Parasitic Wasps

Parasitic Wasps

Donald L.J. Quicke

Advanced Research Fellow in Taxonomy, NERC (Natural Environment Research Council), Initiative in Taxonomy, Imperial College, Silwood Park, Ascot, UK.

CHAPMAN & HALL

London · Weinheim · New York · Tokyo · Melbourne · Madras

Published by Chapman & Hall, 2–6 Boundary Row, London SE1 8HN, UK

Chapman & Hall, 2–6 Boundary Row, London SE1 8HN, UK

Chapman & Hall GmbH, Pappelallee 3, 69469 Weinheim, Germany

Chapman & Hall USA, 115 Fifth Avenue, New York, NY 10003, USA

Chapman & Hall Japan, ITP-Japan, Kyowa Building, 3F, 2-2-1 Hirakawacho, Chiyoda-ku, Tokyo 102, Japan

Chapman & Hall Australia, 102 Dodds Street, South Melbourne, Victoria 3205, Australia

Chapman & Hall India, R. Seshadri, 32 Second Main Road, CIT East, Madras 600 035, India

First edition 1997

© 1997 Donald L.J. Quicke

Typeset in 10/12pt Palatino by Saxon Graphics Ltd, Derby

Printed in Great Britain by Cambridge University Press

ISBN   0 412 58350 X

A catalogue record for this book is available from the British Library

Library of Congress Catalog Card Number: 96–71731

∞ Printed on permanent acid-free text paper, manufactured in accordance with ANSI/NISO Z39.48-1992 (Permanence of Paper).

Dedicated to Mom and Dad

. . . and books that told me everything about the wasp, except why.

Dylan Thomas, *A Child's Christmas in Wales*

Contents

Contents

Preface

The parasitic Hymenoptera are not only a highly successful group of organisms with probably a million or more species; they have been the subject of a huge amount of research over the past 100 or so years. The first truly general account of parasitic wasp development, biology and natural history was provided in *Entomophagous Insects* by Clausen (1940) which includes a wealth of detail, much of which is still of relevance today. This work was followed in 1971 by Dick Askew's *Parasitic Insects*, which similarly summarized much about their biology and gave new insight into their evolution. Much more recently, Charles Godfray's excellent book, *Parasitoids*, dealt extensively with the evolutionary biology of insect parasitoids, and of parasitic wasps in particular. In this book I have only dealt rather superficially with many of these matters. Instead, I have concentrated more on the developmental, physiological, anatomical and molecular details as they relate to the biology of parasitic wasps.

Despite the fact that there are at least as many species of parasitic wasps as there are beetles, and many more than there are of some other major insect orders, there are many areas of their biology that have hardly been investigated. Thus it is not uncommon in major textbooks on insect physiology, cell culture, embryology, nervous systems, toxicology, muscles, flight and many other topics to find only one or two references to parasitic wasps, and often only a few references to the Hymenoptera as a whole – these usually referring to some work on the honey bee, and occasionally an ant or yellowjacket.

Many early studies of anatomy, embryology and physiology reflect a great deal of painstaking work, but they are often in need of revision in the light of new discoveries and making use of modern techniques. Sadly there is little sign of any major trend in that direction; yet (as I hope to point out) there are many interesting problems awaiting detailed investigation, and we have only the scantiest biological information for many major groups.

One subject I have almost entirely avoided is the extensive empirical and theoretical body of work on host/parasitoid population dynamics. This is adequately covered in many other books. Also, I have avoided going into much detail on biological control and integrated pest management, except where other aspects of biology, such as sex determination, have a direct bearing on it. Apart from these, I have attempted to provide at least some references to as many kinds of investigation that parasitic wasps have been subjected to so that readers will gain a better impression of the utility of these insects in research fields other than their own, and hopefully might stimulate some new ideas. Of course, I will have missed some areas, but I hope none too large or important, though I realize that nearly everybody's research is important to themselves. My apologies in advance, therefore, to those whose fields I have overlooked. In two appendices, I summarize host relationships and also provide brief notes on some of the most frequently studied parasitic wasp species that I hope will be of use to students who are suddenly faced with a plethora of new names and associations.

Acknowledgements

I am extraordinarily grateful to my wife, Tanya, for her tolerance of our house being strewn with papers during the writing of this book, and to our new Great Dane pup, Harley, for largely resisting the probably enormous temptation to chew up said papers. My thanks are due to Dick Askew, Andy Austin, Hasan Basibuyuk, Nando Bin, Douglas Dahlman, Chris Darling, Rachel Kruft, Nuncio Isidoro, Anne Le Ralec, Terry Newman, Andy Polaszek, Habibur Rahman, Alex Rasnitsyn, Anna Rivero-Lynch, Richard Stouthamer, Mike Strand and Nathalie Volkoff for allowing me to reproduce some of their excellent photographs and micrographs; John Noyes for allowing me to make use of his table of host associations; and Anne Le Ralec, Mike Fitton, Alan Hook and Hasan Basibuyuk for allowing me to refer to some of their unpublished observations. I would also like to say a special thanks to Mike Fitton, James Cook, Charles Godfray, Mike Sharkey, Dan Gerling, Anne Le Ralec, Alex Rasnitsyn, Nathan Schiff, Nina Trandem, Mike Tristem, Jack Werren, Nando Bin, Carmen Gimeno, Mick Day and Bob Wharton for stimulating discussions, and to the many others who have offered useful suggestions or provided answers to my many and various questions. Rachel Kruft and Frank Wright helped with microscopy and photography.

1

Introduction to the parasitic Hymenoptera

1.1 INTRODUCTION

While some groups of organisms such as mammals, birds, butterflies, and even the social Hymenoptera, have consistently attracted a lot of popular as well as academic interest, the parasitic wasps have been greatly underworked and they have gained very few converts among the ranks of amateur entomologists. This is in spite of the fact that they are not only ubiquitous but also perhaps the most diverse group of insects, rivalling even the beetles in terms of their total number of extant species, and that, as a whole, they are of enormous ecological and economic importance (LaSalle, 1993). Why, then, have they been relatively neglected and failed to achieve the level of interest amongst amateur entomologists that, for example, the beetles have? Certainly they are no less interesting intrinsically, nor are they difficult to collect, and many are very attractively coloured. Probably the answer lies in the fact that they have for a very long time been considered a 'difficult group'; that is, they have been too difficult to identify to allow somebody to get to grips with them within an acceptable period of time. This has been a vicious circle, for since relatively few people have tended to persevere with their study, they have remained more poorly worked with the result that even today many major questions concerning their biology and taxonomy remain unanswered. Even for Great Britain, whose insect fauna is probably better known than that of any other part of the world, there are few key works, and identifying a randomly collected parasitic wasp is likely to be a time-consuming business that may ultimately prove insoluble, at least until a thorough taxonomic revision of the group has been carried out. Add to this the fact that a good day's collecting in Britain in the summer might yield many more than 100 species and the reasons why parasitic wasps have been shunned become starkly obvious. Fortunately, there is a growing number of good comprehensive works on both the taxonomy and biology of Hymenoptera, including the parasitic species (Waage and Greathead, 1986; Gauld and Bolton, 1988; Goulet and Huber, 1993; Hanson and Gauld, 1995) that will, with some practice, allow the correct identification of most individuals to family or subfamily level. This opens the way for amateurs and professionals to get to grips with the order.

The parasitic Hymenoptera are arguably one of the most important insect groups, with many species playing valuable roles in pest control as well as maintaining the diversity of natural communities. LaSalle and Gauld (1991) emphasized their high position in food chains and have suggested that this makes them particularly liable to extinction and also likely to include many keystone species. That many species of parasitic Hymenoptera successfully regulate their host populations so that both they and their hosts usually exist at very low population densities means that they may be particularly vulnerable. Many successful biocontrol programmes have stemmed from locating natural enemies in the native habitat of an introduced pest species. Sometimes these successful biological control agents have been quite difficult to find simply because they and their hosts do not exist at high population densities where they are endemic. Habitat loss through deforestation, etc., could well be leading to extinctions of parasitoids that in the event of future pest outbreaks would be immensely valuable. No one can say how many described species are no longer extant. The degree of host specificity will necessarily play an important role in this and it is interesting to relate this to the growing body of evidence that at least some groups of parasitoids are more host specific in the tropics (**Chapter 10**). Thus it could be that extinctions of parasitic Hymenoptera are intrinsically more likely to occur in tropical ecosystems – the very ones that are probably under most threat at present.

In the same way that a great deal of what we know about mammalian physiology comes from detailed studies of a selected few laboratory animals (rats, mice, etc.), so too does much of what we know about parasitic wasps. Further, particular species have been focused on for particular types of investigation and only a handful could be said to be well studied across a wide range of fronts. A quick comparison of the classification of Hymenoptera in Table 11.1 with a list of frequently studied species provided in **Appendix A**, will show that the majority of families do not have even a single representative that is commonly cultured for experimental purposes. Thus a great deal of what we know about parasitic wasp behaviour, development and physiology is based on a few representatives, and the generality of many findings cannot be taken for granted. Indeed, given the diversity apparent in almost all well studied aspects of parasitic wasp biology, we may expect that many novel strategies remain to be uncovered.

Godfray (1993) made a major contribution to the parasitoid literature by providing an up-to-date and comprehensive summary of their ecology and evolutionary biology. In this book I have tried to take a somewhat different and complementary approach to Godfray's by concentrating rather on what is known about their anatomy, biology and physiology, although **Chapter 10** deals with various ecological features, especially community ecology and diversity. I have not gone into extensive detail about external morphology as this has been adequately covered in many other general and specialist books and papers (Imms, 1935 and subsequent editions; Chapman, 1982; Gauld and Bolton, 1988; Naumann, 1991; Goulet and Huber, 1993; Hanson and Gauld, 1995). **Chapter 2** explores hymenopteran sex and genetics, which affect so many aspects of parasitic wasp biology, and **Chapter 3** draws attention to probably the most informative single feature of a parasitic wasp's life history: whether or not the host is allowed to continue developing following parasitization. **Chapters 4, 5** and **6** deal with developmental and adaptive features from gametogenesis through to the adult, largely taking a systems approach. Throughout, I have attempted to put observations into both a phylogenetic and a comparative framework, and to suggest areas where future studies might be particularly rewarding. **Chapter 7** deals with the physiological interactions between parasitic wasps and their hosts; **Chapter 8** is an account of parasitic wasp behaviour, principally covering host location and discrimination, and parasitoid sexual behaviour. **Chapter 9** is a short one that points out some of the non-physiological ways in which potential hosts avoid or reduce the likelihood of being parasitized. Finally, **Chapter 11** provides a summary of hymenopteran classification, taxonomy and phylogeny, and illustrates how various techniques can be useful for resolving problems from species to suborder level.

1.2 PARASITOIDISM AMONG OTHER INSECTS

Parasitoids, by definition, kill their hosts and often have rather a short period when they are actually acting as a true parasite. That their hosts are almost invariably killed means that they can have profound effects on host population dynamics. In the older literature, parasitoids were often referred to simply as parasites but this term is potentially confusing, and so several alternative terms have been proposed from time to time, such as protelean insect parasite, carnivoroid and parasitoid. Of these, the last term has become the most widely adopted, and the others effectively dropped. Throughout this book I have stuck to the more traditional use of the terms 'parasitic Hymenoptera' and 'parasitic wasp' rather than less euphonious 'parasitoid' equivalents to refer to that grade of Hymenoptera that are predominantly parasitoids. This grouping is neither a monophyletic entity nor composed entirely of parasitoids: a number have evolved from being parasitoids into predators or phytophages (see below). Despite these differences, the taxa involved have generally maintained many morphological, physiological and behavioural connections with their parasitoid ancestors and so their consideration here can be justified.

Parasitoidism is known in members of seven insect orders – the Hymenoptera, Diptera, Coleoptera, Lepidoptera, Trichoptera, Neuroptera and Strepsiptera – but of these, conservatively, 80% of parasitoid species belong to the Hymenoptera (Figure 1.1). The so-called parasitoid Neuroptera belonging to the family Mantispidae include predators of eggs in spider egg sacks and some wasps (Askew, 1971); and in the Strepsiptera, only the Mengeidae are true parasitoids and these attack Thysanura.

Although the hosts of parasitic Hymenoptera are almost invariably killed by their parasitoids, there are a few exceptions. In some cases an incompletely eaten-out host larva may retain the ability to carry on some degree of movement for quite a while after the parasitoid has emerged. In an extreme case – the attack of lepidopterous larvae by braconids of the genus *Microplitis* – the hosts may carry on moving and feeding for up to two weeks following parasitoid larval emergence, but they always die without pupating (Strand *et al.*, 1988; Ritter and Johnson, 1991). The situation with parasitoid Diptera is not always the same. English-Loeb *et al.* (1990) noted that larvae of the arctiid host of the tachinid, *Thelairia bryanti*, frequently survive to pupate and ultimately to emerge as viable and fertile

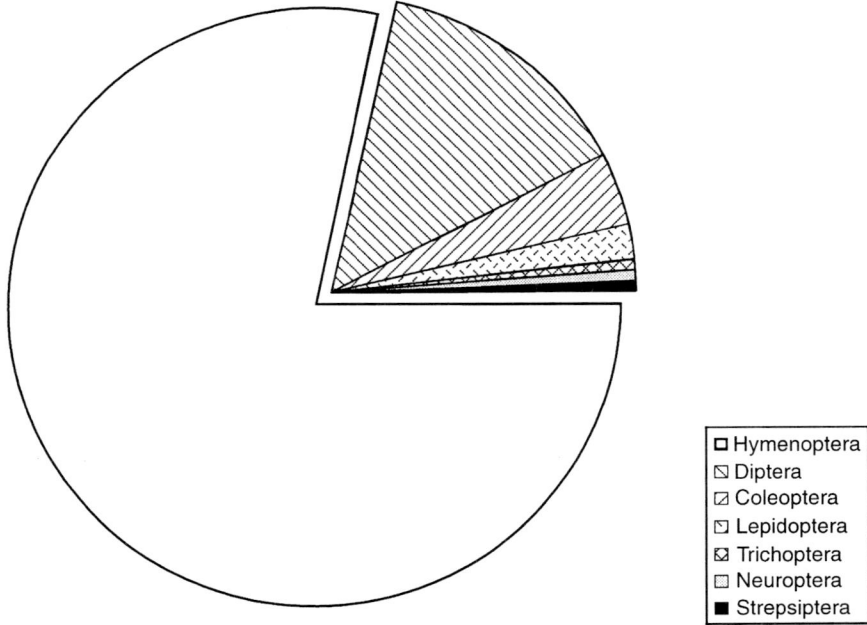

Figure 1.1 Relative number of parasitoid species in different insect orders. (Data from Eggleton and Belshaw, 1992.)

adult moths, and some other tachinids have similarly been reported to leave their hosts alive. Survival of the hosts of parasitic Hymenoptera to the point where they are able to reproduce appears to be restricted to a very few instances in which adult or late nymphal insects are attacked. For example, Timberlake (1916) describes a case involving the coccinellid beetle, *Olla abdominalis,* in which an adult survived to be parasitized by two generations of the braconid wasp, *Dinocampus coccinellae.* Triltsch (1996) found that another ladybird, *Coccinella septempunctata,* can even continue to reproduce when attacked by the same parasite, provided that the temperature is high enough. Kitamura (1988) noted that when *Haplogonatopus apicalis,* a member of the parasitic aculeate family Dryinidae, attacks 4th or 5th instars of the plant-hopper, *Sogatella furcifera,* a few females of the latter can still deposit their eggs.

Eggleton and Belshaw (1992, 1993) compared the evolution and characteristics of parasitoidism in the Hymenoptera with those found in other insect orders, and revealed a number of major differences between them. In the case of the Hymenoptera, it seems that parasitoidism evolved only once, probably in a relative of the Orussidae (see below), and that this was a result of a transition from an endophytic, mycophagous life style as evidenced by putative close relatives of the orussids such as the Siricidae and the Xiphydriidae. In contrast, parasitoidism in the Coleoptera is likely to have evolved at least 14 times, and in the Diptera at least 21 times (Figure 1.2). Further, in the case of the Diptera, the principal route to parasitoidism appears to have been from saprophagy, but also with a significant number having evolved from predatory relationships.

Non-hymenopteran parasitoids do a number of things that hymenopteran ones either do not do, or do only rarely (Eggleton and Belshaw, 1992; Mellini, 1994). Some flies, for example, are effectively parasitoids of snails, woodlice, and centipedes – groups that are not attacked by any parasitic Hymenoptera. (One parasitic wasp, a proctotrupid of the genus *Phaneroserphus,* is possibly a parasitoid of a lithobiid centipede; Newman, 1867.) It is interesting, too, to note that no parasitic wasps attack apterygote insects, and that the more ancient insect orders are utilized as hosts far less than the most recent ones, although the reason for this is not yet known. Of considerable importance is the way in which hosts are located in the various groups. Probably one of the most significant features of the Hymenoptera lies in the fact that they are generally very precise about where they oviposit. Many parasitoid flies (23 families) and beetles (11 families) and even parasitoid Lepidoptera (Jordan, 1926; Ishii, 1990) and Neuroptera simply scatter their eggs in their host's environment and then they rely either on chance encounters between the host and their eggs or on the host-finding capabilities of their 1st instar larvae. In contrast, the great majority of parasitic Hymenoptera place their eggs on, in or very close to a particular host individual (Table 1.1).

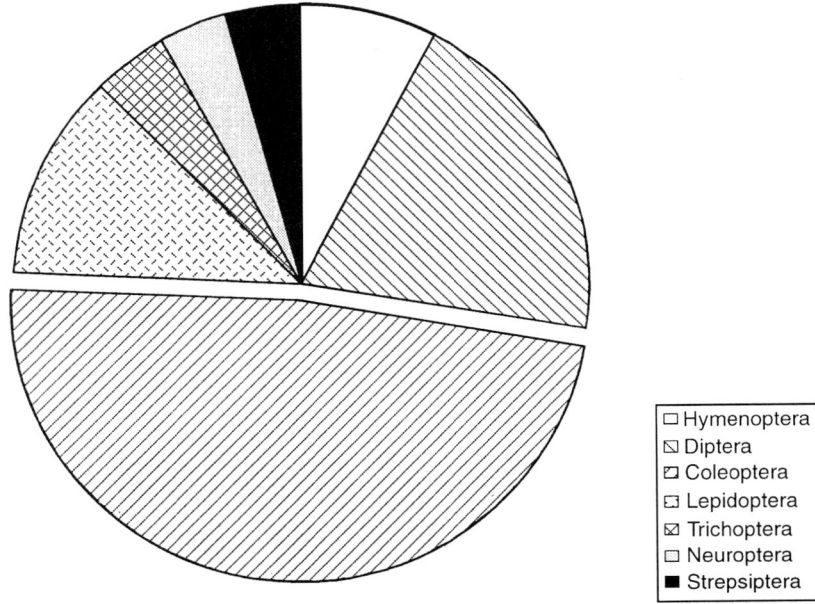

Figure 1.2 Relative number of origins of parasitoid life styles in different insect orders. (Data from Eggleton and Belshaw, 1992.)

More liberal scattering of eggs or their placement away from hosts does occur, for example, in Trigonalyidae, Eucharitidae and Perilampidae, and also in the ichneumonid subfamily Eucerotinae, most of which have host-seeking planidial larvae, but these are very much the exceptions among the Hymenoptera and represent specializations rather than the ancestral biology.

Table 1.1 Estimates of numbers of species of insect parasitoids in which host searching is carried out by the adult or the larva (from Eggleton and Belshaw, 1992)

Parasitoid order	*No. of species in which host searching is carried out by:*	
	adult	*larva*
Hymenoptera	66 500	500
Diptera	7900	8600
Coleoptera	10	3590
Neuroptera	0	50
Lepidoptera	0	10

It is particularly interesting to compare the strategies and biologies of hymenopteran parasitoids with their dipteran equivalents, since both orders include many parasitoid taxa (Figure 1.1). Parasitoid Diptera show quite different host relationships to their hymenopterous counterparts, and this has been attributed to both their adult morphology and their ancestral biology (Eggleton and Belshaw, 1992; Mellini, 1994). Many Diptera have larvae that live and thrive in very toxic environments such as in dung, rotting carcases and other less than pleasant milieus. Gauld *et al.* (1992) suggest that this difference may also have a significant effect on the ability of the different orders to tolerate secondary plant compounds in hosts feeding on toxic plants, and so reflect the host ranges and diversity of them, especially in the tropics. It is interesting to note that this difference may also be reflected in the relative ease with which parasitoid Diptera can be cultured *in vitro* in the absence of insect components, whereas in endoparasitic Hymenoptera there frequently appears to be a requirement for insect factors (**Chapter 5**). Mellini made an extensive comparison of the biologies of the two groups and some of the more noticeable differences are summarized in Table 1.2. Again he drew particular attention to the careful selection of oviposition site and to the consequences of ovipositor morphology in the Hymenoptera (**Chapter 6**). This can account for the fact that not only have parasitic wasps been far more successful in attacking concealed hosts, but also, because they usually penetrate their hosts' bodies with their ovipositor, they can discriminate between unparasitized and previously parasitized hosts, they can inject venoms to paralyse or modify host physiology, and they can even drink host body fluids that ooze from the wound (host-feeding: **Chapter 8**). In addition, the ability to paralyse hosts means that many Hymenoptera can develop as ectoparasitoids, whereas this is rare among Diptera.

1.3 A BRIEF EVOLUTIONARY HISTORY AND PHYLOGENY OF THE HYMENOPTERA

The Hymenoptera appear to have originated around the mid to late Triassic and fossils from this period are known from Middle Asia (Fergana, Uzbekistan) and Australia (Victoria) (Jell and Duncan, 1986; Rasnitsyn, 1988; **Chapter 11**). The earliest hymenopteran fossils belong to the group known as sawflies, the Symphyta, and the oldest fossils show many similarities with the extant sawfly family Xyelidae. During the early Jurassic, 180–200 million years ago, the order showed its first major diversification and the earliest known parasitic groups make their first appearance in the fossil record. These include members of the extant Megalyridae and the extinct Karatavitidae and Ephialtitidae. That these extinct families were parasitoids is, of course, conjecture based upon their morphological features such as possession of a wasp-waist

Table 1.2 Comparison of general biological features of hymenopteran and dipteran parasitoids (interpreted from Mellini, 1994)

Hymenoptera	*Diptera*
Ovipositor capable of piercing or penetrating a hard substrate	Ovipositor typically incapable of piercing
Adult mandibles enable egress from tough, concealed pupation site	Adults lack mandibles, so restricted to sites where legs and ptilinum sufficient to enable escape
Capable of host-feeding to gain proteins for continued egg maturation	Feeding usually limited to predominantly sugary secretions such as honeydew or nectar
Arrhenotokous, so capable of adjusting offspring sex ratio to type of host resource	Amphigonous – sexes suffer differential mortality in small hosts
Less fecund	More fecund
Koinobiont endoparasitoid species usually oligophagous	Koinobiont endoparasitoid species usually polyphagous
Wide range of host stages potentially parasitized by Hymenoptera	Mostly attack late instars of mobile, exposed hosts
Superparasitism often avoided	Superparasitism seldom avoided
Mostly either endophagous or ectophagous with host paralysed	Mostly endophagous, cannot paralyse host

and a long thin ovipositor. During the first half of the early Cretaceous, the order went through a second major diversification, with the appearance of some 17 new families of 'Parasitica' and several parasitic aculeate families, about half of which are still extant. Not a single one of the known extinct hymenopteran families survived beyond the end of the Cretaceous. Thus family level diversification of the Hymenoptera must have been quite a rapid event. For this reason, uncovering the precise phylogenetic relationships between hymenopteran families and super-families has proved a hard problem.

1.4 ORIGIN OF PARASITISM IN THE HYMENOPTERA

As ancestral Hymenoptera (that is, the Symphyta excluding the parasitic sawfly family Orussidae) are phytophagous, it is interesting to speculate how carnivory and parasitoidism could have evolved in the order.

Handlirsch (1907) was one of the first to discuss the evolution of para-sitoidism and his proposal has become generally accepted. In his view, a primitive horntail sawfly (Siricoidea or similar), which would normally have lain eggs in wood, shifted to lay in or on the larvae of beetles dwelling in the same microhabitat, though he did not go into explicit detail of what steps might have been involved. Subsequently, two modi-fied and more explicit hypotheses have been developed based on Handlirsch's idea that the evolution of parasitoidism would involve a siri-cid-like ancestor with a relatively long ovipositor, and a host organism such as beetle larvae living in the same niche. In one scenario, a wasp could be seen to gain advantage if she selected an oviposition site near to the eggs of another insect, since these would provide a rich food source for her own offspring, supplementing the less nutritive plant material. After a time, a dependency could evolve together with various behav-ioural and biological specializations permitting the parasitoid to develop on increasingly large and valuable hosts.

An alternative scenario suggests itself when one considers the biology of the only extant parasitic family of sawflies, the Orussidae. In the horn-tails (Siricidae and Xiphydriidae) phytophagy proper has given way to mycophagy, the fungi in question being the means of obtaining nutrient from wood. In both the horntails (Siricidae) and wood wasps (Xiphydriidae), this association with xylotrophic fungi has become a true symbiosis in that the female wood wasp carries fungal arthrospores, in a pair of specialized abdominal sacs, which she inoculates into the tree at the same time as she oviposits (Francke-Grosmann, 1967; Stillwell, 1965; Spradbery, 1973). The fungi develop on the wood and it is this more digestible, fungus-infested wood that the larval horntails and wood wasps consume. Eggleton and Belshaw (1992) postulated that the transition from phytophagy to parasitoidism was crucially linked with the involvement of fungi in the ancestors of the parasitic Hymenoptera.

An interesting corollary of the shift from primary phytophagy to mycophagy or carnivory concerns the dietary chemistry of the wasps. Plant, fungal and insect diets vary greatly in their compositions of fatty acids, sterols, amino acids, carbohydrates, vitamins, etc. Parasitic wasps have had to adapt to these changes and, in so doing, their own physiolo-gies and biochemistry have undergone marked changes. Some of these are discussed in **Chapter 5**.

It should be noted that Handlirsch's basic idea did not win favour with everyone, and that various alternative hypotheses were put forward, though few of these have gained much support. Malyshev (1968), in par-ticular, stating that 'Handlirsch's hypothesis must be abandoned as com-pletely unrealistic', emphasized that such a sudden and dramatic shift in biology, from wood-feeding to carnivory or parasitoidism, would be unlikely to work on behavioural and physiological grounds. He also pointed out that there appear to be no intermediate stages around today

– that is, he would have expected extant woodwasps also to be gaining advantage by selectively ovipositing in sites near suitable packets of additional animal food. Instead, Malyshev developed an argument based on the observable biologies of extant Hymenoptera, taking into account their supposed instincts, and the important role played by female secretions that modify the larval environment. In his preferred scenario, the Apocrita would have evolved from relatives of gall-forming sawflies, perhaps the ancestors of the Cephoidea, though nothing is known about the biology of these extinct forms. The argument put forward by Malyshev is that the female's secretions cause the plant to produce a localized, nutrient-rich source of plant tissue that would lead to a sedentary larva and would also be attractive to similar, perhaps closely related sawflies that would evolve through stages as inquilines and carnivores, and finally give rise to parasitoids in the same way as can be observed among various extant groups of Hymenoptera (and even Diptera) belonging to a number of different families. As Malyshev's hypothesis suggests that the sister group of the Apocrita was not xylophagous, as in the siricoidean sawflies, but, rather, phytophagous forms living on green plant tissue, resolution of the true phylogeny of the Hymenoptera should enable us to determine whether his or the more widely accepted Handlirschian view is correct. As will become apparent in **Chapter 11**, we are still some way away from the desired level of phylogenetic certainty.

Irrespective of the actual origin of parasitoidism within the Hymenoptera, the transition to a parasitic way of life occurred more or less coincidentally with several morphological changes, the most notable of which being the acquisition of a wasp-waist – a narrowing of the body between the 1st and 2nd abdominal segments. Because the conspicuous body regions behind the head do not correspond to the thorax and abdomen, they have been given different names that can sometimes be confusing. Thus, the middle body region comprising the thorax and 1st abdominal segment is now generally referred to as the mesosoma, and the posterior region is termed the metasoma. In virtually all parasitic Hymenoptera, the metasoma has a very flexible articulation with the mesosoma, which has probably been highly significant in the ability to manipulate the ovipositor. It should be noted, however, that the earliest known fossil parasitic wasps, the Ephialtitidae, did not have such mobile metasomas as present taxa, and so presumably oviposited in similar situations to their waistless sawfly ancestors (Rasnitsyn, 1980).

1.5 OVIPOSITION: A CRUCIAL BEHAVIOUR

A key feature in the evolution of the order has been the precise placement of eggs at sites where there is a rich food source (Gauld and Bolton, 1988), and this is also often associated with modification of the larval food source. All hymenopteran families have two glands associated with the ovipositor system: the acid and alkali glands, so called because of their general histochemical properties. In parasitic and aculeate wasps, the acid gland is generally referred to as the poison or venom gland since it is the principal source of host-modifying and/or pain-inducing secretions. The alkaline gland is also known as the Dufour's gland, first recognized in the social aculeates, and its main role in most groups seems to be connected with pheromone production (**Chapter 6**), though its function in sawflies remains to be discovered. Robertson (1968) postulated that, in the ancestral Hymenoptera, the acid and alkali glands served to produce lubricants to aid oviposition, but without doubt they collectively evolved to produce other sorts of secretions early in the radiation of the order (Gauld, 1991). Even some members of the Xyelidae, the most basal family of Hymenoptera, may produce a cecidogenic secretion (see below), and in the more recent siricid horntails these glands produce secretions that promote the development of a symbiotic fungus that helps to render the wood their larvae feed on more digestible (Spradbery and Kirk, 1978).

1.6 TRANSITION TO SECONDARY PHYTOPHAGY

A considerable number of lineages have evolved from being parasitic on insect or other arthropod hosts, to becoming secondarily phytophagous. In many of these cases, the relationships bear some similarity to parasitism or predation in that the most frequently attacked plant parts are seeds, which may be consumed on a one-wasp, one-seed basis (i.e. akin to parasitoidism), or a one-wasp, several-seeds basis, thus being essentially seed predators. Seeds are well-defined packages of highly nutritious food. The second form of secondary phytophagy that appears to have evolved directly from parasitic habits is gall formation, which is considered separately below. In many cases, there is a growing body of evidence that at least 'seed predation', or 'seed parasitoidism' (as it is sometimes called), actually involves the stimulation of some plant cell proliferation; that is, a sort of incipient gall formation.

One possible scenario for the evolution of phytophagy involves attacking a primary phytophagous host that feeds either within a gall or a seed but which is not large enough to supply the entire needs of the developing parasitic wasp larva. Thus, a clear advantage would accrue to a parasitoid that under these circumstances could consume and utilize the comparatively nutritious plant material that surrounds the primary host. Several possible examples are known, such as those among the Eurytomidae (Varley, 1937). A more extreme case of an association that may have become fixed at this partially phytophagous stage is provided by an Australian species of the ichneumonid genus *Poecilocryptus*. This species has been reared from coccoid-induced galls, and there has now

been confirmation of the initial presumption, based on its enormous larval mandibles (Short, 1978), that it may consume plant material in addition to the coccoids. A similar situation is found in *Grotea*, a genus of ichneumonids, which are parasitoids of solitary bee larvae but which, after
consuming a larva, then consume the pollen loaf of that cell and may go
on to consume the contents (larva and pollen) of another cell
(Slobodchikoff, 1967). This type of relationship has been taken a stage further by the gasteruptiids that oviposit on solitary bee eggs, which the
newly emerged gasteruptiid larva consumes prior to eating the pollen
store (Malyshev, 1964).

Seed predation has evolved in the doryctine braconid genus, *Allorhogas*.
Some *Allorhogas* species are parasitic on pyralid moths in grass stems and
sugar cane but at least one species, *A. dyspistus*, appears to be completely
phytophagous, consuming seeds of the legume, *Pithecellobium tortum*,
though the wasp probably also induced a gall in the *Pithecellobium* seeds
(Macedo and Monteiro, 1989; Marsh, 1991). How such a transition
occurred is not known, and it may be that other species displaying more
intermediate stages remain to be discovered.

1.7 TRANSITION TO CECIDOGENESIS

Many Hymenoptera are cecidogenic; that is, they induce galls in plants,
and their larvae feed exclusively on this modified plant tissue.
Cecidogenesis has evolved among several families of 'Parasitica', of which
probably the best known are the true gall wasps of the family Cynipidae.
Interestingly, whilst many of the gall-forming Hymenoptera from the
north temperate regions belong to the Cynipidae, this family is poorly
represented in Australia, where the majority of gall-forming
Hymenoptera belong to various families of the Chalcidoidea, especially
the Pteromalidae, Torymidae, Eurytomidae and Eulophidae (Naumann,
1991). In addition to the well known cecidogenic taxa among the
Cynipidae and Chalcidoidea, a number of independent instances of the
evolution of phytophagy have been coming to light recently in families
that have for a long time been considered to be entirely parasitic. Within
one braconid subfamily, the Doryctinae, cecidogenesis has evolved at least
twice: once in the genus *Monitoriella* (Wharton, 1993; Infante *et al.*, 1995)
and also apparently in members of the endemic Australian braconid subfamily Mesostoinae which produce stem galls on *Banksia* (Austin and
Dangerfield, cited in Infante *et al.*, 1995).

Galls may result either from the action of the secretions from the
ovipositing female wasp, as in the sawflies studied to date, or from the
wasp's larva, as in the cynipid gall wasps. As stated above, one of the
most characteristic features of the Hymenoptera is the use of the ovipositor not just for laying eggs but also as a means of passing chemical

secretions (venoms) to the host or its substrate in order to bring about some modifications that will be advantageous to the egg's development. This is apparent even in the most primitive of extant hymenopterans, the xyelid sawflies, one species of which, *Xyela pallicaulis*, produces galls on various *Pinus* species in the eastern United States (D.R. Smith, 1970). Galls are also produced by many Tenthredinidae. In the well-known cynipid gall wasps, gall formation is initiated by salivary secretions of the newly hatched wasp larva, but little is known about the induction of galls in most other families apart from some gall-forming sawflies. It is interesting to note that an increasing number of cases of physiologically active salivary secretions is coming to light among the parasitoid taxa (**Chapter 7**) and this might give a clue to the origin of cecidogenesis in this family. Phylogenetic analysis of the Cynipoidea by Ronquist (1994a) suggests that the gall-forming Cynipidae evolved from parasitoids of wood-boring insects. Ronquist proposed that the ancestral parasitoid may have shifted from wood-boring and stem-boring hosts to ones living on more nutritious plant material such as seeds, herb stems or galls. From this, the transition to consuming plant material itself may have occurred as a result of premature host death or inadequate host size. Under such conditions, any larval secretion that would cause proliferation of the plant tissue or an increase in its nutritive quality would be selected for.

1.8 EVOLUTION OF INQUILINISM

Galls in particular offer a very attractive microhabitat for insect development since they provide protection and a source of high quality food, the plant's resources having been effectively hijacked by the gall former to its own ends. As Malyshev (1968) said, 'Naturally, the newly formed galls were bound to attract other insects,' and the evolution of inquilines, getting an easy meal, was almost inevitable. An interesting feature of many inquiline relationships is that they often involve pairs of closely related taxa. The frequent close evolutionary relationship of inquilines and their hosts parallels the evolution of parasitism in many closely related pairs – so-called adelphoparasitism or agastoparasitism. This may be partly because closely related taxa will be likely to cue in on the same signals, and of course have similar physiological requirements. In the case of the Cynipidae, a large group of species are inquilines within the galls of other cynipids. This group has been referred to as the 'Synergini', and it has often been considered a polyphyletic assemblage with different species having evolved independently to be inquilines of their relatives (Gauld and Bolton, 1988). However, Ronquist's (1994b) detailed independent phylogenetic analyses based on adult morphology provided strong evidence in support of a monophyletic origin.

Some remarkable cases of parasitic hymenopterans attacking a gall-forming organism are provided by various tetrastichine eulophids which are known to be predators of cecidogenic mites and even of nematode worms (Vereshchagina, 1961; van den Berg *et al.*, 1990). The dietary versatility in tetrastichines is well known, and species have been recorded from more than 100 families of hosts distributed among 10 insect orders (LaSalle, 1994; see below).

1.9 FIG WASPS

One particular group of parasitic Hymenoptera has evolved, or rather coevolved, in a very close association with figs – whose flowers, which are concealed within a structure called a syconium, they fertilize. This association, and the need for wasps in order to achieve fruit-set, has been known since Aristotle and Theophrastus in the fourth century BC, but the details of what the wasps were doing was not worked out until much more recently (Frank, 1984). Figs cannot be pollinated without their pollinating fig wasps (Agaonidae: Agaoninae) and fig wasps cannot reproduce without their correct species of fig. Whilst the pollinating fig wasps are not parasitoids of other insects, there is a considerable similarity between their biology and that of their parasitic relatives. The same applies to several other groups of phytophagous 'Parasitica', such as the various seed-eating species of Eurytomidae, and these are often referred to as seed-parasitoids. For that reason, many workers who are basically interested in the evolution and biology of parasitic wasps have also paid more than passing attention to the fig wasps.

Female pollinating fig wasps enter the syconium through a small hole, the ostiole, and lay their eggs within developing ovarioles, which are stimulated to develop into galls that provide the developing agaonid larva with all the food it needs. The highly modified males (Figure 6.16) are protandrous and newly hatched pollinating fig wasps mate within the fig. The females then collect fig pollen in structures called corbiculae (Ramírez, 1978; Chapter 6). The mated females now leave the fig, through holes chewed in its wall by the males, in search of a newly developing one, which they enter. Precise details of fig wasp biology are difficult to determine, partly because the microenvironment within the fig is highly specialized, and opening the fig for observation is potentially highly disruptive (Ramírez, 1987).

The developing fig is a microhabitat that provides a food resource for a whole guild of other insects, including several other groups of parasitic Hymenoptera in addition to the pollinating fig wasps. These include inquilines in the strict sense, gall formers, and parasitoids and predators of the fig wasps and their inquilines. The taxonomy of these has been confusing, because of the highly modified nature of both female and male pollinating fig wasps (Figure 6.16) and probably also because of convergent or parallel evolution associated with figs.

Comparison of a phylogeny of *Ficus* subgenera with a putative phylogeny for the genera of Agaonidae revealed considerable concordance, suggesting a high degree of co-cladogenesis between pollinating fig wasps and their host figs (Ramírez, 1974; Herre, 1989) – most genera of pollinators being restricted to a particular section of *Ficus*. However, it should be noted that neither of Ramírez's phylogenies were the result of formal cladistic analysis. Wiebes (1982) made a start at such an analysis and produced a new cladogram for the Agaonidae that broadly resembled Ramirez's result but lacked a great deal of resolution in the Blastophaginae. As Wiebes noted, before firm conclusions could be reached about the degree of coevolution, more analyses need to be performed after a lot more data have been accumulated.

As the fig ovules attacked by the pollinating fig wasp will not develop to form seed, it is obvious that there will be conflict between fig and fig wasp: the former must limit the proportion of ovules attacked to the minimum necessary for adequate pollination, and the latter must maximize reproductive output. Ganeshaiah *et al.* (1995) have shown that this will have interesting consequences for wasp ovipositor length. The optimum wasp ovipositor will be that which leads to maximum fecundity, whereas the fig's style length will be selected for increased variance.

1.10 EVOLUTION OF PARASITISM OF ADULTS

Parasitism of adult insects is a strategy that has been successfully adopted by only a few groups of parasitic wasps (see also section 6.50). Almost all parasitoids of adult insects (or spiders) are endoparasitoids; although some Dryinidae, Embolemmidae, Pompilidae and Rhopalosomatidae are, strictly speaking, ectoparasitic, these families belong to the Aculeata. Why this might be is rather a puzzle. Probably the best studied cases of endoparasitism of adult insects occur in the family Braconidae in which this strategy has evolved on at least four separate occasions. The largest group involved is the subfamily Euphorinae, most members of which are endoparasitoids of adult beetles (Shaw, 1988). The second largest group is the Aphidiinae, which attack both nymphal and adult aphids; then there is the small specialist subfamily, Neoneurinae, which oviposit and develop within adult ants (Shaw, 1993). On the basis of phylogenetic analysis, Shaw (1988) postulated that, in the case of euphorines, the transition to parasitism of adult beetles followed from parasitism of beetle larvae living in the same microhabitat and at the same time as their pupae and adults. Such sympatry and concurrency is exhibited by several groups, including chrysomelid and coccinelid beetles, various Heteroptera and Psochoptera, which are all hosts of some euphorines (Sommerman, 1956; Shaw, 1988).

Some euphorines even parasitize the adults of other Hymenoptera, including other parasitic wasps of the families Braconidae, Ichneumonidae and Tiphiidae (Cole, 1959a; Alford, 1968; Shaw, 1988; unpublished observations). Apart from these braconids, adult insect hosts are also attacked by a few species in several other families of 'Parasitica' – for example, members of the pteromalid genus, *Tomicobia* (Graham, 1969), and the eulophid genus, *Phymastichus* (LaSalle, 1990a). Along with the enigmatic braconid genus *Cosmophorus*, these two chalcidoid genera all attack adult scolytid beetles, but whether any special significance can be attributed to this is at present uncertain.

1.11 EVOLUTION OF PREDATORY RELATIONSHIPS

Several parasitic wasps have evolved to be predators rather than parasitoids in that they make use of more than one host to complete their development. In these cases, 'hosts' include eggs and larvae, and even nematode worms or mites that live in colonies. Members of the cryptine ichneumonid genus *Aritranis* consume several mature bee larvae, one after the other. Egg predation, making use of 'hosts' that lay eggs not just in clusters, but rather in protected cases, is a fairly common strategy in parasitic Hymenoptera. The commonest of these may be spider egg cocoons, and there are similar masses such as those of certain weevils (Kerrich, 1969; Chapter 8). Spider egg masses are predated upon by larvae of various parasitic Hymenoptera (Austin, 1985; Fitton *et al.*, 1987; LaSalle, 1990b), including ichneumonids, tetrastichine eulophids, and pteromalids. In some cases the parasitic wasp also kills the female spider; in others, the wasp relies on stealth to avoid detection by the spider. Cockroach oöthecae are hosts of several groups of wasps, notably all evaniids and several members of the chalcidoid family Eulophidae. In some the parasitoid may lay within a single host egg and only move on to consuming others within the oötheca after an initial parasitoid stage, but there is some disagreement in the literature on this matter, and sufficiently detailed observations are wanting in many cases. The life history of the British evaniid, *Brachygaster minutus*, has been described in some detail by Brown (1973).

One particularly unusual association has been described by van den Berg *et al.* (1990) in which a South African eulophid, *Aprostocetus* sp., is a predator of the nematode worm *Subanguina mobilis* which occurs in and presumably induces galls in various Asteraceae. It is not hard to imagine how such a strange relationship could evolve. *Aprostocetus* and related tetrastichines are strongly associated with galls, and although the host in most instances is an insect, the actual type of gall-forming animal is not very important. Another species, *Quadristichus sajoi*, is predatory on cecidogenic mites (Szelenyi, 1941; Vereshchagina, 1961).

1.12 TRANSITION TO ACULEATE STRATEGY AND TO SOCIALITY

Parasitoidism is limited in two important ways. Firstly, the host must be situated (or able to find a situation) where parasitoid development can be completed successfully, and perhaps even more relevantly the hosts have to be at least as large as the parasitoid that is going to develop on them. These two factors were overcome by the evolution of nesting behaviour whereby the wasp provides a suitable shelter for the host and its young, and subsequently the multiple provisioning of a single cell so that enough smaller prey may be accumulated to provide sufficient food for the wasp's larva to complete its development.

Whilst the ovipositor of parasitic Hymenoptera obviously allows them to become a highly successful group of insects, utilizing a wide range of hosts as larval food, it also poses a limitation in its own right. With very few exceptions, the parasitic Hymenoptera use their ovipositors very specifically to lay their eggs in, on or near their hosts. At least primitively, the ovipositors of Aculeata have lost this function and instead they have become specialized for the injection of subduing venom into a prey and, in some cases, for prey transportation. Transportation of prey or other larval food by aculeates has in fact led to, or perhaps depended on, the development of stronger flight than that needed by most parasitoids, and so in the aculeate lineage we see a number of changes in wing venation, flexion lines and flight musculature not seen in their parasitic cousins. Such changes have probably been essential prerequisites for the evolution of sociality in the flying bees and wasps because of the need to carry heavy loads of food over long distances to their nests.

The loss of use of the ovipositor for oviposition has long been regarded as a defining feature of the Aculeata, but recently it has become clear that at least a few aculeates do actually use the ovipositor as such. This occurs in sapygids which oviposit through the cap of their bee host's newly completed cell (Torchio, 1972), and various dryinids which are partially or wholly endoparasitic within plant hoppers. For example, members of the New World dryinid genus *Crovettia*, are polyembryonic, a life history feature that really does necessitate oviposition within the host. Others lay an egg under the host's cuticle, but as the larva grows it pushes its back out so that only the head and anterior part of its body remain internalized. In yet others (for example, some Gonatopodinae) the egg is lain externally but the 1st instar larva penetrates the host and develops as an endoparasitoid (Olmi, 1994). Some chrysidid larvae also feed by partially entering their host bee larva's body. Carrillo and Caltagirone (1970) provided a detailed account of the biology of the elampine chrysidid, *Pseudolopyga taylori*. This species oviposits in nymphs of the lygaeid bug, *Nysius raphanus*, the ovipositor being pushed into the host and the egg being attached to the host's hindgut. It is not clear whether the chrysidid's egg actually passes down the lumen of the small ovipositor or whether it emerges at the ovipositor base, but the former seems more likely. The independent loss of use of the ovipositor in some parasitic non-aculeates means that an adequate definition of the Aculeata is more one of phylogeny than functionality.

Some, perhaps all, amisegine chrysidids and probably also lobosceliine chrysidids are endoparasitoids in phasmid eggs, but at least in the first case, it is known that the female wasp chews her way into the host egg to make the oviposition hole. Day (1979), noting the similarity between amisegine and lobosceliine mandibles, suggests that the latter probably do the same.

Following the publication of genetic considerations by Hamilton (1967), it has been widely accepted that the Hymenoptera are predisposed to the evolution of sociality due to their haplodiploid sex determination system and the asymmetry of relationships between mothers, sisters and brothers that this leads to. As male Hymenoptera are haploid, each egg a given male fertilizes (female eggs) will contain exactly the same set of genes from him but, as in other insects, each egg receives half of the mother's genes. Together this means that Hymenoptera sisters share 75% of their genes with one another (Table 1.3) rather than the normal 50%.

Table 1.3 Relatedness between siblings in the Hymenoptera: percentage genes shared

	sister	brother
sister	0.75	0.5
brother	0.5	0.5

1.13 SECONDARY EVOLUTION OF PARASITISM

The parasitic way of life of most apocritan Hymenoptera has given rise to a number of other strategies ranging from straightforward predation to gall forming, from inquilinism to kleptoparasitism. On several occasions, there have been reversals in way of life, back from non-parasitic ones to parasitic ones. For example, some African spider-hunting wasps (Pompilidae) have secondarily reverted to become true koinobiont parasitoids. The ancestral behaviour for pompilids is believed to be one in which the wasp selects spiders that naturally live in retreats, and lays a single egg per host (Brothers, 1975). More advanced species transport (drag) their host to a prepared and sometimes quite sophisticated cell. However, a few such as *Homonotus* and *Notocyphus* have completely lost the need for their hosts to be concealed, and they lay on free-living spiders such as lycosids which are only temporarily paralysed by the wasp venom and recover quickly after the pompilid has oviposited (Grout and Brothers, 1982; Wasbauer, in Hanson and Gauld, 1995).

2

Sex and genetics

2.1 HAPLODIPLOIDY AND SEX DETERMINATION

The French apiarist, Dzierzon, first proposed in 1845 that drone bees originate from unfertilized eggs, and workers and queens from fertilized ones, but it was not until some time afterwards that this became generally accepted and was found to apply to virtually all other Hymenoptera. Male Hymenoptera are therefore haploid whilst females are diploid – a system known as haplodiploid sex determination. Unmated female Hymenoptera lay unfertilized eggs that develop into males; and mated females can lay either unfertilized eggs that will develop into males or fertilized ones that develop into females. Whilst not totally restricted to the Hymenoptera, haplodiploid sex determination is found in only a few other groups of organisms, notably in most thrips (Thysanoptera) and in monogonont rotifers.

The genetic mechanism of sex determination in the Hymenoptera that underlies the haplodiploidy has only been worked out for a few taxa, and (surprisingly) at least two different systems are involved. The only system which is known with any degree of certainty is referred to as complementary sex determination (CSD). In the cases demonstrated thus far, CSD involves a single sex determination locus and is therefore referred to as single-locus CSD, though some or even many cases probably involve multiple loci (Luck *et al.*, 1992). Alternatives to the single and multiple locus CSD models centre around genic balance hypotheses (Cunha and Kerr, 1957; Cook, 1991). These different models, and the evidence in support of them, are discussed in greater detail below.

2.2 COMPLEMENTARY SEX DETERMINATION (CSD)

A single-locus CSD system has been shown to exist in a number of hymenopterans (Table 2.1) collectively representing a wide range of taxa (Cook, 1991, 1993b; Cook and Crozier, 1995). In this system, individuals that carry only a single allele at a sex determination locus develop into males and those that are heterozygous develop into females. Haploid individuals necessarily carry a single allele and consequently always develop into males. Allelic sex determination occurs in sawflies (e.g. Smith and Wallace, 1971) and in some Ichneumonoidea and Aculeata; therefore, given some phylogenetic hypotheses of the Hymenoptera (see, for example, Figure 11.1b), it may represent the plesiomorphic condition in the order.

There are essentially two ways of determining the number of sex alleles. Separate lines can be produced that carry known alleles and then the production of diploid males in matings between pairs of lines can be tested until the population has been thoroughly sampled. This exhaustive approach has so far only been applied to the braconid, *Habrobracon*. The other way makes use of the proportions of matched matings – that is, matings in which the female shares one of her sex alleles with her haploid male partner and so will produce at least some diploid male offspring. If the species has single-locus CSD the proportion of matched matings (Θ) is related to the number of sex-determining alleles (K) according to the formula $\Theta = 2/K$ (Ross *et al.*, 1993). This makes a number of assumptions such as equal allelic frequencies, an ability to detect all matched matings (i.e. diploid males have reasonable viabilities) and, in wild populations, monoandry. This approach has been applied to all other cases in Table 2.1. In the case of *Habrobracon*, it is believed that there are nine different alleles within the population. The numbers of sex alleles detected in a single population may, however, be a great underestimate of the total number within a species because of sampling/founder effects (Ross *et al.* 1993; Cook and Crozier, 1995). Thus most diploid individuals will be heterozygous and so develop into females.

One way to detect a single-locus CSD system is by inbreeding, which rapidly leads to the loss of alleles and so increases the likelihood of obtaining diploid males. Inbreeding experiments in *Habrobracon* and other taxa with single-locus CSD causes a detectable increase in the proportion of males after only a few generations. On the other hand, Cook (1991) inbred the doryctine braconid, *Heterospilus prosopidis*, for 15 generations and obtained no increase in the frequency of male production. This test rules out the possibility that *Heterospilus* has a single-locus CSD system, and was most unexpected since phylogenetic analyses of the Braconidae suggest that the Braconinae and Doryctinae are closely related (Quicke and van Achterberg, 1990).

In *Habrobracon*, the sex determination locus has been found to be linked to the fused gene (*fu*) which affects the development of the legs, wings and antennae, and this has been of considerable use in genetic experiments (Whiting, 1943). At the molecular level, there is as yet no explanation of how CSD might work, but there are two main hypotheses. One states that feminization of wasp tissue requires the presence of a heterodimer of the

Table 2.1 Hymenopteran taxa for which single locus CSD has been demonstrated and estimates of numbers of alleles involved

Species	Family	Estimated number of alleles	Reference
Habrobracon hebetor/ juglandis	Braconidae	9	Whiting, 1943, 1960
Diadromus pulchellus	Ichneumonidae	*c.* 15	Periquet *et al.*, 1993
Solenopsis invicta	Formicidae	*c.* 15	Ross and Fletcher, 1985
Solenopsis invicta	Formicidae	10–86*	Ross *et al.*, 1993
Bombus terrestris	Apidae	≥ 24	Duchateau *et al.*, 1994
Apis mellifera	Apidae	19	Adams *et al.*, 1977

* Introduced North American populations have fewer sex determination alleles than native South American ones, presumably reflecting a founder bottleneck.

gene product from the sex determination loci. The other suggests that the RNA products of each sex allele may be defective in some way, and that a fully functional feminizing product can only be assembled if two different, complementary alleles are present. In both cases, the default condition would be the production of male tissue. Some evidence in support of this sort of model comes from careful examination of chimaeric haploid individuals in which, it has been claimed, cells at the borders between two blocks of genetically different haploid tissue were feminized (Whiting *et al.*, 1934). This could be interpreted as being due to the diffusion and mixing of gene products (possibly peptides) between the two tissue types such that heterodimers could be formed only in the border regions.

2.3 NON-CSD SYSTEMS

It is certain that many wasps do not have single-locus CSD or even CSD involving a number of loci (multilocus CSD). This conclusion stemmed originally from the observation that in many species inbreeding does not lead to the production of any diploid male zygotes (Stouthamer *et al.*, 1992). Indeed, there are many species that routinely show sib mating over hundreds of generations, and even mother–son mating is standard in some species (e.g. Hobbs and Krunic, 1971). Such mating systems could not exist with CSD. Cook (1993a) tabulated parasitic wasp species in which repeated inbreeding does not lead to an increase in sex ratio and therefore could not have single-locus CSD – the list included seven chalcidoids, one cynipid and the bethylid, *Goniozus nephantidis*. Another line of evidence comes from cytogenetic studies, notably by Stille and Dävring (1980) on the cynipid, *Diplolepis rosae*, and of Stouthamer and Kazmer (1994) on the microbe-associated thelytokous strains of three chalcidoid species, namely *Trichogramma pretiosum*, *T. deion* and *T.* near *deion*. In these cases, the development of female offspring from unfertilized eggs involves gamete duplication in which the two anaphase nuclei produced by the first cleavage division of the embryo fail to separate and so restore the cell to diploidy. Because this gamete duplication leads to complete homozygosity, as also shown by isoenzyme studies, the production of females rules out any form of CSD.

However, whilst it is clear that there must be some other mechanisms operating in the Hymenoptera, there is far less evidence regarding the nature of such mechanisms. Cook and Crozier (1995) summarize the relevant information. Essentially, two alternative models to CSD have been proposed. One is the genic balance model of Cunha and Kerr (1957); the other is a nucleo-cytoplasmic balance model proposed by Crozier (1971). These are illustrated schematically in Figure 2.1.

The genic balance model is applied most easily to normal diploid species in which sex is determined by the balance between the number of copies of a sex chromosome (or gene or genes) relative to the number of autosomes. In haplodiploid organisms, such as the Hymenoptera, this simple version of the model does not work since the ratio of masculinizing or feminizing genes to the rest of the genome is the same. However, the model can be modified by postulating that masculinizing genes do not act additively, whereas feminizing ones do; and by postulating that a single male gene complement is dominant with respect to a single comple-

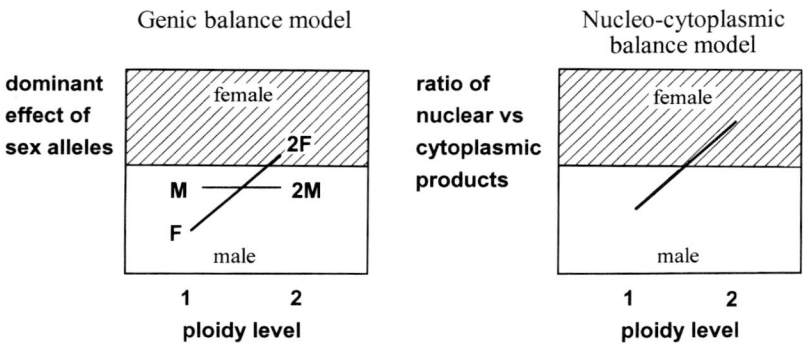

Figure 2.1 Genic balance and nucleo-cytoplasmic balance models of non-CSD sex determination in parasitic wasps.

ment of feminizing genes (Figure 2.1a). Thus a haploid individual with one copy of masculinizing and one of feminizing genes will become a male (because male is dominant), but since any number of extra copies of masculinizing genes have no greater effect than a single copy, diploid individuals, triploids, etc., will become females because the effects of the extra copies of the feminizing genes are additive and come to dominate the male ones.

The nucleo-cytoplasmic balance model proposes that sex is determined by the ratio between the amount of products of particular nuclear versus particular cytoplasmic genes, and that this ratio is affected by the ploidy level (Figure 2.1b). Neither of the above models can account for the existence of diploid males that are known to occur in many species of Hymenoptera. In the absence of other models, the presence of diploid males may be taken as strong evidence in favour of CSD in that taxon (but see below).

2.4 DIPLOID MALES

Inbreeding in taxa that have single-locus CSD will lead to the production of an increasing proportion of zygotes that are homozygous at the sex determination locus. Diploid males were identified a long time ago in the highly studied braconids, *Habrobracon brevicornis*, *H. hebetor* and *H. juglandis* (Speicher and Speicher, 1938, 1940; Cook, 1991). The taxonomic distribution of known diploid males in the Hymenoptera was surveyed by Stouthamer *et al.* (1992). Their list included five ants, six bees, three sawflies and six ichneumonoids, but only one chalcidoid, suggesting that single locus CSD may be common within the Ichneumonoidea + Aculeata clade, but rare or absent in most other groups. However, such conclusions may be premature given that there are no data for or against the occurrence of diploid males in the majority of superfamilies, let alone families. As in most other Hymenoptera in which diploid males have been detected (mostly aculeates), diploid male parasitic wasps tend to have lower viabilities than haploid ones. In some cases most diploid male individuals die early in development (Petters and Mettus, 1980), but in the ichneumonid, *Diadromus pulchellus*, they have viabilities comparable with normal haploid males.

Cook (1991) used a recessive red-eyed genetic mutant to investigate the viability of diploid males in *Habrobracon hebetor* (Figure 2.2). Starting with a heterozygous stock for both sex determination allele (A_1, A_2, ... A_n) and eye colour (B, the dominant black-eye allele and b, the recessive red-eye allele), the members of an all-male brood of a single female were isolated and reared while their mother was maintained cool and alive. The presence of red-eyed males in a brood indicated that the mother was heterozygous for both the eye colour allele and the sex determination allele.

Red-eyed sons were then mated with their mother, which yields an F_2 generation with both red-eyed and black-eyed diploid males as well as both red-eyed and black-eyed haploid males. Crosses between the fertile black-eyed haploid males and red-eyed F_2 females will yield an F_3 generation in which all black-eyed males are diploid. Impaternate diploid offspring can also be produced artificially by feeding females with a colchicine solution: the eggs so produced are tetraploid, but diploidy is often restored at the first cleavage division. In *Habrobracon*, about 5% of such offspring are males.

Much of the early work on diploid males was carried out using members of the braconid genus *Habrobracon*, primarily by Whiting and co-

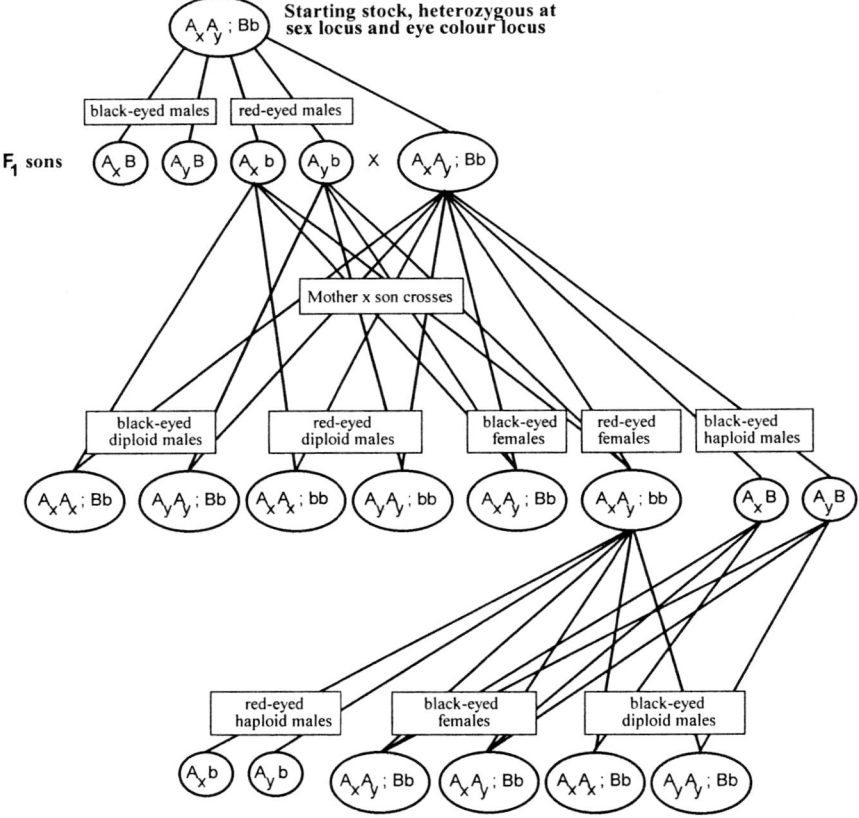

Figure 2.2 Breeding system for producing visually recognizable diploid males in a single locus CSD species in which a visible mutant line is available. In this case, *Ax* and *Ay* represent two alleles at the sex determination locus; and *B* and *b*, respectively, are dominant and recessive eye-colour alleles found in *Habrobracon*.

workers. In recent years most work on the occurrence of diploid male parasitic wasps has focused on the ichneumonid, *Diadromus pulchellus*, and the work of Periquet, El Agoze and co-workers. One of the first surprising findings from this latter work was that diploid males occur at quite high frequencies (9%) in wild populations in the environs of Tours, France (Periquet *et al.*, 1993). Diploid males in this species were routinely produced using a similar breeding strategy to that of Cook (1991), this time making use of a recessive mutant yellow strain. In *Diadromus*, diploid males were found to have essentially the same viability as haploid ones, and the external morphology of both is similar except that diploid males are bigger, resembling the females (El Agoze and Periquet, 1993). In *Habrobracon*, diploid and haploid males appear very similar, but the diploid ones can usually be told apart by the density of the very small setae (microtrichia) on the wing membrane. Each of the microtrichia arises from a single cell, and the size of these cells in diploid males is approximately twice that of those in haploid individuals, resulting in approximately half the density on the wing membrane.

In both *Habrobracon* and *D. pulchellus*, diploid males have been shown to produce diploid sperm (El Agoze *et al.*, 1994), and spermatogenesis and diploid sperm ultrastructure has been described in the latter (Chauvin *et al.*, 1987). In both cases diploid male sperm are motile and appear functional outwardly in that they reach the female's spermatheca after mating. In *Habrobracon* they have a low success in fertilizing (penetrating) eggs (MacBride, 1946) but fertilizations by diploid sperm do occasionally occur leading to triploid individuals. However, no triploid females are ever obtained from mating diploid male *Diadromus* with normal females (El Agoze *et al.*, 1994); instead, the few females that result from crosses involving diploid males are themselves diploids. The implication of this is that, in *Diadromus*, diploid males may at least occasionally carry out successful meioses and produce haploid sperm, though other interpretations cannot be ruled out. Eggs fertilized by diploid sperm apparently suffer 100% mortality during early development. This observation contrasts sharply with observations on several other parasitic Hymenoptera.

2.5 POLYPLOIDY

Apart from diploid males, which are in a sense polyploid, polyploidy has been detected in very few parasitic wasps. Triploid females occur in *Habrobracon* when diploid males, which produce diploid sperm, occasionally successfully fertilize an egg. Polyploid individuals in the pteromalid, *Nasonia* (as *Mormoniella*) *vitripennis*, can be either male or female, and in this case the critical factor in sex determination appears to be whether or not the egg is fertilized, not the ploidy level (Whiting, 1960). Unmated triploid females have a low fecundity but can give rise to both diploid and haploid male offspring, the former being slightly rarer – possibly due to the fact that to obtain a viable diploid egg, exactly two copies of each chromosome needs to go to the correct daughter nucleus during meiosis.

2.6 GYNANDROMORPHS AND MOSAICS

Gynandromorphs are chimaeric individuals with some part or parts of their body male and the rest female. They are quite common in many insect groups and have been a source of some interest among lepidopterists, for example. In non-haplodiploid taxa, such as Lepidoptera, they result from the loss of one of the sex chromosomes, usually during one of the early zygotic cell divisions. In the haplodiploid Hymenoptera, they are also usually presumed to arise from errors in cell division during early development, but because the error required would be the loss of a whole haploid chromosome set, they are perhaps of less frequent occurrence in Hymenoptera, though they are relatively common among ants (Halstead, 1988). Clark and Gould (1972) provided some evidence in *Habrobracon* that some mosaics may arise by post-cleavage fertilization – that is, synkary occurs after the first cleavage division of the oöcyte's pronucleus. This is also consistent with the finding of Petters and Grosch (1976) that the frequency of gynandromorphs in *Habrobracon juglandis* can be increased by cold-shock of newly laid eggs. This abnormal fertilization process was subsequently employed by Petters (1977) to construct an embryological fate map for *Habrobracon* (Chapter 5). Among the parasitic Hymenoptera, it has been noted that gynandromorphs occur occasionally in a number of species (Halstead, 1988), and they have sometimes been helpful for associating males and females in groups that have extreme sexual dimorphism (for example, mutillids) which are seldom either reared or found *in copulo.*

In *H. juglandis*, gynandromorphs occur quite frequently in culture and are perhaps especially common in the ebony stock, in which they occur at a rate of about 4% (Grosch, 1988). Experiments with gynandromorph individuals of this species have shown that behaviours such as recognition of mates and hosts and the ensuing attempts to either copulate or sting are entirely dependent on the sex of the head, or rather, the brain in a sex mosaic individual and not on the sense or reproductive organs (Clark and Egen, 1975). The same is found for other sex-linked behaviours.

2.7 CYTOGENETICS AND KARYOTYPIC VARIATION

Unlike many other groups of insects, there have been comparatively few cytogenetic studies on parasitic Hymenoptera. When Crozier (1975) surveyed available information on chromosome numbers, data were available for only 20 or so species representing just 11 families. Gokhman and Quicke

(1995) listed chromosome numbers for nearly 150 species, collectively representing six superfamilies and 20 families, and an updated version is presented in Figure 2.3. Nevertheless, this still leaves many families and more than half the superfamilies uninvestigated. What are the reasons for this dearth of information? Probably the main reason is that chromosomes can most easily be investigated in immature stages, tissues such a prepupal nerve cord and cephalic ganglia being particularly suitable because they contain sufficient numbers of dividing cells and these cells are almost entirely diploid (or haploid in males) rather than polyploid as in most other somatic larval tissues. Such tissues require identifiable immature stages to be available and in the case of parasitic Hymenoptera this would normally mean that the insect must be in culture (e.g. Strand and Ode, 1990). Further, some of these tissues are difficult to prepare, making reliable results difficult to obtain. In this respect, sawflies and social Hymenoptera are much easier to investigate since many of the former have easily recognizable larvae, and the latter live in colonies from which larvae or pupae can easily be obtained. It has been shown that good chromosome preparations can often be obtained from ovaries of adult female parasitic wasps, especially in recently eclosed individuals, including both mitotic and meiotic plates (Gokhman, 1989; Gokhman and Quicke, 1995). For most groups this means that it is now possible to investigate the cytogenetics of natural populations, and this in turn means that sibling species may well be found to be more common than was previously expected. Unfortunately, post-eclosion cell divisions are rare in some taxa, and other sources of chromosome spreads will be needed for these.

The total range of n values for the Apocrita is 1–46. Within the 'Parasitica', n is known to range between 3 and 17, the smallest chromosome number having been found to date in the chalcidid, *Brachymeria intermedia* (Hung, 1986) and in several members of the aphidiine braconid genus, *Aphidius* (Quicke, unpublished observations), whilst the largest value is known for a couple of ichneumonids and braconids. The majority of chromosome counts available at present are for members of just two superfamilies, the Chalcidoidea and Ichneumonoidea, which have modal values of 5 and 10, respectively. The picture may be somewhat different once more data on other superfamilies have been collected. Chromosome number appears to have some value in higher level phylogeny. Preliminary data for diapriids, scelionids and bethylids suggest that the plesiomorphic chromosome number for the Apocrita is close to 10 and consequently the smaller number in most chalcidoids (except for Encyrtidae and Eurytomidae) may represent a synapomorphic reduction. Most taxa of parasitic Hymenoptera have karyotypes dominated by metacentric and sub-metacentric chromosomes, though within the Gasteruptiidae and Cynipoidea there are more telocentric and acrocentric ones.

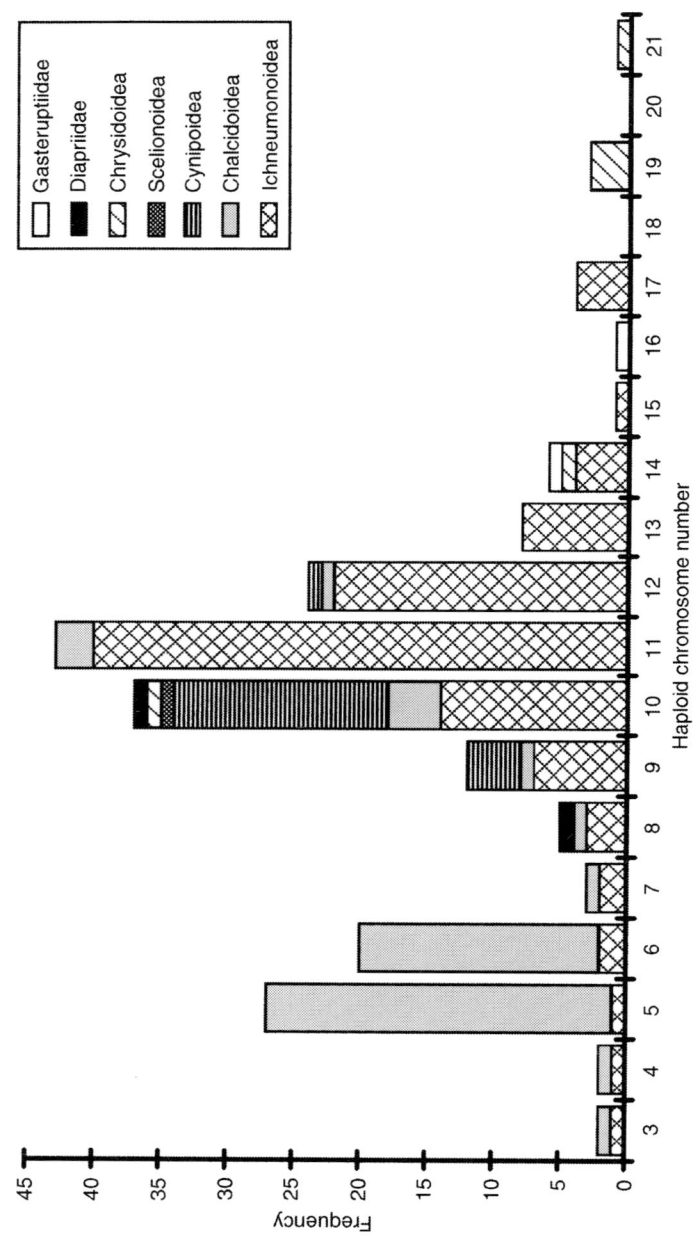

Figure 2.3 Frequency histogram of haploid chromosome numbers in parasitic wasps. (Updated since Gokhman and Quicke, 1995, in accordance with Quicke and Gokhman, in press, and unpublished observations.)

Parasitic wasp chromosomes can be differentially stained to show banding patterns, and these may facilitate recognition of particular chromosomes from others of similar size and morphology. Odierna *et al.* (1993) have illustrated G-banding patterns in somatic metaphases of the aphelinid, *Encarsia berlesei*, and their work provided some evidence for asynchronous chromosome condensation.

Chromosome number has demonstrated potential for revealing cryptic speciation (Gokhman 1991; Gokhman and Quicke, 1995). Several examples are now known among the ichneumonids in which apparently good chromosomal species have remained indistinguishable on the basis of external morphology despite intensive study. Chromosome polymorphisms appear to be rare among parasitic wasps though very few have been studied in detail in the field (Gokhman and Quicke, 1995). Gokhman (1989) has provided evidence for the long-term persistence of a chromosomal polymorphism in the ichneumonid *Tycherus bellicornis*, probably resulting from a translocation and subsequent non-disjunction of chromosomes. Possible chromosome number polymorphisms have been suggested in the cases of two ichneumonids, *Icheumon extensorius* and *I. suspiciosus* (Gokhman, 1993). Both species were revealed to comprise two classes of individuals, one with a haploid number of 12, the other 13. In both species the two chromosomal forms were apparently morphologically indistinguishable, but the possibility that each represents a pair of sibling species rather than one with variable chromosomal races perhaps seems likelier since intermediate chromosomal forms were rare or non-existent.

The number of chromosomes can also be deduced from linkage data if sufficient polymorphic loci are known, and studies in taxa such as *Habrobracon* and *Nasonia*, for which many mutant strains are available, have shown this to be so. RAPD analysis has the ability to reveal large numbers of genetic markers and so might also be useful in this respect. In one such study, Kazmer *et al.* (1995) found that RAPD markers in the aphelinid, *Aphelinus asychus*, fell into four linkage groups. Unfortunately, the karyotype of this species is unknown at present, but the related *A. mali*, and other aphelinids studied, have *n* values of 5 or 6. Unless *A. asychus* is atypical, this could suggest that Kazmer *et al.* may, by chance, have failed to pick up alleles from one or more of the chromosomes.

2.8 RECOMBINATION RATES

There are virtually no data on recombination rates in parasitic wasps, whether derived from genetic studies or from direct observations of crossover in meiosis, and estimates for the Hymenoptera as a whole are not much better (Burt *et al.*, 1991). Rössler and DeBach (1973) reported on the occurrence of chiasmata in the aphelinid, *Aphytis mytilaspidis*, as revealed by staining meiotic chromosomes in oöcytes. The observed rate was 0.9 (61/66) chiasmata per pair of chromosome arms, or approximately 5 per cell.

2.9 GENOME SIZE

Genomic size has only been investigated in five species of parasitic wasp, collectively representing only two superfamilies. For each group, two different approaches have been used. By means of Feulgen staining and quantitative cytometry, Rasch *et al.* (1975, 1977) calculated the haploid genomic size of the braconid wasps, *Habrobracon juglandis*[1] and *H. serinopae*, both to be 0.15–0.16 $\times$ 10^{-12} g DNA, and that of the pteromalid, *Nasonia* (as *Mormoniella*) *vitripennis*, to be 0.33–0.34 $\times$ 10^{-12} g, corresponding approximately to 1.4 $\times$ 10^8 and 3.0 $\times$ 10^8 base pairs for the ichneumonoid and the chalcidoid respectively. Bigot *et al.* (1991) employed DNA reassociation kinetics to calculate genome sizes of the ichneumonid, *Diadromus pulchellus*, and the eupelmid, *Eupelmus vuilleti*. This approach gave the considerably larger estimates: 1–2 $\times$ 10^9 and 10 $\times$ 10^9 base pairs, respectively. However, in both studies, the chalcidoids (represented by Pteromalidae and Eupelmidae) had larger apparent genome sizes than the ichneumonoids (Braconidae and Ichneumonidae). Further studies will be needed to show which of the estimates of genome size are closer to the truth in terms of absolute magnitude, and to see whether or not there is a real difference between the superfamilies.

2.10 GENOMIC MOLECULAR ORGANIZATION

Garguilo *et al.* (1988) have made a detailed study of the rDNA genes of various *Aphidius* species (Braconidae). They presented restriction maps of these and were able to use them to discriminate between four closely related taxa. Bigot *et al.* (1991) used DNA reassociation curves to investigate the overall composition of DNA in the ichneumonid, *Diadromus pulchellus*, and the eupelmid, *Eupelmus vuilleti*. In both, highly repetitive satellite DNA accounted for some 15–25% of the genomic DNA, with a further 26–42% comprising moderately repetitive sequences. Rather surprisingly, there were large differences between males and females in both species in terms of complexity, repeat frequency and number of subcomponents of the moderately repetitive DNA which the authors attributed to sex-linked differences in amplification in somatic tissues. The satellite DNA in both species comprises, largely or entirely, a single family of sequences (Rojas Rousse *et al.*, 1993), the structure of which was elucidated by sequencing multiple clones. Palindromic and A-rich sequences in both taxa suggest

[1] *Habrobracon juglandis* has been previously synonymized with *H. hebetor*; however, the genus is in need of thorough revision and we therefore prefer to keep data reported for *H. juglandis* separate, as this is less likely to lead to a loss of information.

that hairpin-like structures could be formed and that these may play a role in heterochromatin condensation. Gargiulo *et al.* (1988) have made a detailed study of the 28S rDNA genes of various *Aphidius* species (Braconidae) and present restriction maps of these which enable the four closely-related species to be unambiguously distinguished.

The B-chromosome of *Nasonia vitripennis*, which mediates the paternal sex ratio effect (see below), has been shown to contain three different families of tandemly repeated DNA sequence (Eickbush *et al.*, 1992). These authors suggest that two conserved palindromic sequences could be involved in its function or amplification.

Hung (1985) described the only example known to date of a tandem gene duplication in a parasitic wasp. In his case, it was shown that both males and females of the trichogrammatid, *Trichogramma marylandense*, displayed identical isozyme banding patterns for malic enzyme characteristic of heterozygous individuals. Other enzymes behaved normally and the males in question were not diploid, so they were able to conclude that each haploid genome contains two malic enzyme loci, each fixed for a different allele.

2.11 RETROELEMENTS

In many insects, retrotransposons – similar to retroviruses of vertebrates but lacking a protective protein envelope – are widely distributed, and in many cases, one to many copies of the retroelement become included within the germ line genome of the host insect. These retroelements fall into several families, based on sequence homologies. Bigot *et al.* (1994), using Southern blots, surveyed a number of parasitic Hymenoptera for the presence of so-called mariner-like retrotransposons inserted in the genome and found them to be present in virtually all of the taxa examined. Because Hymenoptera are haplodiploid and so recessive lethal genes are eliminated in the haploid male genomes, it was concluded that the insertion sites must be selectively neutral. Burke *et al.* (1993) demonstrated the presence of R1 and R2 non-LTR retrotransposons (a family of retroelements comprising two open reading frames and lacking the long terminal repeats, or LTRs) in the pteromalid, *Nasonia vitripennis*. Varricchio *et al.* (1995) have shown that, in common with many other insects, a proportion of the 28S rDNA genes of the aphidiine braconid, *Aphidius ervi*, are interrupted by a gene sequence that has a high degree of homology with members of the R1 retrotransposon virus family, suggesting that some these viruses have become endogenous in the past. All the insertions are at the same position in the 28S gene, indicating either that the insertions happened a very long time ago (more than 300 million years) or, if they occur more regularly, that the site specificity of the insertion is a feature of the retroelement. The first of these possibilities seems unlikely because the endogenous, non-functional virus gene insertion would have mutated so greatly over such a long period that it would probably not be recognizable. Further, since Hymenoptera are haplo-diploid, it might be imagined that selection against insertions that make a ribosomal gene non-functional will be particularly great in the male, and so might be expected to be eliminated over such a long time. However, some retrotransposons are known to have sequence-specific endonucleases, which could account for the constancy in the positions within the genome of multiple independent insertion events. Probably the endogenous retroelements represent a mixture of preferential and continuing insertions, horizontal and longitudinal transmission, but many more of these sequences need to be obtained and analysed before any firm conclusions are drawn. The distributions of particular integrated retrotransposon sequences nevertheless have great potential for revealing hymenopteran phylogeny.

A particularly exciting discovery is that the selfish B-chromosome of *N. vitripennis*, known as *NvPSR* (see below), includes several endogenous Gypsy/Ty3 retrotransposons, referred to as the *NATE* or *Nasonia* Transposable Element (McAllister, 1995). This Gypsy element is the first to be found in the Hymenoptera. More interestingly, while it is not present in any of the *N. vitripennis* autosomes, it is apparently present in those of other *Nasonia* species, suggesting that *NvPSR* may have originated from its sibling species and been transmitted to *N. vitripennis* through hybridization.

2.12 DOSAGE COMPENSATION

Whilst it is well known that the oögonia have twice as many chromosomes and, correspondingly, twice as much DNA as do spermatogonia in Hymenoptera, the situation for somatic tissues is far less clear. Cells in many somatic insect tissues are polyploid and the same is true for many tissues in both adult and immature Hymenoptera. White (1954) wondered whether somatic cells in male and female Hymenoptera would have the same levels of ploidy, or whether the number of endopolyploidy cycles occurring in diploid females might be half the number in males, so that the final DNA content in the two sexes would be the same. The answer has been revealed by Rasch *et al.* (1977), who used Feulgen-staining and microdensitometry to show that the amount of DNA in some adult male parasitic wasp tissues is the same as that found in equivalent adult female wasp tissues even though the germ line tissue in the two sexes differs by a factor of two (Figure 2.4). This may indicate that there is an optimum, sex-independent ploidy level for each of the various somatic tissues. Wahrman and Zhu (1993) have shown that cells from male and female *Nasonia* (as *Mormoniella*) *vitripennis* maintained in culture largely retain their original ploidy levels, though many aneuploids (25–30%) were also present, as in many cell cultures.

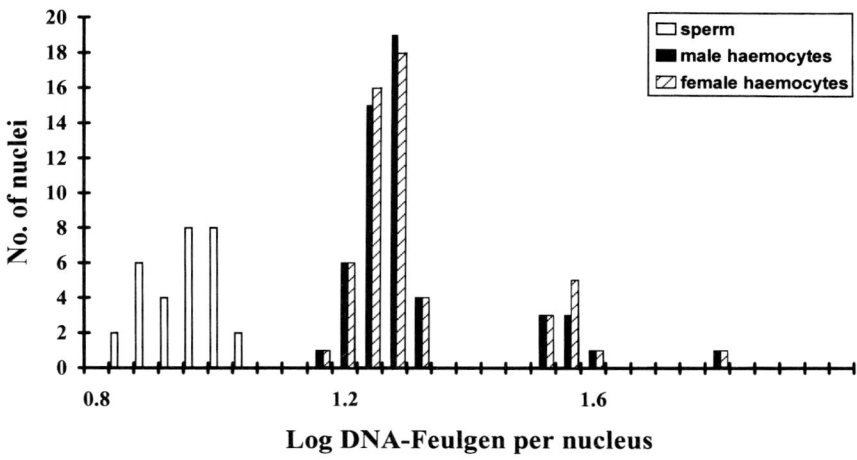

Log DNA-Feulgen per nucleus

Figure 2.4 DNA contents of individual nuclei of sperm and haemocytes from males and females of the braconid, *Habrobracon serinopae*, as measured using the quantitative Feulgen reaction. Haemocytes from both haploid males and diploid females contain similar levels ($2n$ and $4n$) of DNA. (Redrawn from Rasch *et al.*, 1977.)

2.13 GENETIC DIVERSITY

Alloenzyme electrophoretic studies have been extensively employed to examine levels of genetic diversity within populations and species of organisms. The Hymenoptera as a whole have proved a fertile area for this type of study because a number of isoenzyme investigations within the order have produced evidence that in general they display lower levels of heterozygosity than do most other insects, though as for many other features most data come from non-parasitic taxa. For example, Berkelhamer (1983) tabulated mean heterozygosity for 51 hymenopterans, but only one of these, the braconid '*Opius' juglandis*, belonged to the 'Parasitica'. Both Lester and Selander (1979) and Berkelhamer (1983) reported mean heterozygosity levels of approximately 0.036 compared with typical values for entirely diploid insects of 0.135. Explanations for this low level of genetic diversity centre around the putative effects of haplodiploidy (e.g. Pamilo *et al.*, 1978). The three main hypotheses are:

1. Increased selection against deleterious alleles which occur in males without the buffering effects of less deleterious ones.
2. Reduced effective population size and resulting increase in the rate that alleles would be expected to become fixed in the population.
3. Increased genetic linkage due to lower levels of recombination.

Suggestions that these low levels of heterozygosity might be associated with sociality (Graur, 1985) have been largely discounted since various studies had shown that they also characterize most parasitic species and some sawflies (Woods and Guttman, 1987). However, not all allozyme studies on Hymenoptera have revealed low genetic diversity. For example, Sheppard and Heydon (1986), who investigated three species of tenthredinoid sawflies, found levels that were quite comparable with those of diploid species. Boato and Battisti (1996) found the same for seven species of the pamphiliid genus, *Cephalcia*.

Packer and Owen (1992) have also challenged the view that Hymenoptera have lower than average heterozygosities by providing a very detailed study of heterozygosity in 172 species as reported in the literature (Figure 2.5). They concluded that, whilst overall levels of heterozygosity in the order may be low, many studies have not studied those enzyme loci that are most typically heterozygous. Thus, P_v, the proportion of loci detected with a given enzyme staining system that were variable within a species, was highest for guanine deaminase (EC 3.5.4.3; Gda), β-galactosidase (EC 3.2.1.23; Bgal), catalase (EC 1.11.1.6; Cat), acetylhexosaminidase (EC 3.2.1.52; Aha), amylase (EC 3.2.1.1; Amy), NADPH diaphorase (EC 1.6.99; Diap), esterase (Est), glucosephosphate isomerase (EC 5.3.1.9; Gpi) and phosphoglucomutase (EC 5.4.2.2; Pgm), each of which was variable in at least 40% of taxa (loci) examined (Figure 2.5). However, heterozygosity for several enzymes appears to be variable across the order. Thus, whilst two-thirds of the nine sawflies investigated for catalase were found to be heterozygous for this enzyme, none of the six 'Parasitica' examined were, though the coverage of different studies on allozymes of parasitic wasps was too sporadic to permit much in the way of a detailed statistical analysis. Packer and Owen (1992) suggested that various aspects of enzyme expression and ease of staining following electrophoresis may be in part responsible for the relatively few studies that have employed the most heterozygous systems. Thus, some enzymes show rather low levels of activity in adult insects compared with larval and pupal stages (e.g. Aha and Bgal) whilst others seldom give clear results (e.g. Cat and Gda). Further investigations of large numbers of loci among the parasitic taxa are clearly needed before firmer and more comparable estimates of heterozygosity will be possible.

Isoenzyme analysis of the aphidiine braconid, *Aphidius ervi*, has been used to examine changes in heterozygosity and effective population size in laboratory lines (Unruh *et al.*, 1983). Each generation was started with 30 females, but the observed per generation decline in heterozygosity ($\hat{H}_2$) of 0.037 corresponded to an effective founder population of only nine mated females. This would result in an halving of heterozygosity in only 18 generations (for this species, approximately 9 months in culture). It was suggested that this effect was most probably due to consistent inequality

		Tpl				
		Pep				
		Mdh				
		Ldh				
		Lap				
		Idh				
		Iddh				
		Hk				
	Xdh	G6pd				
	Sod	Gapd				
Umpk	Pk	Enol	6Pgd			
Skdh	Pgam	Dia	Mpi	Pgm		
Pgk	Gp	Akp	Me	Gpi		
Guk	Ark	Ao	Hbdh	Est		
Glud	Ald	Acp	G3dp	Diap		
aGlu	Aku	Acon	Fum	Amy		Gda
Fdpd	Adk	Aat	Ala	Aha	Cat	Bgal

| 0 | 10 | 20 | 30 | 40 | 50 | 60 |

$$P_V$$

Figure 2.5 Enzymes tabulated according to the proportions of loci examined across the Hymenoptera that were heterozygous within a species. (Redrawn from Packer and Owen, 1992.)

in the reproductive contributions of individual females. The consequences of this observation for culturing parasitoids for biocontrol are quite considerable (Stouthamer *et al.*, 1992).

Hung *et al.* (1988) used isoenzyme analysis to investigate whether arrhenotokous strains of the ichneumonid, *Mesochorus nigripes*, had higher or lower levels of heterozygosity than thelytokous ones. If there was stronger selection in haploid individuals, then it would be expected that thelytokous forms would be able to maintain higher levels of heterozygosity, though this might be counterbalanced if the thelytoky involved marked differences in clonal fitness. Hung *et al.* found mean heterozygosities for the arrhenotokous and thelytokous strains of 0.187 and 0.103, respectively. These did not differ significantly and so their study provided no evidence that selection in haploid individuals was higher. The possibility that other factors may be responsible for the low levels of heterozygosity in many Hymenoptera, perhaps connected with the phylogeny of the order, has to be considered.

Another technique that is gaining ground for assessing interspecific genetic diversity, and as a corollary as a tool for recognizing and detecting different strains of a species, is random amplified polymorphic DNA (RAPD) analysis (Landry *et al.*, 1993) (Chapter 11). Edwards and Hoy (1995) used this technique to detect variation within isofemale lines of the aphidiine braconid, *Trioxys pallidus*, and this procedure has been used in a number of other similar situations. At first sight, its ability to reveal many putative gene markers would make it seem ideal for identifying the origins of parasitoids after the release of different stocks in biocontrol programmes, or for assessing lineage purity. However, a word of caution was introduced by Kazmer *et al.* (1995), who studied RAPDs in the aphelinid, *Aphelinus asychus*. Bands from heterozygous parents were found to be inherited 1:1 by haploid sons, but offspring were also found to display some 16% of non-parental bands, meaning that RAPDs can only be used safely in combination with extra information on allelic inheritance, etc.

2.14 RELATIONSHIP BETWEEN HOSTS AND HETEROZYGOSITY

Nemec and Starý (1983) studied allozyme (esterase) polymorphism in several aphidiine braconids in relation to host specificity and alternation of hosts. Having shown that the allozymes detected in a monomorphic strain of *Aphidius ervi* were not influenced by the species of aphid used as the host, they then showed that transferring polymorphic strains of several species from their original host aphid to their alternate hosts caused a rapid loss of alleles. Thus it may be concluded that hosts exert different selection pressures on different alleles, or perhaps on linked alleles. The natural alternation of hosts in many aphidiine species might therefore play a significant role in the maintenance of high levels of heterozygosity. The apparent strength of such differential selection complicates the use of electromorphs for defining biotypes and may even necessitate a new definition. Following from this, Nemec and Starý (1984) proposed that the level of heterozygosity displayed by a parasitoid species developing in different hosts will indicate which host is ancestral for the species. The concept, referred to as the 'population diversity centre' hypothesis, suggests that the host in which a parasitoid displays the highest level of heterozygosity will be the one from which it has subsequently radiated out to accumulate additional hosts. Thus far, data in support of this come from studies of two aphid/aphidiine braconid systems (Nemec and Starý, 1984; Powell, 1994). It would be interesting to know whether this is a general phenomenon of parasitic Hymenoptera, or restricted either to some koinobionts or even just to aphidiines.

2.15 HERITABILITY OF BEHAVIOURS AND LIFE HISTORY PARAMETERS

It has been known for a long time that different biotypes, strains or cultures of a given species of parasitic wasp can differ markedly from one another in behavioural and life history characters. Ruberson *et al.* (1989b) surveyed intraspecific variation in morphological, behavioural, reproductive and physiological parameters across a range of taxa, with a strong bias towards species involved in biocontrol. Amongst others, intraspecific variation has been described in such features as fecundity, sex ratio production (Parker and Orzack, 1985; Ruberson *et al.*, 1989a; Orzack and Gladstone, 1994), facultative sex ratio response (see below), sequence of progeny sex allocation (Wajnberg, 1993), parasitization efficacy (e.g. Mimouni, 1992), gregariousness (Legner, 1991), developmental duration (Weseloh, 1986), response to host cues (Prevost and Lewis, 1990), host age preference (Ruberson *et al.*, 1989a), longevity (Ruberson *et al.*, 1988), temperature tolerance (Ruberson *et al.*, 1989a), fecundity (Mimouni, 1992; see p. 306) and locomotor activity (e.g. Fleury *et al.*, 1994). For example, Veerkamp (1980) showed that different strains of the widely studied *Drosophila* parasitoid, *Leptopilina heterotoma* (referred to as *Pseudeucoila bochei loc. cit.*) differ in copulation duration, offspring sex ratio and the percentage of host mortality they cause. In the braconid, *Microplitis croceipes*, experiments with isofemale lines over a number of generations have shown that the wasps' response to volatile allelochemicals is strongly influenced by familial origin (Prevost and Lewis, 1990). In a detailed investigation of 41 isolines of the mymarid, *Anagrus delicatus*, Cronin and Strong (1996) demonstrated significant genetic variability in fecundity, time spent in a host patch, number of ovipositions per patch, and ovipositor and hind tibia length. Of the variables investigated, significant heritability was demonstrated for fecundity ($h^2 = 0.47\pm0.16$; Figure 2.6) and ovipositor length ($h^2 = 0.36\pm0.17$). Geographical variation in 'virulence' (resistance against encapsulation), sex ratio, host selection and diapause features have been found in the braconid, *Asobara tabida*, a parasitoid of *Drosophila* spp. (Kraaijeveld and van Alphen, 1994, 1995; Kraaijeveld *et al.*, 1995), and demonstrated corresponding geographical variation occurs in the host (Kraaijeveld, 1994). This variation can provide many opportunities for investigating the costs of defence and virulence and their underlying genetics and physiology.

Artificial selection has successfully been employed to alter sex ratio in the chalcidoids, *Nasonia vitripennis* and *Dahlbominus fuliginosus* (Wilkes,

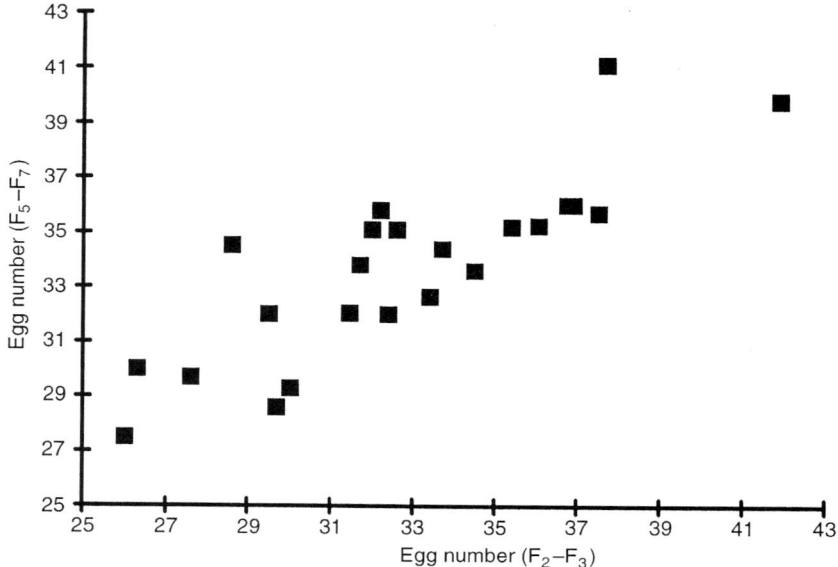

Figure 2.6 Correlation between mean fecundity in early (F_2–F_3) generations and late (F_5–F_7) generations of 41 isolines of the mymarid, *Anagrus delicatus*. (Redrawn from Cronin and Strong, 1996.)

1964). In the latter, the high male-producing selection line appears to have been the result of selecting for males that produce an abnormally high proportion of non-functional sperm with a sinistrally coiled head (Chapter 4), and so was perhaps best regarded as selecting for male infertility.

2.16 MATING STRATEGY AND VIRGINITY

Nearly all work on parasitic wasp mating strategy comes from laboratory investigations, and only a small amount from field studies. One of the few 'field' studies is that of Antolin and Strand (1992) who investigated the mating strategy of the braconid, *Habrobracon hebetor*, an idiobiont ectoparasitoid of stored grain moths – *Plodia interpunctella* in their study. Males of this wasp aggregate at the tops of piles of corn where they compete for mates, most females dispersing from their emergence sites before being mated. *H. hebetor* has single-locus sex determination (section 2.2), which means that inbreeding will tend to produce diploid males and therefore be particularly disadvantageous. As the mating system of this species greatly reduces the likelihood of sib–sib mating, there is likely to be only a very low level of diploid male production under 'field' conditions. Interestingly, *H. hebētor* is an example of a species in which searching for males is incompatible with searching for hosts since the females hunt for hosts beneath the surface, rather than on the surface of the stored produce where the males are. Therefore, there may be a trade-off between the two behaviours. Guertin *et al.* (1996) found that approximately 10% of females do not mate at all and they suggest that this may represent a split sex ratio strategy.

The fact that virgin female Hymenoptera can still pass on their genes to the next generation by laying unfertilized, male-producing eggs means that, unlike many other organisms, there is likely to be less selective pressure to find a mate. Thus it might be expected that, especially among wasps that emerge late in the season when males are relatively scarcer, a proportion of females will remain virgins and lay all male-producing eggs. Although many studies have shown that virginity is uncommon in most parasitic Hymenoptera (Godfray and Hardy, 1993; Hunter, 1993), many deal with taxa that have overlapping broods, and none so far have followed mating status of females throughout a whole season. More work is needed to assess the true degree of virginity in parasitic wasp populations.

Hardy and Cook (1995) showed that in the bethylid, *Goniozus nephantidis*, a gregarious parasitoid of dermestid beetle larvae, many broods contain no males at emergence. As this species appears to rely entirely on sib-mating at the emergence site, the females from such broods are likely to remain virgins and so, in turn, lay all-male broods. Such broods, given that this is a typical local-mate competition species, would be highly detrimental because the all male broods are unlikely to get any matings. Failure for broods to produce males might have been due to pre-adult mortality of the one or few males in each brood – some 28% of *G. nephantidis* young die before emergence. However, that this is not the case was apparent because all-female broods constituted a similar proportion of those broods in which there was no mortality. Explanation of this unlikely observation may require more detailed study of the actual mating system of wild *G. nephantidis* populations in which sib-mating might plausibly not be such a strict rule. Because of the effectively near-zero fitness of obligately sib-mating wasps that fail to produce a male in any given brood, it had been postulated that virginity in such taxa would be far more strongly selected against than it would in out-breeding species (Godfray, 1988). Godfray and Hardy (1993) surveyed evidence of virginity in 24 species of parasitic Hymenoptera and related this to mating structure (though because of the sparsity of information on actual mating structure in parasitic wasps, these authors used gregariousness as a probable correlate of sib-mating). Contrary to expectations, they found no evidence to support the idea that the frequency of virginity was related to mating structure, and possible reasons for this were discussed by Heimpel (1994).

2.17 EFFECT OF MATING STATUS ON OVIPOSITION BEHAVIOUR

Several studies have examined the effect of mating on oviposition behaviour in parasitic wasps. Virtually all have shown that virgin females attack fewer hosts in a given time than do mated ones of an equivalent age (e.g. Michaud, 1994). There is also a tendency, at least in some species, for virgins to avoid superparasitism more than mated individuals (Michaud, 1994). When a female that has already laid eggs as a virgin is then given the opportunity to mate, there is usually no problem, and after mating she assumes oviposition behaviour characteristic of females that mated directly after emergence. However, there are a few exceptions and in some taxa, females appear not to respond to their own mating status (see below; Cole, 1981; Ueno, 1995).

2.18 PROTANDRY

In almost all species of parasitic wasp, males emerge a day or more before females – a phenomenon referred to as protandry. The advantage of protandry is presumably far greater in those species in which females only mate once: a late-emerging male is likely to encounter predominantly already mated females, and he is genetically doomed as he is not likely to get any matings with these. Since the majority of parasitic wasp females only mate once (Chapter 8), protandry is probably particularly common in this group of insects. As always, there are some exceptions; for example, in the heteronymous autoparasitoid, *Encarsia deserti* (Gerling *et al.*, 1987). However, no one seems to have systematically investigated whether protandry is less developed in taxa with polyandrous females.

Even though males are typically smaller than females among the parasitic Hymenoptera, protandry generally seems to reflect a shorter pupal phase for males, rather than any gender difference in the period of larval development (Eller *et al.*, 1990).

2.19 SPERM COMPETITION

There have been few studies on sperm competition in the Hymenoptera and very few among parasitoid species. This dearth no doubt in part reflects the reluctance of females of many parasitoid species to mate more than once (section 8.33). Of the studies that have been carried out, the majority have found some (and sometimes pronounced) first-male precedence, which at first might seem surprising given the simple spherical nature of the spermatheca in most or perhaps all parasitic wasps. Using eye colour mutants of the pteromalid, *Nasonia vitripennis*, Holmes (1974) showed a strong first-mating dominance: first-mating (virgin prior to the experiment) males fathered 98% of daughters on average. However, in the same species, van den Assem and Feuth-de Bruin (1977) found that the second male fathered a constant proportion of offspring, indicating that there was no precedence for either the first or second male's sperm. Van den Assem and Feuth-de Bruin's result for this species was also found by Beukeboom (1994) who used the effects of a sex-ratio distorting B-chromosome as a marker (section 2.33). In the pteromalid, *Lariophagus distinguendus*, van den Assem *et al.* (1989) found some evidence of first-male precedence, whilst El Agoze *et al.* (1995) showed strong first-male sperm precedence in the ichneumonid, *Diadromus pulchellus*. The technique of El Agoze *et al.* was interesting in that it made use of the fact that diploid males of *D. pulchellus* are almost completely sterile – though significantly, when females are mated with diploid males, the total number of progeny remains about the same as in normal matings, the difference being that diploid male matings yield almost entirely male offspring. El Agoze *et al.* showed that if a female is mated consecutively with a diploid and a normal male, then the progeny produced consisted almost entirely of males. This was true whether the inter-mating interval was 5 minutes or 24 hours. First-male precedence was also found using genetically marked stocks in the aphelinid, *Aphytis melinus*, by Allen *et al.* (1994). First-male precedence in all of these cases could result from the spermatheca having been largely filled by the first insemination so that only a less than maximal amount of the second male's sperm can gain access (Holmes, 1974). Consequently, one might predict that the proportion of daughters fathered by the first male would be reduced if that male had been partly exhausted of mature sperm through previous matings, as has been demonstrated in several species (section 4.5).

Most of the previous studies did not take into account the time interval between matings, and it is now clear that this can be an important factor. Working with eye colour mutants of the eulophid, *Dahlbominus fuscipennis*, Wilkes (1966) showed a strong first-male preference if the interval between matings was greater than 24 hours, but approximately equal paternities if the inter-mating interval was shorter. In these cases, the absence of a clear precedence when inter-mating interval was short could be interpreted as a result of mixing of sperm in the female reproductive tract before they reach the spermatheca, or perhaps once inside the spermatheca it takes a while for sperm to become properly aligned. The opposite trend was observed by Martínez-Martínez *et al.* (1993) with genetically marked stock of the opiine braconid, *Diachasmimorpha longicaudata*. This species, too, showed no particular first- or second-male preference when the inter-mating interval was short (less than 6 minutes), though the order of release of sperm from the spermatheca indicated some ordering, with second males fertilizing a higher proportion of eggs laid soon after the mating, but fewer later on. When the inter-mating period was approximately 24 hours, the proportion of eggs fertilized by the second male was markedly higher (75%). Given the simplicity of the spermatheca in parasitic Hymenoptera, it is difficult to envisage how the proportion of paternities achieved by the second-mating male could increase with the inter-mating interval. Leakage of sperm from the spermatheca over time seems unlikely, given the apparent control the female has over sperm release in most species; and similarly it seems unlikely that the first male's sperm would suffer a markedly higher rate of death than the second male's, given the potential longevity of sperm in other species. The observation that, in some species, second males that mate soon after the first can achieve nearly equal levels of paternity might provide an explanation for the post-copulatory courtship behaviour seen in many species (Chapter 8).

2.20 SEX RATIO

Although females of a given species have a life-time optimum sex ratio, many species lay different sex ratios at different times of their lives. Sometimes, fertilized females produce significantly more males soon after mating, and the proportion of females increases gradually thereafter (Génieys, 1925; Donaldson and Walter, 1984). However, in many and perhaps the majority of species, females can and do lay a high proportion of fertilized eggs from immediately after mating, and later in a female parasitoid's life, at least under laboratory conditions, females tend to lay increasing proportions of male eggs (for example: King, 1987; Tillman, 1994; Brodeur and McNeil, 1994). This gradual, or in some cases rapid, increase in male production could be due to the female running out of sperm, or at least running out of viable sperm, and so may be out of her control rather than being adaptive. Whether female parasitic wasps in the wild ever run out of viable sperm remains to be demonstrated.

Using iso-female lines, Orzack and Gladstone (1994) have shown that sex ratio in the pteromalid, *Nasonia vitripennis*, is under partial genetic control with a heritability of between 0.05 and 0.15. Sex ratio heritability was found not only for the sex ratio of eggs laid on normal host pupae, but also for those laid on pupae which had previously been parasitized by another female and which would normally elicit an increased proportion of males in the brood because of the reduced level of sib-mating that would normally result (see below).

Seasonal variation in sex ratio has been noted by Godfray and Shaw (1987) for the eulophid, *Eulophus larvarum*, with the summer generation having an even sex ratio whilst the spring generation is more female biased, and by Allen *et al.* (1994) in the aphelinid, *Aphytis melinus*, with autumn broods having a higher proportion of males. Both species therefore show the same trend. In the case of *E. larvarum*, the between-brood sex ratio variance also greatly increased in the second generation. As the spring generation parasitoid pupae stay together and the species displays a high level of local mate competition, the female-biased sex ratio in that generation is to be expected. Godfray and Shaw suggested that, in the absence of any evidence of a difference in brood size between the generations, the greater proportion of single sex broods and higher proportion of males in the summer generation could be explained either by high levels of virgin oviposition or by strong intersexual competition within broods such that females producing mixed broods will be at a disadvantage.

Precise sex ratios are known to be laid by several gregarious parasitoids (section 2.25) but recently an interesting case of precise sex ratio production has been described for the megaspilid, *Dendrocerus carpenteri*, a solitary hyperparasitoid of aphids parasitized by aphidiine braconids (Chow and Mackauer, 1996). In this species, females decide on the sex of the first egg to be laid in a host patch according to host quality, but the subsequent egg is of the opposite sex irrespective of host quality. This behaviour ensures a one-to-one sex ratio and therefore perhaps maximizes the probability of each daughter being fertilized.

2.21 SEX ALLOCATION DEPENDING ON HOST TYPE AND SIZE

It is very well known that many parasitic wasps show a wide range of intraspecific size variation and, further, that males tend to be smaller than females, which probably reflects the fact that size is not such an important determinant of fecundity in males as in females (Godfray, 1993). Since parasitic Hymenoptera can control the sex of each individual offspring, it follows that they often make offspring sex decisions according to the size (or quality) of the host resource. For idiobiont taxa (Chapter 3) whose hosts do not develop further after parasitization, females are particularly likely to lay male eggs in or on smaller hosts and female eggs in or on larger ones. Similar decisions are also made by some koinobionts (that is, taxa whose hosts can still develop and grow after parasitization) and this is especially likely in those species whose host's growth potential after parasitization is rather limited (Tanaka *et al.*, 1992).

Whilst it is easy to understand how both ecto- and endoparasitoids of exposed hosts might adjust their sex ratio in accordance with host size, it is more surprising that the same response is shown by at least some parasitoids of concealed hosts: they cannot see or physically measure their hosts. Urano and Hijii (1995) described two such cases involving idiobiont braconid ectoparasitoids of subcortical coleopteran larvae (Cerambycidae, Curculionidae, Scolytidae). In both *Atanycolus initiator* and *Spathius brevicaudis*, the sizes of emerging wasps was positively correlated with host size.

2.22 SIB–SIB, MOTHER–SON MATING AND LOCAL MATE COMPETITION

Outbreeding populations will always tend to evolve towards equal investment in male and female offspring. In situations where the investment in both sexes is equal, or virtually so, this means that sex ratios will approximate unity (Fisher, 1930). However, as the mating system tends towards inbreeding, the predicted sex ratio becomes more female biased. If hosts are distributed in patches that are attacked by only small numbers of female parasitoids, and the offspring of these mate within the patch without dispersing, then inbreeding between the offspring of these foundresses will be high and will be inversely related to the initial number of parent females that oviposited in the patch. In these mating systems, the individual's competition for mates is restricted to those present within the patch, and hence is referred to as local mate competition (LMC) (Hamilton, 1967). The importance of LMC is that it accounts for most of the strongly female-biased sex ratios found among parasitic Hymenoptera. In extreme cases, as found in many gregarious parasitic wasps, female offspring always mate with one of their own brothers, a situation known as sib-mating.

Sib-mating is rare in solitary parasitoids but may be common in gregarious ones or in parasitoids which have strongly aggregated hosts, such as many egg parasitoids (Askew, 1968; Waage, 1982). In many gregarious species, the adult female will lay a very strongly female-biased brood, sometimes with only one (e.g. Klein *et al.*, 1991) or two male eggs (e.g. Braman and Yeargan, 1989). These males are typically protandrous; that is, they emerge before their sisters and will mate with each sister in turn as she emerges. In some cases, mating may even take place within the host remains. A similar situation exists with pollinating fig wasps. The mating strategies of the female parent might serve to reduce the effects of inbreeding through off-patch or multiple matings.

LMC is expected to lead not only to an extremely female-biased sex ratio, but also to low sex ratio variance. Thus in an extreme situation involving small broods in which the normal optimal number of males per brood is only one, production of two males is detrimental because it means effectively one fewer female offspring; but production of no males leads to no fertilized females, which in strictly sib-mating species would mean zero fitness. Morgan and Cook (1994) provided evidence for low sex ratio variance in the gregarious LMC bethylid, *Laelius pedatus*, which lays the most precise sex ratio of any known parasitic wasp. Nearly all broods contained exactly one male, and the proportion that lacked males was very close to what would be expected from the probability of mortality.

Evidence for the predicted effects of LMC on sex ratio was reviewed by Hardy (1994). Whilst a reasonable body of evidence was in qualitative agreement with expectations, the review highlighted the need for more studies of mating structure in wild populations. Hardy concluded that in most cases there is probably a significant amount of away-from-patch mating.

Mother–son mating appears to be normal for some eulophids. Balfour-Browne (1922) and Hobbs and Krunic (1971) described how an unmated female of *Melittobia acasta* which has laid a small batch of male eggs will stay with these throughout their 20 days development, and mate with one of her resulting sons upon their eventual emergence. The now mated wasp will then lay a large batch of eggs which will mostly develop into daughters. In *Habrobracon*, females mate preferentially with their own sons – time to copulation and number of courtship attempts being lower than with brothers or unrelated males. This seems odd since *Habrobracon* is an outbreeding species (Guertin *et al.*, 1996); however, it may be related to increased colonization potential.

2.23 DETERMINATION OF PRIMARY SEX RATIO

Primary sex ratio is the sex ratio of the zygotes of a female. In the case of parasitic wasps in which sex is determined by whether or not an egg is fertilized as it is laid, it is the ratio of unfertilized to fertilized eggs laid. If all eggs develop to maturity, then the primary sex ratio will be reflected in the sexes of the adults emerging. However, very often there is differential mortality of male and female offspring, so the sex ratio of emerging adults, the secondary sex ratio, will differ (often markedly) from the primary ratio. Differential survival can manifest itself at any stage from embryo to pupa.

It is often very difficult to determine primary sex ratio in parasitic wasps because of differential survival, or, in the case of polyembryonic or other gregarious species, because of differential egg division or differential siblicide (Chapter 5). In some cases, it may be possible to determine primary sex ratio because of behavioural differences during ovipositing fertilized versus unfertilized eggs – for example, in the position where they are laid – or through movements that are associated with the fertilization process itself (Cole, 1981; Suzuki *et al.*, 1984; van Dijken and Waage, 1987; LaSalle, 1990a; section 2.24). However, in most situations no obvious difference exists, and so other approaches to determining primary sex ratio have to be taken. These can be divided into direct cytological tests and indirect assessment of sex-specific mortality.

As an example of the first of these, van Dijken (1991) determined primary sex ratio of the cassava mealybug parasitoid, *Apoanagyrus lopezi* (as *Epidinocarsis*), cytologically by staining and counting metaphase chromosomes in recently deposited parasitoid eggs using lacto-acetic orcein. In principle, there is no reason why other chromosome staining methods should not be just as reliable. However, the ease with which chromosomes can be stained in eggs varies considerably between taxa: the technique may be more applicable to some than to others, and may be more difficult in species with large, yolky eggs. In solitary parasitoids, haplodiploidy also allows the assessment of sex-specific mortality since unfertilized females lay only male eggs. Thus, male mortality can be calculated by observing virgins ovipositing and rearing their offspring. Armed with this information one can back-calculate primary sex ratio, and the female-specific mortality, from that observed at emergence from mixed broods. Of course, this is not a direct method like the cytological one, and it is therefore open to several possible sources of error, such as host quality effects, and considerable care in experimental design may need to be exercised to minimize these.

2.24 SIGNS OF FERTILIZATION

In general it is not possible to detect whether a wasp has fertilized a given egg without examining it (van Dijken, 1991) or rearing it through. Fertilization state can be assessed in some heteronomous parasitoids because male and female eggs are laid in different places (Walter, 1986), but in most other cases there is no sex difference in oviposition site. A very nice example of sex-related ovipositional behaviour occurs in the eulophid, *Phymastichus coffea*, which parasitizes adults of the coffee berry borer, *Hypothenemus hamperi*. In this species, the wasp lays female eggs in the abdomen of the host and male ones in the thorax (LaSalle, 1990a). The heteronomous aphelinid, *Coccophagus bartletti*, lays female eggs if the wasp inserts her ovipositor only once into an host, but male ones if the ovipositor is inserted twice. Unfortunately, not all ovipositor insertions in this species are associated with oviposition. In the eulophid, *Pediobius foveolatus*, it is not possible to tell if an individual egg has been fertilized, but unmated females take longer to lay each male egg than do mated females which predominantly lay fertilized, daughter-producing ones (Hooker and Barrows, 1989).

Interestingly, at least in the pimpline ichneumonids, *Itoplectis maculator*, *I. naranyae* and *Pimpla nipponica*, fertilization can be inferred with a high degree of accuracy from the female's movements during oviposition (Cole, 1981; Ueno, 1995). When unfertilized eggs are laid, the lower ovipositor valves are usually moved to-and-fro continuously without interruption from the commencement of oviposition to deposition of the egg. When fertilized eggs are laid, the to-and-fro movement of the lower valves is always interrupted for 1–2 seconds just prior to the egg being seen to pass along the ovipositor, presumably allowing time for sperm to be released on to it as it passes the spermathecal duct. Since fertilized eggs are always associated with the paused behaviour, but about 10–20% of pauses gove rise to unfertilized eggs, the fertilization process is not 100% efficient – perhaps due to sperm depletion, imperfect control of sperm release or reduced/suboptimal sperm quality. Further, Ueno noted that both behaviours, continuous and interrupted, were displayed in similar proportions by both mated and virgin wasps, and concluded that in these species females were unable either to remember or determine their matedness status; that is, virgin females still seemed to be trying to lay fertilized eggs. It seems quite likely that such interruptions are not restricted to this one species, and may be widely distributed among the order, in which case it might provide a useful tool in many investigations.

2.25 SEX RATIO IN GREGARIOUS BROODS

The optimal sex ratio that a wasp should lay in either a gregarious or quasi-gregarious brood will be determined by a number of factors, including the degree to which LMC takes place and consequent foundress number and clutch size effects, the relative fitness gain from producing male versus female offspring, the resource size, etc. (Godfray, 1993). Optimum sex ratio has been extensively studied in the gregarious pteromalid, *Nasonia vitripennis*, which is a typical LMC species. Males of this species, which are brachypterous and flightless, emerge before the females and position themselves near an emergence hole in their fly puparium host from which females will subsequently emerge. Ovipositing *Nasonia* adjust the sex ratio of their brood according to how many other females are searching for hosts on the same patch in accordance with LMC theory (Werren, 1983).

Braman and Yeargan (1989) studied sex allocation in the scelionid, *Trissolcus euschisti*, when presented with egg broods of its host, the spined soldier bug, *Podisus maculiventris* (Pentatomidae), that had already been attacked to various extents by a conspecific female. Scelionids are unusual among the parasitic wasps in that they tend to lay male eggs at the beginning of the egg-laying sequence (Waage, 1982). Therefore if a female encountered a host egg cluster that had already been partially attacked by another scelionid it was predicted that it might adjust its sex ratio towards a larger proportion of males in agreement with LMC theory (Hamilton, 1967; section 2.22). However, no such effect was found in the case of *T. euschisti*.

Pickering (1980) investigated sex ratios in the gregarious ichneumonid, *Pachysomoides stupidus*, an idiobiont ectoparasitoid of polistine wasps. The ichneumonid female lays from one to 31 eggs on a single host. Her young start feeding on the host's abdomen, and as they grow they progressively eat their way towards the host's head. However, as they grow, the confining wall of the host's nest cell mean that only the more competitive can continue to grow. Those that are excluded from feeding pupate early and give rise to small individuals. Indeed in *P. stupidus*, individuals of either sex range in size by a factor of 12, as measured by dry weight. Female larvae are believed to have a competitive advantage over their brothers, and so if the female wasp were to lay broods with a one-to-one sex ratio on each host, the larger the brood, the greater will be the proportion of males that will fail to attain their maximum potential size, and some may fail to develop fully. Pickering examined the sex ratio from a large number of broods (Figure 2.7) and found that brood sex ratio was significantly over-dispersed. It is proposed that the female wasp, by laying either predominantly male or predominantly female broods, equalizes the fitness and numbers emerging of both sexes, though the theoretical basis for this assumption requires further investigation (Godfray, 1993).

Polyembryonic taxa necessarily produce gregarious broods. In most species, these are strongly female biased in agreement with LMC theory. In polyembryonic encyrtids, large broods resulting from oviposition of a single egg will be either all male or all female, and usually, within a species, the size of such broods is roughly similar. When mixed sex broods occur, it is believed that they almost always arise from deposition of one male and one female egg, but the emerging adults have a far from equal sex ratio, and are strongly female biased. This appears to result from selective siblicide (fraternicide), mediated by special precocious larvae (section 5.9). It is not yet clear whether this represents a selfish female behaviour or is actually also advantageous to the male genotypes, though probably the latter is the case. Surprisingly, in an unidentified polyembryonic species of the encyrtid genus, *Copidosoma*, Rössler and DeBach (1972) found that some 'all-male' broods actually contained a single female. Moreover, many broods were unisexual and so the offspring would have to disperse from the emergence site to find mates. The presence of a single female in otherwise all-male broods is difficult to explain because, assuming the different sexes originated from different eggs, it requires a mechanism for the differential mortality of daughters in mixed broods, or the production of non-polyembryonic female-producing eggs.

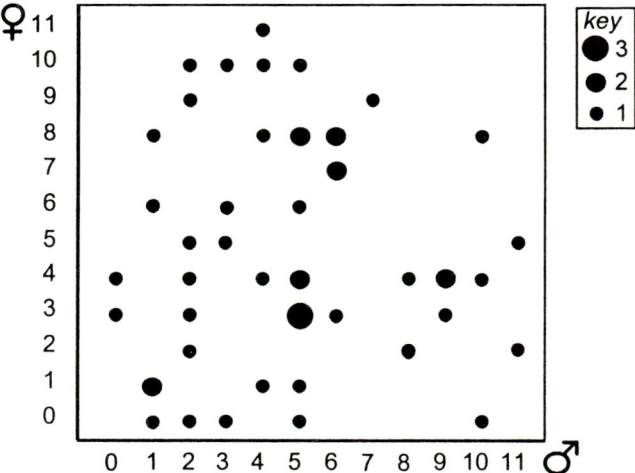

Figure 2.7 Distribution of sex ratios in broods of the ichneumonid, *Pachysomoides stupidus*, an ectoparasitoid of polistine wasps. (Redrawn from Pickering, 1980.)

2.26 HETERONOMY AND HETEROTROPHY

In some parasitic wasps belonging to the chalcidoid family Aphelinidae, males and females have different developmental biologies and host relationships (Flanders, 1964) – a system known as heteronomous parasitism (Walter, 1988a). In all such cases, females develop as primary endoparasitoids of homopteran nymphs (whiteflies or scale insects), whilst males develop in a number of different ways depending on the species, and a fairly complex though self-explanatory terminology has been developed to describe these. The different known relationships are summarized in Table 2.2.

Heteronomous biologies are known in about 10 genera of Aphelinidae. In most species, the males are obligate hyperparasitoids of the larvae or pupae of their conspecific primary parasitoid females, a relationship referred to as obligate heteronomous autoparasitism. In these cases, unfertilized haploid eggs will not develop as primary parasitoids (Walter, 1983b). In some instances, the pupa of a closely related species may also be used as a host for male eggs, as in the case of *Encarsia smithi* and its sibling species, *E. opulenta* (Nguyen and Sailer, 1987). Such species are referred to as facultative autonomous autoparasitoids. In heteronomous alloparasitoids, males never parasitize their conspecific females but instead develop as hyperparasitoids of quite unrelated primary parasitoids, including other aphelinids as well as various encyrtids or eulophids (Williams and Polaszek, 1996). Among heteronomous aphelinids, it is interesting to note that members of a single genus can show a variety of different host rela-

Table 2.2 Summary of male biologies in heteronomous aphelinids (based on Walter, 1988a)

Host relationship	Host	Terminology
Primary ectoparasitoids	Usually the same species of scale insect or whitefly as the conspecific, endoparasitic female	Diphagous parasitoids
Primary endoparasitoids	Lepidoptera eggs	Heterotrophic parasitoids
Hyperparasitoid*	Conspecific male or female larvae or pupae	Obligate heteronomous autoparasitoids
Hyperparasitoid*	Male or female larvae or pupae of conspecific or other parasitoids	Facultative heteronomous autoparasitoids
Hyperparasitoid*	Male or female larvae or pupae always of other parasitoid species	Heteronomous alloparasitoids

* Some are endoparasitic, though most are ectoparasitic.

tionships. Thus within *Coccophagus*, there are species which are diphagous parasitoids (e.g. *C. hemera, C. bartletti*) and obligate (e.g. *C. semicircularis*) and facultative (e.g. *C. lycimnia, C. atratus*) autoparasitoids.

Some obligate autoparasitoid aphelinid species lay either a fertilized or an unfertilized egg in their host whitefly, but whilst female eggs develop normally, the male egg completes embryogenesis and then enters a quiescent phase (inhibited hatching) which is probably analogous to diapause or the protracted 1st instar stage of many more typical koinobionts (Walter, 1983b). The solitary male larva can remain in this quiescent state for weeks or months and hatching will only take place when and if a female egg has also been laid and has developed to a suitable stage. A particularly interesting example is provided by *Pteroptrix* (as *Archenomus*) *orientalis*. In this species, females lay either a single egg or two eggs in a single host (Viggianni and Garonna, (1986)/1987; Viggianni, 1988). If two are laid, one is always a female and the other a male. As above, female larval development is rapid, whereas the male egg hatches but remains as a quiescent first instar larva until its sister has reached the appropriate prepupal phase for parasitization. In the case of the genus *Casca*, male eggs are laid in the embryo of a female, but the latter is quickly killed and the male is secondarily a primary parasitoid of the coccid – thus although the male egg will not develop in the primary host, the resulting larva will (Flanders, 1966).

Heteronomous autoparasitism has consequences for sex ratios in natural populations, and also poses the interesting question of how and why did this behaviour evolve. Several scenarios for the evolution of heteronomous biologies in the Aphelinidae have been proposed over the years (Zinna, 1961, 1962; Flanders, 1967; Walter, 1983a). As regards the sex ratio consequences of autoparasitoids, Hunter (1993) studied a population of *Encarsia pergandiella* in New York. Females, for some unexplained reason, showed a distinct preference for ovipositing in whitefly that already contain a developing female primary *Encarsia*, even though unparasitized hosts were far more numerous and would have yielded daughters with presumably a higher resulting fitness gain. Further work on other species would show whether this is a general tendency in autoparasitoids.

2.27 CYTOPLASMIC INCOMPATIBILITY

Many insects, and at least some other arthropods, possess symbiotic microorganisms which are responsible for the occurrence of cytoplasmic incompatibility strains (Rousset *et al.*, 1992). In these cases, when crosses occur between individuals carrying different strains of the incompatibility organism, the egg is fertilized but syngamy (fusion) of the paternal and maternal nuclei fails to occur, and the resulting embryo is consequently haploid. In most insects these haploid embryos fail to develop, but in haplodiploid species they develop successfully into males (Breeuwer and Werren, 1990).

Within the parasitic Hymenoptera, cytoplasmic incompatibility agents are best studied in the pteromalid, *Nasonia vitripennis*. Inbreeding various eye colour mutant stocks and wild type stocks can rapidly lead to the formation of variously incompatible lineages (Conner and Saul, 1986). As in other insects, incompatibility manifests itself by an inability to cross with other strains – mating and syngamy occur but no daughters are produced. Treatment of stocks with antibiotics such as tetracycline has complicated effects (Richardson *et al.*, 1987). Females reared from hosts that had been injected with antibiotic were incompatible with males that would otherwise have been compatible; these females mated with incompatible males obviously produce only sons, but the latter are surprisingly compatible with females that would otherwise have been incompatible. When a female mates with an incompatible male, his sperm enter her eggs but the sperm nucleus degenerates into a tangled mass of chromatin and fails to fuse with the female pronucleus (Ryan and Saul, 1968). The reversal of incompatibility is due to the elimination of rickettsia-like bacteria that are endosymbionts of the wasps. In *Nasonia* the cytoplasmic incompatibility organism has been shown by sequence analysis of 16S rDNA to belong to the genus *Wolbachia* of the alpha-Protobacteria (Breeuwer *et al.*, 1992). In order for a cross to be compatible, both the male and female wasp must carry the same *Wolbachia* strain or the male must be without *Wolbachia*. Although *Wolbachia* are present in the testes of infected males, they are not transmitted in the sperm (see, for example, work on *Drosophila* by Bressac and Roussets, 1993); however, the sperm themselves must be modified in some way that identifies them as having come from a male carrying a different *Wolbachia* strain and renders them incapable of fertilizing an egg unless repaired. *Wolbachia* in the egg can then make the sperm's DNA functional if they are of the correct strain, but the precise mechanism is not known. Thus far, two strains of *Wolbachia*, A and B, have been found in *Nasonia* (Perrotminnot *et al.*, 1996). Other evidence for cytoplasmic incompatibility factors has come, for example, from the detailed crossing experiments between various populations of *Trichogramma* by Pinto *et al.* (1991) (section 11.10).

2.28 PARTHENOGENESIS – THELYTOKY AND DEUTEROTOKY

A widespread type of reproduction in the parasitic Hymenoptera is a form of parthenogenesis in which unfertilized females produce all-female broods, termed thelytoky (Table 2.3). In some other taxa, unfertilized eggs may give rise to either females or males: this is termed deuterotoky. In fact, many so-called thelytokous species are strictly speaking deuterotokous by this definition, but male production is rare. The mechanisms involved may be very different from those that give rise to the cyclical deuterotoky in taxa such as many cynipid gall wasps. Thelytoky may be characteristic of a species as a whole or more usually it occurs in particular races (e.g. Hung *et al.*, 1988; Aeschlimann, 1990; Hunter *et al.*, 1993). It is clear that several different forms of thelytoky exist in different wasps. These include cyclical thelytoky, which is common in gall wasps (Cynipidae) and also occurs in the ichneumonid *Sphecophaga vesparum* (Donovan, 1991), the eurytomid, *Tetramesa grandis* (Phillips and Emery, 1919, as *Harmolita grandis*), and in the braconid, *Microctonus brevicollis* (Künckel d'Herculais and Langlois, 1891), though there is no evidence that the same mechanism is involved in these various examples. In the gall-forming eurytomid, *Trichilogaster acaciae-longifoliae*, equal numbers of each sex are produced when it develops in *Acacia longifolia* var *floribunda*, but 99% of offspring are female when reared from *A. longifolia* (Nobel, 1940). In the latter case, it is not known whether the ovipositing female adjusts the sex ratio of the eggs laid or whether the male progeny suffer an especially high level of developmental mortality in the one host. In many cases thelytoky results from infection with certain rickettsia-like bacteria belonging to the genus *Wolbachia*, which in other species mediate sexual incompatibility (section 2.27), but in other cases no such organism is involved. Sometimes thelytokous lines have been found to result from crossing different bisexual strains of a parasitoid. Each of these will be considered in more detail below.

Cornell (1988) considered what ecological factors may select for thelytoky but was unable to find supporting evidence for any of them. This is not so surprising when it is realized that in most thelytokous taxa investigated to date the thelytoky is mediated by selfish symbiotic microorganisms, and Cornell did not distinguish between these and intrinsic factors.

2.29 CYTOGENETICS OF THELYTOKY

A detailed discussion of the possible karyological mechanisms of thelytoky was provided by Rössler and DeBach (1973). Thelytokous egg production may or may not involve meiosis; if meiosis is absent then it is termed apomixis, otherwise it is termed automictic thelytoky. Discounting apomictic reproduction in which there is no reduction in chromosome number, there are a number of distinct forms of automictic thelytoky each of which involves restoration of the diploid chromosome number by some mechanism or other. Rössler and DeBach describe six fundamentally different thelytokous mechanisms, some of which encompass several distinct variants. Some mechanisms maintain heterozygosity, others yield only homozygous offspring. Under most circumstances, all these mechanisms will lead to homozygous lines in due course, though within a population there may be many lines, each fixed for different alleles. Even rare maintenance of sex in such thelytokous populations will help to avert complete homozygosity (section 2.32).

Table 2.3 Numbers of thelytokous species known from parasitic Hymenopteran superfamilies and families (modified after Luck *et al.*, 1992, and taking into account Day, 1971; Young, 1990; and Weinstein and Austin, in press)

Superfamily/Family	Number of thelytokous species reported	
Trigonalyoidea	1	
Trigonalyidae		1
Proctotrupoidea	6	
Proctotrupidae		5
Pelecinidae		1
Platygastroidea	1	
Platygastridae		1
Cynipoidea	> 2000	
Cynipidae		> 2000
Others		?
Chalcidoidea	121	
Aphelinidae		38
Encyrtidae		23
Eulophidae		9
Eupelmidae		2
Eurytomidae		9
Leucospidae		1
Mymaridae		9
Pteromalidae		3
Signiphoridae		4
Torymidae		3
Trichogrammatidae		20
Ichneumonoidea	32	
Braconidae		21
Ichneumonidae		11
Chrysidoidea	6	
Bethylidae		1
Dryinidae		5

Very little is known about which mechanisms may apply to the various thelytokous parasitic Hymenoptera, even though karyological studies are not particularly difficult in species with alecithal eggs. In the cyclically thelytokous cynipid gall wasp, *Neuroterus baccarum*, Suomalainen (1950) provided some evidence that the mechanism involves abnormal meiotic division I with the products of meiotic division II giving rise to cleavage nuclei. The consequences are that the offspring would be mosaics and will remain heterozygous for loci that have undergone crossover. Sanderson (1988) found the same for several other gall-forming cynipids.

The cytology of egg production in both thelytokous and arrhenotokous strains of the aphelinid, *Aphytis mytilaspidis*, have been described by Rössler and DeBach (1973); that of the ichneumonid, *Venturia canescens*, by Speicher (1937); and that of the microorganism-induced thelytoky in *Trichogramma* species by Stouthamer and Kazmer (1994). In all these, events are indistinguishable from normal through the first meiotic division, but diploidy is restored in the oviposited egg by a failure of the chromosomes to segregate at the first meiotic division followed by normal mitosis, though the precise cytological events can be difficult to follow. When thelytokous females of *Trichogramma* mate, they are capable of laying fertilized eggs, and in these the first meiotic division progresses normally with no doubling of chromosome number. Parthenogenesis in the aphelinid, *Encarsia formosa*, has been shown to be brought about by affecting the anaphase of the first mitotic division following meiosis. In normal gamete formation, the haploid cells formed after meiosis undergo two further mitotic divisions so as to form four haploid cells from each meiotic product. In *Wolbachia*-infected cells the first mitotic division does not lead to cell division, and the number of chromosomes is therefore doubled back to the diploid number, resulting in complete homozygosity. In this way, *Wolbachia*-infected females produce half as many diploid eggs as do conspecific uninfected individuals.

The thelytoky found in some parasitoids, however, appears to differ from that found in the *Wolbachia*-infected *Trichogramma* investigated by Stouthamer and Kazmer. For example, in the aphidiine braconid, *Lysiphlebus cadui* (Colin Denholm, personal communication), clones of females occur. These include heterozygous clones, thus excluding the possibility of gamete duplication at the first mitotic division.

2.30 GENE FLOW BETWEEN THELYTOKOUS AND ARRHENOTOKOUS STRAINS

Many species are known from both arrhenotokous and thelytokous populations, sometimes occurring sympatrically but, more commonly, allopatrically. In some cases males of the arrhenotokous race will not show any interest in thelytokous females; for example, in Aeschlimann's (1990) work on the mymarid, *Anaphes diana*, and similarly attempts to hybridize biparental European strains of the braconid, *Meteorus pulchricornis*, a parasitoid of the gypsy moth, *Lymantria dispar*, with uniparental East Palaearctic ones were unsuccessful, indicating that their reproductive strategies effectively render them reproductively isolated (Fuester *et al.*, 1993). In contrast, Stouthamer and Kazmer (1994) used genetic markers to show that, in thelytokous *Wolbachia*-infected strains, *Trichogramma* species when mated with conspecific males incorporated the male genome in their offspring at approximately the same frequency as did females of conspecific normal sexual (arrhenotokous) strains. Cytogenetic studies showed that this occurs because thelytoky in *Wolbachia*-infected *Trichogramma* involves so-called gamete duplication, a process in which the haploid chromosome number of the egg at the first post-meiotic mitotic anaphase is doubled by the failure of the nuclei to segregate. Gamete duplication would normally result in purely homozygous offspring, but it is clear, at least in *Trichogramma*, that should a sperm from a male fuse with the egg before fusion of the first mitotic nuclei, it can prevent that fusion and a normal heteroparental zygote will develop. Thus, Stouthamer and Kazmer have shown that *Wolbachia*-induced thelytoky does not necessarily mean that populations will tend to homozygosity and that gene flow between thelytokous and arrhenotokous individuals can occur.

2.31 THELYTOKY RESULTING FROM CROSSING STRAINS

Thelytoky sometimes originates in experimental crosses between different strains of the same species of parasitoid or between different but closely related species (Luck *et al.*, 1992). Examples of the first kind occur when strains of the pteromalid, *Muscidifurax raptor*, are crossed (Legner, 1987), also from particular crosses between members of geographically widely separated strains of the aphidiine braconid, *Aphidius colemani* (Tardieux and Rabasse, 1988), and occasionally when some strains of the braconid, *Habrobracon hebetor*, are crossed (Speicher, 1934). Crosses between closely related species leading to thelytoky have been reported for two pairs of species of *Trichogramma* (Nagarkatti, 1970; Pintureau and Babault, 1981). In Nagarkatti's example, female *T. perkinsi* were crossed with males of an unidentified *Trichogramma* species (D-67) from Colombia, parasitic on *Diatraea*. In the cross, several hybrid females were obtained but only one appeared to be thelytokous.

2.32 DEUTEROTOKY

Deuterotoky is a form of parthenogenesis in which both males and females develop from unfertilized eggs. This can be part of a cyclical parthenogenetic system as found in cynipid gall wasps or in the ichneumonid, *Sphecophaga vesparum*, but it can also occur in normally thelytokous taxa, and in the latter situations males are typically very rare. Allozyme analysis of some such species has shown that at any one site there can be several distinct clonal lineages (section 2.29). This is very much in agreement with the complementary sex determination mechanism (CSD) that appears to be normal for braconid wasps. In taxa employing CSD, gamete duplication (which would lead to homozygosity) could not be involved because individuals that are homozygous at the sex-determining locus (or loci) develop into males.

2.33 SEX-RATIO DISTORTERS – B-CHROMOSOMES AND OTHER FACTORS

The best-known selfish B-chromosome among the parasitic Hymenoptera is found in some populations of the pteromalid, *Nasonia vitripennis*. This chromosome, commonly referred to as a paternal sex ratio (*PSR*) chromosome, is inherited through males. When the sperm fertilizes the egg, the *PSR* causes the rest of the chromosomes from the sperm to degenerate, thus converting the initially diploid female zygote to a haploid male one that will still carry the *PSR* chromosome (Werren *et al.*, 1981). *PSR* males produced slightly larger (all-male) broods than normal males, probably due to smaller size and thus reduced sib–sib competition (Beukeboom, 1994). *PSR* males were also found to develop slightly faster than normal ones. Beukeboom and Werren (1992) noted that, in nature, populations of *N. vitripennis*, which parasitizes dipterous pupae, normally display sib-mating, giving rise to a highly subdivided population structure. Theoretical and experimental results show that the occurrence of *PSR* will be favoured in situations where there is a higher foundress number, or where some *Nasonia* females also carry a female-favouring sex ratio distorter referred to as *MSR* (maternal sex ratio). The nature of the *MSR* factor, which occurs in some females in many populations of *Nasonia vitripennis*, is currently unknown except insofar as it is another cytoplasmic factor (Skinner, 1982). Unlike *Wolbachia*-infected individuals, the female offspring of *MSR*-infected females seem to result from fertilized eggs, and therefore *MSR* presumably acts by influencing fertilization rate.

A similar instance of a genetic factor giving rise to loss of one chromosome set from a fertilized egg, and consequently leading to male offspring, is reported by Hunter *et al.* (1993), who described an unusual and weakly heritable trait in the heteronomous autoparasitoid whitefly parasitoid, *Encarsia pergandiella* (Aphelinidae). Normally in *E. pergandiella*, diploid (fertilized) eggs are lain in the primary host whitefly whereas unfertilized eggs are lain in conspecific female larvae. Unfertilized male eggs will not normally develop within an unparasitized primary host. However, in populations derived from a strain collected at Ithaca, New York, some males were found to develop as primary parasitoids: approximately 50% of primary male matings resulted in at least some primary male offspring. Cytological investigation showed that these apparently primary parasitoid males actually originated from fertilized eggs, but that after fertilization one set of chromosomes became over-condensed and subsequently lost, converting an initially diploid (female-producing) egg into a haploid (male-producing) one. This chromosomal loss was also found to be caused by a supernumerary *PSR* chromosome. The fact that the female wasp would be unaware of the presence of the *PSR* within her mate's genome explains why she lays the egg directly into the primary host just like other fertilized female-producing eggs. Perhaps also because it is the paternal genome that is lost, the now haploid egg can complete development as if it is a female one, though more will need to be known about the mechanisms that determine why fertilized eggs can, and unfertilized ones cannot, develop as primary parasitoids. Whereas in *Nasonia vitripennis* the loss of one haploid chromosome set is caused by a supernumerary B chromosome, Hunter *et al.* were unable to detect any extra chromosomes in the case of *E. pergandiella*.

2.34 SEX-RATIO DISTORTERS – *WOLBACHIA, ARSENOPHONUS* AND OTHERS

A particularly interesting aspect of sex determination in many disparate groups of parasitic wasps is the role of endoparasitic microorganisms that distort sex ratios. Typically, in these groups, untreated females produce all female broods. Whilst some species are known in the wild only from these uniparental forms, others show natural variation with some populations being uniparental, thelytokous, whilst others are sexual.

It was discovered some while ago that rearing the females of some thelytokous strains, such as some thelytokous members of the chalcidoid genera, *Encarsia, Muscidifurax, Ooencyrtus* and *Trichogramma*, at elevated temperatures led to the formation of males (Flanders, 1942c; Wilson and Woolcock, 1960; Legner, 1985). A similar effect can be achieved by feeding many thelytokous female hymenopteran parasitoids with antibiotics such as tetracycline, which often causes them to produce male offspring (Stouthamer, 1991; Stouthamer *et al.*, 1990; Zchori-Fein *et al.*, 1992b) within a few days of treatment. This effect has become known as curing, and it may be permanent or just temporary. Collectively, these observations suggested that the thelytoky might be associated with some form of microorganism.

Microorganisms within insect cells can be revealed with the bacterial stain lacmoid, followed by bleach treatment, or by using the double-stranded DNA stain, DAPI (4′,6-diamidine-2-phenylindole-dihydrochloride). Eggs of field-collected thelytokous species and races of several species examined in this way often appeared spotted (Figure 2.8b), whereas those from naturally sexual races lacked the dark-staining microorganisms in their cytoplasm (Figure 2.8a). Similarly, staining of eggs from thelytokous and cured strains of *Trichogramma* showed that eggs of the former were infected with numerous bacterial cells, whilst these were absent from the latter (Stouthamer and Werren, 1993). Although these observations provide strong evidence that thelytoky in many parasitic wasps is indeed the result of infection with a microorganism, it has so far proved impossible (despite many efforts) to culture these endosymbionts or artificially inoculate uninfected females. Therefore final positive proof that they alone are the causal agents of thelytokous parthenogenicity will have to wait, though the evidence that they are is now virtually overwhelming.

Molecular genetic studies of the 16SrDNA gene (Stouthamer *et al.*, 1993) and of the bacterial cell-cycle gene, *ftsZ* (Werren *et al.*, 1995), have shown that the microorganisms responsible for causing uniparentality in *Muscidifurax uniraptor* and *Trichogramma deion* belong to the rickettsia-like Proteobacteria (Figure 2.9a), within which they constitute a closely related group. Further, they appear to be closely related to similar organisms that are responsible for post-zygotic cytoplasmic incompatibility in a wide range of other insects, and also feminization in some crustaceans. The organisms responsible for both parthenogenesis and cytoplasmic incompatibility are placed together in the same genus, *Wolbachia*. Species of *Wolbachia* are widely distributed among insects as a whole and also occur in a range of other invertebrates. In some – for example, in *Drosophila* – they have no apparent effect at all; but in many species in a wide range of orders they have been shown to be responsible for incompatibility, and this role has been estimated to affect up to 10% of all insects. In more than 20 species of parasitic Hymenoptera they have been shown to be responsible for parthenogenesis. *Wolbachia* are normally transmitted from female to offspring intra-ovum; consequently they also occur in males in most if

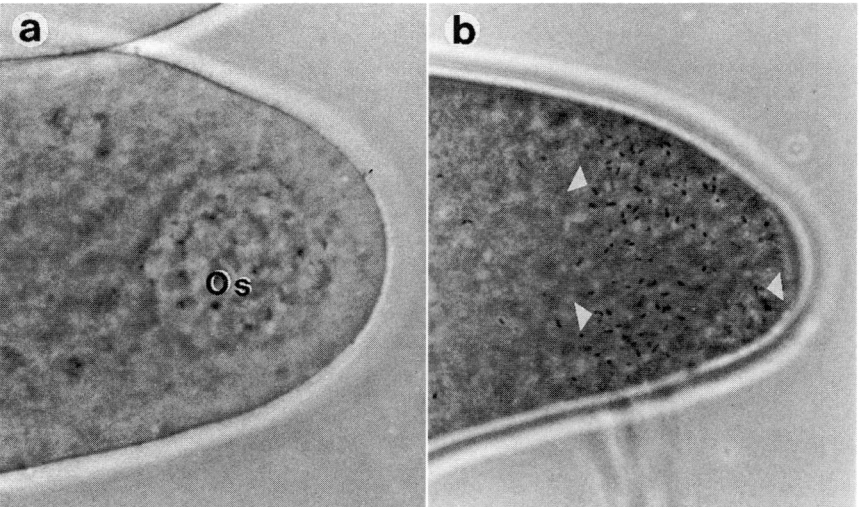

Figure 2.8 Micrographs of anterior poles of eggs from two strains of *Trichogramma deion* (Trichogrammatidae), showing (a) absence in a strain from Seven Pines, California, and (b) presence (white arrows) in a strain from Sanderson, Texas, of DAPI-stained *Wolbachia*, which are responsible for inducing thelytokous reproduction. Os, oösome. (Courtesy of Richard Stouthamer.)

not all insects that carry them (Bressac and Rousset, 1993), but are not transmitted in sperm. The aggregation of *Wolbachia* in the region of the oösome (Figure 2.8), which is part of the egg that will develop into the pole cells and thence the germ line, may be a significant part of their vertical transmission. Ovarian cells of *Habrobracon* also contain bacteria (Cassidy and King, 1972), but this genus is not known to show either cytoplasmic incompatibility or thelytoky, except for one instance when two particular different strains were crossed (Speicher, 1934) and therefore the possibility that this microorganism may be involved in some sort of compatibility system cannot be excluded.

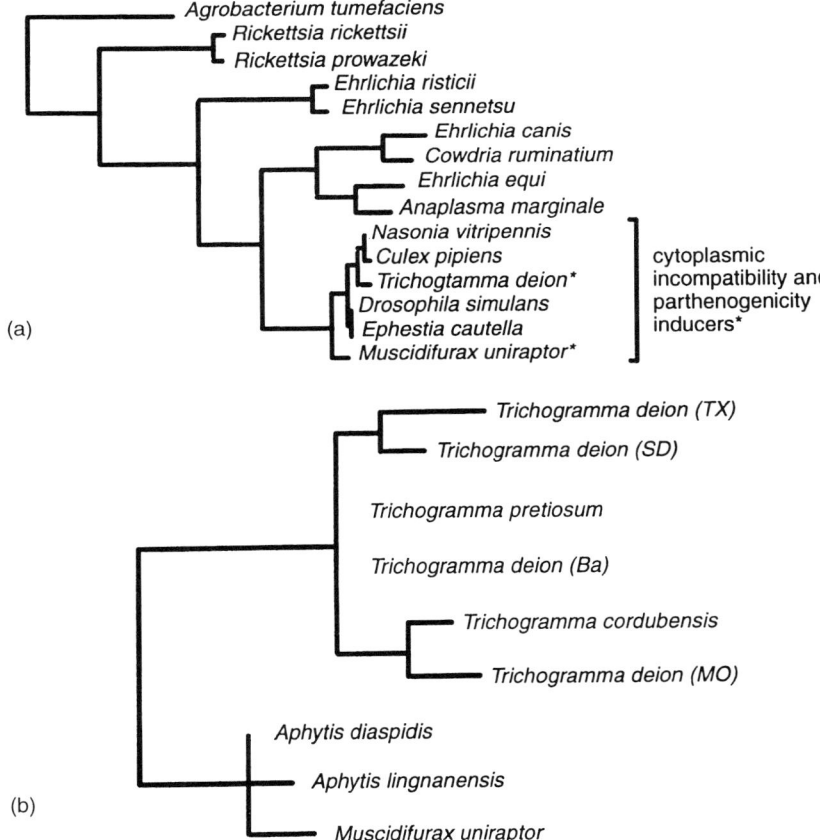

Figure 2.9 Phylogenies of *Wolbachia* and related endosymbiotic microorganisms as derived from analysis of the 16s rRNA gene. (a) Relationship to other Protobacteriaceae (redrawn after Stouthamer *et al.*, 1993). (b) Consensus, maximum parsimony phylogeny of parthenogenesis-inducing *Wolbachia* from various *Aphytis*, *Trichogramma* and *Muscidifurax* strains and species (redrawn from Zchori-Fein *et al.*, 1995.)

Another bacterium, *Arsenophonus nasoniae*, a member of the Enterobacteriaceae, is responsible for all-female broods in *Nasonia vitripennis* and is consequently known as 'son-killer' (Huger *et al.*, 1985; Skinner, 1985; Werren *et al.*, 1986; Gherna *et al.*, 1991). *Arsenophonus* only occurs at low frequencies in wild populations; it is transmitted longitudinally and, in instances of superparasitism, horizontally. The bacterium is injected by an infected female into the host Diptera pupa, and re-infects any female larvae present, but for some unknown reason it kills male ones. Godfray (1993) notes that the unusual reproductive biology of some strains of the mymarid, *Caraphractus cinctus*, might be attributable to a similar son-killing factor.

In the pteromalid, *Muscidifurax raptor*, there is some evidence that sex ratio can be distorted by the presence of a microsporidian parasite that is transmitted from mother to offspring via the ovaries (Antolin, 1992).

2.35 LONG-TERM EFFECTS OF *WOLBACHIA*-MEDIATED THELYTOKY

In some parthenogenetic taxa, it has proved impossible to establish sexual lines even after curing with antibiotics so as to produce males, because the males so produced either do not produce functional sperm or in some cases will not even mate. Such examples are found in species in which thelytoky has become fixed and it can be assumed that, since males would not have been needed for very many generations, genes for otherwise important male traits such as mating behaviour and functional sperm will have progressively accrued mutations until they are no longer functional, providing an effective prezygotic reproductive barrier between thelytokous and arrhenotokous strains. For example, Zchori-Fein *et al.* (1992b) showed that cured males of *Encarsia formosa* produce sperm, but that these sperm are non-functional. Further, only a small percentage of the males produced showed any attempt to mate with virgin females, though as expected no sperm were subsequently found in the females' spermathecae. Zchori-Fein *et al.* also found that cured females appeared to produce more offspring than untreated controls but the result was not statistically significant. In further work on the same species, Stouthamer *et al.* (1994) found a significant *Wolbachia*-induced depression of fecundity, but failed to detect any effect in the pteromalid, *Muscidifurax uniraptor*. In an interspecies comparison, Volkoff and Daumal (1994) found that the number of mature eggs at emergence was smaller in thelytokous species than arrhenotokous ones, but the number of species considered was rather low and so care needs to be exercised in extrapolating from this observation. Certainly, this aspect of thelytoky requires wider investigation, as it would be surprising if there were not some cost to carrying the *Wolbachia* load.

2.36 HORIZONTAL TRANSMISSION OF *WOLBACHIA*

Pinto *et al.* (1993) used allozyme variation to investigate the relationships of a number of strains of several closely related species of *Trichogramma*, including five thelytokous and 15 sexual strains of *T. deion*. Their results showed that the thelytokous strains did not form a single monophyletic group, and therefore that *Wolbachia*-induced thelytoky had arisen on at least three separate occasions among the 20 strains investigated. Horizontal transmission of *Wolbachia* in the field seems the likeliest explanation. A similar conclusion was reached by Zchori-Fein *et al.* (1995) who carried out a parsimony analysis of the 16S rDNA sequences of the parthenogenesis-inducing *Wolbachia* of two *Aphytis* species, *Muscidifurax uniraptor* and seven *Trichogramma* species, and showed that those from the *Trichogramma* formed a monophyletic group and that those from the two *Aphytis* species differed in only two positions (Figure 2.9b). Given that the two *Aphytis* species belong to widely different species groups, it was suggested that the high degree of similarity between their 16S rDNA sequences probably means that they also acquired their *Wolbachia* via horizontal transmission.

2.37 THELYTOKY AND BIOCONTROL

Many parasitic wasps are known to have thelytokous races. Aeschlimann (1990) and Stouthamer (1993) have considered the relative advantages and disadvantages of using normal sexual arrhenotokous strains versus asexual thelytokous ones in biocontrol. Firstly, such considerations may hinge upon any difference in fecundities or searching efficiencies between the strains. Most available data sets reveal no significant differences in these parameters, so what other factors may be relevant? Thelytokous strains may have the intrinsic advantage that, since they produce no male offspring, all of their progeny go on to attack new hosts directly. This factor may also be important in mass rearing, and at low population densities, where sexual strains may be disadvantaged because females may find it difficult to find mates. As some thelytokous strains are 'curable' by antibiotic or heat treatment, they can in principle be used to test experimentally the relative advantages and disadvantages of sex in biocontrol.

2.38 SEX DETERMINATION AND BIOCONTROL

As noted by Stouthamer *et al.* (1992), the successful rearing of parasitic wasps for biocontrol and the success of any introductions may well be influenced by the sex determination system of the wasp concerned (see above). Specifically, those taxa that have a single-locus complementary sex determination (CSD) mechanism tend to fare less well, firstly because when attempts are made to mass-produce them from small starting populations there will be a rapid increase in the proportion of diploid males, generation upon generation, as a result of the inbreeding. These effects will not be randomly distributed across the parasitic Hymenoptera since single-locus sex determination appears to be largely restricted to the Ichneumonoidea (Cook and Crozier, 1995), though there are certainly some exceptions to this, and at least some ichneumonoids do not have CSD. Stouthamer *et al.* (1992) hypothesized that this could be responsible for the relatively lower success rate of ichneumonoids in biocontrol programmes compared with other parasitic groups such as the Chalcidoidea.

2.39 GENETIC MARKERS AND MUTATIONS

Parasitic Hymenoptera are ideal subjects for the detection of visible mutants because recessive alleles will be unmasked and therefore visible in the haploid males. Compared with many groups of organisms, relatively few externally visible genetic markers seem to have been recognized and cultured in parasitic wasps, with the marked exceptions of those found in a few much-cultured taxa such as *Nasonia*, *Dahlbominus* and *Habrobracon*. By 1935, some 90 visible mutants had been described for last of these – just over half cropping up naturally in culture, the others artificially induced, mostly by X-rays. So it seems likely that scarcity of reported mutants in other taxa simply reflects the amount of bulk rearing and the interests of hymenopterologists rather than any lack of mutations. A list of mutant stocks of *Dahlbominus fuscipennis* was provided by Baldwin *et al.* (1964) and for *N. vitripennis* by Saul *et al.* (1965). Using such induced genetic markers, it has been possible to construct a preliminary gene map for *Nasonia vitripennis* (Saul, 1990).

The eulophid, *Dahlbominus fuscipennis* and the pteromalid, *Nasonia vitripennis* (often referred to as *Mormoniella vitripennis* in this literature), and various braconids of the genus *Habrobracon* have been intensively used in research on the mutagenic effects of radiation, chemicals and even ultrasound (Whiting, 1943, 1954; Baldwin, 1969, 1972; Grubbs and Conner, 1976). It is interesting to note that, for a while, *H. hebetor* was quite a popular tool for assessing the mutagenicity of various substances and environmental factors (e.g. von Borstel and Smith, 1977; Petters *et al.*, 1983). Mutations have even been investigated in *Habrobracon* that were carried out into space in Biosatellite II (von Borstel *et al.*, 1968).

In the absence of naturally occurring visible genetic markers that may be useful for experimental studies, it is quite simple to create useful ones in the laboratory. Radiation and various chemicals, especially mitomycin C and ethyl methanesulphenate, have been widely employed to induce mutations, notably in *Habrobracon* and *Nasonia*. Much of this work has been adequately summarized by Grosch (1988). Most induced (or at least, detected) mutations are recessive lethals – often temperature-dependent ones – and Grosch noted that visible external mutations do not arise often enough for the production of dose–effect curves.

A few naturally occurring visible mutants have been found and used in other parasitic wasps. For example, in the opiine braconid, *Diachasmimorpha* (as *Biosteres*) *longicaudatus*, a dark metasomal form has been studied by McInnis *et al.* (1986), who provided evidence that the black body form was the result of a single recessive allele; Rössler and Debach (1972) have made use of a single-locus eye-colour mutant in the aphelinid, *Aphytis* sp.; and in the microgastrine braconid, *Microplitis croceipes*, Li and Steiner (1995) described a clear wing-colour mutant (*cw*) apparently controlled by a single recessive allele. Whiting (1954) showed that considerable similarities exist between X-ray induced eye-colour mutants in two confamilial parasitic wasps, *Nasonia vitripennis* and *Pachycrepoides dubius*, suggesting that the underlying genetic mechanisms and metabolic pathways could be similar or homologous.

2.40 MUTATION RATE

Since Hymenoptera are haplodiploid and males carry only a single copy of each gene, they have no genetic buffering to protect them from recessive lethal mutations. This property has been extensively used in studies of chemical and radiation induced mutations. It also provides an opportunity to examine the natural rate of recessive lethal mutations. Smith and Shaw (1980) used the number of adult wasps of each sex emerging in mixed-sex and all-male broods of four gregarious species of the braconid genus, *Apanteles*, to obtain a maximum likelihood estimate of mutation rate. As expected, haploid males suffered a higher mortality than the diploid females, and suggested a mutation rate per genome per generation of 0.035 – a value close to that found for non-haplodiploid insects.

2.41 GENETIC TRANSFORMATION

Transgenic natural enemies offer the possibility of improving pest control by modifying any of several aspects of the control agent's biology. This will undoubtedly be a rapidly expanding area of endeavour once suitable transformation techniques have been developed. The first successful transformation involving a parasitic wasp has been achieved by injecting adult females of the braconid, *Cardiochiles diaphaniae*, with the plasmid *phsopd* in which the parathion hydrolase gene of the bacterium, *Pseudomonas dimuta*, had been inserted (Presnail and Hoy, 1996). Out of 38 injected wasps, one subsequently displayed a hybridization pattern indicating that the plasmid had integrated with the wasp's nuclear genome. If this procedure proves to work with other systems, it may soon be possible

to produce transgenic parasitic wasps with various added resistance genes
– for example, to pesticides or to host defence reactions.

2.42 RESISTANCE TO PESTICIDES AND USE IN INTEGRATED PEST MANAGEMENT (IPM)

Whilst there are many undesirable problems associated with chemical pest
control that make biological control far more desirable, there are many
cases in which biological control alone has failed to achieve an economically acceptable level of pest control. Under such circumstances, the application of limited quantities of chemical pesticides at times when the biological control is starting to fail can provide a cost-effective and ecologically tolerable alternative. An interesting example of the importance of pesticide resistance has been described by Steinberg *et al.* (1987). During the
decade leading up to their study, *Pteroptrix smithi* had gradually replaced
another aphelinid, *Aphytis holoxanthus*, as the major parasitoid of Florida
red scale in Israel. Laboratory studies showed that, according to a number
of criteria, *A. holoxanthus* was superior to and should have outcompeted *P.
smithi*. However, its failure to do so in the field most likely resulted from its
greater susceptibility to the widely used insecticide, malathion.

Persistence of insecticide resistance in parasitoids in the absence of
selection for resistance has obvious implications for IPM programmes
when there may be periods of several years between applications of particular pesticides. Baker (1995) demonstrated that there was considerable
persistence of resistance to malathion in field-collected populations of two
parasitic wasps: the pteromalid, *Anisopteromalus calandrae*, and the braconid, *Habrobracon hebetor*. Although not all such studies have found such
a high degree of genetic stability, most have shown reasonably high levels indicating that temporarily stopping pesticide application is not likely
to lead to a dramatic increase in parasitoid susceptibility.

The possibility of selecting for strains of parasitoids that are resistant to
pesticides is therefore attractive. However, many studies of artificial
selection for resistance in insects in general have shown that under many
selection regimes the resistance achieved is polygenic; that is, it is the
result of the actions of many separate genes each contributing just a little
to the overall resistance of the species. This contrasts markedly with the
typical observation of resistance that develops in wild populations of pest
species, in which it is often only one or two genes that confer a high level
of protection against a given pesticide. Such discrepancies appear to
result primarily from the strength of the selection pressure involved and
the typical practical constraints on the size of laboratory populations used
in selection experiments.

Havron *et al.* (1987a) developed a methodology to determine the susceptibility of small parasitic wasps such as egg parasitoids to various

insecticides. The previous lack of suitable techniques had undoubtedly hampered the screening of such fragile insects, which are not amenable to standard techniques that can be applied to larger species. Their technique uses a controlled exposure of the wasps to pesticide in sucrose solution. Although the procedure leads to a systemic rather than topical exposure, and a consequent reduction in the accuracy of the applied dose per individual, it should yield reliable data on relative toxicities. Subsequently, Havron *et al.* (1987b) developed a novel breeding and artificial selection regime for increasing resistance of parasitic Hymenoptera to insecticides which makes use of their haplodiploid sex determination mechanism (Havron and Rosen, 1988). Selection pressures against deleterious alleles are far greater in male Hymenoptera than they are in females because in the hemizygous state their effects are not buffered by the presence of a second, less deleterious version of the gene. Their selection regime is based entirely on males (Figure 2.10), a procedure facilitated in the case of their experiments with *Aphytis* wasps because their pupae could readily be sexed by means of morphological features. The mating sequence and selection procedure was carried out for more than 30 generations, with the concentration of insecticide (malathion) being increased incrementally at various times through the programme.

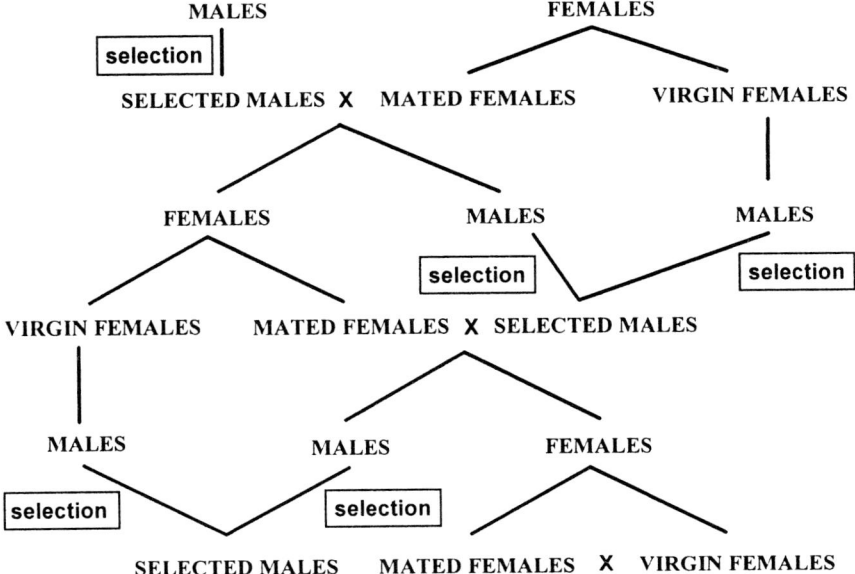

Figure 2.10 Protocol for rapid selection of insecticide resistance in parasitic wasps that takes advantage of hemizygous nature of males. (Redrawn from Havron *et al.*, 1987a.)

2.43 POPULATION SIZE AND BIOCONTROL SUCCESS

Rosen (1988) emphasized the desirability of making large and genetically diverse introductions of potential biocontrol agents. As an example he noted the lack of success obtained when his research group attempted to select the aphelinid parasitoid, *Aphytis holoxanthus*, artificially for increased resistance to the organophosphate insecticide, azinphosmethyl. The failure was attributed to the fact that the initial population of *Aphytis holoxanthus* imported for control experiments was too small, and therefore had effectively passed through a genetic 'bottleneck' (Havron and Rosen, 1988). A related problem with parasitic Hymenoptera is that the effective population size of a culture is considerably less than the number being used to start each new generation (section 2.13). In consequence, the longer a species is maintained in culture before release, not only will the level of genetic diversity in the introduced inoculum be lower, but also the sex ratio (and diploid male production) is likely to be increased in species that have complementary sex determination (Stouthamer *et al.*, 1992). This can be largely overcome by maintaining a number of separate lines, each of which will act as a reservoir for sex determination and other alleles (Cook, 1993c). This has been extended by Boush and Hopper (1995) who calculated the number of mildly inbred lines that are needed to maintain most common alleles in the total population. For arrhenotokous taxa such as parasitic wasps, the number is about 25.

3

Life history strategy: idiobionts and koinobionts

3.1 INTRODUCTION

One of the most conspicuous features of an insect parasitoid is whether or not its larvae develop as endoparasitoids or as ectoparasitoids. However, this only explains some aspects of parasitic wasp biology. Haeselbarth (1979) and Askew and Shaw (1986) drew attention to another aspect of parasitoid biology that has become widely appreciated for its explanatory ability: that is, whether or not the host insect continues to develop further after it has been parasitized. Those parasitoids whose hosts do not develop further are referred to as idiobionts whilst those whose hosts carry on their development for at least a while post-parasitization are called koinobionts. Some idiobionts develop completely within a host egg or in a host pupa; others attack a mobile larval stage, but this is almost always paralysed permanently at the time of parasitization so that it can no longer move. Koinobionts typically attack larvae, often early instars, or eggs in the case of egg–larval and egg–pupal parasitoids. In practice there is a complete spectrum between idiobiont to koinobiont strategies, and there are exceptions to virtually all of the various correlated biological features listed in Table 3.1. Gauld and Hanson (1995) summarized the adaptive suits of idiobionts and koinobionts: idiobionts tend to evolve morphological adaptations that are associated with gaining access to hosts, whereas koinobiont parasitoids have usually had to adapt physiologically to host defences. Of course, this is a gross simplification, and both idiobionts and koinobionts may have special morphological and physiological adaptations, but because koinobionts have generally a more protracted parasitic phase they do tend to show a higher degree of physiological adaptation. Interestingly, not only do many biological characteristics correlate with a parasitic wasp's life history strategy, but it is also noticeable that so too does the ease with which parasitic wasp larvae can be cultured *in vitro* –

the great majority of successful culturings have involved idiobionts, including both endo- and ectoparasitic ones (Chapter 5).

Of the two strategies, the idiobiont is perhaps the most heterogeneous since it includes species that prevent their larval hosts from developing further by paralysing them, and other species which complete their development within a host's egg or pupa. In the latter two cases, the critical feature that unites these strategies is that the host stage attacked is immobile and is quite often in a protected situation anyway.

3.2 ECTOPARASITISM VS. ENDOPARASITISM

Most koinobionts are endoparasitoids, and many idiobionts are ectoparasitoids, but a few exceptions occur – notably polysphinctine ichneumonids, which are highly specialized parasitoids of spiders; tryphonine and some adelognathine ichneumonids, most of which attack sawflies; some eulophids (several *Eulophus* and *Elachertus* spp.) (Askew and Shaw, 1986); and a few bethylids (e.g. Peter and David, 1991). In the case of tryphonine ichneumonids, adaptations of the egg (Figure 4.7d–f) and larvae mean that they can feed on their still mobile and developing hosts from the outside, and can even remain with the host during ecdysis, though the

Table 3.1 Biological and life history traits associated with idiobiont and koinobiont strategies

Life history strategy	
Idiobiont	*Koinobiont*
• Ectoparasitism	— Endoparasitism
• Host concealed	— Host exposed
• Generalists	— Specialists
• Long-term host paralysis	— No or temporary host paralysis
• Rapid larval development	— Slow or delayed larval development
• Large eggs	— Small eggs
• Females carry few mature eggs at any one time	— Females carry many mature eggs at a given time
• Synovigeny	— Pro-ovigeny
• Host-feeding frequent	— Host-feeding uncommon
• Oösorption	— No oösorption
• Long adult life span	— Short adult life span
• Host stage attacked larger than wasp	— Host stage attacked often smaller than wasp
• Wasp may choose sex of egg to match host size	— No such relationship
• Mostly diurnal	— Diurnal or nocturnal
• Sexual dimorphism often pronounced	— Sexual dimorphism absent or less pronounced

majority attack late instars and so do not need to. It is fairly obvious, however, that endoparasitism is likely to be most strongly associated with koinobionts.

Being ectoparasitic on active mobile hosts presents several special problems. For example, the host may try to physically damage or dislodge the developing parasitoid, e.g. by biting it or by scraping it off. Not surprisingly, therefore, koinobiont ectoparasitoids generally place their eggs, and develop, in places on the host's body where they are least at risk. The polysphinctine ichneumonids, for example, oviposit between the cephalothorax and abdomen of their spider hosts, and the developing wasp larva forms a 'saddle' that the spider cannot reach. Many tryphonine ichneumonids that attack caterpillars oviposit anterodorsally on them so that the host larva cannot turn around and bite them. The presence of an ectoparasitoid larva on a host may also make the latter more conspicuous and thus liable to be noticed and attacked by predators such as birds. In many cases, therefore, the host lives in a rather concealed situation, or is exposed mostly at night or soon wanders off to produce a cocoon. When this is not the case, as with some ectophagous eulophids, the parasitoid has a very rapid development.

Idiobiont endoparasitoids are effectively restricted to attacking the egg or pupal stage of the host. Whether parasitoids of adult insects should be regarded as idiobionts or koinobionts is debatable, but given that the size of the host resource is fixed at the time of parasitization and that the host does not go on to construct a safe retreat, they are probably best regarded as idiobionts despite the fact that their hosts remain active and may feed.

3.3 CONCEALED VS. EXPOSED HOSTS

Apart from the eggs attacked by egg parasitoids, the majority of hosts of idiobionts are concealed in some way at the time they are attacked. A greater proportion of koinobionts parasitize exposed hosts, such as exophytic larvae (Hawkins *et al.*, 1990). Hosts of many koinobionts are mobile and fairly free-ranging, whilst those of many idiobionts tend to be sedentary.

3.4 GENERALIST VS. SPECIALIST

It has been noted for a long time that idiobionts in general have broader host ranges than do koinobionts, though there are many exceptions within both groups (Askew and Shaw, 1986). Several factors conspire to make the host ranges of koinobionts narrower in general than those of idiobionts. Some of these have to do with physiological interactions with hosts; others relate to differences in the speciation processes that occur in the two groups. Whilst it may at first seem that host range will be narrower for endoparasitoids than for ectoparasitoids, because the former have to overcome the physiological defence mechanisms of the host (e.g. encapsulation), it has become apparent that endoparasitism does not mean the same for idiobionts as it does for koinobionts. Koinobiont endoparasitoids have to avoid host defences for a prolonged period of time, but endoparasitic idiobionts tend to attack hosts at stages of their development (e.g. eggs and pupae) that are less capable of suppressing parasitoids physiologically, and of course the parasitoids are not constrained to develop slowly. Further, endoparasitic idiobionts develop rapidly themselves but curtail host development so they have no reason to allow the host to maintain its defence systems.

As an independent test of Askew and Shaw's hypothesis, Sheehan and Hawkins (1991) compared host ranges of the ichneumonid subfamilies Metopiinae and Pimplinae, which are koinobiont and idiobiont endoparasitoids of immature Lepidoptera, respectively. In this study, numbers of host taxa were regressed against the number of collections, thus accounting for the effects of uneven sampling. As predicted, the idiobiont pimplines were found to attack more host species – in fact, on average, 2.7 times more. Shaw (1994) compared the host ranges of an ecto- and endoparasitic group of idiobionts, members of the pimpline ichneumonid tribes Ephialtini (the genus *Scambus*) and Pimplini, respectively; and of an ecto- and endoparasitic group of koinobionts, the polysphinctine pimplines and the rogadine braconid genus *Aleiodes*, respectively. The polysphinctines all attack spiders; *Aleiodes* species are endoparasitic within lepidopterous larvae; members of the Pimplini are very largely endoparasitic within lepidopterous pupae; and ephialtines are ectoparasitoids of a wide range of host groups, including sawflies and lepidopterous larvae. The main determining factor in host range in these groups was not whether the parasitoid is endo- or ectoparasitic but whether it is a koinobiont or an idiobiont. The idiobionts appear to attack almost any host of suitable size that is present in their searching niche. In some cases, the searching niche may include only one or a few hosts, but in other cases there may be many, and all (or almost all) are attacked. In contrast, the koinobiont species have considerably narrower host ranges. Shaw therefore cautiously suggested that the relatively broad host ranges of idiobionts are not the result of some slow accretion processes, whereby new hosts are added one by one to the existing set of potential hosts, but rather that idiobionts from the very beginning are able to attack, and do attack, a wide range of hosts that occur in their searching environment. Koinobionts, on the other hand have had to evolve much more precise behavioural and physiological relationships with their hosts. Individuals that oviposit into or on a host for which they are not adapted will normally fail; at the least an egg will be wasted and at worst, as in the case of polysphinctines which attack spiders (a potentially dangerous host), the wasp may lose its life and therefore any further reproductive potential. Selection on koinobionts will therefore be for specialist host acceptance cues. Extra host species will only be recruited slowly into a parasitoid's host range, and the probability of expanding to a new host will be determined by both the frequency with which that host is encountered (opportunity) and probably the relatedness of the new host to the existing hosts, since closely related hosts may require more similar adaptations in the parasitoid, though there are plenty of examples in which a parasitic wasp may attack two or more very distantly related hosts.

3.5 SYNOVIGENY VS. PRO-OVIGENY

Synovigenic insects continue to mature eggs through part or all of their adult life, whereas strictly pro-ovigenic taxa emerge into adulthood with their full lifetime complement of mature eggs. Between these extremes are so-called pro-synovigenic species that emerge with a good number of mature eggs but can mature more as their original supply of mature eggs becomes depleted. Most koinobiont parasitic wasps are pro-ovigenic; a higher proportion of idiobionts are synovigenic. This correlation may stem from a generally greater degree of dispersion of hosts of idiobionts compared with those of koinobionts, idiobiont hosts generally being more mature (excepting those of egg parasitoids) and therefore having had more time to disperse, and from the comparatively rapid development of idiobionts which favours the evolution of larger, more yolky eggs.

3.6 OÖSORPTION

Flanders (1950) pointed out the strong correlation between being synovigenic and displaying oösorption, and therefore oösorption also correlates with being an idiobiont. This is dealt with in more detail in Chapter 4. Synovigenic species can be divided into two groups: those in which ovulation is under internal control (internally induced), and those in which it is not (externally induced ovulation). In the latter, oösorption is effectively obligatory because egg-maturing continues irrespective of the availability of hosts.

3.7 ADULT LONGEVITY

Adult longevity is closely related to how easily, how frequently and in what numbers a parasitoid normally locates hosts. If hosts tend to be hard to find and to occur singly, then adult wasps will tend to live longer in order to have time to encounter a sufficiently large number of hosts. It is clear from a wealth of literature on culturing parasitic Hymenoptera that idiobionts are typically longer lived than koinobionts, but data from such rearings have also to be treated with caution and the actual longevities stated should not necessarily be taken as representative of longevity in the field, for which there is remarkably little good data (Chapter 10).

Koinobionts often attack exposed hosts and early stages that have not had much time to disperse. Attacking the former means that their handling times may be much shorter than with many idiobionts that may perhaps have to drill through some tough substrate just to reach a single individual. Thus, in general terms, koinobionts may be able to find and deal with a large number of hosts within a comparatively short time after emerging and so may use up a large proportion of their egg load quite soon.

Gauld and Gaston (1994) have suggested that longevity may be greater in the tropics, especially for koinobionts, because their hosts may be scarcer there and consequently the wasps will have to live longer in order to find enough. However, most of the evidence for this is rather anecdotal and there is a real need to obtain reliable data on longevity in the field for a wide range of parasitic wasps from both temperate and tropical habitats.

3.8 LARVAL DEVELOPMENTAL PERIOD

Blackburn (1991) surveyed life span and fecundity in 474 species of parasitic Hymenoptera, and found a significant positive association between being koinobiont and both pre-adult life span and pupal period. As pointed out by Gauld and Bolton (1988), idiobionts by and large deal with their hosts as if they were simply pieces of meat. As such they usually consume them rapidly so as to prevent them going off, though in fact it seems that many of those that develop more slowly probably produce some sort of antifungal or antibacterial secretions. Shaw (1994) noted that in the case of the gregarious ectoparasitic braconid, *Histeromerus mystacinus*, the host remained sweet-smelling throughout the period that the wasp larvae slowly consumed it. The problem of host putrefaction is no less a concern for endoparasitic idiobionts; for example, rupture of the host gut would be expected to release large numbers of bacteria. Führer and Willers (1986) have shown that the larvae of the ichneumonid, *Pimpla turionellae*, an idiobiont endoparasitoid of various Lepidoptera pupae, secrete antibacterial and antifungal compounds from their anus as well as substances that inhibit melanization in damaged host tissues. Thus whilst endoparasitic idiobionts may not have evolved quite such complex physiological interactions with their hosts as koinobionts have, they nevertheless often possess highly evolved systems to protect themselves from various sorts of damage.

3.9 RELATIVE SIZES OF PARASITOID AND HOST

Some features are necessarily fixed for idiobionts and koinobionts. For example, the host of an idiobiont cannot grow after parasitization and so must provide all the necessary resources for its parasitoid. The size of an adult idiobiont is therefore necessarily never more than that of its host. For koinobionts, there is no such necessary relationship. Whilst they are often larger than their host at the time of parasitism and so make use of the host's ability to feed and grow, there are plenty of examples in which the adult parasitoid is much smaller than the host, as is the case with gregarious species.

Many parasitic wasps are highly variable in adult size, and this is particularly true of idiobionts. A fairly extreme example is provided by the pimpline ichneumonid, *Pimpla turionellae*, in which Aubert (1959) found females that ranged in size from 3 to 15 mm. This means that they can adjust their development to the amount of available food and so make use of a large range of potential hosts, in terms of the size range both within a host species and also between potential host species, other things being equal. It is also well documented that female size is strongly correlated with fitness, more so than male size is (Godfray, 1993). Thus if a parasitic wasp can assess the value of an individual host at the time of oviposition, it can select the offspring sex that will maximize its fitness; in other words, it can choose to lay a higher proportion of male eggs in small hosts since their final small size will mean less of a loss of fitness than it would for a daughter.

3.10 HOST PARALYSIS

Some koinobionts temporarily paralyse their hosts, probably in most cases in order to enable the female wasp to oviposit successfully, but by definition this paralysis must disappear fairly rapidly in order to allow the host to continue to fend for itself. Sometimes, this temporary paralysis gives the wasp the opportunity to host-feed as well as to oviposit (Jervis and Kidd, 1986). Host paralysis is displayed by virtually all ectoparasitic idiobionts – not just those attacking larval stages but also those whose larvae feed externally on prepupae and pupae.

3.11 HOST-FEEDING

Host-feeding is widespread among the parasitic Hymenoptera and has been surveyed in detail by Jervis and Kidd (1986). The great majority of taxa that host-feed are idiobionts, and in comparison the behaviour is almost absent among koinobiont species. This difference seems largely to reflect the fact that most koinobionts are pro-ovigenic and so have no need to consume protein-rich host haemolymph. Idiobionts, on the other hand, are predominantly synovigenic; they produce large, protein-rich eggs and so need to take in extra proteins in order to be able to continue maturing more oöcytes. Some of the few host-feeding koinobionts, and the best studied, are members of the braconid subfamily Rogadinae, such as various *Aleiodes* spp. (Shaw, 1983). These wasps are endoparasitoids of lepidopterous larvae, but females carry few mature eggs at any one time and their eggs are yolk-rich. This unusual combination may reflect their phylogenetic position in that they are one of the few endoparasitic and koinobiont groups within a large clade of ectoparasitic idiobionts. Further, synovigenic species often host-feed and this is likely to cause some unnecessary harm to a host that has to go on developing and behaving normally. This may either favour non-concurrent host-feeding (section 8.19) or limit it to those cases where the host can readily withstand the extra stress.

3.12 DIURNAL VS. NOCTURNAL

Huddleston and Gauld (1988) pointed out that, at that time, there were no known nocturnal idiobiont parasitic Hymenoptera, and that all known ichneumonoids displaying the 'ophionoid' facies characteristic of nocturnally active taxa are koinobionts (Chapter 6). Since then, some braconine braconids that are almost certainly idiobionts have been shown to be attracted to lights at night, and are therefore suspected as being nocturnal, or at least crepuscular (Quicke, 1992). Unfortunately, little is known of their hosts. One member of this group, *Aphrastobracon flavipennis*, is known to attack a carnivorous *Eublemma* moth larva that feeds on lac insects, under whose resin it hides, suggesting that its nocturnal behaviour is also associated with greater host accessibility at night, but other apparently nocturnal members of the group attack concealed hosts.

In another instance of nocturnal activity by an idiobiont, Völkl and Kranz (1995) have recently reported that in the megaspilid, *Dendrocerus carpenteri* (a hyperparasitoid of aphids via an aphidiine braconid primary host), nocturnal host searching and parasitization occur in addition to more normal diurnal host seeking. It was shown that *D. carpenteri* is more active at night if its egg-load is high, and so nocturnal host searching may be a way of making good a low level of success during the daytime, perhaps due to adverse weather conditions.

3.13 EGG AND PUPAL PARASITISM

Whilst most idiobionts are ectoparasitoids of larval hosts, those that develop on a single host egg are all endoparasitoids, as also are almost all that attack pupal hosts. However, despite being endoparasitic, they do not share many features in common with koinobiont endoparasitoids. For example, their development is not integrated with the host's physiology or endocrinology; hosts at these stages have poor encapsulating abilities, and the host does not have to be enabled to continue its development. Therefore these hosts can be treated more or less as simple bags of food, just like the paralysed larval hosts of other idiobionts. Host ranges of egg and pupal parasitoids are typically broad and constrained more by the parasitoid's host location behaviour rather than host defences. This is not to say that there are no host-imposed constraints. Respiration, for example, may be problematic, and in the case of pupal endoparasitoids there are potential risks from the putrefaction of unconsumed host tissue. As with ectoparasitic idiobionts, these endoparasitic ones tend to develop rapidly, or at least to complete their larval development quickly. This will reduce the effects of any decline in host quality that will occur naturally or as a result of infection by microorganisms.

Hirose (1994) showed that the size of the egg parasitoid guild of lepidopterous eggs in Japan was largely a function of the size of the lepidopteran egg, with larger eggs supporting larger guilds of parasitoids (Figure 3.1). Apparently the reason behind this is simply that small eggs do not provide a sufficient resource for the complete development of the larger species of egg parasitoids such as the eupelmid, *Anastatus japonicus*, which is more than five times larger than the smaller trichogrammatid egg parasitoids.

3.14 EGG–LARVAL AND EGG–PUPAL PARASITISM

A number of koinobionts belonging to a diverse array of groups of parasitic wasps have evolved to oviposit into host eggs but delay completion of their development until the host is much further developed, sometimes

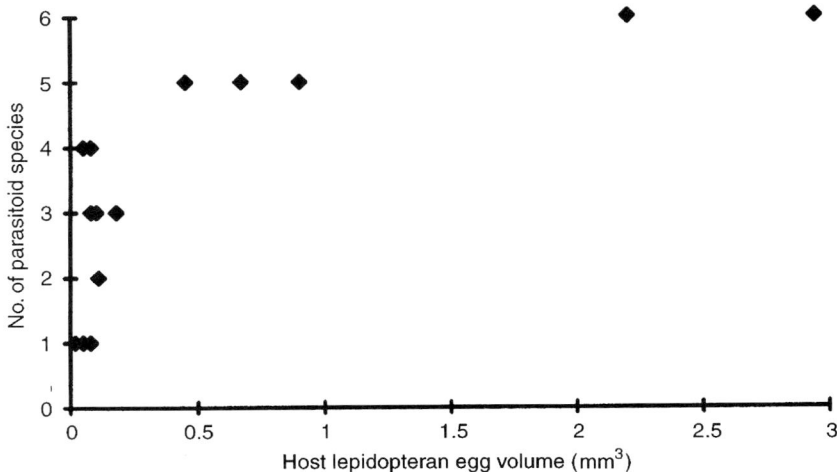

Figure 3.1 Relationship between host egg size and number of egg parasitoid species for 14 species of Japanese Lepidoptera. (Redrawn from Hirose, 1994.)

not starting feeding until the host has pupated. Usually, however, the parasitoid egg hatches soon after oviposition, and it is the 1st instar larva that spends a long time before developing any further. In some cases, for example in *Ibalia* (Ibaliidae) and *Chelonus* (Braconidae), this means the 1st instar parasitoid stage may last for many months. Clausen (1954) provided a survey of the occurrence of egg–larval and egg–pupal parasitism, and an updated list of taxa involved is presented in Table 3.2. What is immediately surprising is that, for the family-level groups in which egg–larval/pupal parasitism occurs, this is exclusively the case in only about half of the groups. Further, whilst a very large proportion of hymenopterous parasitoids in general attack Lepidoptera or Coleoptera, these host groups are relatively uncommon amongst egg–larval/pupal parasitoids, which rather seem to be associated with dipteran and hymenopterous hosts.

In contrast to many idiobiont egg parasitoids, egg–larval/pupal parasitoids nearly always lay in a host which is at an advanced stage of development (Clausen, 1954). However, there appears to be considerable variation in exactly how the parasitoid gains access to the inside of the host larva. Several studies, all dealing with the braconid subfamily Cheloninae, illustrate the possible range of variation. In *Chelonus inanitus*, the egg is laid under the chorion but outside the actual egg cell of the host (Rechav and Orion, 1975). The 1st instar parasitoid larva penetrates the yolk of the host egg and then the haemocoele of the host embryo before it finally hatches. Cox (1932) showed that *Ascogaster carpocapsae* would lay into host eggs at all stages of development except for the most advanced black spot stage, but would never lay in such a position that it would be within the developing host embryo. However, Boyce (1936), working on the same species, disagreed and stated that histological sectioning of host embryos clearly showed that the *Ascogaster* egg could also be laid within the developing embryo. In both *Chelonus annulipes* and *Phanerotoma flavitestacea* the egg is normally laid in the yolk of the host, and the 1st instar larva enters the host haemocoele at 3.5 days, before the host's midgut is united (Wishart and van Steenburgh, 1934; Hawlitzky, 1972). However if the host egg is more than 4 days old, the *Phanerotoma* will lay direct into the embryo (Hawlitzky, 1979).

Several ovo-larval parasitoids have been shown to be facultatively capable of attacking newly emerged host larvae that may still be near the egg mass (Cowan, 1979), though clearly some are much more specific and have only ever been observed to oviposit in the host's egg stage.

3.15 OFFSPRING SEX DETERMINATION

Another consequence of an idiobiont's host not growing any more is that the parasitoid can potentially assess its size and so make decisions about

Table 3.2 The occurence of egg–larval or egg–pupal parasitism

Parasitoid family group	Species/genus	Host order	Reference
Ibaliidae	Perhaps most, but at least one attacks early instar larvae too	Hymenoptera	Spradbery, 1970a
Aulacidae	All species for which observations exist	Hymenoptera, Coleoptera	Skinner and Thompson, 1960, cited in Gauld and Bolton, 1988
Monomachidae	All species for which observations exist	Diptera	
Platygasteridae	Many but details of biology only available for a few representatives	Diptera	Parnell, 1963
Encyrtidae	Several genera	Lepidoptera, Hemiptera, Coleoptera, Araneae	
Chalcididae	*Chalcis* spp.	Diptera	Cowan, 1979
Eulophidae	*Entodon ergias*	Coleoptera	Beaver, 1966
Braconidae – Cheloninae	All known species	Lepidoptera	Shaw and Huddleston, 1991
Braconidae – Opiinae	A few species	Diptera	
Braconidae – Alysiinae	Several genera of Dacnusini	Diptera	Clausen, 1954; Mook, 1961
Braconidae – Helconinae	Probably some *Diospilus*	Coleoptera	Parrott and Glasgow, 1916; Alauzet, 1987
Braconidae – Ichneutinae	Some species (Ichneutini)	Hymenoptera	Zinnert, 1969; Sharkey and Wharton, 1994 – but some evidence to contrary as regards other tribes
Braconidae – Microgastrinae	A few species but representing several species groups	Lepidoptera	Shaw and Huddleston, 1991; Naumann, 1991
Ichneumonidae – Collyriinae	All known species	Hymenoptera	Salt, 1931
Ichneumonidae – Stilbopinae	All species for which observations exist	Lepidoptera	Hinz, 1981
Ichneumonidae – Campopleginae	One *Campoletis* sp.	Lepidoptera	Clausen, 1954
Ichneumonidae – Ctenopelmatinae	Some Pionini	Hymenoptera	McConnell, 1938
Ichneumonidae – Diplazontinae	A few species	Diptera	Rotheray, 1981a

whether or not it is worth ovipositing on, or whether it might be more suitable for development of a male or a female offspring. The ability of female hymenopterans to control the fertilization of their eggs, combined with their haplodiploid sex determination system, gives them the ability to determine the sex of each individual offspring. Many species have been noted to lay unfertilized eggs on small or otherwise low quality hosts; males of a large proportion of hymenopterans are smaller than corresponding females (Gauld and Fitton, 1987) and therefore can develop successfully on hosts that would only support the development of an undersized and presumably less fecund female. Size has a larger effect on the fecundity of females than males in most parasitic Hymenoptera (Heinz, 1991), and it has been shown on many occasions that idiobiont parasitoids selectively lay male eggs on small hosts – presumably because being small does not disadvantage a male offspring as much as it would a female one.

It is obvious that idiobiont wasps that attack exposed hosts, such as trichogrammatids attacking insect eggs or bethylids that attack dermestid beetle larvae, could easily make an assessment about its size, but it is far less clear how such individual host size assessment could take place in those parasitoids that attack concealed hosts with which only their ovipositors come into contact. Nevertheless, Urano and Hijii (1995) have shown that two braconid parasitoids of beetle larvae that live under bark through which the wasps drill do in fact lay a higher proportion of male eggs on smaller host individuals. Presumably, there are chemical or mechanical cues about host size, or perhaps age, that can be detected by the wasp's ovipositors, but no details are known as yet.

Of course, assessing host size is at least potentially possible for idiobionts, since by definition the host will not be doing any more growing after parasitization, but in most cases koinobionts are unlikely to be able to predict how much resource a host chosen at the time of parasitization will eventually yield. For this reason, koinobionts do not normally make decisions about the sex of particular offspring in relation to host size, but many idiobionts do. However, Tanaka *et al.* (1992) also showed that sex allocation in the koinobiont braconid, *Cotesia kariyai*, is determined by the stage (and therefore size) of its host caterpillar, *Pseudaletia separata*. The wasps lay the same number of female eggs in late 3rd instar hosts as they do in late 5th instar ones, but lay more males in the latter. The explanation put forward is that as males, more so than females, respond to reduced food by maturing to a smaller size, the wasp uses the potential extra (but not accurately predictable) food resource provided by the larger host for males.

4

Preimaginal development: from gametogenesis to syngamy

4.1 SPERMATOGENESIS

The early stages of gametogenesis in insects are essentially similar in both sexes. Testes and ovaries are compartmentalized, respectively, into a series of seminiferous tubules or ovarioles (Büning, 1994). (The seminiferous tubules of males are often referred to as follicles, but these are equivalent to ovarioles rather than the follicles of female ovaries.) Germ-line cells – oögonia and spermatogonia – divide to form the oöcytes or spermatocytes. The developing oöcytes and sperm are surrounded by a layer of cells which are attached basally to a basal membrane. In males, each seminiferous tubule contains a succession of different zones characterized by different stages of spermatogenesis. The spermatogonia or primordial germ cells each become surrounded by a layer of somatic cells and these form a cyst in which the spermatogonium divides repeatedly so as to produce from 64 to 256 spermatocytes. (The cysts of male testes are equivalent to the follicles of female ovaries.) Each spermatocyte then divides meiotically to give rise to four spermatids, which mature to form the sperm.

Nearly all studies of spermatogenesis in the Hymenoptera deal with aculeate taxa and only a few limited studies have dealt with the process in parasitoids. The most detailed published study to date is that of Hogge and King (1975) on *Nasonia*; and Newman and Quicke (in preparation) have investigated spermatogenesis in various braconids. In general among insects, spermatogenesis is complete by the time of eclosion if the adult life span is short, or it may continue for at least a while after eclosion in longer-lived species. Preparations from a number of wild-caught parasitic wasps indicate that perhaps most belong to the latter category, but there are clearly many exceptions. For example, Gerling and Legner (1968) showed that in the pteromalid, *Spalangia cameroni*, the testes were depleted of sperm by the last day as a pupa, and the testes a few days after emergence were empty deflated bags. Spermatogenesis in *Trichogramma brassicae* was also found to be completed by the time of emergence (Chihrane and Laugé, 1994). It would be interesting to know whether continued spermatogenesis in the adult is correlated with either longevity or mating system (section 4.5). In many species, mature sperm are stored in a distinct seminal vesicle (Chapter 6), but in others they appear simply to fill the more or less uniformly tubular vas deferens.

A number of key stages in spermatogenesis in the braconid, *Aleiodes*, which possesses fairly typical-looking mature sperm, are illustrated in Figure 4.1. In the secondary spermatocyte, the acrosomal complex and axoneme (mobile organelle of the tail) both develop next to the nuclear envelope, the acrosome at the part adjacent to the cyst wall and the tail diametrically opposed (Figure 4.1a). The axoneme develops from one member of a centriolar pair. Next to the centriole, an electron-dense patch assembles which will become the centriolar adjunct, a structure of uncertain function but which apparently contains RNA. The mitochondria become clustered in one part of the cytoplasm and start to fuse (Figure 4.1b). The characteristic structure that results is called the nebenkerne, which will eventually form the two separate, elongate mitochondrial derivatives that lie parallel to the axoneme in the tail of the mature sperm. The axoneme grows out at right angles from the centriole/nuclear envelope complex before the nucleus starts to change shape and elongate (Figure 4.1c). Nuclear elongation is not a simple process, but rather involves a number of special features. The nucleus develops a 'dorso-ventral' polarity with fibrous chromatin accumulating on one side (Figure 4.1d). Microtubules associate differentially around the nuclear envelope and the nucleus is then drawn into an elongate gutter shape that is crescent-like in transverse section (Figure 4.1e). In *Aleiodes*, rough endoplasmic reticulum can be seen looped within the concave part of the gutter-shaped nucleus, perhaps indicating that this is an active site of protein synthesis. Finally the concavity becomes occluded from the anterior and posterior poles and the stretched nucleus of the mature sperm is now completely circular in section (Figure 4.1f).

Spermatogenesis appears to be adversely affected by high temperature in some taxa (Yang *et al.*, 1993). *T. brassicae*, for example, when kept at 44°C for 6 hours, exhibits pycnosis in primary spermatocytes with resultant complete sterility. This thermo-sterilization might explain the failure of this species under extreme climatic conditions (Chihrane and Laugé, 1994).

4.2 STRUCTURE AND ULTRASTRUCTURE OF MATURE SPERM

With the exception of bees and ants, there have been very few studies of sperm in the Hymenoptera, and very few indeed for parasitic taxa (Jamieson, 1987). Until recently, sperm ultrastructure had been described

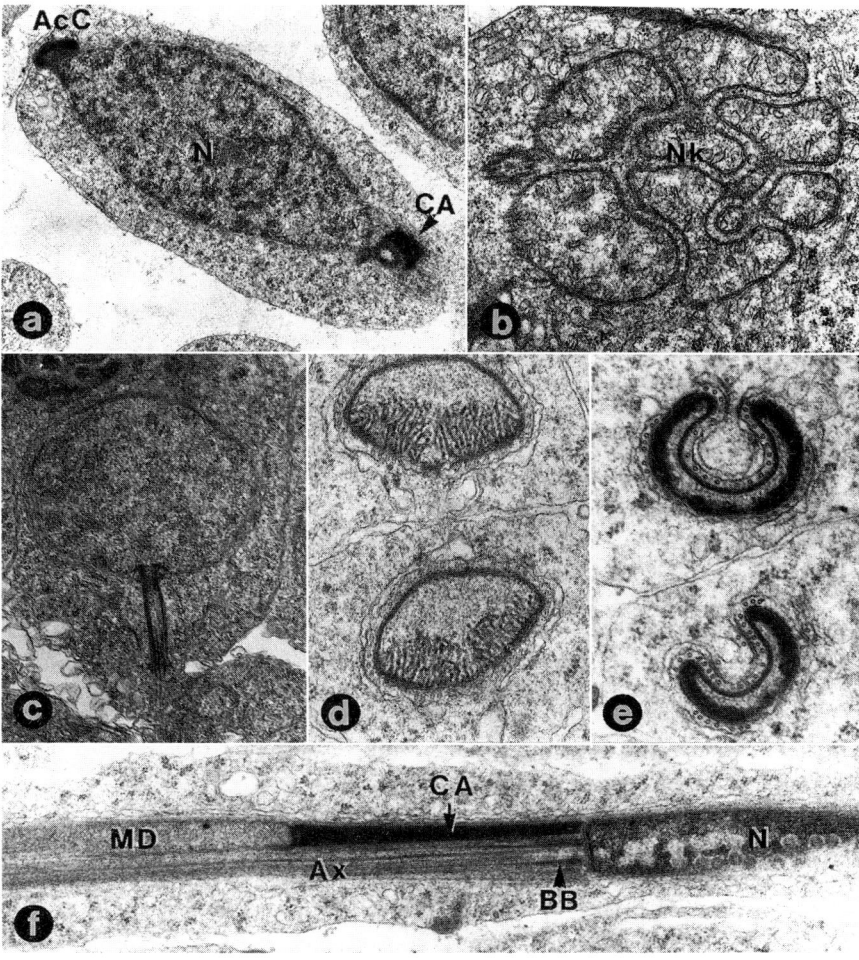

Figure 4.1 Features of spermatogenesis in the braconid, *Aleiodes coxalis*: (a) secondary spermatocyte showing development of acrosomal complex (AcC) and centriole with centriolar adjunct (CA), both next to the nuclear envelope; (b) nebenkerne (Nk) – the aggregated and fused mitochondria that will form the pair of mitochondrial derivatives of the mature sperm; (c) axoneme beginning to grow from centriole attached to nuclear envelope; (d,e) two transverse sections showing successive stages of nuclear elongation; (f) longitudinal section through nearly mature sperm showing posterior of nucleus (N), axoneme (Ax), basal body (BB), centriolar adjunct (CA) and mitochondrial derivative (MD).

for a few chalcidoids (e.g. Hogge and King, 1975) and for the ichneumonid, *Diadromus pulchellus*, including sperm from diploid males (Chauvin *et al.*, 1987). A survey by Quicke *et al.* (1992b) revealed considerable variation in terms of gross structure and ultrastructure throughout the Hymenoptera. All sawflies studied to date, from xyelids to siricids, have their mature sperm grouped into spermatodesmata in which the acrosomes and anterior ends of the nuclei of all the sperm from one testicular cyst are embedded in a gelatinous extracellular matrix and which swim with synchronous tail movements (Figure 4.2a; unpublished observation). In contrast, apocritan spermatozoa are, with few exceptions, solitary and in none are well-formed spermatodesmata passed to the female. Whether orussids also produce spermatodesmata is uncertain from Cooper's (1953) description, and nothing is known about any of the basal apocritan families Stephanidae, Megalyridae and Trigonalyidae.

The typical Hymenopteran sperm has an elongate nucleus and a normal motile tail with a pair of mitochondrial derivatives. Total sperm length in most groups ranges approximately between 40 and 250 μm, though especially short and highly modified sperm occur in a large clade of braconid wasps (Quicke *et al.*, 1992b). At the anterior of the sperm is the acrosome. The basic structure of this seems to be fairly constant within the parasitic and aculeate Hymenoptera, with a cone-shaped acrosomal cap surrounding a cavity within which there is an electron-dense rod, the perforatorium (Figure 4.2c); this sometimes intrudes into the anterior end of the nucleus, which is concave to accommodate it. In some groups, such as most ichneumonoids and some parasitic aculeates, the acrosome and anterior end of the nucleus is often surrounded by a thick and sometimes complex matrix (Figure 4.2c,d). Nothing is known of the function of this structure, or for certain whether it is intra- or extracellular. In mature *Aleiodes* sperm, the nucleus abuts posteriorly on one side to a basal body which appears to be derived from the centriole and extends posteriorly as the axoneme, and on the other side to a mitochondrial derivative and the centriolar adjunct, the latter being interposed between the nucleus and the second mitochondrial derivative (Figure 4.1f). Whether this is true of all other parasitic wasp taxa is at present uncertain. In most groups, the mitochondrial derivatives run along one side of the axoneme, but in many Chalcidoidea they follow a spiral course around a similarly spirally twisted axoneme (Figure 4.2b). In these the nucleus is also often spirally twisted, though not always in the same direction (Lee and Wilkes, 1965), and this has been related to function (see below).

4.3 SPERM POLYMORPHISM

Sperm polymorphism has been receiving considerable attention of late with the possibility that different morphs, even apparently abnormally

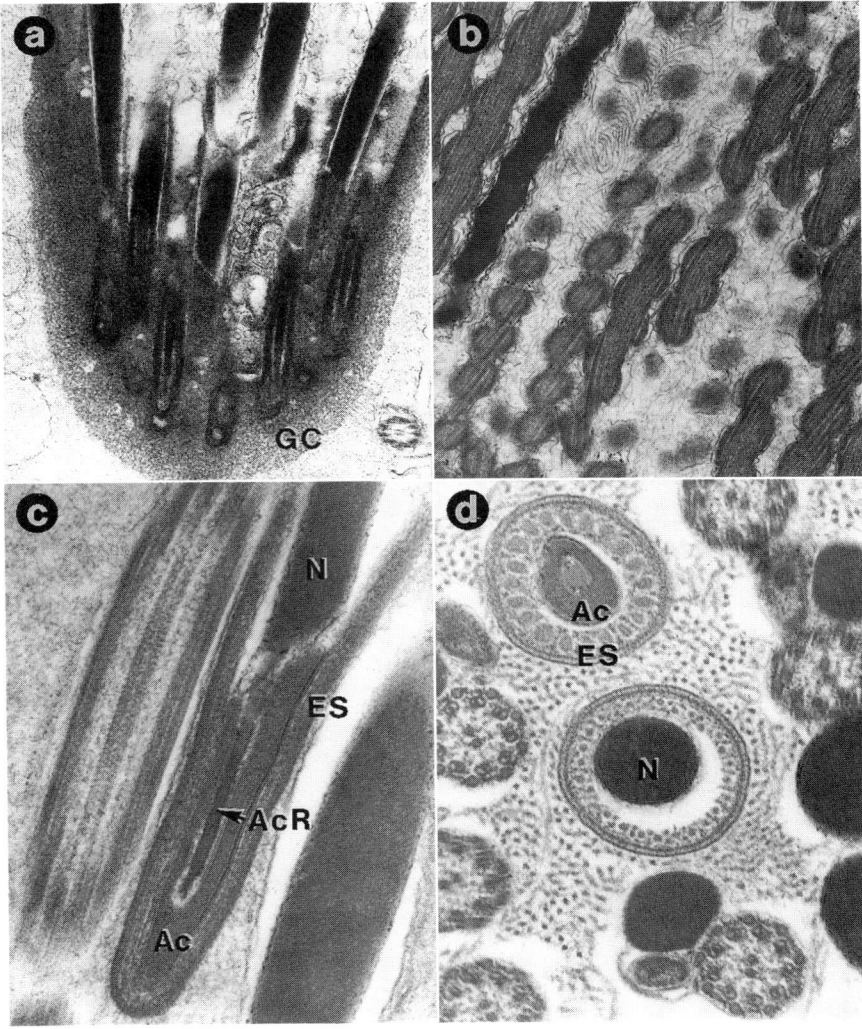

Figure 4.2 Features of some mature hymenopteran sperm: (a) longitudinal section through head end of spermatodesmata in the sawfuly, *Xyela julii* (Xyelidae), showing the gelatinous extracellular cap (GC); (b) spirally twisted nuclei and tail regions of *Encarsia* (Aphelinidae); (c) transverse section through anterior end of sperm of *Netelia* (Ichneumonidae) showing acrosome (Ac), acrosomal rod (AcR) or perforatorium, and extracellular sheath (ES); (d) transverse sections through various levels of *Metopius* sperm (Ichneumonidae) showing complex anterior (extracellular?) sheath (ES) surrounding anterior of nucleus (N) in the middle section, and in the top section, the acrosome (Ac).

formed ones, may each play their own role – if not in fertilization, then in sperm competition. Sperm polymorphism does not appear to be very obvious or common in most parasitic wasps, but it has been reported for two chalcidoids, *Dahlbominus fuscipennis* (Lee and Wilkes, 1965) and *Trichogramma brassicae* (Chihrane and Laugé, 1994). *D. fuscipennis* apparently regularly produces more than one sperm morph, including both length and coiling variants. Among the sperm removed from the spermathecae of freshly mated females, a proportion have the nucleus spiralled in the opposite direction to the axoneme whilst in the majority both sperm regions coil in the same direction (Wilkes and Lee, 1965). All individuals examined possessed sperm of these two types. Wilkes and Lee proposed that the rarer, sinistral sperm morphs, which have the nucleus coiling in the opposite direction to the axoneme, were able to locate and enter the micropyle of the egg but were unable to penetrate the vitelline membrane and therefore actually to fertilize the egg. It was suggested that the sinistrally coiled sperm were responsible for haploid male production in this species but no strong evidence was presented to show that normal male production actually involves these atypically coiled sperm. Until further evidence to the contrary is obtained, they could just as easily be regarded as abnormal, dysfunctional sperm which simply fail to do their intended job. Indeed, Wilkes (1964) had shown that selection experiments for high sex ratio led to males that produced a higher proportion of abnormal sperm than those of the base population. An interesting consequence of this could be that females of *D. fuscipennis* might not distinguish between sperm types, and so would release dysfunctional sperm with the purpose of fertilizing an egg. The presence of a significant proportion of dysfunctional sperm would therefore cause a potentially disadvantageously high proportion of male offspring. In *Trichogramma*, the proportion of atypical sperm seems to be increased by exposure to high temperature during the pupal phase (Chihrane and Laugé, 1994), but far less is known about this system than that of *D. fuscipennis*.

4.4 SPERMATOPHORES AND SPERM BUNDLES

It is not absolutely clear whether parasitic wasps always produce spermatophores. Flanders (1945) reported their occurrence in the braconid, *Macrocentrus ancylivorus*, and Gordh and Hendrickson (1976) noted a capsule, which they believed to be a spermatophore, in the female genital tract of the ichneumonid, *Bathyplectes anurus*, that had been fixed immediately after mating. El Agoze *et al.* (1995) reported seeing several spermatophores in females of another ichneumonid, *Diadromus pulchellus*. Wilkes (1965) reported that the sperm of the eulophid, *Dahlbominus fuscipennis*, formed themselves into a compact bundle in a pocket just beyond the seminal vesicle (Figure 6.1a), and that during mating a single

bundle was transferred to the female. The sperm were described as forming a viscous mass but without any apparent surrounding membrane. The
speed with which sperm in some species reach the spermatheca and the
lack of references to spermatophores or sperm bundles in most species
suggests that most species simply ejaculate a more or less fluid semen.
This leaves us wondering about the function of the large accessory glands
of all male parasitic Hymenoptera (Chapter 6) which undoubtedly produce considerable amounts of secretion. In this respect, it is interesting to
note that the sperm of the pteromalid, *Spalangia cameroni*, are stimulated
to swim vigorously when contacted by accessory gland secretion (Gerling
and Legner, 1968), so in this species packaging into a spermatophore
would be a waste of effort.

4.5 INSEMINATION POTENTIAL

Male insemination potential is very important in species that have low sex
ratios and are routinely subject to local mate competition. For example, a
male will fail to achieve maximum fitness if it cannot inseminate all the
females with which it has the chance to copulate. This problem has been
intensively investigated in the eulophid, *Colpoclypeus florus*, by Dijkstra
((1986)/1987). Some males were found to be capable of fertilizing up to 15
females, which is roughly the maximum number that a male in this LMC
species is likely to encounter in broods. However, an individual's insemination potential was also found to be strongly correlated with its size –
larger males being able to inseminate more females – and a similar effect
has been found with other taxa (Barrass, 1961).

Gordh and DeBach (1976) examined the insemination potential of
males of the aphelinid wasp, *Aphytis lingnanensis*. They found that males
can mate approximately 25 times during their lifetime and produce
between 150 and 170 offspring (i.e. daughters). However, the number of
eggs fertilized declined with the number of previous matings, indicating
that sperm availability may be limiting. Wilkes (1965) found a similar lifetime mating potential for the eulophid, *Dahlbominus fuscipennis*. Nadel and
Luck (1985) examined male sperm depletion in the pteromalid,
Pachycrepoideus vindemiae, and showed that the numbers of daughters produced by successive females mated with the same male in rapid succession declined rapidly and almost linearly with rank in mating sequence.
By the ninth or tenth female, the male had virtually run out of mature
sperm that could be transferred. Similarly, Schlinger and Hall (1960)
found that the first female inseminated by a male *Praon palitans*
(Braconidae: Aphidiinae) produced a one-to-one sex ratio, but the seventh
produced only one daughter to 58 sons. The braconid, *Habrobracon lineatellae*, however, was found by Laing and Caltagirone (1969) to have a far
lower insemination potential, and although a male would mate with more

than 30 females it succeeded in fertilizing only the first three of these. Similar low insemination potentials have been found for various other species (Simmonds, 1953).

As was noted above, many parasitic wasps continue spermatogenesis for long periods post-eclosion and so the fact that a male may run out of sperm if he has the opportunity to mate with a lot of females in quick succession does not necessarily signify the end of his reproductive life – though in the case of *Nasonia*, only an incomplete recovery in insemination potential was apparent after a rest of one or two days (Barrass, 1961).

Ramadan *et al.* (1991) made the useful distinction between full and partial inseminations. In the case of the opiine braconid, '*Biosteres*' *vandenboschi*, a male kept with an excess of virgin females managed to inseminate 10 of them fully, and partially inseminated a further dozen. In the case of *Pachycrepoideus vindemiae* the male was able to replenish his sperm stocks rapidly, and if matings were separated by an interval of about 30 minutes he could supply a maximal sperm load to each female, at least up to the ninth in sequence. It has been suggested that the inclusion of more than one male in large broods of the gregarious sib-mating bethylid, *Goniozus nephantidis*, may be a strategy to counter limiting insemination potential of a single male (Hardy and Cook, 1995). This would be particularly likely if the females all emerged in a short period of time.

The fact that many male parasitic wasps will happily continue mating (up to several hundred times in *Nasonia*!) even though they have long since run out of sperm might be an example of spiteful behaviour, because females they have mated with will often accept no further mates, or at least not for a while (van den Assem, 1986).

4.6 OÖGENESIS

Oögenesis has been described in detail for rather more parasitic wasps than has spermatogenesis, but compared with other insect orders the process of oögenesis is poorly known within the Hymenoptera as a whole (Büning, 1994). Indeed, modern techniques have hardly been applied even for the better studied aculeate groups. Much of what we know stems from early publications, some dating back into the last century, though these often have to be treated with some caution, and some aspects may need to be reappraised in the light of recent investigations. Most studies have concerned oögenesis in various Ichneumonoidea. Speicher (1936) described in detail the karyological aspects of oögenesis from oöcyte to the stages of embryogenesis in the braconid *Habrobracon* and much of their findings were confirmed and enhanced by the excellent study of Cassidy and King (1972). Amos and Salt (1974) provided an atlas of oöcyte formation in the campoplegine ichneumonid, *Venturia* (as *Nemeritis*) *canescens*. Klag and Bilinski (1993) described the process of vitellogenesis in two

other ichneumonids, the ectoparasitic tryphonine, *Cosmoconus meridionator*, and the endoparasitic banchine, *Lissonota catenator*, and in particular
considered the formation of the oösome, a characteristic structure visible
under the light microscope (Figure 2.8), formed at the posterior end of the
egg in many insects. They showed that one component of the oösome is
derived from mitochondria in the nurse cells and transported into the egg;
the other main component appears to develop within the oöcyte cytoplasm itself. The oösome in most insects is clearly an important structure
because its obliteration leads to the development of an infertile individual,
but this is apparently not the case in the parasitic Hymenoptera, or at least
not in the ichneumonid, *Pimpla*, though obliteration does apparently prevent formation of pole cells (Achtelig and Krause, 1971). A detailed electron microscopical survey of oögenesis in the braconid *Cardiochiles nigriceps*, was provided by Davies *et al.* (1986) paying particular attention to the
fibrous layer of the chorion. In other groups there have been fewer
detailed studies. Krainska (1961) described basic cell changes and some
histochemical aspects of oögenesis in the gall-forming cynipid, *Cynips folii*;
and King and Richards (1969) carried out a light and electron microscopic
study of oöcyte development in the pteromalid *Nasonia vitripennis*; Le
Ralec (1991, 1995) presented scanning and transmission electron micrographs of various stages in egg development in species representing several families and related these to egg type – for example, whether the eggs
are hydropic or not (section 4.12).

Oögenesis follows a fairly typical form in most parasitic Hymenoptera
and the following summary is collated from various studies and summarized in Figure 4.3. At the anterior ends of the ovarioles is a region referred
to as the germarium, where stem and pre-follicle cells divide continuously. Each division gives rise to a replacement stem cell and a cell referred to
as a cystoblast which continues to divide to form a cluster of so-called cystocytes – the germ cell cluster. In *Nasonia*, cystocyte division continues
until a 16-cell cluster is formed, one cell of which will become the oöcyte
and the other 15 its associated nurse cells (trophocytes). In other taxa there
may be more or fewer nurse cells produced; for example, in *Habrobracon*
there are 31 nurse cells. The nurse cells and the oöcyte are connected by
intercellular bridges and it is through these connections that yolk materials synthesized in the nurse cells are transported to the oöcyte in the
process of vitellogenesis. The whole cystocyte/oöcyte complex is initially
associated with a group of pre-follicle cells (20 in *Habrobracon*) which
group and divide to surround it. The follicle cells develop and differentiate, with those surrounding the nurse cells forming a thin squamous layer,
and those around the oöcyte, which is at the posterior end of the cluster,
forming a more cubic or columnar epithelium. The nurse cells rapidly
enlarge as the cluster passes along the ovary, and so too does the oöcyte.
The nurse cell nuclei undergo several rounds of endomitosis until they

have very high ploidy levels, presumably associated with the need to produce large amounts of protein during vitellogenesis, at least in many taxa.
Follicle cells also undergo endomitosis.

Vitellogenesis in insects typically involves production of a polypeptide
(vitellogenin) in the female's fat body, its release and subsequent sequestering by follicle cells and final storage in the oöcyte as vitellin, the major
yolk protein. Three different types of vitellin production have been identified among insects. In group 1 taxa, which includes most insects, vitellogenin is produced as a large polypeptide, but then cleaved before secretion into the haemolymph by the fat body cells. Hymenoptera have been
considered to belong to a unique group (group 2) in that the vitellogenin
is not cleaved before secretion. However, recent work by Takadera *et al.*
(1996) on vitellogenesis in tenthredinoid sawflies shows that at least some
Hymenoptera display an essentially group 1 type of vitellogenesis.
Nothing is known about the biochemistry of vitellin production in any of
the parasitic wasps and so the distribution of group 1 and group 2 vitellogenesis in the order is potentially an interesting area for further study.
Apparently the hydropic, alecithal eggs of some parasitic wasps lack
vitellin completely (Tilden and Ferkovich, 1987).

Following completion of vitellogenesis, the follicle cells secrete an extracellular matrix that surrounds the oöcyte and condenses to form the
vitelline membrane. A group of follicle cells from the anterior end of the
nurse cell cluster migrates through to the anterior end of the oöcyte
(Figure 4.7a), where it completes the vitelline membrane and chorion of
the micropylar region. There is then a marked change in the secretory

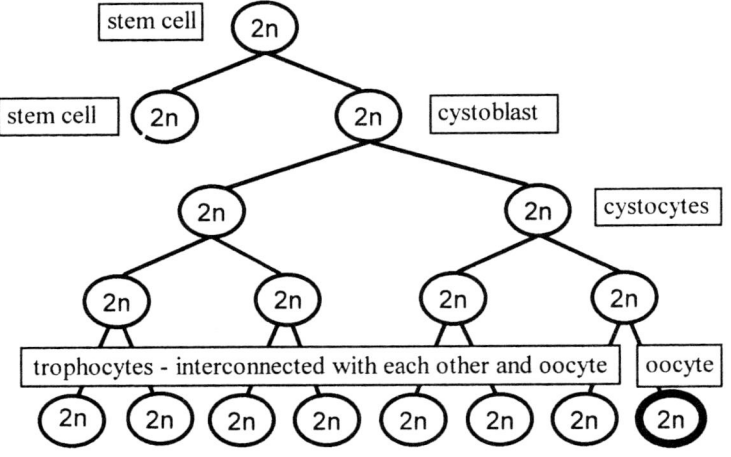

Figure 4.3 Cell divisions from the stem cell (primary oögonia) stage which occurs
in the germarium through to formation of the nurse cell cluster and oöcyte.

products of the follicle cells and the subsequent secretions form the chemically and physically distinct outer layer of the mature egg, the chorion.

4.7 MEIOSIS

In insects, the process of meiosis occurs within the oöcyte. In parasitic Hymenoptera, uterine eggs are usually in a prolonged first meiotic metaphase which may not proceed to anaphase until oviposition (Rössler and DeBach, 1973). The second division usually occurs upon oviposition, and usually takes only an hour or so. Only one of the four haploid nuclei that result can become part of the zygote (Figure 4.4). Meiosis in the oöcyte is not accompanied by cell division so after the two meiotic divisions there will be four nuclei – the oöcyte nucleus and three others, which are termed polar nuclei or polar bodies. Some or all of the polar nuclei usually degenerate or are eliminated soon after the first cleavage division of the zygote (e.g. Rössler and DeBach, 1973). However, in a number of taxa, including some parasitic wasps, they continue to play an active role in embryonic development (Tremblay and Caltagirone, 1973; section 5.4).

4.8 ACCESSORY NUCLEI

In addition to the polar nuclei, and not to be confused with them, oöcytes of a number of Hymenoptera have a number of rather enigmatic

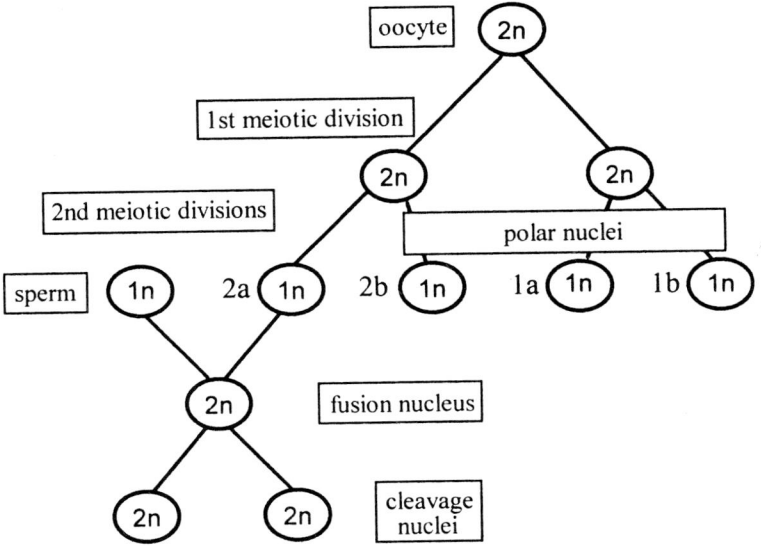

Figure 4.4 Stages from fertilization to the initiation of embryogenesis.

organelles, the accessory nuclei, which bud off the main nucleus before vitellogenesis is complete. Among the parasitic Hymenoptera, accessory nuclei have been reported in the braconids, *Cotesia* (as *Apanteles*) *glomerata* (King and Fordy, 1970) and *Habrobracon juglandis* (Cassidy and King, 1972; King and Cassidy, 1973); the ichneumonids, *Ophion luteus* (King and Fordy, 1970), *Cosmoconus meridionator* (Bilinski, 1991b), *Pimpla turionellae* (Meng, 1970), *Lissonota* sp. and *Coleocentrotus* sp. (Klag and Bilinski, 1994); and the chrysidid, *Chrysis ignita* (Szklarzewicz *et al.*, 1993). They also occur in sawflies and various bees and social wasps. They have not yet been observed in chalcidoids or other 'microhymenoptera' but it is not certain that they are absent from these groups. Accessory nuclei have a characteristic internal structure, and several studies suggest they contain large amounts of RNA (Bilinski *et al.*, 1993). Accessory nuclei are also known in Mallophaga and Diptera, and probably in Mecoptera. Differences in their structure and behaviour suggest that those of the Hymenoptera, Mallophaga and Diptera have independent origins (Büning, 1994), but not enough is known about those of the mecopteran, *Panorpa*, to proffer an opinion.

In most insects, the first meiotic division is not completed until after the egg is laid, and often the egg is at metaphase of the first division at the time it is laid. Details have been described for a few parasitic wasps (e.g. Speicher, 1936). In some cases the first meiotic division is completed within the adult wasp before oviposition; the second division and, in the case of fertilized eggs, fusion of the sperm nucleus with the egg nucleus occur soon after oviposition. However, Rössler and DeBach (1973) showed that in the thelytokous aphelinid, *Aphytis mytilaspidis*, the first meiotic division is only completed some three hours after the egg has been laid.

Klag and Bilinski (1994) showed that in the ichneumonid, *Coleocentrotus soldanskii*, more than one of the cystocytes in the germ cell cluster undergo meiosis, but only one of these differentiates to form the oöcyte. Nevertheless the other cells (potential pro-oöcytes) that undergo meiosis are initially distinguishable from the other developing nurse cells, though they do subsequently become transdetermined to form normal nurse cells, and the accessory nuclei resulting from their meiotic division eventually degenerate. Klag and Bilinski suggest that the transdetermination may be the result of a chemical gradient set up within the germ cell cluster.

4.9 EGG MATURATION

Some parasitic wasps emerge as adults with their full complement of mature eggs and no more maturation occurs. These are referred to as pro-ovigenic (section 3.5). Some others emerge with no mature eggs and have to mature them as they go along; these are referred to as synovigenic. Between these extremes are many examples with intermediate strategies.

There is a strong tendency for synovigenic species to be idiobionts (Chapter 3) and also to host-feed (section 3.11), but neither correlation is perfect. The encyrtid, *Leptomastix dactylopa*, for example, is synovigenic but does not require protein in order to mature more eggs (Zinna, 1960).

Not all parasitic wasps emerge as adult females with mature eggs ready for oviposition. Those that need to feed first, in order to mature their eggs, are referred to as anautogenous, as opposed to autogenous ones which can lay eggs without feeding (Englemann, 1970; Jervis and Kidd, 1986). In many cases, protein food does not appear to be necessary, the wasp only requiring a source of carbohydrates such as glucose, sugar or honey (Labeyrie, 1960). Mao and Kunimi (1994) provide a nice example: the chalcidid, *Brachymeria lasus*, a pupal endoparasitoid of the tea tortrix moth, *Homona magnanima*, does not start to lay for the first 3–10 days after eclosion, and the number of mature eggs remains very low unless the wasps can feed (Figure 4.5). An even longer pre-oviposition period is found in *Bracon cephi*, females of which do not lay for about the first 3 weeks of their total life expectancy (in the laboratory) of 4 weeks (Nelson and Farstad, 1953). Females of the ichneumonid, *Exeristes comstockii*, will not mature eggs on a sugar diet, but can be stimulated to do so by feeding with the juvenile hormone mimic, farnesylmethyl ether (Bracken and Nair, 1969). This observation is particularly interesting since it is widely assumed that host-feeding is necessary in order to supply proteins needed for egg maturation – the underlying assumption being that many synovigenic parasitoids do not carry enough protein or other food with them into the adult stage to meet their full reproductive potential. However, Bracken and Nair found that when stimulated by a juvenile hormone mimic, *Exeristes* females were able to produce as many mature viable eggs as controls that had been allowed to host-feed. Clearly in this case, and under the laboratory conditions employed, the wasps did have a sufficiency of protein reserves at emergence, and the results suggest that host feeding may be a means of obtaining juvenile hormone and so perhaps of regulating egg maturation in accordance with host availability. Indeed, in *Exeristes* the juvenile hormone mimic was most effective when administered orally. In *E. comstockii*, vitamins are also important for egg maturation (Bracken, 1966). Without vitamins in the diet, an adult female *Exeristes* will lay a maximum of 0.5 eggs per day, declining to nearly 0 after about 20 days. Addition of vitamins leads to a dramatic increase in fecundity after a few days' lag, with wasps laying nearly two eggs a day. Given this result, it is not unlikely that many studies of lifetime fecundity of parasitoids in which the females have only been fed sugar-water may grossly underestimate the parasitoid's true potential.

Nobel (1940) showed that the gall-forming eurytomid, *Trichilogaster acacia-longifoliae*, has a full complement of mature eggs after it has chewed its way out of the gall, but that if its pupae are dissected out of the gall prior

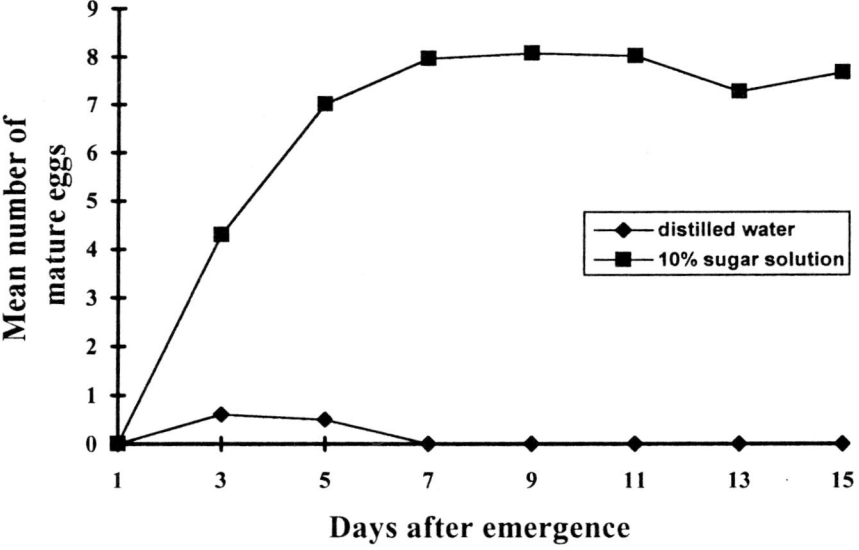

Figure 4.5 Effect on the number of mature eggs carried by females of the chalcidid, *Brachymeria lasus*, fed with distilled water or sugar water. (Redrawn and modified after Mao and Kunimi, 1994.)

to emergence the eclosing adult wasps have no mature eggs, leading Nobel to suggest that the chewing acts in some way as a trigger for egg maturation, and perhaps the ingestion of some chemo-stimulant may also be involved. It would be interesting to know how many regular host-feeding species either do not require the protein, or possibly also require vitamins, or make use of host hormones taken in at the same time.

4.10 CHORION

The chorion is the outer protective shell or membrane of the insect egg, and is secreted by the follicle cells. In ectoparasitoids, it may serve principally to resist desiccation and protect the egg from pathogens, whilst in endoparasitoids it is the interface between parasite and host. In many taxa it has to be highly elastic since it has to permit considerable distortion as the egg is squeezed along the egg canal of the ovipositor, and for those taxa with hydropic eggs (section 4.12) it has to permit the egg to swell very considerably once inside the host. Given these various requirements, it is not surprising that it is often far from a simple membranous egg covering and may have a very convoluted appearance as well as being multilayered (Mouzaki and Margaritis, 1994; Le Ralec, 1991; Figure 4.6a,b).

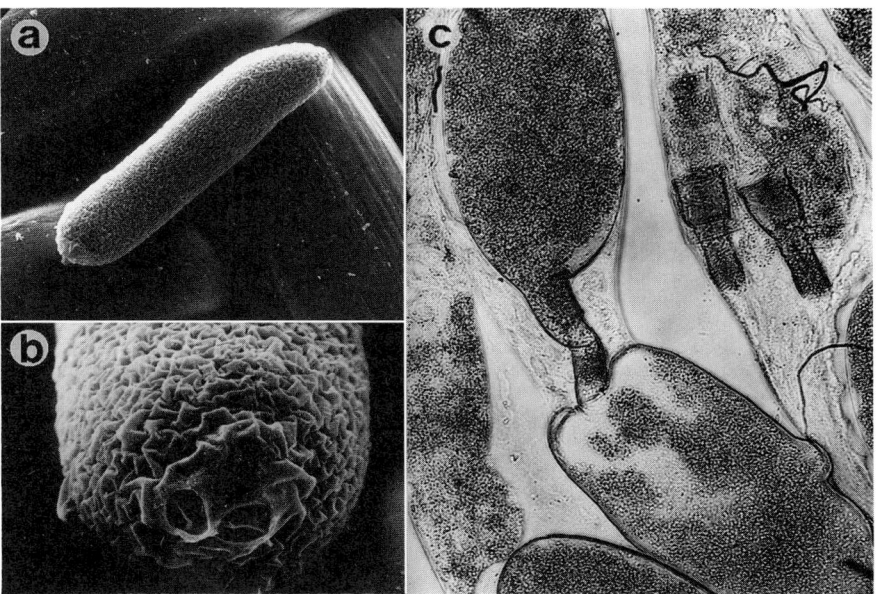

Figure 4.6 Ovarian egg features: (a,b) hydropic egg of the aphidiine braconid, *Ephedrus plagiator*, showing highly corrugated chorion and micropyle (courtesy of Anne Le Ralec); (c) dissection of ovary of the encyrtid, *Leptomastix dactylopii*, showing two intact, dumb-bell shaped eggs and a remnant of the mid-region following oösorption (courtesy of Anna Rivero-Lynch).

Chorionic structure in parasitic wasps has been investigated by both electron microscopical and histochemical procedures for a number of species, including the pteromalid, *Nasonia vitripennis* (King *et al.*, 1969b) and the braconid, *Cotesia glomerata* (King *et al.*, 1969a), and for a range of taxa by Le Ralec (1991), including aphidiine braconids, an encyrtid, a megaspilid and a eucoilid. The stages of chorion synthesis have been described in detail for the ichneumonid, *Campoletis sonoriensis*, by Norton and Vinson (1982) and for the phytophagous eurytomid, *Eurytoma amygdali* (Mouzaki and Margaritis, 1994).

Chorionic structure varies considerably: in some it is thin and has a simple structure; in others it is thick, obviously multilayered and sometimes with trabeculae (canal-like structures). The chorion of some *Eurytoma* species (Eurytomidae) is densely covered in hard dark spinules (Arthur, 1961), but the function of these is uncertain since they are not present in other taxa with similar biologies. There is even a suggestion that this feature could be intraspecifically variable (Askew, personal communication). In some endoparasitoids the chorion has a distinctly fibrous outer layer, attached to an underlying layer furnished with protuberances that is important in avoiding encapsulation (Davies and Vinson, 1986). In the braconid, *Cardiochiles nigriceps*, the fibrous layer appears to be polysaccha-ride-rich as it was removable by driselase but not by protease (Davies *et al.*, 1986). The latter authors noted that similar-looking flocculent coatings are present in several other ichneumonids and braconids that oviposit in lar-val hosts, but that it is not present in parasites of eggs or pupae (*Nasonia*, *Telenomus*, *Trichogramma*). However, their examples of non-larval para-sitoids all belonged to either the Chalcidoidea or Platygastroidea rather than the Ichneumonoidea, and therefore the distribution of this coating type might be taxonomically constrained; further, the ichneumonoids studied are also ones that have polydnaviruses or virus-like particles (sec-tion 7.11). Having said this, however, the endoparasitic ichneumonids and braconids that do possess a flocculent outer chorionic layer have certainly evolved independently from ectoparasitic ancestors, so presumably their coatings did too.

4.11 DIFFERENT EGG TYPES

As with many other aspects of wasp biology, the eggs have been studied rather less than those of many other groups of insects. Indeed, in Hinton's (1981) three-volume monograph on insect eggs, the section on the Hymenoptera is only six pages, compared for example with 77 for Hemiptera and 55 for Coleoptera.

The general morphology of parasitic wasp eggs has been surveyed by Hagen (1964). Eggs of most taxa are ovoid and many have a narrow pedi-cel at the end that enters the ovipositor first. In the case of many taxa belonging to various families whose hosts are concealed deep within wood, the narrow pedicel may be extremely long (Figure 4.7b), and is associated with the passage of the egg along the long, narrow ovipositor (Hagen, 1964; Spradbery, 1970b). Cooper (1953) noted that in the parasitic sawfly, *Orussus sayi*, some eggs may in fact be longer than the entire female wasp (16.9 mm compared with 13.7 mm) – necessarily, the pedicels of these eggs are folded back upon themselves in the ovarioles (see Figure 6.1g). In this and biologically similar species, the egg has necessarily to dis-tort considerably during the process of oviposition: the narrow pedicel emerges from the apex of the ovipositor before all the egg has entered the ovipositor's anterior end; the bulk of the contents of the egg is then squeezed down the part within the ovipositor and the apex of the pedicel correspondingly expands as cytoplasm flows into the part emerging from the ovipositor apex. In the kleptoparasitic ichneumonid, *Pseudorhyssa ster-nata*, which threads its ovipositor down the boring already made by its rhyssine ichneumonid primary host but lays a larger egg, the egg chorion

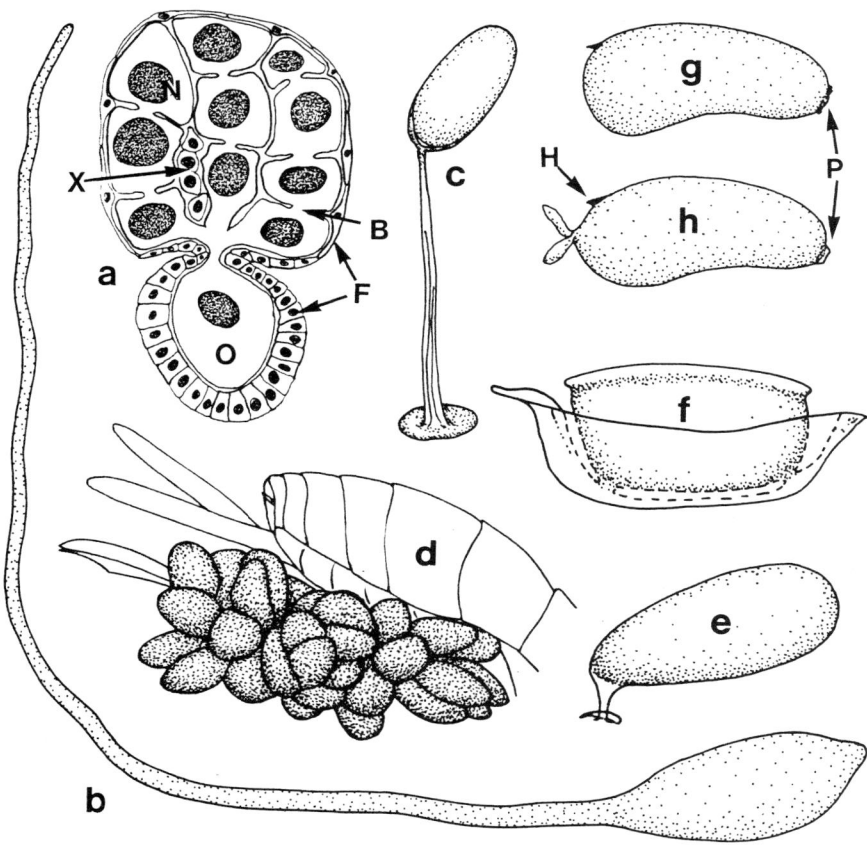

Figure 4.7 Features of oögenesis and mature eggs. (a) Previtellogenic ovarian follicle of the braconid, *Habrobracon*, showing nurse cells (N) and oöcyte (O) connected by intercellular bridges (B), and follicle cells (F) including a group (X) migrating through nurse cell cluster to anterior (micropylar) end of oöcyte (redrawn after King and Cassidy, 1973). (b) Egg of *Orussus sayi* (Orussidae) (based on Cooper, 1953). (c) Laid egg of the ichneumonid, *Euceros frigida* (redrawn from Tripp, 1961). (d) Anchored eggs of the tryphonine ichneumonid, *Polyblastus macrocentrus*, held ready for oviposition at the base of the ovipositor (redrawn from Kasparyan, 1981). (e) Ovarian egg of a tryphonine ichneumonid. (f) Egg of the tryphonine ichneumonid, *Exenterus* (based on Mason, 1967). (g) Newly laid egg of the dryinid, *Dicondylus americanus*, and (h) after development, showing the hook (H), pigmented adhesive ring (P) and, in (h), a pair of outgrowths reported to be from the egg but possibly of embryonic origin (see section 5.11) (redrawn from Giri and Freytag, 1989).

is conspicuously transversely striate (Spradbery, 1970b) which may be an adaptation for passing down an exceptionally thin egg canal. Eggs of dumb-bell shape (Figure 4.6c), such as those produced by encyrtids and tanaostigmatids, serve a similar function to the elongate pedicel found in other groups, the posterior part of the egg being a bulbous expansion. During oviposition, the body of the egg emerges from the ovipositor first, and then the bulb, the yolk having been temporarily displaced into the bulb before subsequently re-entering the body of the egg.

The number of mature eggs carried by a female is inversely correlated with egg size (Iwata, 1960a), and also generally with ovariole number, though there are some marked exceptions as in the aphidiine braconids. In both the Braconidae and Ichneumonidae, synovigenic idiobiont taxa that normally carry 10–50 mature eggs generally have between two and eight ovarioles per ovary; whilst many pro-ovigenic koinobiont groups that carry several hundred mature eggs have more than 10 ovarioles per ovary (Iwata, 1960a,b). However, the correlation is not perfect and within the Braconidae – for example, members of the koinobiont 'microgastroid' lineage – there are only two ovarioles. In the aphidiine braconids, with only one or two pairs of ovarioles, there may be as many as 1000 mature eggs at eclosion. Whether this represents an evolutionary limitation is not certain. Within a species, the number of eggs carried by a female is often positively correlated with her size, so size can be used as an estimator of female reproductive potential or quality.

4.12 HYDROPIC AND ANHYDROPIC EGGS

Two functional types of the chorion that surrounds mature eggs have been described for parasitic wasps by Flanders (1942a), anhydropic and hydropic, though these terms are often applied more loosely to the whole egg. Anhydropic eggs have a large enough quantity of yolk for subsequent embryonic development; hydropic eggs have insufficient yolk for the embryo to develop and so food has to be absorbed from the host through the chorion (Ferkovich and Dillard, 1987). The chorions of anhydropic eggs are the typical form among insects – they are tough, relatively thick and impermeable. Hydropic eggs occur widely in the parasitic Hymenoptera and have clearly evolved on numerous separate occasions, though they are restricted to endoparasitic taxa. Once laid inside the host, hydropic eggs usually swell greatly over a period of hours to days (Schlinger and Hall, 1960). The change can be enormous: in some cases (for example, euphorine braconids) hydropic eggs can swell to more than 1000 times the volume they were in the ovary (Jackson, 1928; Smith, 1952).

Hydropic eggs have little yolk, which is predominantly composed of lipids (Le Ralec, 1995). They are also typically supplied with numerous ribosomes and mitochondria which presumably play an important role in their growth within the host. These organelles appear to originate in the nurse cells and are passed to the oöcyte through the anterior intercellular bridge (nutritive pore) during oögenesis (King *et al.*, 1971; Le Ralec, 1995). The structure of the hydropic chorion of the microgastrine braconid, *Cotesia glomeratus* (as *Apanteles*), was described by King *et al.* (1969a) who showed that in comparison with anhydropic chorions, as found for example in the pteromalid, *Nasonia vitripennis*, that of *Cotesia* is very reduced in thickness and complexity. A detailed comparative scanning electron microscope survey of eggs of several families of parasitic wasps by Le Ralec (1991) showed that the hydropic chorions of the braconid subfamily Aphidiinae and of the eucoilid, *Leptopilina*, are highly convoluted whilst that of the encyrtid, *Epidinocarsis lopezi*, is smooth but still very flexible. This compares with the rigid and smooth chorions of the anhydropic taxa investigated.

The yolk of anhydropic eggs, in addition to having numerous lipid droplets, is very rich in proteins, of which vitellin is the major component. The appearance of the proteinaceous component in the yolk of hydropic eggs appears to depend upon whether or not the wasps host-feed (Le Ralec, 1995). Host-feeding taxa have a large number of protein bodies in the yolk which appear similar to those of many other insect eggs. However, the protein inclusions had quite different forms in the two non-host-feeding taxa that produce anhydropic eggs examined by Le Ralec: the megaspilid *Dendrocerus carpenteri*, and the eucoilid, *Leptopilina boulardi*. Whether this is actually related to the source of proteins used in yolk production cannot be stated with certainty as there is a possibility that it reflects some phylogenetic effect, since no eggs of host-feeding eucoilids or megaspilids were examined. Further comparative work on this would be very interesting.

4.13 STICKY EGGS

Le Ralec (1991) showed that in the eucoilid, *Leptopilina boulardi*, an endoparasitoid of *Drosophila* larvae, the long pedicel of the egg often adheres firmly to host tissues, especially to the host's gut. It is not certain whether this stickiness serves a similar function to that of the *Asobara* eggs (Braconidae) studied by Kraaijveld (1994) (see below and Section 7.19). Similar observations have been made for a wide range of other taxa, including the braconid, *Cotesia glomerata* (King *et al.*, 1969a). Eggs with an anchor also occur in some endoparasitic ichneumonids, including some Anomaloninae and Tersilochinae (Iwata, 1960a; Gauld, 1976). In this case the anchor is used to fix the end of the egg into gut wall tissue, serving to protect that part of the egg from being encapsulated and so allowing the parasitoid embryo to develop and to hatch. A similar effect is achieved by some strains of the alysiine braconid, *Asobara tabida*, whose eggs are sticky and thus tend to attach themselves to or embed themselves in host tissue despite having originally been laid in the host's haemocoele (Kraaijeveld, 1994; Kraaijeveld and van Alphen, 1994).

4.14 ANCHORED EGGS

Most parasitic wasp eggs are roughly ovoid, with or without a narrow pedicel, but some groups produce more highly modified structures. For example, members of the ichneumonid subfamily Tryphoninae are characterized by the production of eggs furnished with an anchor-like structure derived from the chorion that is used to attach them to their hosts (Figure 4.7d,e) (Kasparyan, 1981). Tryphonines are interesting in that they are ectoparasitic koinobionts, a rather rare combination. Their anchored eggs are thus an adaptation to prevent them from being dislodged by the host until the parasitoid's larva ecloses (Mason, 1967). The placement of the egg is therefore critical as the still mobile and active hosts would be able to remove or destroy it if it were placed within reach of their mandibles. In many the anchor is a simple wedge-shape and when inserted through the host's cuticle it prevents the egg from becoming detached. In other cases, such as in the genus *Exenterus*, the anchor is sometimes highly modified to hold the egg body more or less flush with the host's cuticle; the anchor stalk may be duplicated and the anchor a deep boat-shape that almost completely encompasses the egg body (Figure 4.7f) (Morris, 1937; Mason, 1967). Some Tryphoninae – for example, *Netelia* and *Phytodietus* – have eggs with a rather different form of anchor comprising a coiled tail which is inserted through the host's cuticle, though whether this is homologous to the other anchors requires further investigation.

Eggs of some dryinids are furnished with a hook-like process and a brown adhesive ring (Figure 4.7g,h) (Giri and Freytag, (1986)/1989), but the precise mechanism by which these help to hold the egg between the tergites of the homopteran nymph hosts has not been determined. Simple anchors in the form of a simple, long pedicel are also present in the eggs of various tetrastichine eulophids (Zinna, 1955), and they serve virtually identical functions to those of the Tryphoninae.

A different and non-homologous form of anchor is present on the eggs of the ichneumonid genus, *Agriotypus*, members of which are ectoparasitoids of the aquatic larvae of caddis flies (Trichoptera). In these, the anchor is not a modification of the chorion, but is an extra-chorionic structure secreted by the female (Clausen, 1931a). Another extrachorionic anchor-like modification has been described by Neser (1973) in the eulophid, *Euplectrus* sp. In this wasp, the structure is used to glue the egg to the host, whose cuticle is not broken.

Members of another ichneumonid subfamily, the Eucerotinae, lay stalked eggs on leaves (Figure 4.7c). In fact they lay huge numbers of such eggs because the chance of any of them surviving is very small. First instar larvae of eucerotines are planidial and after hatching they wait for a passing caterpillar or larval sawfly, to which they then attach themselves and on which they remain until the larva pupates (Varley, 1964). The planidial larvae feed a little from their host but will not develop any further unless their carrier-host is parasitized by a tachinid fly or ichneumonid.

4.15 RESPIRATORY ADAPTATIONS OF EGGS

Eggs in the Encyrtidae are highly modified and have been extensively studied. They are dumb-bell shaped, the larger part of which will contain the embryo, whilst in some species the other end contains a structure referred to as the aeroscopic plate. This highly specialized structure is associated with parasitoid respiration, and it has been described in great detail by a number of workers (Maple, 1937, 1947; Zinna, 1960; Nénon and Biossangama, 1985). Only the part containing the oöcyte/embryo is inserted into the host, the other remains protruding through the host's cuticle. The aeroscopic plate consists of a spongy hydrophobic reticulum, and provides a pathway for gas exchange. The earlier instars of the parasitoid larva are correspondingly modified, being metapneustic with a single caudal pair of spiracles (Zinna, 1960; see also section 5.16). Inside the host, the parasitoid larva remains with the caudal spiracles in close proximity to that part of the aeroscopic plate. In the case of *Ooencyrtus*, an egg parasitoid of the family Encyrtidae, the stalk which remains protruding from the host egg after oviposition has been shown to be used by the parasitoid as a marker of whether a potential host has been previously parasitized (Takasu and Hirose, 1988; Chapter 8).

The eggs of the phytophagous eurytomids, such as the almond wasp, *Eurytoma amigdali*, have processes at both ends. The one at the posterior pole is somewhat flattened. Mouzaki and Margaritis (1994) have suggested that this pole may remain outside the almond when the egg is lain and may therefore play a role in egg respiration. It is clear that many more detailed ultrastructural studies of parasitic wasp eggs need to be carried out in order to determine the full range of their respiratory adaptations.

4.16 THE MICROPYLE AND FERTILIZATION

The basic structure of mature parasitic wasp eggs is like that of other insects in that the oöcyte is surrounded by a vitelline membrane, outside of which is the chorion. At one end of the chorion is the micropylar region, where channels allow the sperm access to the egg cell itself. In some taxa, the micropylar region is distinctly pigmented, though more often it is inconspicuous. In common with most insects, the micropyle in the relatively few parasitic Hymenoptera that have been reported is at the anterior end of the egg – that is, the end which is towards the anterior of the wasp inside the ovary (Bronskill, 1959; Rotheram, 1973; Le Ralec, 1991; Quicke, personal observation of Stephanidae). Rather surprisingly, a posterior location has been reported for the secondarily phytophagous eurytomid, *Eurytoma amygdali* (Zarani and Margaritis, 1994). Whether the posterior micropylar appendage of *Eurytoma* is homologous with the anterior ones of other Hymenoptera is unclear: although Zarani and Margaritis provide a detailed study of its morphogenesis in *Eurytoma*, there are no comparable studies of other wasps. Nor is it known whether a posterior micropyle occurs in any other parasitic wasps, but if the posterior position is associated with the development of a respiratory process at the anterior end, i.e. the end that may be left protruding from the host (as is proposed for *Eurytoma*), then it might be expected to occur also in other groups whose eggs have a specialized respiratory structure.

There have been few detailed morphological studies of the micropylar region in the parasitic wasps. The micropyle of *Nasonia vitripennis* was described by King (1962a); those of *Pimpla turionellae* (Ichneumonidae), *Cotesia congregata* (Braconidae) and *Coelichneumon rudis* (Ichneumonidae) were illustrated by Bronskill (1959), Beckage and DeBuron (1994) and Tarasco (1995), respectively. Le Ralec (1991) presented scanning electron micrographs of the micropylar region of eggs of the aphidiine braconid, *Ephedrus plagiator* (Figure 4.6b), and the eucoilid, *Leptopilina boulardi*. Considerable differences between these suggest that micropyle structure may be of potential use in phylogenetic studies in the parasitic Hymenoptera, as it is in various other insect groups.

It is possible that in some parasitic wasps there is no micropyle. El-Agoze (1985) reported that sperm of the ichneumonid, *Diadromus pulchellus*, are able to penetrate the egg over a large area, and that a fertilization cone forms at the point of entry. Presumably this could only happen if the chorion was very soft.

4.17 POLYSPERMY

As in most animals, polyspermy – the penetration of a single egg by more than one sperm – is rare in Hymenoptera. Speicher (1936) noted it in two out of 150 fertilized eggs in *Habrobracon*, and Wilkes (1965) found only four instances out of 254 eggs in the eulophid, *Dahlbominus fuscipennis*. As females of the latter species use nearly all the sperm transferred to them by their mate to produce female offspring, it seems likely that the low level of polyspermy reflects more the controlled release of single sperm from the spermatheca rather than the barriers in the egg. Perhaps sperm use in less conservative species, which lay large numbers of eggs in a single oviposition, may result in higher instances of polyspermy. King (1962a) suggested that since the micropyle is not positioned in the exact centre in some taxa, such as *N. vitripennis*, fertilization may depend on the egg's precise orientation within the common oviduct as it passes the opening of the spermathecal duct. Based on the topology of the *Nasonia* egg, King predicted that approximately one-quarter of eggs would remain unfertilized even if the female always released sperm. This possibility was reiterated by Grosch (1988) in relation to braconids.

4.18 OÖSORPTION

Oösorption – the resorption of mature eggs – is known to occur in at least some members of seven insect orders: Thysanura, Dermaptera, Orthoptera, Heteroptera, Diptera, Coleoptera and Hymenoptera (Bell and Bohm, 1975; Chapman, 1982). Many parasitic wasps are known to reabsorb eggs under conditions of stress or the absence of hosts (Quezada *et al.*, 1973). Egg resorption is frequently associated with idiobiont wasps that produce large, yolky, anhydropic eggs and host-feed (Flanders, 1950), and has a wide distribution among parasitic wasp families.

Detailed investigations of oösorption have been provided for *Nasonia vitripennis* (Hopkins and King, 1964), the eulophid, *Chrysocharis pentheus* (Sugimoto *et al.*, 1983), the aphelinid, *Coccophagus bartletti* (Walter, 1988b), the braconid, *Psyttalia* (as *Opius*) *concolor* (Stavraki-Paulopoulou, 1966), and the encyrtid, *Leptomastix dactylopii* (Rivero-Lynch, 1994) (Figure 4.6c). Oösorption takes place within a few days of egg maturation if hosts are not available. King (1963) fed the microtubule inhibitor colchicine to the pteromalid, *Nasonia vitripennis*, to prevent further egg maturations, and so was able to show that oösorption commences 2–3 days after egg maturation if hosts are not available. King and Ratcliffe (1969) noted that the terminal (fully developed) egg in each ovariole is resorbed if it has not been laid within about 24 hours of maturation. However, as pointed out by Walter, colchicine has wide-ranging effects and such results have to be treated with caution. Egg resorption in the aphelinid, *Aphytis melinus*, results when the wasp is prevented from host-feeding even though it is supplied with water and honey (Collier, 1995). Host-feeding must also, therefore, have a role in supplying metabolic requirements, and in the absence of hosts for food, the wasp has to recycle some of the nutrients contained within their mature eggs.

4.19 EGG SECRETIONS

There is some evidence that parasitic wasp eggs may secrete compounds that adversely affect competitors (see Section 7.22).

5

Preimaginal development: from embryo initiation to pupa

5.1 INITIATION OF EMBRYOGENESIS

Whilst embryogenesis in most insects is triggered by fertilization, in haplo-diploid organisms and in parthenogenetic taxa in general there has to be a second process such that unfertilized eggs will also develop when required. In many parasitic wasps, the egg undergoes a very considerable distortion when it passes down the narrow egg canal of the ovipositor. For example, in the case of the parasitic sawfly, *Orussus sayi*, Cooper (1953) calculated that during oviposition the egg, which has a maximum diameter at the bulbous end of 0.4 mm in the ovary (Figures 4.7b, 6.1g), must be squeezed through a long egg canal with an internal diameter of only 0.04 mm, resulting in an increase in surface area from 7.4 mm^2 to 10.3 mm^2 – an increase of approximately 40%. Went and Krause (1973, 1974a,b) and Went (1982) have studied experimentally the effects of changing egg architecture on embryogenesis in the ichneumonid *Pimpla turionellae*. Ovipositional distortion can be mimicked by passing unfertilized eggs removed from a female wasp through the narrow orifice of a glass micropipette, and this process leads to cleavage just as in natural 'oviposition'. Mechanical inducibility of embryogenesis appears to be widespread among parasitic Hymenoptera: King and Rafai (1973) showed that ovarian eggs of the pteromalid, *Nasonia vitripennis*, could be stimulated to develop by means of pressing with a needle; Vinson and Jang (1989) demonstrated that distortion and certain hypertonic saline treatments could also induce development in the ichneumonid, *Campoletis sonorensis*; and von Borstel (1960) described an interesting mutant of *Habrobracon hebetor* whose eggs did not develop when oviposited because they emerged at the base of the ovipositor rather than being distorted by passing along it.

Perhaps surprisingly, fertilization is not an adequate stimulus for development, at least in *Pimpla turionellae*: Went and Krause (1974a) removed fertilized eggs from the base of the ovipositor before distortion, but these would not develop further. Went (1982) provided evidence that, rather than activation, ovipositional distortion of the egg is more akin to derepression. In many cases, host factors do not normally appear to provide adequate stimuli either, as transplantation studies involving *P. turionellae* and *N. vitripennis* have shown (Salt, 1965; Went, 1982). There are some exceptions: Greany (1986), in an attempt to culture microgastrine braconids *in vitro*, showed that ovarian eggs of *Cotesia marginiventris* would sometimes undergo embryogenesis and hatch if wasp ovaries were transferred intact into one of his culture media which contained host fat body material. This seems to suggest that the initiation of embryogenesis might depend at least in part on chemical stimuli from the host, but other factors cannot be totally excluded.

Distortion certainly does not appear to be important in the case of tryphonine ichneumonids, since in this group only the egg anchor (section 4.14) passes down the lumen of the ovipositor and embryogenesis has often already commenced within the female. In many tryphonines an egg, or sometimes a whole mass of eggs (genus *Polyblastus*), may be held suspended from the base of the ovipositor awaiting laying (Figure 4.7d). Also, in tryphonines, females have sometimes been found to exhibit egg dumping when kept in a tube in the absence of hosts (Shaw, 1995). One possible reason for this may be that if the eggs are not laid in a given time, embryogenesis might proceed too far, and Shevyrev (1912) has observed that sometimes they will actually hatch and start consuming their own mother from within, thereby killing her. Such viviparity, whilst not universal in tryphonines, is nonetheless quite common in some genera such as *Aderaeon*, *Dyspetes*, *Netelia* and *Polyblastus* (Kasparyan, 1981). Distortion may also be less important in taxa with small, alecithal eggs, such as many koinobionts. Flanders, for example, noted that in the braconid, *Macrocentrus*, eggs are approximately the same size as the ovipositor lumen, so distortion in these is presumably comparatively small, and the same is probably true of many other taxa.

In some sawflies, and perhaps the majority, eggs can be stimulated by distilled water to begin embryogenesis (Naito, 1982; Sander and Feddersen, 1985). In this system, if mature eggs can be dissected from female sawflies and placed on a filter paper soaked in distilled water, cleavage ensues within a few minutes, making this an ideal system for the study of sawfly chromosomes and embryogenesis. Further, Hatakeyama *et al.* (1994) has shown that this system can be used to investigate the importance of the location of sperm within the egg for successful fertilization. They found that sperm injected at the anterior, micropylar end of the egg would sometimes fuse with the oöcyte nucleus and produce a diploid insect. In contrast, injection into the posterior end never produced diploids but did sometimes produce haploid–haploid chimaeras. There have been only limited attempts to activate parasitic wasps artificially using distilled water or hypotonic media. Sander and Feddersen found that placing mature eggs dissected from the ichneumonid, *Venturia canescens*, in distilled water caused an initial activation, but that embryogenesis failed at the blastoderm phase. If the distilled water stimulus was combined with mechanical stress (partial flattening under a coverslip), then embryogenesis proceeded normally and a viable larva could be obtained.

Egg activation in the ichneumonid, *Pimpla turionellae*, can be induced by the ionophore A23187, which very selectively increases the permeability of biological membranes to the divalent cations Ca^{2+} and Mg^{2+} ions (Wolf and Wolf, 1988). In some instances injection of the ionophore was sufficient to lead to complete embryogenesis, with the successful hatching of a 1st instar larva, thus demonstrating that the release of Ca^{2+} from intracellular stores is a sufficient stimulus for embryogenesis, and presumably under normal oviposition conditions such a release is caused by egg distortion. Importantly, in some eggs in which the injected A23187 remained localized, development was also localized, and so it may also be presumed that, in this wasp, distortion of more or less the whole egg is essential for the initiation of successful development.

5.2 EMBRYOLOGY

In all cases studied to date, the orientation of embryonic development is already determined in the ovary of the wasp – as in other insects, they follow Hallez's Law; that is, the anterior end of the egg (towards the wasp's head) will develop into the anterior of the embryo (Hallez, 1886). Bilinski (1991b) has shown that in the ichneumonid, *Cosmoconus*, the antero-posterior and dorso-ventral axes are already defined in the mature oöcyte as indicated by the positions of the main and polar nuclei, the oösome, the accessory nuclei and lipid and yolk droplets.

Most work on embryogenesis in parasitic wasps was conducted before 1970, and the great majority of workers used light microscopical techniques exclusively. Nevertheless, a great deal of information was obtained, revealing a wide range of variation. Embryological fate maps showing the ultimate destinies of particular cell regions in the embryonic blastoderm have been produced for the honey bee and the braconid, *Habrobracon* (Petters, 1977). These were found to be essentially similar to one another as well as to the *Drosophila* fate map. In *Habrobracon*, Petters made use of mosaics produced using the ebony mutant and cases in which fertilization goes wrong with nuclear fusion occurring between the sperm nucleus and one of the post-cleavage nuclei.

Most embryological studies concern the Chalcidoidea and Ichneumonoidea, with very little work having been published on members of any of the other parasitic superfamilies. Within the Chalcidoidea, detailed descriptions of embryogenesis have been provided for *Eurytoma aciculata* (Eurytomidae: Ivanova-Kasas, 1958), *Nasonia vitripennis* (Pteromalidae: Bull, 1982), *Trichogramma evanescens* and *T. chilonis* (Trichogrammatidae: Gatenby, 1917, and Tanaka, 1985a,b, respectively). As regards the Ichneumonidae, detailed descriptions of embryogenesis have been provided for *Pimpla turinellae* (Bronskill, 1959), *Diadegma* (as *Angitia*) *vestigialis* (Ivanova-Kasas, 1960), *Mesoleius tenthredinis* (Bronskill, 1964), and in the Braconidae for *Aphidius* and *Ephedrus* (Ivanova-Kasas, 1956), *Habrobracon juglandis* (Amy, 1961), *Coeloides brunneri* (Ryan, 1963) and *Cotesia congregata* (Beckage and DeBuron, 1994). Embryogenesis has only been described in detail for a handful of parasitic aculeates such as the dryinid, *Hoplogonatopus atratus* (Abe and Koyama, 1991).

5.3 POLYEMBRYONY

Polyembryony is a process in which a single egg, instead of developing straight away into a single embryo, divides a number of times to form a number of separate cells, each of which finally develops into an embryo (Ivanova-Kasas, 1970). Polyembryony, as a normal reproductive strategy rather than a teratology, is rare in insects; it is best known in Strepsiptera and, in particular, the parasitic Hymenoptera (Table 5.1). Within the Hymenoptera, polyembryony occurs in four families – Braconidae, Dryinidae, Platygastridae and Encyrtidae (Clausen, 1940), each in a different superfamily – and it is clear that it has evolved on at least four separate occasions. Polyembryonic broods originating from a single egg are always composed of just one sex, but mixed-sex broods can result when more than one egg is laid in a single host. Some polyembryonic broods comprise only a few individuals, as in at least one platygastrid (Leiby and Hill, 1923), but in some encyrtids many hundreds – or in some *Copidosoma* species even more than a thousand – separate embryos may develop through the repeated division of a single zygote (Silvestri, 1906). In the polyembryonic braconid, *Macrocentrus ancylivorus*, only one adult ever emerges from the host caterpillar, the embryonic development of all other embryos having been suppressed by the first one to become a larva.

Most details of polyembryonic development have been presented for several encyrtids (Silvestri, 1906; Leiby, 1922; Ivanova-Kasas, 1970; Baehrecke and Strand, 1990). However, the last of these, which documents development in *Copidosoma floridanum*, points to a number of inconsistencies in terminology and some mistakes in earlier works, and consequently care should be exercised in their interpretation. In polyembryonic encyrtids, the egg divides into a number of morulae. These are ensheathed in a syncytial membrane derived from polar bodies which forms a trophamnion and develops an intimate relationship with host tissues – the trophamnion cells produce microvilli that penetrate host cells

Table 5.1 Incidence of polyembryony in the Hymenoptera

Taxa	Family	Reference
Most species	Encyrtidae	Leiby, 1922; Doutt, 1947; Cruz, 1986; Wang and Laing 1989
Macrocentrus gifuensis; M. ancylivorus	Braconidae	Parker, 1931; Daniel, 1932
Platygaster zosine; other species reported are synonyms	Platygastridae	Marchal, 1904; Hill, 1926; Gauld and Bolton, 1988
Aphelopus spp.	Dryinidae	Buyck, 1949

to increase the uptake of nutrients (Cruz, 1986). Development of *C. floridanum* embryos is apparently synchronized with host development since it depends on host endocrinology (Strand *et al.*, 1991; Chapter 7). Ligation of the host behind the head blocks development of *C. floridanum* embryo and serosa, but the effect is not simply reversed by the application of juvenile hormone analogues (Baehrecke *et al.*, 1992).

5.4 EMBRYONIC MEMBRANES

Many insect embryos develop within a single layered cellular membrane that probably serves a variety of functions depending on the group. A considerable diversity of such membranes has been observed within the parasitic Hymenoptera. The main function in many groups is presumably protective, but in some the membrane clearly has a trophic function; that is, it is responsible for absorbing nutrients from the host and passing them to the developing embryo within.

Tremblay and Caltagirone (1973) summarized the variety of extra-embryonic membranes in parasitic Hymenoptera by proposing four groups: the true serosal membrane (serosa), a pseudoserosa, a deuteroserosa and a trophamnion. The true serosa, or its equivalent in parasitic Hymenoptera, derives from extra-embryonic cells of the germ band (the ichneumonids *Diadegma* and *Mesoleius*) or from the blastoderm cells forming the dorsal cap (the ichneumonid *Pimpla*). During embryogenesis, a group of cells may grow out from either of the above regions to spread all over the surface of the embryo, forming a complete outer membrane. In the braconid, *Cardiochiles*, the serosal membrane develops from cells at both the anterior and posterior poles of the egg some 14 hours after oviposition (Pennacchio *et al.*, 1994b). It comprises both discrete cells at the poles and a syncytium in the abdominal region: only the discrete polar cells give rise to teratocytes (see below). Experiments involving the transfer of eggs between different artificial media at various stages in serosal development suggested that, at least in *Cardiochiles*, the serosal cells have a nutritive function. In some cases, the 'serosal' membrane originates simultaneously over the whole blastoderm in a process called delamination, and the resulting membrane is termed a pseudoserosa. In such cases the blastoderm cells divide to produce two layers, the outer of which is composed of distinctive cells. Delamination to form a pseudoserosa occurs, for example, in the ichneumonid, *Mesochorus*, and in the aphidiine braconid, *Ephedrus*. The deuteroserosa, *sensu* Tremblay and Caltagirone (1973), is derived from cells that were initially internal and associated with the yolk, but come to the surface and spread to form a membrane as a result of a considerable distortion of the growing blastoderm. Deuteroserosa apparently also occur in *Trichogramma*.

Beckage and DeBuron (1994), using both scanning and transmission electron microscopy, showed that in the microgastrine braconid, *Cotesia congregata*, what was previously believed to be a single-cell thick serosal membrane actually comprised two membranes: the true serosa immediately below the chorion and an amnion which delaminates towards the inside from it. The amnion was found to remain attached to the embryo for a while after the serosa had dissociated to form teratocytes, and so could play a significant role in host/parasitoid immunological interactions. In the opiine braconid, *Diachasmimorpha* (as *Biosteres*) *longicaudata*, a koinobiont parasitoid of Diptera, the serosal cells have been shown to secrete polypeptides, and so they probably play some role in host regulation (Lawrence, 1990a).

The serosa proper, pseudoserosa and deuteroserosa share the fact that they are derived entirely from cells of the embryo, i.e. from cleavage nuclei, and so are genetically identical with it. The trophamnion, on the other hand, is formed from the polar nuclei, usually by themselves but sometimes after fusing with one of the cleavage nuclei (Tremblay, 1966). The trophamnion is thus a specialized membrane that serves to help the developing embryo and 1st instar larva to obtain nutrients from its host. Afterwards, the trophamnion may disintegrate into a number of cell clusters which may also serve as food for the growing parasitoid larva (Nahif and Madel, 1990). The distribution of trophamnion possession through the parasitic Hymenoptera has been summarized by Whitfield (1992) who noted its presence in at least some members of the Ichneumonidae, Braconidae, Cynipoidea and Scelionoidea. As there is strong evidence that the ancestral members of both the Ichneumonidae and Braconidae were ectoparasitoids, and further, as there appears to be no close relationship between the Cynipoidea and Scelionoidea, it can be presumed that a trophamnion has developed independently within the parasitic Hymenoptera on at least four separate occasions.

5.5 TERATOCYTES

Teratocytes – embryo-derived cells that develop in the host independently from the embryo – form from the dissociated cells of the serosal membrane which normally surrounds the developing embryo in parasitic wasps (Figure 5.1). In these cases, instead of these extra-embryonic serosal cells simply being sloughed off and dying, they absorb nutrients from the host's body (Alford, 1968; Gerling and Orion, 1972; Vinson and Lewis, 1973; Volkoff and Colazza, 1992). Teratocytes usually swell considerably after separation, and 10-fold changes in diameter (Sluss and Leutenegger, 1968) or 100-fold increases in volume (Zhang *et al.*, 1994) are not uncommon. Consequently they can be very conspicuous when a parasitized host is dissected (Figure 5.1b); hence their alternative name of giant cells. They do not divide, but they do undergo multiplication of their DNA, giving rise ultimately to cells with very high ploidy levels. Teratocytes are known to occur with certainty in only three families of parasitic wasps, collectively representing just two superfamilies: the Platygastroidea (Gerling and Orion, 1972; Volkoff and Colazza, 1992) and the Ichneumonoidea (Zhang *et al.*, 1994). The single supposed teratocyte that was reported for the trichogrammatid, *Trichogramma brasiliensis*, by Voegele *et al.* (1974) may be a somewhat different entity, and Dahlman (1990) postulated that it could be derived from the single combined polar body. Hellqvist (1994) reported that teratocytes may also occur in the diapriid, *Synacra* sp., but his observations were made through the host cuticle and so require confirmation. Within the Ichneumonoidea, teratocytes have a limited distribution, occurring only in the braconid subfamilies Aphidiinae, Cardiochilinae, Cheloninae, Euphorinae, Helconinae, Meteorinae and Microgastrinae (Shaw and Huddleston, 1991), whilst in the Platygastroidea they occur in both scelionids and platygastrids. Within the Braconidae, there is a general consensus of opinion that the Cardiochilinae and Microgastrinae are closely related, as are the Euphorinae and Meteorinae, and so the distribution of teratocytes in this family probably reflects no more than four separate evolutionary events, and perhaps even fewer.

Tremblay (1966) showed that in the aphidiine braconids, *Lysiphlebus* and *Diaretiella*, the teratocytes ultimately derive from the largest cell of the eight-cell-stage blastomere. Subsequent work by Tremblay and Calvert (1971, 1972) showed that this cell is actually the result of fusion of one of the eight-cell-stage blastomeres with the polar nuclei. This large cell divides on the surface of the embryo to form a layer of large polyploid cells. In *Cardiochiles nigriceps*, Pennacchio *et al.* (1994b) have shown that teratocytes develop only from the serosal membrane at the poles of the embryos; the remaining serosal cells surrounding the wasp's abdominal region form an anucleated syncytium that remains attached to the 1st instar larva. The differences in the developmental origins of teratocytes in different braconid

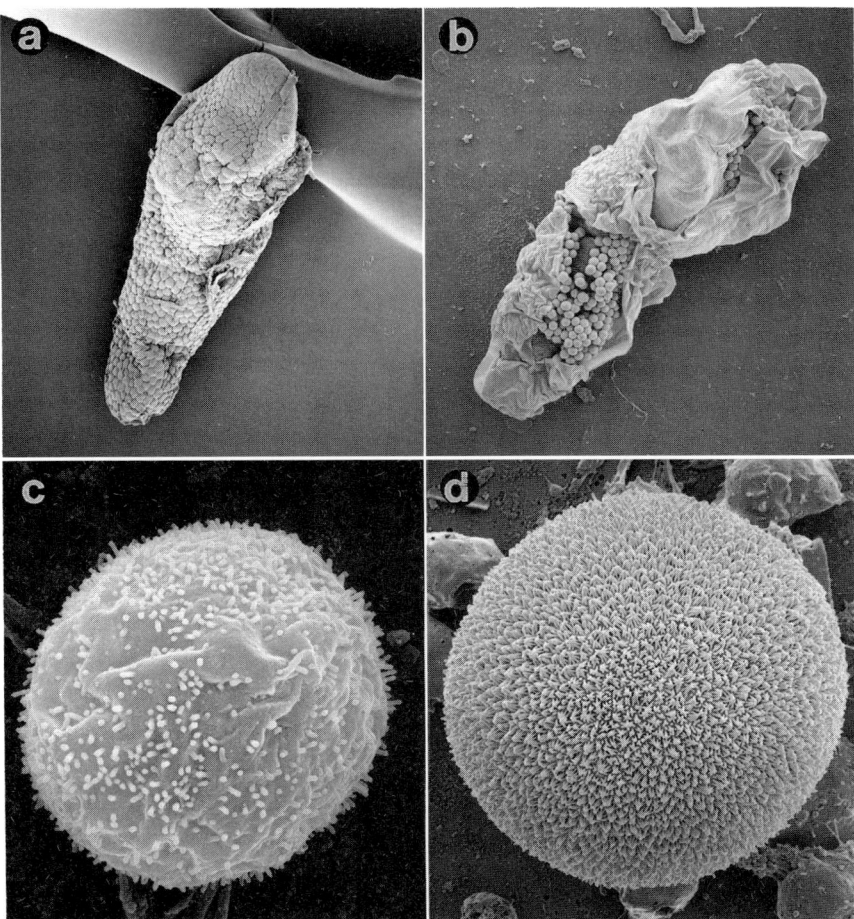

Figure 5.1 Serosal membrane and teratocytes of the braconid, *Microplitis croceipes*: (a) embryo 36 h old, showing individual cells of serosal membrane; (b) hatching embryo 40 h old, showing dissociation of serosa into individual cells which will develop into teratocytes; (c) newly dissociated teratocyte, showing sparse short microvilli; (d) teratocyte more than a day after release, showing very dense microvillar covering. (Courtesy of Douglas Dahlman. Reprinted from International Journal of Insect Morphology and Embryology, Zhang *et al*, 1994, with kind permission of Elsevier Science Ltd.)

subfamilies, and even between different tribes of the Aphidiinae, suggest that even within this family they may not all be homologous, and this might also be reflected in their function (Wharton, 1993).

Upon hatching of the embryo in the host, those cells comprising this membrane divide and separate into clusters and finally into individual cells. For many years it was believed that these cells acted simply as a way in which the parasitoid larva could acquire host tissue as food. As shown by Volkoff and Colazza (1992), their number may decline as the parasite larva develops, perhaps but not necessarily because the latter consumes them. It has now been shown that, at least in some cases, teratocytes are involved in endocrinological interactions with the host (Zhang *et al.*, 1992; Pennacchio *et al.*, 1994b,c), and these are discussed in greater detail in Chapter 7 (see particularly section 7.6).

Detailed electron microscopical studies have been carried out on tera-tocytes and their development in a number of species of parasitoids, including the braconids *Cardiochiles nigriceps* (Pennacchio *et al.*, 1994b,c), *Dinocampus* (as *Perilitus*) *coccinellae* (Sluss and Leutenegger, 1968), *Aphidius matricariae* (Tremblay and Laccarino, 1971), *Cotesia* (as *Apanteles*) *kariyai* (Tanaka and Wago, 1990), and *Microplitis croceipes* (Zhang *et al.*, 1994). The surfaces of teratocytes soon after separation have few microvilli (Figure 5.1c), but that of mature ones is densely covered with microvilli (Figure 5.1d). The fatty acid and amino acid compositions of teratocytes of two euphorine braconids, *Leiophron uniformis* and *Peristenus stygius*, were investigated by Cohen and Debolt (1984). The possibility that not all tera-tocytes of a given species are equal in terms of function may also have to be considered since Volkoff and Colazza (1992) showed that those of the scelionid, *Trissolcus basalis*, comprise three different size classes.

5.6 HATCHING

Clausen (1932) reported that humidity was the most important factor in determining the time of hatching of eggs of the koinobiont ectoparasitoid tryphonine ichneumonids. In groups with planidiform larvae, eggs are laid away from the host; the minute 1st instar larvae hatch soon after lay-ing and then either await their primary host or move off actively in search of one.

Members of the Trigonalyidae have a particularly unusual biology. They are usually hyperparasitoids of tachinid flies or ichneumonids, or parasites of vespids (Weinstein and Austin, 1991) but occasionally they are primaries on sawflies (Weinstein and Austin, 1995). Trigonalyids lay very large numbers of eggs along the edges of leaves (inserted into the leaf tis-sue). The eggs are consumed by the secondary host, a lepidopteran or sawfly larva, and are stimulated to hatch by the environment of the cater-pillar's gut, but they usually only complete development if this is subse-quently parasitized or eaten. Clausen (1931b) showed that eggs could be stimulated to hatch in mild potassium hydroxide, but only if the chorion had been damaged. The implication is that mastication by the secondary host larva is an essential part in the process, in addition to the host's gut chemistry. A more thorough set of experiments was conducted on eggs of the Australian species, *Taeniogonalys venatoria*, by Weinstein and Austin (1995). Eclosion was most successfully induced *in vitro* by exposing eggs to saline (pH 7.5), acid saline (pH 6.0) and the enzyme, cathepsin, in acid saline.

In a few instances, a parasitic wasp larva hatches away from its feeding site on the host, and has to gain entrance by chewing its way inside. A remarkable variant on this theme is found in the aphelinid genus, *Eretmocerus*. In one species which attacks the whitefly, *Bemisia tabaci*, Gerling *et al.* (1990, 1991) showed that the parasitoid egg is laid underneath the whitefly nymph, and that the newly emerged parasitoid larva bites a small hole in the ventral cuticle of its host, through which the parasitoid larva enters. However, host epidermal tissues form a capsule around the *Eretmocerus* larva, isolating it from the host's haemolymph. Unlike normal encapsulation processes, this capsule appears to be advantageous to the parasitoid, keeping it functionally external to the host and its cellular defences. Other factors presumably released into the host seem to cause a general tissue breakdown by the time that the capsule that surrounds the *Eretmocerus* also disintegrates, allowing it to complete its feeding.

5.7 GENERAL LARVAL MORPHOLOGY

Apart from a number of isolated descriptions, Parker (1924) provided the first really detailed survey of parasitic wasp larvae though his study was restricted to the Chalcidoidea. In addition to describing general appearance, he also illustrated details of such features as the larvae musculature, central nervous system, gut and glands. Although mature parasitic wasp larvae are generally similar from family to family, the basic and perhaps most widely distributed type usually being referred to as hymenopteriform (Figure 5.2a), early instars – and especially the 1st instar – vary considerably in their complexity and external features. Several different names have been coined for these larval types though they are by no means all-inclusive nor are some of them clearly defined. Including the hymenopteriform type, Clausen (1940) recognized 14 different 1st instar larval forms: hymenopteriform (Figure 5.2a), polypodiform (Figure 5.2b), vesiculate (Figure 5.2b), caudate (Figure 5.2c), eucoiliform (Figure 5.2e), sacciform (Figure 5.2f), cyclopiform (Figure 5.2h), mymariform (Figure 5.2i), planidial (Figures 5.2j, 5.3), encyrtiform, teleaform, agriotypiform, microtype and mandibulate (Figure 5.4). In most of these cases, the parasitoid larva undergoes hypermetamorphosis from specialized early instar forms to more or less uniformly hymenopteriform later ones. In some cases, the larva may pass through three distinct morphological forms. For example, scelionids of the genus *Trissolcus* have a teleaform 1st instar, sacciform intermediate one and hymenopteriform final stage (Volkoff and Colazza, 1992), and some *Anaphes* species (Mymaridae) have a mymariform 1st instar followed by sacciform 2nd to 4th instars

(Boivin *et al.*, 1993). In some braconids, 1st and last instar larvae are mandibulate but the 2nd, fluid-fading, form is not. More gradual transitions also occur: for example, some koinobiont parasitoids (e.g. agathidines) hatch as caudate larvae with little or no obvious anal vesicle, but the balance shifts throughout the instar until at its conclusion the larva is markedly vesiculate, with an inconspicuous tail (Quednau, 1970; Odebiyi and Oatman, 1972); these agathidine larvae also come under the polypodiform category (Figure 5.2b).

Vesiculate larvae occur in some braconids (Shaw and Huddleston, 1991) and in some banchine ichneumonids – the vesicle has often been suggested as having a principally respiratory function (section 5.16) but may also be involved in nutrition and excretion (section 5.14), and in physiological interactions with the host (Chapter 7). Caudate larvae occur in many groups (e.g. Ichneumonoidea, Proctotrupoidea, Cynipoidea and Chalcidoidea) and this has undoubtedly evolved on many independent occasions (Eastham, 1929). The exact nature of the tail may not be the same in all cases. Thorpe (1932b) noted that the tails of some ichneumonid larvae were supplied with tracheae and so were also reasonably presumed to play a role in respiration. Larvae of tryphonine ichneumonids, which develop externally on mobile larval hosts, have many long setae in their early instars and possess a distinct suctorial disc at the apex of the abdomen (Clausen, 1932) (Figure 5.2g). Polysphinctine ichneumonids which are koinobiont ectoparasitoids of spiders have polypodiform larvae which, in addition to the ventral 'feet', have retractile, dorsal, wart-like processes that are used to grip the host between its cephalothorax and abdomen, and are used later in their development for holding the spider's silk as the ichneumonid larva completes its feeding (Nielsen, 1923; Clausen, 1940; Fitton *et al.*, 1987). Roskam (1986) described a strange V-shaped structure associated with an early stage of the platygastrid, *Platygaster betularia*, but its exact nature is unknown and no such feature has been observed in any other platygastrids. It would be interesting to know exactly what Roskam had seen.

Larval anatomy has been described for a number of groups (e.g. Eastham, 1929), but most work has concentrated on the structure of the head, which has been particularly important in the systematics of the Ichneumonoidea. Vance and Smith (1933) provided a general terminology for head sclerites. Short (1952) described the head musculature and skeleton for several taxa, and Cutler (1955) has described it in great detail for the pteromalid, *Nasonia vitripennis*. The heads of ichneumonoids are well sclerotized and have provided many characters for phylogenetic analysis. In contrast the heads of chalcidoids usually have relatively little in the way of sclerotized structures, apart from the mandibles – some of this reduction may reflect the loss of cocoon spinning (Flanders, 1938; Cutler, 1955; section 5.27). As noted by Hagen (1964) there is considerable variation in the nomenclature for larval head structures, which results from attempts to homologize larval features with those of adults – a task

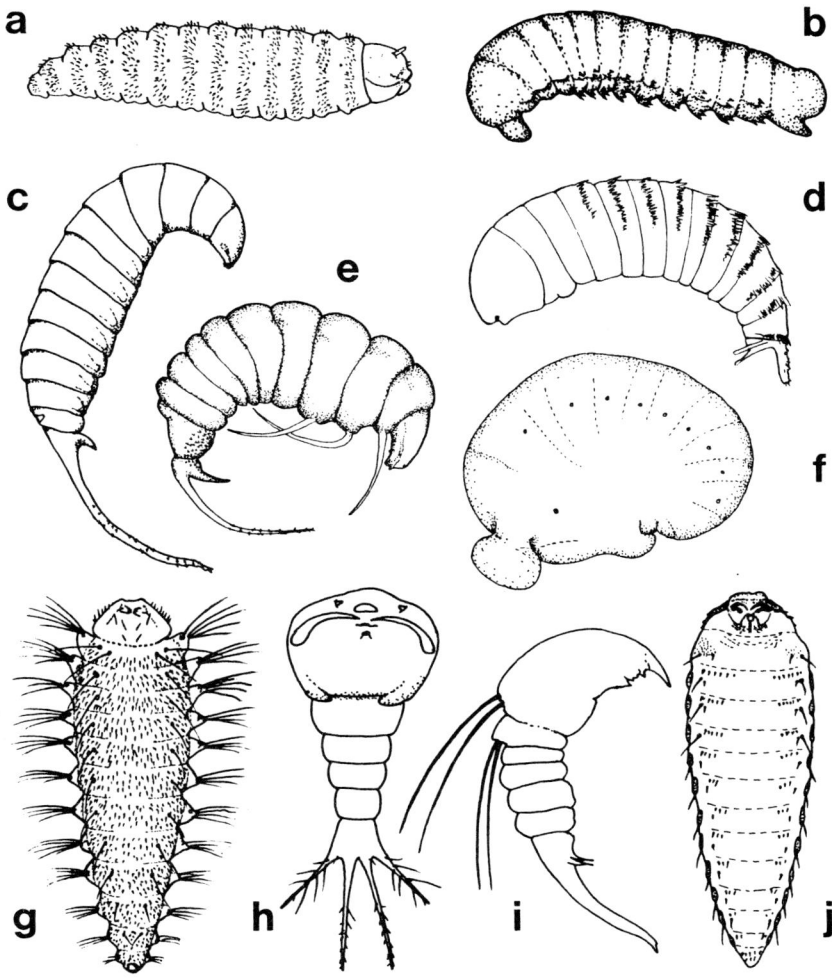

Figure 5.2 Various larval forms in the parasitic Hymenoptera. (a) Hymenopteriform final instar of the exothecine braconid, *Phanomeris phyllotomae* (redrawn after Dowden, 1941). (b) Vesiculate and polypodiform 3rd instar larva of the agathidine braconid, *Agathis pumila* (redrawn from Quednau, 1970). (c) Caudate larva of the ichneumonid, *Habronyx heros* (redrawn after Porcelli, 1988). (d) Stylized aphidiine braconid (redrawn after O'Connell, 1989). (e) 1st instar of the eucoilid, *Aganaspis pelleranoi* (redrawn after Ovrusky, 1994). (f) Sacciform larva of an unidentified gonatopodine dryinid (redrawn after Ponomarenko, 1975). (g) 1st instar larva of the tryphonine ichneumonid, *Exenterus abruptorius* (redrawn after Morris, 1937). (h) 1st instar cyclopiform larva of *Platygaster* sp. (Platygastridae). (i) 1st instar mymariform larva of the mymarid, *Anaphes* sp. (redrawn from micrograph of Boivin *et al.*, 1993). (j) 1st instar plannidial larva of the eucerotine ichneumonid, *Euceros frigidus* (redrawn from Tripp, 1961).

that has not proved easy in many cases. The teleaform 1st instar larvae of scelionids resemble the mymariform ones of some Mymaridae and this has been postulated as indicating a relationship between the two groups (Rasnitsyn, 1980). However, as discussed by Gibson (1986a), this could represent convergent evolution since there appear to be some differences in detail between the larvae in the two families and both families have similar biologies, being egg parasitoids.

5.8 PLANIDIAL LARVAE

Members of several families of parasitic insects produce free-roaming, host-seeking 1st instar larvae. These have been classified into two morphological categories: planidia and triungulins (Clausen, 1940; Askew, 1971). Notable among these are the oil beetles (Meloidea) and rhipiphorids, which have triungulin-type larvae. Planidial larvae are found in Diptera (Acroceridae) and, among the Hymenoptera, in members of the chalcidoid families Eucharitidae and Perilampidae (Figure 5.3) (Heraty and Darling, 1984; Heraty and Barber, 1990; Darling, 1992) and in eucerotine ichneumonids (Tripp, 1961; Varley, 1964) (Figure 5.2j). Whilst it is certain that planidial larvae have evolved independently in the Ichneumonoidea and Chalcidoidea, it seems probable that they have a common origin in the two chalcidoid families, despite their different biologies. Eucharitids are parasitoids of ants, and adult females deposit their eggs into plant tissue in places where host ants are likely to occur. The planidia hatch and apparently attach to almost any passing insect: if this happens to be an ant, or an insect that ends up in an ant's nest, it has a good chance of finding an ant larva which, following pupation, will be its host. Perilampids have a similar way of life in that they lay large numbers of eggs in or on leaves, buds, etc. and the resulting planidia attach indiscriminately to passing insects. In this case, however, the ultimate hosts are wood-boring larvae and the chance of an individual planidium ending up near a host is extremely small. Nevertheless, Heraty and Darling (1984) have provided evidence that these two families are sister groups or perhaps even that the Perilampidae are paraphyletic with respect to the Eucharitidae. Partly this evidence comes from a study of the detailed morphology of the planidial larvae in the two groups (Heraty and Darling noting nine synapomorphies) and partly from life history, though adult morphology is far less conclusive.

The hyperparasitic ichneumonid *Euceros* relies on its planidiform 1st instar larva finding an individual of its secondary host, a sawfly or lepidopteran larva, so as to reach its primary host (Varley, 1964). In *E. unifasciatus*, the chance of getting its eggs eaten by an appropriate, or rather potential, secondary host is enhanced by its selection of oviposition site, which is just basal along a pine shoot to the eggs of its secondary host, the sawfly, *Neodiprion*. The 1st instar *Euceros* larva parks itself on a secondary

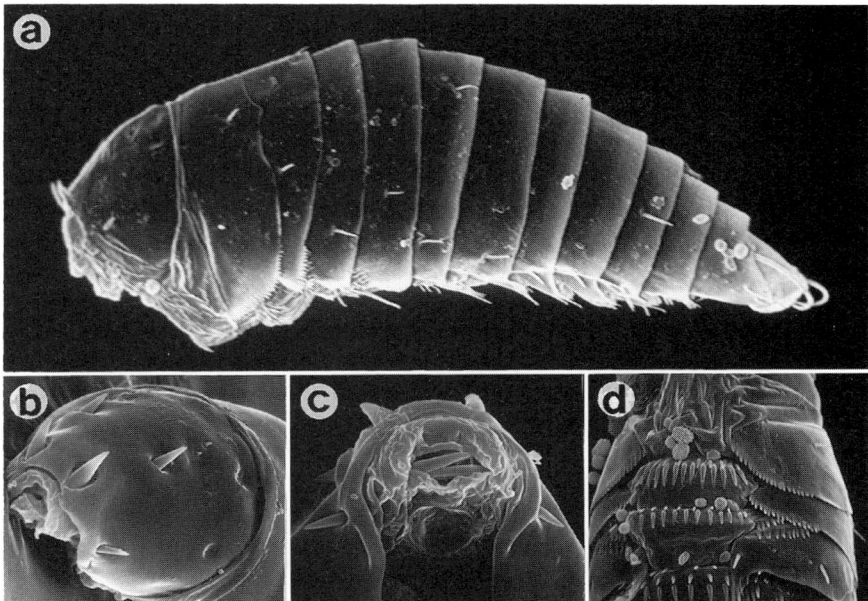

Figure 5.3 Planidial larva of *Perilampus hyalinus* (Perilampidae): (a) lateral habitus, total length 0.2 mm; (b) dorsum of cranium, showing recurved spines and sensilla; (c) ventrum of cranium, showing mouthparts; (d) ventral part of body, showing sternal spines and serrate posterior margins of tergites. (Courtesy of Chris Darling.)

host larva but does not develop further until the host has spun its cocoon, and only then if it has been parasitized by a suitable primary host.

5.9 PRECOCIOUS OR KILLER LARVAE

Whilst some polyembryonic encyrtids produce typical broods composed of a single type of larva (Wang and Laing, 1989), several others have been known for a long time to produce two distinct larval morphs within a single host (Silvestri, 1906; Doutt, 1952; Cruz, 1981). The normal larvae develop slowly and eventually give rise to adult males and females. The second type, variously called asexual, teratoid or precocious larvae, are distinctly serpentine in appearance, have very large mandibles and develop rapidly but do not reach pupation. They do not moult and apparently lack some internal organs (Tremblay, 1991). Usually an all-female or mixed-sex brood of the polyembryonic encyrtid, *Copidosoma floridanum*, contains 10–15 such precocious individuals, but in *C. sosares* each brood normally seems to produce just one precocious larva, though a small percentage have more (Hardy, 1996).

Several roles have been postulated for these precocious larvae. Silvestri (1906) suggested that they help to break down host tissues for future consumption by their siblings, but whilst Silvestri's proposal remains a possibility, evidence now shows that their main role may be quite different. One possibility is that they serve to protect their siblings from competition or attack by conspecific non-siblings (i.e. conspecific superparasitism) or allospecific competitors (i.e. multiple parasitism) (Cruz, 1986; Cruz *et al.*, 1990). By carefully examining the composition of precocious larvae through development, Cruz *et al.* showed that there are in fact two distinct types of precocious larvae in the polyembryonic encyrtid, *Pentalitomastix* (as *Copidosomopsis*) *tanytmema*, as had been previously suggested by Doutt (1947). The smaller of the two forms develops early and only in small numbers, whereas the larger morph develops about 12 days later and in larger numbers. Neither morph appears to undergo ecdysis. Cruz *et al.* suggest that the sequential production of two morphs of precocious larvae means that the parasitoid's brood will be protected from superparasitism and multiparasitism for a larger proportion of its developmental period than it would be if it were only to produce the one type.

A more sinister role has also been demonstrated. By injecting vitally-stained male polygerm of *Copidosoma floridanum* into hosts containing female broods, Grbíc *et al.* (1992) have shown that these precocious larvae, which are predominantly female, attack male polygerms significantly more than they do female ones (Figure 5.4). They concluded that an important role of precocious larvae in *Copidosoma floridanum* is to bias the sex ratio of the emerging wasps towards females. The precocious larvae can be seen as self-sacrificing sisters which help to resolve the sexual conflict between mother and offspring in favour of the daughters. Significant in this is the fact that the size of emerging wasps is related to brood size/host resource combined with the normal asymmetry in genetic relatedness inherent in haplodiploid organisms. Thus, both males and females would benefit by increasing the proportion of their own sex in the emerging brood but females benefit more because they are less closely related to their brothers than their brothers are to them. Further, Grbíc *et al.* noted that female size declines with increasing brood size more than male size does and so females stand to lose more through being in large broods relative to males. Whether this selective fratricide is true of precocious larvae in all those encyrtids that have them is not known, and it is still to be determined whether the brothers and sisters in mixed broods are in conflict or whether the actions of the precocious larvae are also in accord with maximizing the fitness of the male clone.

5.10 NUMBER OF LARVAL INSTARS

Data on the number of larval instars in the parasitic Hymenoptera is hard to come by (Hagen, 1964) and therefore there are many groups for which

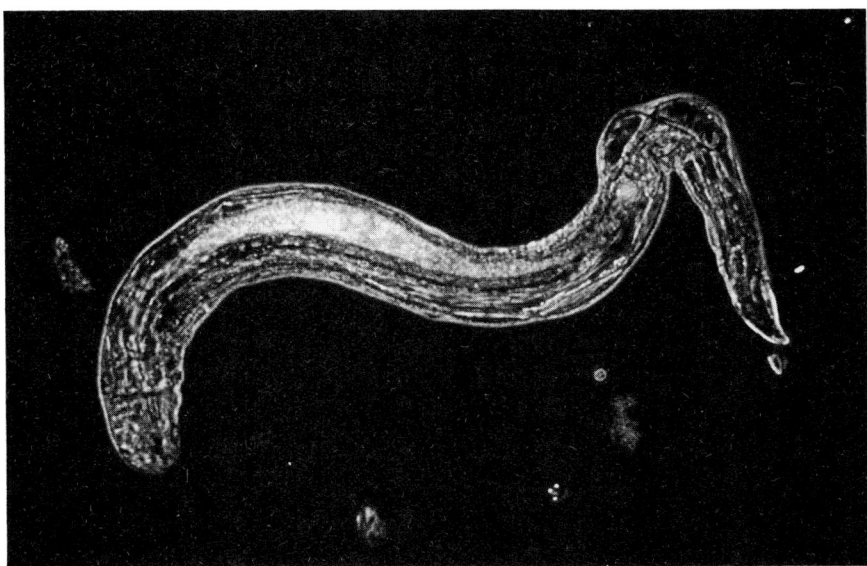

Figure 5.4 Precocious larva of the polyembryonic encyrtid, *Copidosoma floridanum*, that has consumed vitally stained polygerm of male conspecifics. (From Grbic *et al.*, 1992, reprinted with permission from Nature, McMillan magazines, © 1992.)

there are no data at all. Further, by no means all records are reliable, and one should consider carefully the type and quality of published observations before accepting the conclusions. Sometimes the appearance of a given instar may be quite different early versus late in its development, and such differences have often been assumed to reflect the existence of additional instars. Sometimes larval endoparasitoids moult upon leaving the host's body to feed upon it externally or pupate, leaving the cast skin in the exit hole and so making it difficult to observe. Only in a few instances have such careful morphometric studies, applying Dyar's law (Dyar, 1890), been conducted as those of Brothers (1972) on mutillids, O'Donnell (1987) on aphidiine braconid larvae and Volkoff *et al.* (1995a) on *Trichogramma*.

From what is known, the plesiomorphic number of larval instars in parasitic Hymenoptera appears to be five (Evans and Eberhard, 1970), though some sawflies have up to eight. However, many groups display reduced numbers, three being the normal minimum. When reduced numbers of instars do occur, they are nearly always associated with endoparasitism, though there appear to have been no suggestions as to why this should be (Hagen, 1964). In the extremely small mymarids and trichogrammatids, some observers reported only one or two instars (e.g. Meyerdirk and Moratorio, 1987; Hutchison *et al.*, 1990) but others have reported higher numbers for the same or closely related taxa. Volkoff *et al.* (1995a) presented a detailed study of *Trichogramma cacoeciae*, showing conclusively that in this species there is only one instar, and these authors explain why other less thorough and careful studies might have been misleading.

Generally the number of larval instars is fixed within a species. Sexual dimorphism in the number of larval instars has been reported by Broodryk and Doutt (1966) in the heteronomous aphelinid, *Coccophagoides utilis*. Ramadan and Beardsley (1992) reported that in the opiine braconid, *Diachasmimorpha tryoni*, an endoparasitoid of Diptera larvae, 1st instar larvae may undergo supernumerary moults, especially when very young hosts are attacked. The significance of supernumerary instars, if they do occur, is unclear but it may be connected with avoiding the host's encapsulation response. However, Lathrop and Newton (1933) interpreted similar data for another opiine, *Opius melleus*, as indicating that superparasitism had occurred, and that the extra 1st instar parasite larval head capsules found in hosts reflects the outcome of competition. Further, Lathrop and Newton found at least some hosts with multiple 1st instar parasitoid larvae, proving that superparasitism does occur in this species.

5.11 LARVAL FEEDING AND DIGESTIVE SYSTEM

In contrast to sawflies whose guts are intact throughout the whole of their development, those of virtually all Apocrita remain incomplete until they are about to pupate (Figure 5.5a). Thus in the Apocrita, the midgut (ventriculus) and the hindgut (proctodaeum) both end blindly but abutting one another. Fusion of their lumens usually occurs at the end of the final instar and this allows expulsion, in the meconium, of undigested food accumulated throughout development together with nitrogenous waste. This is seen as one of the principal apocritan adaptations since it allows the larva to avoid fouling its environment which, in the case of parasitoids, would almost certainly lead to a rapid reduction in host quality and viability. Even in the phytophagous groups of Apocrita such as the cynipid gall wasps and various chalcidoids, the larval gut remains divided up to the prepupal stage when the undigested plant material is at last voided. The only exceptions to this appear to be opiine and alysiine braconids which do not void the meconium until adult eclosion, and various bees and the Gasteruptiidae in which earlier fusion occurs. Members of the Gasteruptiidae are kleptoparasitoids of solitary bees whose pollen stores they consume, and in these the midguts and hindguts become functionally contiguous early in the final instar so that the faeces are excreted as a series of pellets over a period of time rather than in a single mass. This may indicate that pollen has a

significantly greater amount of indigestible residue than the largely liquid diet of true parasitoids. Gall wasps and others with similar biology
appear to have circumvented this problem by expelling a single large
meconium and probably by regulating the physiology of the plant tissue
on which they feed to be as waste-free as possible.

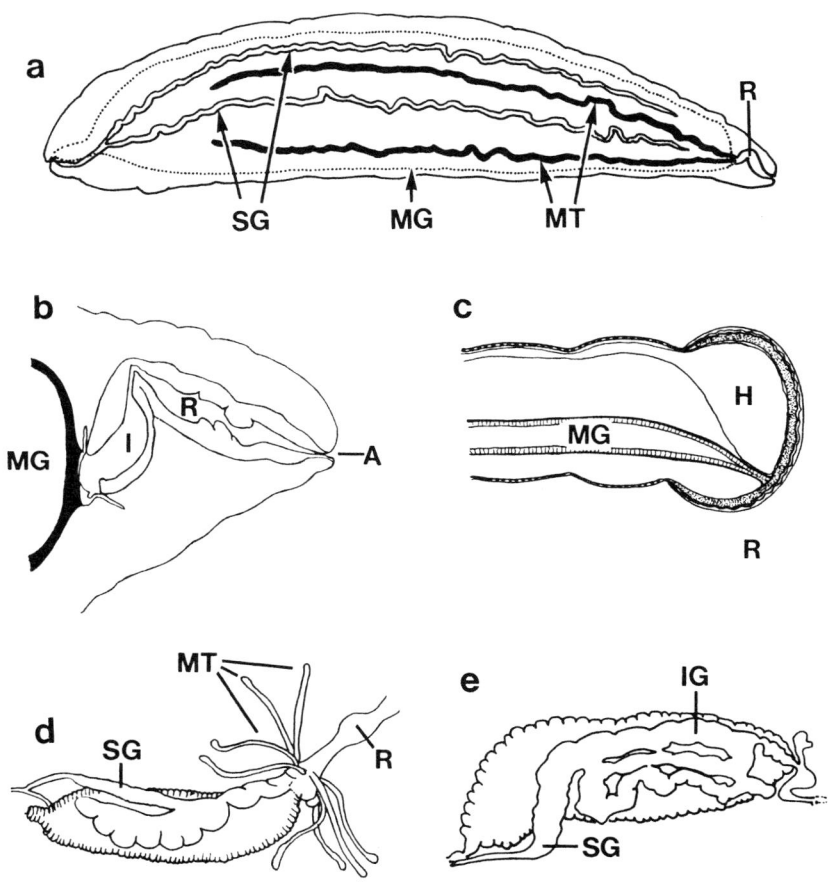

Figure 5.5 Larval internal anatomy features. (a) Gut, salivary glands and iliac
glands. Sagittal section through (b) hindgut of *Pimpla turionellae* (redrawn after
Führer and Willers, 1986); (c) anal vesicle of a microgastrine braconid (redrawn
after Thorpe, 1932a); (d,e) gut, Malpighian tubules, iliac and labial (salivary)
glands of the encyrtids, *Encyrtus* sp. and *Microterys* sp., respectively (redrawn after
Flanders, 1938). A, anus; H, heart; I, ileum; IG, iliac gland; MG, midgut; MT,
Malpighian tubules; R, rectum; SG, salivary glands.

Larval gut morphology has been described in detail for a range of taxa. In many parasitic wasps the midgut swells considerably as a result of ingestion of host haemolymph and tissue fragments and it may constitute as much as 90% of the larva's volume. Eastham (1929) described a disintegration of oesophageal cells in a proctotrupid larva, with the cells breaking off into the midgut lumen. Eastham clearly states that this is not part of peritrophic membrane formation, and suggests that it may be concerned with digestion. Whether any similar process occurs in other parasitic wasps is not known and more observations, especially using modern techniques, are needed, including on other proctotrupids. There is growing evidence that the larval hindgut and associated glands in many endoparasitoids are physiologically active, often appearing to contain secretory material (Christiansen-Weniger, 1994), and may be involved in physiological interactions with the host (Chapter 7). In common with many other larval tissues, the midgut epithelial cells of *Habrobracon* are quadri- or octo-nucleate as a result of incomplete mitotic divisions (Grosch, 1951).

Detailed observations of feeding in parasitic wasp larvae, especially of early stages and of endoparasitic taxa, are extremely difficult and it is not surprising that there have been few detailed studies, especially in the latter cases. Observations made *in vitro* may help to overcome some of these difficulties in the future but to date such observations have hardly been made. The general consensus is that many endoparasitic wasp larvae are believed initially to feed largely on fluids (e.g. Klomp and Teerink, 1978; Mazanec, 1990), though probably most end up feeding on macerated host tissues (e.g. Polaszek, 1986). In young larvae, host fluids are sucked into the alimentary canal by a stomodaeal pump, and muscles can often be seen to raise and lower the roof of the pharynx rhythmically, thereby sucking in gulps of host haemolymph (e.g. Osborne, 1960). In the proctotrupid, *Phaenoserphus viator*, feeding in the 4th and last instar is apparently still largely suctorial, as by this stage the larval mandibles have become very reduced in size and remain motionless (Eastham, 1929). Eucoilids have distinctive (eucoiliform) 1st instar larvae that have very small mandibles and it has often been assumed that this type of larva is also adapted for fluid feeding (e.g. Meyers and Deonier, 1993). However, that of *Tribliographa daci*, a parasitoid of *Ceratitis* fruit flies, has been shown to attach itself to host tissues (probably mostly individual cells) by a pre-oral sucker and to use the mandibles to shred these into tiny fragments, which are then sucked in (Cals-Usciati *et al.*, 1985). This arrangement involving a pre-oral sucker may be fairly common in ectoparasitic species but is apparently rare in endoparasitoids.

The functional feeding morphology of the 1st instar larva of the mymarid, *Anaphes victus*, has been investigated by Nénon *et al.* (1995a). In some groups, particularly egg parasitoids, the early instar larvae are virtually immobile and sacciform – that is, they lack obvious segmentation – and it is believed that these may absorb much or all of their nutrients across their cuticle (Hagen, 1964; Hutchison *et al.*, 1990; Nénon *et al.*, 1995a). The larval cuticle of the campoplegine ichneumonid, *Venturia canescens*, is impermeable even to water, according to Salt (1966), who also provided some evidence that food could be absorbed through the rectum (via the anus). Edson and Vinson (1977) found evidence from ligation experiments with radio-labelled sugars and amino acids that the anal vesicle (everted rectum) of the braconid *Microplitis croceipes*, was important in the uptake of simple food molecules; the lipid, triolein, was not absorbed by the rectum but was absorbed through the cuticle. However, in some endoparasitoids, the larval cuticle is very thin and may lack the normal waxy layer. These may be permeable to both lipophilic and hydrophilic molecules such as trehalose. The next step is to determine the relative importance of the various possible nutrient uptake pathways *in vivo*.

Abe and Koyama (1991) described an adaptation of the embryo of the dryinid, *Hoplogonatopus atratus*, a parasitoid of the small brown plant hopper, *Laodelphax striatellus* in Japan, in which a pair of lobes develop at the anterior end of the embryo and grow towards the host's intestine. The authors suggest that these may serve a nutritive function analogous to the trophamnion found in some other wasp embryos. This structure has also been studied in some detail by Ponomarenko (1971, 1975) who had shown that it persists throughout most of larval development in gonatopodines (Figure 5.2f) and had postulated that it was this relatively benign way of obtaining nutrition from the host that enabled its host to remain active for a long time whilst the dryinid larva grows.

Gauld (1988a) noted that many koinobiont endoparasitoids do not complete consuming their hosts until the latter have voided their gut contents, and with them any potentially harmful plant materials. Interestingly, at least some pimpline ichneumonids that are idiobiont pupal endoparasitoids are at risk from release of bacteria from the host's gut, which might cause septicaemia when it is finally ruptured (Willers, 1980), and the larvae of these wasps produce bactericidal compounds that are believed to counter the risk (Führer and Willers, 1986; see also Chapter 7). Evidence to date suggests that these may be produced by a modified secretory part of the hindgut (Figure 5.5a).

Cannibalism, in terms of eating conspecifics as a major dietary component as opposed to simply killing them off to eliminate competition, has been described in a few parasitic wasp species (e.g. Minkenberg, 1990) and it is likely that this will be found to be more widespread as systems are studied in detail. Hobbs and Krunic (1971) illustrated it in the case of the gregarious pteromalid, *Pteromalus venustum*, a parasitoid of bee larvae. In this species, younger larvae will start to consume less mobile fully fed conspecific larvae or prepupae once the host runs out.

A remarkable instance of a pupal hymenopteran parasitoid being able to feed is related by Austin (1984). In this case, the wasp, *Ceratobaeus* sp., is an Australian scelionid parasitoid of spider eggs. Austin observed that the contents of the spider egg were not all consumed at the time of the parasitoid moulting to the pupal stage, but continued to be ingested afterwards. Whilst not unknown in other insects, feeding by the pupal stage has not been observed in any other Hymenoptera, parasitic or otherwise, though it is quite possible that it could occur in other species and has simply been overlooked.

5.12 LARVAL DIETARY PHYSIOLOGY AND METABOLISM

The nutritional requirements of parasitic wasp larvae have been most thoroughly studied in the few species that can be readily cultured in or on artificial media, and this means principally idiobiont taxa (Chapter 3). Several studies have shown that parasitic wasp larvae selectively deplete their hosts of particular resources (Thompson, 1982, 1983; Cloutier, 1986). In addition to the nutritional requirements of parasitic wasps, there have been a few investigations of the subsequent metabolism of food substances, and in particular of amino acids and lipids.

Much of what we know about the lipid, carbohydrate and amino acid requirements, and metabolism of parasitic wasp larvae comes from the work of S.N. Thompson and colleagues, who have concentrated on three idiobiont ichneumonid species: the pimpline pupal endoparasitoid, *Itoplectis conquisitor*, and two tryphonine ectoparasitoids, *Exeristes comstockii* and *E. roborator*. Thompson (1979) showed that larvae of *E. roborator* reared on lipid-free artificial diet did not require carbohydrate, but that its presence greatly enhanced larval growth rate (Figure 5.6), suggesting that it was far more efficient for the wasp larva to convert carbohydrates into fatty acids than to make them through other biochemical pathways.

Traditional deletion techniques have been used by Thompson (1981a) to determine the essential amino acid requirements of four *in vitro* cultured parasitoids: *Brachymeria lasus* and *B. ovata* (Chalcididae), *E. roborator* (Ichneumonidae) and *Pachycrepoideus vindemiae* (Pteromalidae). All were found to have an absolute requirement for 10 amino acids (arginine, histidine, isoleucine, leucine, lysine, methionine, phenylalanine, threonine, tryptophan and valine). These ten were the same as found for the honey bee and for most other insects. In *E. roborator*, the minimum dietary concentrations of amino acids and carbohydrates were found to be interrelated. With 6% amino acids, no glucose was necessary as the wasp larva was able to synthesize carbohydrates from the available amino acids. However, wasp larvae could not survive on lower amino acid concentrations unless they were also provided with a carbohydrate source. Yazgan (1972) found that females of the ichneumonid, *Itoplectus*, were particularly susceptible to

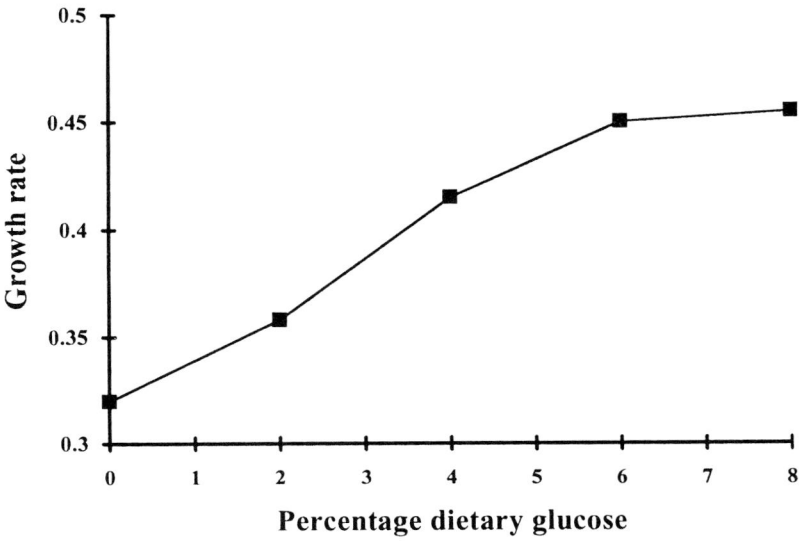

Figure 5.6 Effect of level of carbohydrate in diet (as glucose) on the larval growth rate of the ichneumonid, *Exeristes roborator*, reared on artificial medium. (Redrawn after Thompson, 1979.)

reduced amino acids, and this may mean that females have a need for higher quality hosts.

Parasitic wasps can reasonably be classifiable into lipid conformers and lipid regulators. Fatty acid profiles of the larvae of thirty parasitic wasp species were examined and compared with those of their hosts by Thompson and Barlow (1974). Most contained predominantly laurate, myristate, palmitate, palmitoleate, stearate, oleate, lineolate and lineolenate. Some parasitic wasp taxa had their own characteristic fatty acid profiles and in this respect are like virtually all other insects. However, what was particularly interesting about Thompson and Barlow's findings was that some ichneumonids had fatty acid profiles that were almost indistinguishable from those of their hosts. This reinforced earlier findings by Thompson and Barlow (1970) that the fatty acid profile of *Itoplectis conquisitor*, an ichneumonid endoparasitoid within various lepidopterous pupae, depends on the particular host species attacked (Figure 5.7). Thompson and Barlow have suggested that whether a wasp is a lipoconformer or regulator could influence the range of suitable hosts for a given species, lipid conformers more readily being able to utilize novel hosts. (See also Chapter 6.)

No insects can manufacture their own steroids *de novo*; they have to rely entirely on other food sources to obtain these essential compounds

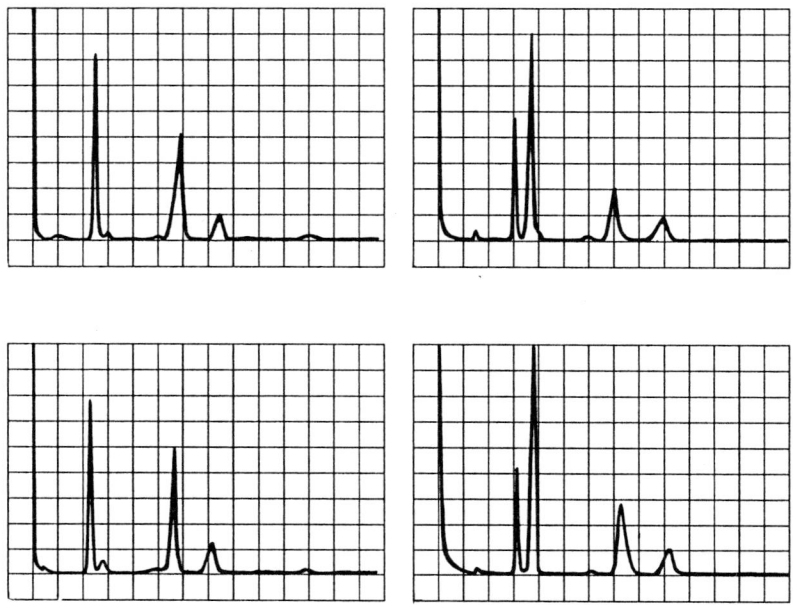

Figure 5.7 Fatty acid profiles of larvae of the ichneumonid, *Itoplectis conquisitor* (lower row) and those of two of its hosts, *Galleria mellonella* (upper row, left) and *Ostrinia nubialis* (upper row, right), upon which they were reared. (Redrawn after Thompson and Barlow, 1974.)

(Thompson, 1981b). In insects the dominant steroids, just as in vertebrates, are C_{27} compounds such as cholesterol, which they need. Plants, however, produce predominantly C_{28} and C_{29} compounds, referred to as phytosterols. These have an additional alkyl group at the C_{24} position which insects remove to obtain the C_{27} compounds they need. Within the Hymenoptera, the phytophagous sawflies, as with other insects, possess the necessary enzymatic pathways to convert the phytosteroids to C_{27} compounds – mainly cholesterol, but desmosterol in the Xiphydriidae (Svoboda *et al.*, 1995; Schiff and Feldlaufer, 1996). This ability appears to have been lost in the 'higher' Hymenoptera, including social bees and ants, even though some of these have become secondarily phytophagous. It would be very interesting to know something about the steroid chemistry of orussids, and of the specialist phytophagous gall wasps (Cynipidae), though it seems likely that the ability to dealkylate phytosterols was lost soon after the evolution of the parasitoid way of life and the consequent availability of C_{26} sterols in the diet.

Several studies have shown that parasitoid venoms may play an important role in regulating host physiology to the advantage of the parasitoid larva's nutrition. Venom from the idiobiont ectoparasitoid, *Bracon mellitor*, causes an increase in the free amino acid composition of its host (boll weevil) larvae, and a concomitant decrease in soluble proteins (Guerra *et al.*, 1993). Morales Ramos *et al.* (1995) have shown that in the case of another boll weevil ectoparasitoid, the pteromalid, *Catolaccus grandis*, both female venom and 1st instar larval saliva effect changes in host amino acid composition; whilst Rivers and Denlinger (1995) found that venom of the pteromalid, *Nasonia vitripennis*, leads to an accumulation of lipid in the fat body of its host Diptera pupa. Such effects may be widespread and may be responsible for increasing the dietary efficiency and developmental rate in the parasitoid. Many other changes in host metabolism have been observed in parasitized hosts (e.g. Thompson, 1982; Hawlitzky and Boulay, 1986), though it is seldom possible to pin down the exact causes. In general, the parasitoid-induced changes often seem to be involved in freeing nutrients into the host haemolymph so that they can be more efficiently taken up. The utilization of such compounds appears to be directly responsible for the high food conversion efficiency of parasitic Hymenoptera. Thus Howell and Fisher (1977) showed that *V. canescens*, a koinobiont endoparasitoid of *Anagasta* larvae, has a food conversion efficiency of 61% compared with 20–40% for most other animals. The ability to selectively induce the mobilization of host compounds for ingestion also means that for much of the parasitoid's development it can effectively tailor its diet to be as close as possible to optimal.

In gregarious species, it is often the case that the individual larvae are in competition with their siblings for food, but some experiments on host quality conducted by Hawkins and Smith (1986) have shown that this is not necessarily always the case and that synergistic relationships may also exist. Using the Indian braconid, *Rhaconotus roslinensis*, a gregarious ectoparasitoid of pyralid borers of sugarcane (Cherian and Israel, 1941), Hawkins and Smith showed that the greater the density of parasitoid larvae, the greater was their individual feeding efficiency (Figure 5.8). These authors suggest that this might be the result of the additive effect of digestive enzymes 'injected' into the host by the batch of larvae.

5.13 SEQUESTERING UNUSUAL COMPOUNDS AND METALS

Many parasitic wasps attack hosts that are in some way protected by either their own toxins or toxic compounds sequestered from host plants. Such compounds may limit the ability of parasitic wasps to develop in some hosts, but those that do no doubt have appropriate means for either detoxifying the compounds or sequestering them themselves (section 10.12). However, very little work has been carried out on this. Fung (1988)

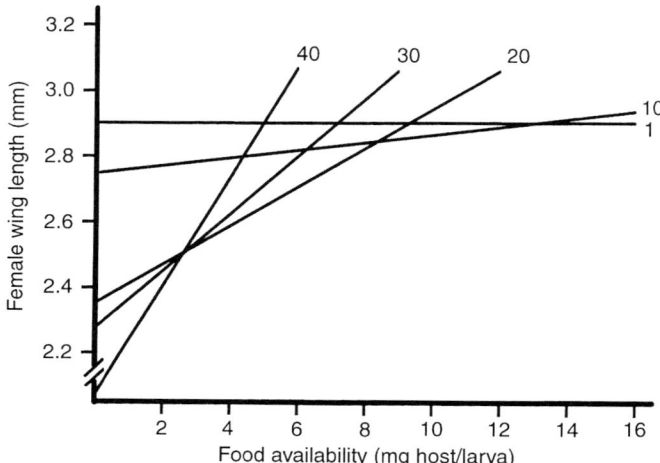

Figure 5.8 Regression lines of adult female size against larval food availability in the gregarious ectoparasitoid, *Rhaconotus roslinensis* (Braconidae), on sugar cane borer, *Diatraea saccharalis*, showing positive relationship between the number of parasitoid larvae feeding on a given amount of resource and the size of the resulting female wasps (numbers at ends of regression lines are numbers of parasitoid larvae per clutch). (Redrawn from Hawkins and Smith, 1986.)

showed that butenolides, a group of compounds that appear to be characteristic of yponomeutid moths and probably provide them with protection from predators, are taken up by all the ichneumonid parasitoids, at least in one species. Benn *et al.* (1979) had demonstrated that the braconid *Microplitis* sp., which attacks the unpalatable day-flying arctiid, *Nyctemera annulata*, contains the same pyrrolizidine alkaloids as its host, which in turn had been sequestered from the moth's larval food plants, *Senecio* sp. It is not known whether these sequestered compounds that are preserved into the wasp's adult stage help to provide protection against predators.

A small amount of research has looked at the uptake of heavy metals by parasitic wasp larvae from contaminated hosts, both as a result of general environmental concern and also from an interest in the pathways that radioactive contaminants might take through organisms and food chains. The braconid, *Habrobracon*, has been the main experimental organism for the radioisotope work, but most of the research involved feeding the isotopes to adult wasps (Grosch, 1988). Radioactive ^{45}Ca and the related ^{90}Sr have been injected into host larvae parasitized by *Habrobracon*, which as expected were subsequently found to be heavily labelled. These alkaline earth metals bind strongly to chromatin and this association was not only detectable in the sperm nuclei of the *Habrobracon*, but also stayed with its chromatin after fertilization and through several mitotic divisions (Steffensen and LaChance, 1960). Given that ^{90}Sr is a major radio-contaminant in some nuclear fall-out, the consequence of the observed tight association with chromatin could be serious.

From the limited data on larval metal biology, it seems that parasitic wasps are active accumulators of several metals including lead, cadmium and zinc whilst at least some of their hosts actively exclude them (Ortel, 1995; Bischof, 1996). It is now becoming apparent that the adults of some parasitic wasps have their mandibles and/or their ovipositors hardened by the incorporation of either zinc or manganese in the cuticle (Section 6.14), and it is therefore likely that they have evolved to accumulate these metals.

5.14 LARVAL EXCRETORY SYSTEMS AND PRODUCTS

A key feature of parasitic wasp larval development is the separation of the mid- and hindguts until just prior to pupation. The meconium is the accumulated waste products and indigestible food that the parasitic larva is unable to void until the midgut and hindgut connect in the last larval instar. However, in opiine and alysiine braconids, at least some taxa do not void the meconium until emergence as adults (Pemberton and Willard, 1918; Salkeld, 1967) and the same applies to a few species of the mymarid genus, *Anaphes* (Boivin *et al.*, 1993). The form of the meconium can be quite characteristic of particular taxa and can help to identify the species of parasitoid that had attacked certain hosts (Flanders, 1942b). In some eulophid parasitoids of leafminers, such as various *Chrysocharis* species, the meconium is produced as a number of separate lumps which the wasp larva arranges in a circle around itself. When the leaf dries up, these 'pit props' help to keep the two sides of the mine apart, protecting the eulophid pupa from being crushed (Viggiani, 1964; Askew, 1971). The main nitrogenous waste components of the meconium appear to be either urates or allantoic acid, and small quadrate or rhombic crystals presumed to be these are scattered throughout the meconial pellets (Salkeld, 1967).

Malpighian tubules are typically reduced in parasitic wasp larvae (Flanders, 1938) and waste nitrogen products are stored in special fat cells until metamorphosis (Salkeld, 1967). Later instars are frequently noted as becoming distinctly yellowish compared with earlier stages and perhaps this indicates a build-up of nitrogen-rich storage compounds. The yellow pigments of at least some wasps are pterins (Leclercq, 1951; Section 6.34), nitrogen-containing uric acid derivatives, and thus adult pigmentation might be a sink for some unwanted larval nitrogenous waste.

Flanders (1938), in considering the origins of the cocoon in chalcidoids (section 5.27), drew attention to a number of previous anatomical studies which showed that in many parasitic wasp larvae there are both normal-looking salivary glands (labial glands) and a second pair of elongate glands arising from the pyloric region adjacent to the usually reduced true Malpighian tubules. These were referred to by Flanders as iliac glands but apparently homologous structures found in ichneumonoids have traditionally been regarded simply as an especially well-developed pair of Malpighian tubules (Salkeld, 1959; Führer and Willers, 1986). In some taxa, for example in proctotrupids (Eastham, 1929), there is a third, shorter gland that runs posteriorly, and it is likely that this has a quite different function from that of the other two hypertrophied tubes. In some chalcidoids these glands are joined to the salivary glands (by a 'protoplasmic' bridge), the combined organ being referred to as the ilio-labial gland (Figure 5.5d,e; Flanders, 1938; Zinna, 1960). The exact nature of this connection needs to be investigated using modern microscopical techniques. In any case, it seems clear that these iliac glands are responsible in many 'microhymenoptera' for the production of the fluid material that is released from the anus and hardens to form a cocoon-like structure (section 5.27). If the iliac glands are essentially modified Malpighian tubules it would be interesting to know whether the cocoon material of these chalcidoids is particularly rich in nitrogenous compounds and thus potentially a nitrogen waste dump.

Wigglesworth (1953) suggested that the anal vesicles of early instars of some braconid larvae may function in water/ion balance, and Lewis (1970) noted a pair of well-developed tubules associated with the anal vesicle of the braconid, *Microplitis croceipes*, which might have a secretory function – these iliac glands may be modified Malpighian tubules. Experimental evidence for this role came from Edson and Vinson (1976) who collected materials from anal vesicles of *M. croceipes*, in a drop of saline whilst the rest of the wasp larva was isolated in paraffin oil. They found that ammonia, bicarbonate ions and possibly arginine and alanine were accumulated in the saline. Ammonia is also the main nitrogenous excretory product in many aquatic animals, including aquatic insects (Wigglesworth, 1953), and it is energetically advantageous to excrete ammonia compared with urates, etc. Edson and Vinson note that *M. croceipes* larvae orientate themselves with their anal vesicles in close proximity to the host's Malpighian tubules. This arrangement may have evolved so that the waste ammonia and bicarbonate, which are relatively toxic, can be absorbed quickly by the host's Malpighian tubules and so rendered harmless to host and parasitoid alike.

5.15 *IN VITRO* REARING OF PARASITIC WASP LARVAE

There have been a good number of attempts to rear parasitic wasps on artificial diets or in artificial or semi-synthetic media, sometimes with the ultimate objective of providing a simple way of mass rearing for biocontrol purposes, and sometimes in order to gain a better understanding of the physiological requirements of the insects. Three different types of artificial media can usefully be distinguished: totally chemically defined media without any extracts; media containing various non-insect animal extracts or yeast extracts; and media containing various amounts of insect material. There has been little success with the first of these, though the method would have the potential of revealing much about parasitoid larval nutrition. In several cases, addition of host insect material has been found to be essential and this may reflect the need for particular as yet undefined factors, perhaps involved with endocrine or other interactions. The process of *in vitro* rearing is fraught with difficulties – viable eggs have to be obtained, aseptic conditions maintained, dietary requirements supplied, endocrinological environment controlled, and suitable physical conditions created to allow moulting, pupation, cocoon production and emergence, to name just a few. The whole process of developing *in vitro* rearing systems is a bit like cookery: try something and see if it works. Indeed, *in vitro* culture of hymenopteran parasitoids appears to be rather more difficult than culturing parasitoid Diptera, and hymenopteran parasitoids in particular, often seem to require insect-derived (usually host-derived) factors. Thus, in the survey by Nettles (1990), out of nine parasitic species of Diptera cultured *in vitro*, eight could be reared to adult stage in the absence of insect dietary components, whereas only six out of 11 parasitic hymenopterans could.

To date only a few systems have been developed which permit a parasitoid to be reared successfully from egg to adult (Thompson, 1986b; Nettles, 1990). Some of the most successful have involved idiobionts and, in particular, ectoparasitoids, which have the advantage that they are effectively carnivores and are not involved in an intimate physiological and endocrinological interaction with their hosts. The earliest successful attempt appears to be that of Bronskill and House (1957) who reared *Pimpla turionellae*, an ichneumonid idiobiont endoparasitoid of Lepidoptera pupae, on a pork-liver slurry in saline, with 70% reaching adulthood. As regards idiobiont ectoparasitoids, Guerra *et al.* (1993, 1994) and Guerra and Martinez (1994) have successfully reared the braconid, *Bracon mellitor*, and the pteromalid, *Catolaccus grandis*, two important ectoparasitoids of the boll weevil, *Anthonomus grandis*, on agar-based food medium. Female *Catolaccus grandis* were caused to oviposit through a parafilm membrane into chambers (tissue culture plate wells) containing the artificial diet, by smearing the parafilm with the artificial oviposition stimulant, n-hexane. The *in vitro* system produced more wasps per parent female, and a higher percentage of female offspring than standard *in vivo* culturing methods, and the resulting wasps appeared behaviourally and reproductively similar to those obtained from natural hosts. The system is potentially viable economically because of the difficulties inherent with rearing hosts to the 3rd instar for mass culture of both *B. mellitor* and *C. grandis*. Yazlovetsky and Nepomnyashchaya (1989) successfully cultured another braconid, *Habrobracon* (as *Bracon*) *hebetor*, on an artificial nutrient medium, and Yazlovetsky and Ageeva (1995) the elasmid, *Elasmus albipennis*. In the latter case, the amino acid composition of host haemolymph was used to estimate the nutrient requirements of the elasmid larvae (see also Rojas *et al.*, 1996). Thompson (1975) had a high level of success in rearing the ectoparasitic ichneumonid, *Exeristes roborator*, on a medium free of insect material. Whereas Yazgan and House (1970) were able to use agar to solidify their medium, Thompson failed with that method and finally found that incorporating albumin and sterilizing the medium for at least 40 minutes yielded a semi-solid diet on which the *Exeristes* would feed.

Apart from that on *Pimpla* mentioned above, work on idiobiont endoparasitoids has also proved moderately successful, including both pupal and egg parasitoids. *Nasonia vitripennis*, an idiobiont pteromalid that is endoparasitoid within Diptera puparia but feeds externally on the fly pupa within these, was first cultured *in vitro* by Boulétreau (1972) in insect haemolymph plasma. Soon after, Hoffman and Ignoffi (1974) successfully reared the related *Pteromalus puparum*, a common polyphagous endoparasitoid of Lepidoptera pupae, in a medium devoid of insect components, though when *Heliothis zea* haemolymph plasma was included the pteromalid larvae developed almost twice as quickly, reaching three times the weight and producing twice as many successfully emerged adults, compared with those in medium completely free from insect material. Yazgan and House (1970) reared another pimpline ichneumonid, *Itoplectis conquisitor*, an idiobiont endoparasitoid of Lepidoptera pupae, axenically on a chemically defined diet, and this has formed a basis for a number of nutritional studies. Dindo *et al.* (1994) successfully reared a further pupal endoparasitoid, the chalcidid, *Brachymeria intermedia*, on an artificial diet containing veal homogenate.

Most work with idiobiont endoparasitoids has been carried out on egg parasitoids (Hoffman *et al.*, 1975; Grenier, 1994), especially *Trichogramma* species but also scelionids (Volkoff *et al.*, 1992) and encyrtids (Masutti *et al.*, 1993). This work has been facilitated by the fact that several species can readily be persuaded to oviposit directly into artificial eggs. For *T. pretiosum* to pupate successfully *in vitro*, small polar (possibly sugar-like) compounds present in host haemolymph are required (Nettles, 1990).

In contrast to the work with various idiobiont taxa, success with koinobionts has been far more limited and culturing many of these seems to be fraught with difficulties. Greany (1986) described systems that would permit development of two microgastrine braconids, *Cotesia marginiventris* and *Microplitis croceipes*, from post-germ band stage embryos to full-grown 1st instar larvae. Pennacchio *et al.* (1992) succeeded in rearing the related braconid, *Cardiochiles nigriceps*, from egg to 2nd instar; and Ohbayashi *et al.* (1994) brought the ichneumonid, *Venturia canescens*, to the 2nd instar in an insect cell culture medium. A low rate of success was reported by Rotundo *et al.* (1988) for the aphidiine braconid, *Lysiphlebus fabarum*, using a liquid medium to which *Lysiphlebus* teratocytes had been added. This wasp could not be induced to oviposit in artificial substrates so its larvae were dissected out of parasitized natural host aphids with or without their teratocytes. Out of 48 larvae tested, two made it to adulthood.

5.16 LARVAL RESPIRATION

A detailed review of the respiratory adaptations of parasitic Hymenoptera larvae was provided by Clausen (1950) and so this section will only provide a brief overview and point out some more recent findings. As with many areas of traditional physiology, the topic has not received a great deal of attention in recent years, and there is undoubtedly much interesting information that could be obtained using more modern technologies. For most endoparasitoids, the tracheal system of early larval instars is closed and spiracles may be either totally absent or not connected to the tracheal system, which is itself usually filled with liquid. Open and connected spiracles often form only in the last instar; thus in these early stages gaseous exchange must be largely or entirely transcuticular. As will become apparent, there is a lot of evidence that most parasitic wasp eggs and larvae require O_2, but Salt (1966) and Edson and Vinson (1976) have provided evidence that some are at least facultatively anaerobic. Oxygen availability may be rather limited for many endoparasitoid eggs and larvae, and they are often very vulnerable to reductions in oxygen level. This may in fact be one of the reasons why encapsulation by hosts kills them so readily.

Several different types of potential larval respiratory adaptation have been described in parasitic wasps, particularly in the apneustic early stages. Early instar larvae of some endoparasitic ichneumonids and braconids possess a vesiculate swelling at the posterior end of the body, usually referred to as an anal vesicle (Figure 5.5b) which is formed by the evagination of the hindgut (proctodaeum). The vesicle is gradually lost in later instars as the hindgut becomes internalized. As other evidence strongly suggests that ancestral ichneumonids and braconids were ectoparasitoids, the anal vesicles in these two families must have evolved independently on at least two occasions. In both families, anal vesicles are largely occupied by the posterior portion of the midgut (Bledowski and Krainska, 1926), which is closely apposed to the everted rectum and has a moderately well-developed tracheal system, and, in the case of some braconids (e.g. microgastrines), by an enlarged terminal section of the heart (Figure 5.5b) (Thorpe, 1932a), but this is apparently not true of all species (Clausen, 1950; Osborne, 1960). Surprisingly, Edson and Vinson (1976) found no evidence of the anal vesicle of the microgastrine braconid, *Microplitis croceipes*, taking up O_2 as had been widely expected and widely cited. Lewis (1970) noticed a pair of tubes in the anal vesicle of the microgastrine braconid, *Microplitis croceipes*, which it was suggested were Malpighian tubules, and Edson and Vinson provided experimental evidence that, at least in this species, the anal vesicle may have some other functions too. Without doubt, anal vesicle structure in both ichneumonids and braconids would benefit from detailed ultrastructural investigation using electron microscopy. A respiratory function has also been proposed for the tail of caudate larvae (Figure 5.2c) but this seems unlikely, at least in some cases, since Thompson and Parker (1930) found that the tail was occupied largely by fat body, though it also possessed a pair of tracheal trunks. Further, Thorpe (1932b) was unable to find any physiological indication of a respiratory function. The hindgut of some 1st instar alloxystine cynipoids is also modified, forming a particularly enlarged chamber which has also been suggested as most probably having a respiratory role (Hagen, 1964), though it may also be secretory (Nahif and Madel, 1990).

Many mature parasitic wasp larvae are holopneustic, with two thoracic and eight abdominal pairs of spiracles, but considerable variation occurs both within and between families. For example, platygastrids have two thoracic and one abdominal pair (Hill, 1926), signiphorids have one thoracic and three abdominal, those of some encyrtids are metapneustic with a single pair of posterior spiracles (Maple, 1937; Zinna, 1960), and the sacciform and mymariform larvae of mymarids are totally apneustic (Nénon *et al.*, 1995a). Conflicting reports in quite a few cases seem to indicate careless observation rather than actual differences, and so interpretation of published reports needs to take this into account. The single pair of spiracles in metapneusic encyrtids is surrounded by a peritreme that helps to hold the spiracles in place next to the internal part of the aeroscopic plate, part of the encyrtid's highly specialized egg. Associated with these, the encyrtid larva has a specialized pair of glands (iliac or ilio-labial glands) which it has been suggested secrete a hydrofugic liquid that helps to keep the aeroscopic plate and spiracles from filling with water.

Mature larvae of some Encyrtidae and Eulophidae bite respiratory openings in the host's cuticle before they pupate (Zinna, 1960). In the case of *Edovum puttleri*, a eulophid egg-parasitoid which attacks eggs of Colorado beetle, the final instar larva has oddly curved, sickle-shaped mandibles which are used to pierce the host egg's chorion prior to its pupation (Laudonia and Viggiani, 1986; Colazza and Bin, 1992) (Figure 5.9). Some encyrtids pupate within the host while it is still alive and induce the host's tracheal system to anastomose within its pupal membrane, thereby ensuring an adequate level of gas exchange for the developing parasitoid (Clausen, 1950). The detail of what happens was described by Thorpe (1936) for *Encyrtus infelix*. The 4th instar larva, which is amphipneustic, becomes surrounded by a thin membrane made up of host phagocytes and tracheal branches, and the sheath becomes attached to the main tracheal trunks of the host opposite each of the parasitoid's spiracles. Dye injection showed that there is an actual connection between the host's trachea and the lumen of the sheath in which the parasitoid larva is located, and gas bubbles are formed within the sheath opposite the spiracles of the parasitoid larva.

There have been numerous studies, either specific or in passing, on the respiratory systems of parasitic wasp larvae but there have been virtually no actual physiological measurements of respiration rates. Gromyszkalkowska and Grochowska (1992) measured respiration rate in the alysiine braconid, *Polemochartus liparae*, a pupal endoparasitoid of the fly, *Lipara similis*. Fisher (1963) showed that 1st instar larvae of *Venturia* (as *Nemeritis*) *canescens*, consume 0.03–0.06 μl of O_2 per hour. The linear relationship between atmospheric O_2 partial pressure and that of the haemolymph of the host (*Ephestia*) larvae allowed Fisher to compare the O_2 requirements of the different immature stages of *Venturia*. The eggs and 1st instar larvae were found to be particularly susceptible to reduced O_2 level; O_2 consumption was found to increase rapidly with development, and so although the actual requirements of the egg and 1st instar were small compared with that of later stages, their requirement was absolute. This apparent absolute O_2 requirement led Fisher to conclude that oxygen availability might be an important factor in the physiological suppression of supernumerary parasitoids in some species – the greater sensitivity of earlier stages to reduced O_2 agrees well with observations that the older of two competing parasitoids usually has the advantage.

5.17 LARVAL ENDOCRINOLOGY AND ENDOCRINE SYSTEM

As with other insects, parasitic Hymenoptera metamorphosis is regulated by the balance between two hormones: juvenile hormone (JH), which promotes the retention of larval features, and ecdysteroids, which lead to metamorphosis. Several different forms of JH are found in insects. Larvae of parasitic wasps and Diptera appear principally to synthesize JHIII, whereas in Lepidoptera larvae the main ones are JHI and JHII, a difference which facilitates investigation of JH origins in these host–parasitoid systems (Beckage, 1991). From the few studies there have been, it appears that levels of haemolymph juvenile hormone (JHIII) remain high throughout most of the larval development of Hymenoptera, including both parasitic and non-parasitic taxa. Lanzrein and Hammock (1995) showed this to be the case for the endoparasitic braconid, *Chelonus inanitus*, by measuring JHIII metabolism *in vitro* using larval homogenates. Only after the final instar larva had emerged from the host did JHIII metabolism increase markedly, with JHIII-esterase activity becoming a major factor. As would

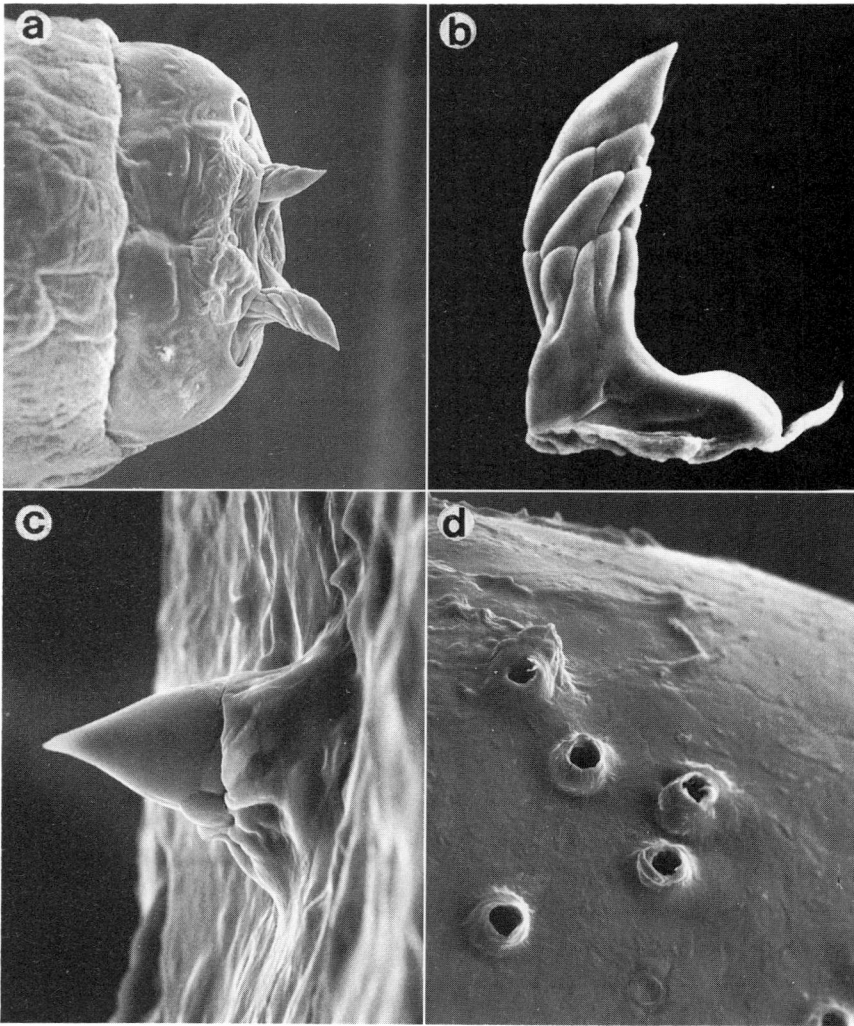

Figure 5.9 Larval respiratory adaptations of the eulophid, *Edovum puttleri*, which use their well-developed diverging mandibles to pierce holes through the host Colorado potato beetle egg – the holes appear to be held open by a hardened wasp larval salivary secretion. (a) Larval head; (b) isolated mandible; (c) mandible pushed through host egg chorion; (d) holes made in host egg chorion. (From Colazza and Bin, 1992.)

be expected, a distinct peak of ecdysteroids occurs immediately before ecdysis (Claret *et al.*, 1978).

The anatomy and cytology of the neuroendocrine system of parasitic wasp larvae has been almost totally neglected and the only morphological study appears to be that of Vagina (1982). Subsequently, Vagina (1987) used several traditional, non-specific histological techniques to stain neuro-secretory material in brain cells of larvae of *Alysia manducator*, a braconid endoparasitoid of the flesh fly, *Calliphora vicina*. Her results showed an accumulation of putative neuroendocrine substances in the first medial group of neurosecretory cells during cold- and heat-induced diapause, and a rapid release of these upon release from diapause. She also found evidence of cyclical accumulation of these compounds during prolonged cold diapause. Within the Hymenoptera, more sophisticated techniques such as immunostaining of specific neuropeptides appear to be limited to work on the honey bee (Kutsch and Breidbach, 1994), though the same procedures would seem to be perfectly applicable to both immature and mature hymenopterous parasitoids. A considerable body of work has been carried out on the interactions between parasitoid and host endocrine systems (for reviews, see Beckage, 1985, 1991). Some of these are discussed in greater detail in Chapter 7.

5.18 DIAPAUSE

Diapause is often important for maintaining phenological synchroniza-tion between a parasitoid and its host. Adult diapause is known for only a few species of parasitic wasps (Chapter 10), though it is probably more widespread than currently appreciated. Larval diapause is very common and some species are also known to remain in diapause for more than a year, so missing out a complete host season (e.g. Morris, 1937). Larvae of many species are known to enter diapause in response to environmental triggers or to host physiological cues, or a combination of the two (for a review, see Tauber *et al.*, 1983; Brodeur and McNeil, 1989b; Christian-Weniger and Hardie, in press). In the simplest cases, diapause may be trig-gered by just one factor (e.g. Coop and Croft, 1990; Neilis *et al.*, 1996), usu-ally day length; in other cases temperature may play a role, and also the responses and physiological stage of the host.

Unlike most diapausing insects, except some Diptera, diapause in many parasitic wasp larvae can also be determined by the ovipositing female – the maternal effect (Saunders, 1965; Fabres and Reymonet, 1991). Whether a female lays an egg that will develop into a diapausing larva depends on photoperiod, temperature experienced and the availability of hosts. Kraaijeveld and van Alphen (1995) showed that the alysiine braconid, *Asobara tabida*, a parasitoid of *Drosophila* spp., may enter diapause to avoid both unsuitable weather (i.e. winter) and unsuitable biotic factors (i.e. increased host ability to encapsulate, which is positively correlated with temperature). It seems likely that the ovipositing female predominantly determines whether her larvae will enter diapause, but the observation by Kraaijeveld and van Alphen that the host species can also have an effect suggests the larvae themselves may be responding to their environment. In *A. tabida*, male larvae are more likely to enter diapause than female ones, again suggesting that there are some intrinsic larval effects. In the pteromalid, *Dinarmus acutus*, an ectoparasitoid of bruchid beetle larvae and a species in which diapause is maternally induced, it is possible to distinguish larvae that will enter diapause from non-diapausing ones by an orange secretion they produce (Reymonet *et al.*, 1987).

Intraspecific variation in diapause induction is becoming known in an increasing number of cases and is normally associated with geographical races. For example, Alvi and Momoi (1994) have shown that one Japanese population of the braconid, *Cotesia plutellae*, from the north, can be induced into diapause by a short photoperiod, whereas larvae from southern populations can not. Another interesting example involves the braconid, *Eubazus semirugosus*, an endoparasitoid of *Pissodes* weevils, including various pest species. *E. semirugosus* from lowland sites in Europe does not require diapause, whereas it is obligatory for a biotype from the mountains, where summer is shorter and where there is no opportunity for it to have more than one generation per year (Kenis, 1994; Kenis *et al.*, 1996). Such variation in diapause is indicative of evolution acting to ensure synchronization of parasitoid and host phenologies. Kenis *et al.* point out the importance of the diapause strategy of particular parasites in the likelihood of successful biocontrol introductions.

The endocrinology of diapause has been investigated most thoroughly in *Nasonia vitripennis* (Schneiderman and Horowitz, 1958; DeLoof *et al.*, 1979) and in *Pimpla hypochondriaca* (as *instigator*) by Claret *et al.* (1978).

5.19 ECTOPARASITIC VS. ENDOPARASITIC LARVAL FEATURES

The great majority of parasitic Hymenoptera develop as either endoparasitoids or ectoparasitoids. Given the microenvironmental, respiration and feeding differences that these strategies usually entail, it is not surprising that the larvae have evolved suites of characters adapting them to a particular strategy. In general, endoparasitoid larvae have flattened or reduced antennae, toothless mandibles and less heavily sclerotized head capsules than ectoparasitoids, but there are many exceptions. In many braconids (Shaw and Huddleston, 1991), ibaliids and at least a couple of tersilochine ichneumonids (Cushman, 1916; Osborne, 1960), the larvae are endoparasitoids during their early instars but emerge from the host before it is fully consumed to complete feeding and pupate externally (Table 5.2). This intermediate strategy has obviously evolved independently in several families, and even on several occasions within a family; it may serve a number of functions, including respiration and avoiding infection from

Table 5.2 Pupation site and cocoon production in selected families of endoparasitic wasps (Ichneumonoidea and various Chalcidoidea excluded)

Family	Pupation site	Cocoon production	External feeding	References
Aulacidae	External	Present	Absent	Skinner and Thompson, 1960, cited in Gauld and Bolton, 1988
Evaniidae†	External		Present	Brown, 1973
Ibaliidae	External	Present	Present	Chrystal, 1930; Spradbery 1970a
Eucoilidae	External*		Present	Meyers and Deonier, 1933
Figitidae (Anacharitinae)	External	None?	Present	Miller and Lambdin, 1985
Ceraphronidae	Internal		NA	Parnell, 1963
Megaspilidae	External	None	Present	Cooper and Dessart, 1975
Trigonalyidae	External	Vestigial or none	Present	Cooper, 1954; Weinstein and Austin, 1995
Diapriidae	Internal	?	Absent	O'Neill, 1973
Pelecinidae	Partially external	None	Presumed absent	Lim *et al.*, 1980
Heloridae	Partially external	None	Absent	Clancy, 1946
Proctotrupidae	Partially external	None	Presumed absent	Eastham, 1929; Hoebeke and Wheeler, 1990; Williams *et al.*, 1992
Scelionidae	Internal	Secreted membrane	NA	Volkoff and Colazza, 1992
Platygasteridae	Internal	Some	NA	Hanson and Gauld, 1995

† Evaniidae are included since they commence development as endoparasitoids of cockroach eggs but they soon emerge to feed externally as predators on the remaining eggs in the oötheca (Brown, 1973).

* Pupation of eucoilids is within the dipteran host's puparium but external to the host pupa.

incompletely consumed host tissues when a single host provides more food than the wasp larva requires.

A few parasitic wasps can develop facultatively as ecto- or endoparasitoids. Evans (1933) noted that as the egg of the alysiine braconid, *Alysia manducator*, is attached just under the host Diptera cuticle, it sometimes gets pulled out of the host during the final moult to the pupal stage. However, it can develop perfectly well as an ectoparasitoid. Evans suggested that this may reflect a relatively recent transition to endoparasitism within the Alysiinae.

5.20 COLD TOLERANCE

Salt's (1959) work on cold tolerance of larvae of *Bracon cephi*, an ectopara-sitoid of the wheatstem sawfly, *Cephus cinctus*, is widely quoted in ento-mological textbooks. Salt found that the haemolymph of hibernating *B. cephi* larvae can contain up to 5 molar glycerol, giving it a freezing point of –7°C to –17°C. However, supercooling may allow this to be depressed to as low as –47°C. Grosch *et al.* (1977) showed that prolonged cold storage (2 weeks at 1–5°C) of 4th (penultimate) instar larvae of a related braconid, *Habrobracon hebetor*, induced supernumerary ovariole production. Cell division continues at such temperatures and it was postulated that the cold storage allowed continued division of presumptive ovariole sheath cells leading to production of a greater than normal number of tubes.

5.21 LARVAL MOVEMENTS

Larval mymarids, which are egg parasitoids of hemipteran eggs, are very active, wiggling all the time apparently in order to keep the egg contents churned up (Whalley, 1969; Meyerdirk and Moratorio, 1987). Larvae of several endoparasitic wasps are known to show a fairly stereotypical path of movements within the host during the parasitoid's development. For example, Führer and Kilincer (1972) showed that larvae of the ichneu-monids, *Pimpla turionellae* and *P. flavicoxis*, are initially located dorsally within the host pupa. As they grow they move anteriorly, then ventrally along the underside of the pupa, and finally complete development in a position close to where they started. Such fixed behaviours may be associ-ated with the need to consume less vital host tissues before more impor-tant ones, but Führer and Kilincer also suggest that it may play a role in the elimination of supernumeraries before they consume too much of the host. Violent movements of final instar larvae are believed to be responsi-ble for the jumping cocoons of some ichneumonids (section 5.26).

5.22 PROTECTIVE BEHAVIOURS OF LARVAE

Nearly mature larvae of the megaspilids appear to be able to defend themselves successfully against both intraspecific and interspecific ter-tiary parasitism by behavioural means. Seven-day-old larvae of *Dendrocerus carpenteri*, a megaspilid hyperparasitoid of aphids via aphidi-ine braconids, are able to defend themselves from tertiary parasitism by another *D. carpenteri* or other hyperparasitoids by violent movements. By the time that the megaspilid reaches its 4th larval instar, it is not only big-ger and morphologically modified compared with earlier instars, having a conical tail and rows of spine-like projections on the body segments, but is also behaviourally different in that it now thrashes violently when prodded. This behaviour, in combination with the seeming body armour, appears to provide a great deal of protection from attack by either conspecific or other hyperparasitoids (Bennett and Sullivan, 1978; Carew and Sullivan, 1993).

5.23 LARVAL COMPETITION

Many 1st instar parasitic wasp larvae have relatively huge mandibles that are especially developed for fighting and killing any other parasitoid larvae or eggs that they encounter (Figure 5.2h), and usually the mandibles of subsequent instars are much reduced by comparison. Little is known for certain about how active this seeking out of potential competitors is, but it is widely accepted that freshly emerged larvae of many species actively search out and attempt to kill competitors. Fisher (1970) notes that when the ectoparasitic eurytomid, *Eurytoma curculionis*, hatches, it roams over its weevil host larva or pupa and bites and kills any others it encounters. Larval sapygids that are kleptoparasites of various bees do the same (Torchio, 1972), and it seems likely that many endoparasitoids will generally do likewise.

The relatively greater mobility of 1st instars compared with later ones often appears to give them a combative advantage. For example, 1st instar larvae of the ichneumonid, *Venturia canescens*, have a distinct advantage over 2nd instars and a greater advantage over 3rd instars (Marris and Casperd, 1996).

5.24 SILK AND SALIVARY GLANDS

The larvae of many parasitic wasps have a well-developed pair of tubular glands which run along the body next to the gut and open into a duct that opens just below the mouth (Figure 5.5a). These are the salivary glands, also called silk or labial glands. In early instars it is presumed that they produce salivary secretions, which in a number of cases have now been shown to have profound physiological effects on their hosts (Chapter 7). In the final instar in aculeates and the Ichneumonoidea, and probably in many less derived groups (e.g. Orussidae, Megalyridae), they are responsible for the production of the silk used in cocoon construction. The silk glands open at a specially modified cuticular structure, the silk press, which is involved in shaping and applying the extruded silk strands.

The silk glands of parasitic Hymenoptera and aculeates appear to differ from those of sawflies in that the secretory units in the latter are clearly removed from the epithelium, and are connected to the gland lumen by quite long ducts (Kenchington, 1972). The morphology, histology, development and degeneration in the pupal stage of the salivary glands in *Habrobracon juglandis* was described by Grosch (1952). This taxon appears to be unique in that the glands fuse again posteriorly so as to form a loop. The cell nuclei were shown to undergo amitotic divisions so that the mature salivary cells each contained four nuclei.

In most chalcidoids and other 'microhymenoptera' (e.g. Platygastroidea), silk production has been replaced by formation of a non-fibrous cocoon which is the hardened product of a secretion of the iliac glands (section 5.13). The silk press is correspondingly lost in these taxa, though they still possess well-developed salivary glands.

5.25 COCOONS

Many parasitic wasps, including nearly all Ichneumonoidea and aculeates, spin a silken cocoon prior to pupating. In some, cocoon production has been completely lost and it is rare or absent among Ceraphronoidea, Chalcidoidea, Stephanoidea Cynipoidea, Platygastroidea and Proctotrupoidea, which normally pupate within the shelter produced by the host. It also appears to be absent in stephanids (Hanson and Gauld, 1995). In the case of proctotrupids, helorids and pelecinids, the wasp larva pupates without forming a cocoon and with its posterior end still embedded in the host remains (references in Table 5.2). This may be the case for various other proctotrupoid families whose biology is not yet fully known (i.e. Austroniidae, Peradeniidae and Vanhorniidae). The cocoons of ichneumonoids and most 'macrohymenoptera' are almost always composed of silk secreted by a pair of labial glands (also called silk or salivary glands) but this is not the case in most 'microhymenoptera'.

5.26 SILK COCOONS AND SILK

Cocoon construction has been described in detail for only a few species (for example: Fulton, 1940; Baccetti, 1958; Cross and Simpson, 1972; Wilson and Ridgway, 1974; Salt, 1977). Cocoons are generally considered to be protective, though what their exact role is has seldom been investigated, and some taxa survive perfectly well without making any cocoon. Hymenoptera pupae are certainly vulnerable to damage, and of course they are subject to parasitization by other insects as well as infection by microorganisms. However, some studies have indicated that parasitic wasp cocoons may be important in preventing desiccation, e.g. in the braconid, *Cotesia glomerata* (Tagawa, 1996). Silk of the South African mutillid, *Pseudomethoca frigida*, is white when first formed, but tans to a golden-brown colour within a few hours provided that there is sufficient moisture in the substrate (Brothers, 1972). This and other observations led Brothers to suggest that the tanning process may be important in water-proofing and in the subsequent prevention of desiccation. Silk cocoons normally comprise several distinct layers, the outer one having a relatively loose network of fibres, and the innermost a smooth surface of tightly woven fibres. Cocoons are typically sub-spherical to ellipsoidal though in some meteorine braconids and campoplegine ichneumonids they are pendulous – presumably an adaptation that makes them less susceptible to predation. The whole construction process can take a long time: for example, in the ichneumonid, *Bathyplectes curculionis*, Cross and Simpson showed that it takes about 30 hours. Probably cocoon construction is fairly consistent within a species except where seasonal or diapause-related variation is found, but in the ichneumonid, *Exetastes cinctipes*, the number of silk layers, whilst modal at three, varies from one to eight (Slovák, 1984).

A considerable number of instances of seasonal dimorphism in parasitic wasp cocoons, including among parasitic aculeates, has been discovered (Schlinger and Hall, 1960; Cross and Simpson, 1972; Shaw and Huddleston, 1991). A typical instance of such dimorphic cocoon production is provided by the ichneumonid social wasp parasite, *Sphecophaga vesparum* (Donovan, 1991). In *S. vesparum*, summer cocoons are thin-walled and white whereas overwintering cocoons are tough and yellow. Indeed, the latter may more than just overwinter, and the adult ichneumonid may not emerge from them for up to four years. Another cryptine ichneumonid parasitoid of social wasps, *Latibulus argiolus*, produces a summer cocoon which remains within the pedotrophic cells of the nest of its host, *Polistes*; it also produces an overwintering one which drops from the host nest and, through violent movements of the final instar larva, hops around until a stable (protected) site is reached (Frilli, 1965b, 1981). The robustness of this overwintering cocoon was further demonstrated by Frilli, who showed that it could protect the *Latibulus* larva for several weeks if submerged in water and also when the cocoon was 'bedewed' with a solution of DDT in petroleum. In some aphidiine braconids (*Pseudopraon* and *Lipolexis*) non-diapausing cocoons are formed inside the mummified host aphid's remains whereas diapausing ones are constructed underneath the aphid cadaver Shuja-Uddin, 1977). Some campoplegine ichneumonid cocoons can jump when disturbed, and Finlayson (1964) proposed that this jumping was probably the result of violent thrashing of the caudal appendage of the final instar ichneumonid larva. Day (1970) compared the hyperparasitism rate in cocoons of *Bathyplectes anurus*, which jumps, with that of *B. curculionis*, which cannot. In a test situation, five times more *B. curculionis* were parasitized than *B. anurus*, demonstrating the probable protective function of this adaptation. Temperature and light also affect jumping behaviour in *B. anurus*, suggesting that it might help cocoons to end up in sites with favourable microclimates as well as protecting it from pseudohyperparasitoids.

The cocoon of the aquatic ichneumonids, *Agriotypus* spp., are specially modified with a long ribbon-like structure that floats freely in the streams in which they live as parasitoids of various caddis larvae (Trichoptera)

(Clausen, 1931a). The ribbon appears to act as an extra organismal gill, allowing gas exchange with a gas bubble that develops within the cocoon (Messner and Taschenberger, 1981). Removal of the ribbon results in death of the *Agriotypus* (Elliot, 1982).

There has been little work actually on the silk produced by parasitic wasp larvae, but that of several sawflies and aculeates has been subject to intense investigation. Among the sawflies, it has been found that cocoons are constructed from three distinct types of fibrous protein: fibrin, polyglycine and collagen, depending on taxonomic grouping (Lucas and Rudall, 1968; Rudall and Kenchington, 1971). The first significant study of parasitic wasp silk was by Baccetti (1958) who analysed the silks of the ichneumonid, *Bathyplectes corvina*, and the braconid, *Apanteles fraternus*. In both it was found to be highly birefringent, composed of protein and polysaccharide, and chromatographically similar to that of silkworms, *Bombyx mori*. Silk of another braconid, *Macrocentrus thoracicus*, was analysed chemically and by X-ray diffraction by Rudall's group, and was found to have an unusual repeat structure unlike that of all but one sawfly; chemically it was a parallel beta-fibrin type protein with 80% of residues comprising serine, alanine and glycine. It would be interesting for more parasitic Hymenoptera silk to be subjected to modern physical and chemical investigation and especially to compare the labial gland silk of ichneumonoids with the non-fibrous iliac gland products of chalcidoids (section 5.27). In at least some species, the silk appears to be supplemented by the secretions of other glands (Shuja-Uddin, 1977) which may fill in the interstices between the fibres.

Parasitic wasp cocoons are not normally orientated at random, nor are the larvae within them or their emergence holes. Salt (1977) showed that the cocoons of the ichneumonid, *Venturia canescens*, which are made within the cocoon of their *Ephestia* moth host, are almost always pointing towards the open end of the latter. Salt went on to consider how the parasitoid larvae could orientate itself for pupation within its own cocoon, since during cocoon construction the wasp larva rotates end on end several times, and as construction progressed it would become progressively isolated from external cues. Surprisingly, Salt found that the silk wall of the cocoon was not woven uniformly and that the wasp larva incorporates orientation marks at the anterior end – distinct dark pigmented zones and a special cross-woven area. Correct orientation of the pupating parasitoid larva may be 'a matter of life or death', especially in those situations in which the cocoon is formed within a confined space, such as a wood-boring. In a preliminary survey of cocoons of a few other ichneumonids, he found orientation marks in virtually every case.

5.27 NON-SILK COCOONS

In many chalcidoids, the cocoon-like structure has a different origin: it is produced from a liquid secreted by the anal end (and occasionally also the oral end) of the final instar parasitoid larva. This liquid flows over the larva and then dries and hardens (Flanders, 1938; Colazza and Bin, 1992; Ceresa-Gastaldo and Chiappini, 1994). At least some scelionids appear to produce a similar type of cocoon (Volkoff and Colazza, 1992), and this may be a synapomorphy linking the Chalcidoidea and Platygastroidea. Flanders (1938) surveyed much of the literature to that date and showed that the secretion of the chalcidoids originates from the iliac glands (section 5.13). These glands, however, appear likely to be homologous with similar glands found in most ichneumonoid larvae that are usually assumed to be a specially modified pair of precociously developed Malpighian tubules, and in ichneumonoids these are not involved in cocoon production. In some chalcidoids these iliac glands anastomose with the salivary glands (labial glands) but the detailed nature of the connection remains uncertain. However, this observation agrees well with those of Flanders, and of Gastaldo and Chiappini, that in some species the cocoon-forming liquid issues from both the mouth and the anus. Some chalcidoid cocoons, such as those of euplectrine eulophids that are constructed outside the host's remains, are apparently composed of silk, but this is released through the anus rather than the mouth as in ichneumonoids. Production of silk by Malpighian tubules (or their derivatives) is not unique to the Hymenoptera: it also occurs in a few Neuroptera and Coleoptera (Imms, 1935). Flanders also noted that the hindgut of chalcid larvae is particularly muscular and is therefore well adapted for expelling the cocoon-forming secretions.

5.28 HOST MUMMIFICATION

Mummification is the process whereby some parasitoids cause their host's cuticle to harden, stretch and dry before the parasitoid larva pupates. Mummies are formed by aphidiine braconids and aphelinids, both of which attack aphids; by rogadine braconids, which attack Lepidoptera larvae (Figure 5.10); and by alomyine ichneumonids, which are specialist parasitoids of hepialid moth larvae (Hinz and Short, 1983). Usually the parasitoid larva pupates within the mummy but in the case of the aphidiine, *Praon*, and its relatives the parasitoid larva escapes from the mummified aphid and pupates beneath it (Starý, 1970).

Mummification often, though not always, involves the final instar parasitoid larva biting a hole in the host's cuticle to allow any remaining fluids to escape, and frequently these substances, and/or secretions of the parasitoid larva, may be used to glue the mummy in place as in the rogadine braconid genus, *Aleiodes* (Shaw and Huddleston, 1991). Usually the parasitoid larva spins a cocoon within the mummified host remains.

Although not a cocoon as such, the mummified remains of aphids parasitized by aphidiine braconids are frequently distinguishable according to whether the aphidiine is going to enter diapause or not. Krespi *et al.*

Figure 5.10 Mummified Lepidoptera larvae. (a) Adult *Aleiodes nigricornis* (Braconidae: Rogadinae) and the mummified host (*Apamea monoglypha*) larval remains from which it emerged. (b) *Acronicta menyanthidis* larvae mummified by *Aleiodes rugulosus*. (Courtesy of Mark Shaw.)

(1994) showed that mummies of non-diapausing *Aphidius rhopalosiphi* are paler and thinner than those of diapausing ones. These differences probably reflect the need for greater or longer-lasting protection for diapausing parasitoids, and result from differences in the cocoon within the mummified host remains.

5.29 IMAGINAL BUD DEVELOPMENT

In the Hymenoptera in general, the adult ectoderm, including the foregut and hindgut and exocrine glands, originates entirely from larval imaginal buds rather than involving any larval cuticular tissues. In many species, imaginal buds of wings, legs and sometimes gonads are visible through the cuticle of final instar larvae as white patches, but there have been few detailed studies of imaginal bud development in parasitic Hymenoptera. For example, I have been unable to find any studies at all on wing bud development though it has been investigated in some detail in the honey bee, along with the development of other thoracic structures, by Daly (1964) and this work may be of some relevance. D'Rozario (1942) and Rakshpal (1945) have studied the development of the male and female genitalia from their imaginal buds (Chapter 6). The paired iliac glands (probably a specialized pair of Malpighian tubules; section 5.13) in many parasitic wasp larvae disintegrate prior to pupation and the adult Malpighian tubules derive from the multiple smaller structures that are usually discernible near the origin of the iliac glands. A detailed electron microscopic study of antennal bud development in the braconid *Habrobracon* (as *Bracon*) *hebetor* was provided by Abbott and Grosch (1987). Antennal bud growth and elongation proceed at a pace during the 3rd and 4th larval instars (*H. hebetor* goes through five larval instars in total), though at this stage the morphological changes are due to rapid cell division rather than any marked increase in the size of the individual cells. The development of segmentation occurs synchronously along the bud, and the final number of segments is apparent from the earliest stages of segmentation.

5.30 TIMING OF METAMORPHOSIS

Many species that overwinter as immature stages in a host or in the cocoon do so as final instar larvae, others as pupae or as teneral adults. In most species from the temperate region, the overwintering phase is in obligate diapause, and the resumption of development will not occur until it has spent a prolonged period at low temperature. Some species may spend two winters in the cocoon as a mature larva (Lathrop and Newton, 1933). In contrast, the aquatic ichneumonid, *Agriotypus*, whose cocoon is formed within a caddisfly case at the bottom of a cold stream, overwinters in the cocoon as a teneral adult. This strategy may be associated with being able to become active early in the year and because if it had not already metamorphosed it might not be possible to complete metamorphosis at the right time due to the low temperature.

5.31 DEVELOPMENTAL ABNORMALITIES

The braconid, *Habrobracon hebetor*, has been used as a test organism for assessing adverse developmental effects of both radiation and putative toxins. For example, Abbott and Grosch (1984) found that triethylamine, chloroform and ether compounds, often employed as (or in) insect anaesthetics, caused delayed development, disruption of cocoon spinning and morphological abnormalities.

5.32 PUPAE

Unlike some sawflies, parasitic wasps do not have a true pre-pupal phase demarked by an ecdysis (Morris, 1937), and thus the pupa is formed directly from the last larval instar. The pupae of all Hymenoptera are typical exarate – that is, the appendages (legs, wings, antennae) are free from the body, making them generally rather vulnerable to damage. Consequently they are usually protected in a cocoon (of either their own or their host's construction) or in some similarly safe place.

An interesting case of pupal dimorphism was described in the ectoparasitic eulophid, *Eulophus larvarum* (Shaw, 1987). These gregarious parasitoids of exposed Lepidoptera larvae pupate naked on the leaf surface around the dead host remains. Non-diapausing pupae are brown and firmly stuck to the leaf, but diapausing (overwintering) pupae are black and much less firmly attached. Some mixed broods also occur, but as Shaw points out, it is not certain whether these are all from the same parent.

5.33 PUPATION SITE

Most koinobiont endoparasitoid hymenopterans, including braconids and various cynipoids, emerge from their host remains prior to pupation. This behaviour may have evolved so as to prevent the parasitoid's pupa/cocoon being in contact with unconsumed host tissues which could become infected with potential pathogens. For example, the totally endoparasitic ichneumonid, *Venturia canescens*, fails to develop into adult wasps if it attacks too large a host. In most cases the parasitoid larva spins a cocoon, though cocoons are absent in many chalcidoids and in a few proctotrupoids. In the Braconidae, in which some 30 endoparasitic subfamilies are known, pupation within the host remains is only known to occur in the Aphidiinae and Rogadinae (which mummify aphids and lepidopterous caterpillars, respectively), in the Opiinae and Alysiinae (which are endoparasitoids within Diptera puparia) and in a very few other rare groups. Several groups of endoparasitoids that emerge from the host prior to pupation go through a short externally feeding phase. Although this is common in braconids (Shaw and Huddleston, 1991), it is virtually unknown in their sister family, the Ichneumonidae, though it apparently occurs in a few tersilochines that are endoparasitoids of beetle larvae (Osborne, 1960). Details of pupation site, external feeding phase, and cocoon formation are summarized in Table 5.2.

6

Adult morphology and adaptations

6.1 REPRODUCTIVE SYSTEMS

Much of our understanding of the basic structure and homologies of the genital elements in the Hymenoptera stems from the detailed investigations of Ross (1945), Oeser (1961), Scudder (1961a,b; 1971), and most particularly those on sawflies by E.L. Smith (1968, 1969, 1970). Few of these workers did much with parasitic taxa, but luckily many aspects of genital morphology are fairly well conserved throughout the order. Most components of the female genitalia have homologous counterparts in the male. Thus the upper ovipositor valve, technically the gonapophyses of the 9th abdominal segment (Table 6.1), is homologous with the male's aedeagus, whilst the ovipositor sheaths are homologous with the parameres; the lower valve homologues are lost in males.

Table 6.1 Some frequently used terms for constituent parts of the hymenopteran ovipositor system

Lower valves of ovipositor	Upper valve of ovipositor	Ovipositor sheaths	References
Ventral stylets	Dorsal stylet		Various
Stylets	Stylet sheath		Copland and King, 1971a
Ventral valve	Median valve	Dorsal valve	Naumann, 1991
Lancets	Lance		Ross, 1945
Lancets	Stylet	Stylet sheath	Various
8th gonapophyses	9th gonapophysis	Gonostylus	Fergusson, 1988
8th gonapophyses	9th gonapophysis	Gonoplac	E.L. Smith, 1968, 1969, 1970, 1972
1st valvulae	2nd valvula	3rd valvulae	Various

6.2 MALE INTERNAL REPRODUCTIVE SYSTEM

The male internal reproductive system of parasitic wasps comprises paired testes, vas deferens, accessory glands and ejaculatory duct which leads to the aedeagus of the external genitalia. The paired testes can be separate (Figure 6.1a) or united within a single scrotal sac (Figure 6.1b) which, if present, is usually sited dorsal to the gut (e.g. Togashi, 1970). Each testis contains a number of seminiferous tubules (follicles) and, in common with many insects, the stage of spermatogenesis is generally more advanced towards the posterior end of these. Spermatogenesis is described in more detail in Chapter 4. The vas deferens (or, if so differentiated, the seminal vesicles) joins to either the proximal or distal ends of a pair of often large accessory glands. The efferent ducts from the accessory glands unite to form the ejaculatory duct.

6.3 SEMINAL VESICLE

Seminal vesicles – swollen regions of the vas deferens specialized for sperm storage – are present in many parasitic wasps, notably Chalcidoidea. In some other groups, such as in many ichneumonoids, the vas deferens is not visibly differentiated externally; however, as Vinson (1969) has shown, the region adjacent to the testes is cytologically differentiated with a more columnar type of epithelial cell, though it is not especially swollen. In most sawflies, except Cephoidea, the seminal vesicles are a convoluted, swollen region (Figure 6.1b) that could be described as a seminal glomerulus (Togashi, 1970), but when clearly differentiated the seminal vesicles of parasitic Hymenoptera (and also of cephoid sawflies) are generally simple bulbous swellings, and differ markedly from those of other sawflies. Togashi proposed a transition series for the various forms of seminal vesicle he observed, with the cephoid type being regarded as relatively plesiomorphic. However, it seems more probable, given what is known of sawfly phylogeny (e.g. Rasnitsyn, 1988), that the uncoiled condition in Cephoidea is derived. In at least some chalcidoids, including the eulophid, *Dahlbominus fuscipennis* (Wilkes, 1965), and the pteromalid, *Spalangia cameroni* (Gerling and Legner, 1968), the seminal vesicle has a valve at its posterior end which opens into a distinct second chamber where sperm bundles are assembled prior to ejaculation (Figure 6.1a). Gerling and Legner made a number of very careful observations on the storage and release of sperm from the seminal vesicles of *S. cameroni*. The sperm were apparently kept in motion by continuous undulations of the walls of the anterior chamber of the seminal vesicle. Manipulation experiments provided evidence that secretions from the accessory gland serve to activate the sperm. Dissections of many other taxa show the sperm to be active within the vas deferens, but it is difficult to be certain that this is not the result of dissection trauma.

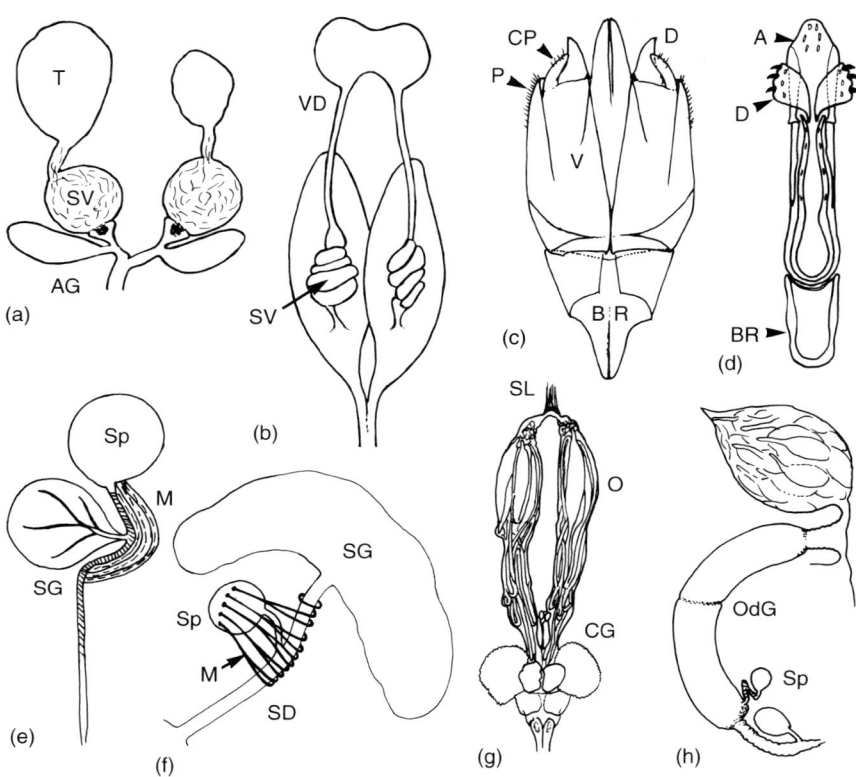

Figure 6.1 Features of the male and female internal and external genitalia. Testes, seminal vesicle and accessory glands of (a) the eulophid, *Dahlbominus fuscipennis*, and (b) the sawfly, *Sirex* sp. Male external genitalia of (c) the braconid, *Acrophasmus*, and (d) the scelionid, *Telenomus sokolovi*. Spermatheca and gland of (e) the pteromalid, *Nasonia vitripennis*, and (f) the ichneumonid, *Diadromus pulchellus*. Female reproductive system features of (g) the orussid, *Orussus sayi*, and (h) the scelionid, *Trissolcus basalis*. A, aedeagus; AG, accessory glands; BR, basal ring; CG, colleterial glands; CO, common oviduct; CP, cuspidal process; D, digitus; G, gland; M, muscles; O, ovaries; OdG, oviduct gland; P, parameres; PD, primary duct; R, reservoir; SD, spermathecal duct; SG, spermathecal gland; SL, suspensory ligament; SP, spermatheca; SV, seminal vesicle; T, testes; V, volsella; VD, vas deferens.(Redrawn after a: Wilkes, 1965; b: Togashi, 1970; d: Javahery, 1968; e: King, 1962b; f: Rojas-Rousse and Palevody, 1981; g: Cooper, 1953; h: Rosi *et al.*, 1995.)

6.4 ACCESSORY GLANDS

The paired accessory glands are a prominent feature of the internal male reproductive system in all male Hymenoptera, and their contents mix with sperm from the vas deferens or seminal vesicle before ejaculation. The accessory glands have quite a complex structure with a stellate lumen (in transverse section) caused by five radial septa composed of muscle and connective tissue (Figure 6.2a; Rojas-Rousse, 1972). This structure appears to be quite conservative since it is also found in aculeates (Wheeler and Krutzsch, 1992). The glandular cells are large and packed with secretory droplets (Figure 6.2). There are no intracellular chitinous secretory mechanisms as found in most exocrine glands, and instead their structure suggests that they are essentially apocrine – that is, the secretion is meant to contain part of the cell rather than just vesicle contents. Several distinct types of secretory granule can usually be distinguished, suggesting that they produce a mixture of products. The cells have large nuclei and extensive golgi apparatus and rough endoplasmic reticulum (Figure 6.2b–d), suggesting that the secretory products are probably proteins, glycoproteins and or polysaccharides, and this is supported by histochemical work on the ichneumonid, *Diadromus*, showing that the glands are rich in acid mucopolysaccharides and proteins (Rojas-Rousse, 1972).

6.5 MALE EXTERNAL GENITALIA

The external genitalia of male parasitic wasps are relatively conserved through the order, though there have been a number of reductions in various parts in particular groups (Gibson, 1986a). The basic structure as described by Snodgrass (1941) is illustrated in Figure 6.3a, and a typical, reduced selionoid type genital armature is shown in Figure 6.3b. The medial aedeagus is equivalent to a penis and is the structure that delivers the male's sperm to the correct point in the female's reproductive tract which, in most species, is probably close to the opening of the spermatheca into the upper end of the vagina or common oviduct. The aedeagus is bordered on either side by the volsella and lateral to these are the usually lengthened parameres. The whole complex is normally united at the base and adjoined to a separate annular structure, the basal ring (or gonobase), though this is lost in the Chalcidoidea and Mymarommatidae (Gibson, 1986a). The volsella on either side usually bears one articulated process called the digitus, and often a second process, called the cuspidal process, which also articulates with the volsella in some taxa and can form a pincer-like structure together with the digitus. When developed, the digitus and probably the cuspidal process appear to be involved with gripping the female during copulation, and this has been shown for some parasitic aculeates which exhibit phoretic copulation (Evans, 1969). However, detailed investigation is lacking for most parasitic hymenopteran families. In contrast to non-parasitic Aculeata, the male external genitalia in the great majority of parasitic taxa are bilaterally symmetrical. Some asymmetry has been found in a number of parasitic aculeates and in a few non-

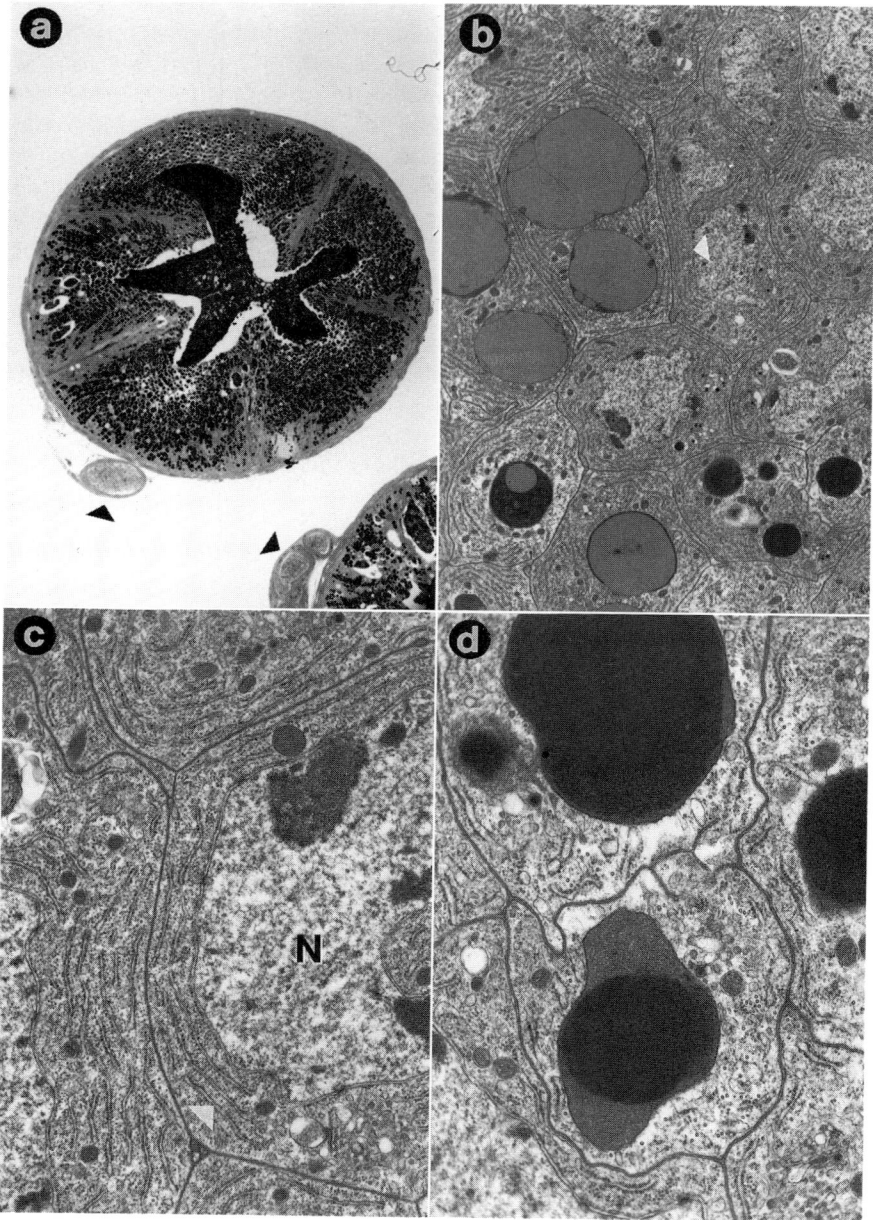

aculeates such as some Trichogrammatidae (e.g. *Ufens* and *Hispidophila*). The significance of asymmetrical genitalia in parasitic Hymenoptera is not yet known though perhaps in some cases it is related to mating posture.

It is well known that many parasitic wasps show enormous intraspecific size variation, with males often being very small and females sometimes very large. Surprisingly few studies seem to have focused on the potential problems that very small males may face, and only a few deal with the effect of size on male fecundity. In most cases they still seem to be quite capable of mating, even with atypically large females (e.g. Chrystal and Skinner, 1931); however, in the pteromalid, *Nasonia vitripennis*, van den Assem and Jachmann (1982) noted that with extreme size discrepancy only about 50% were able to achieve successful genital contact. Similarly, Grosch (1948) reported that very small males of *Habrobracon hebetor* were unable to couple with large conspecifics because their genitalia could not reach. The ejaculate size of small males (or at least, the number of sperm involved) is quite probably smaller than that of large ones in most cases (e.g. Dijkstra, (1986)/1987; Chapter 4). In one very interesting study, Ramadan *et al.* (1991) compared the fecundities (expressed as mating potential) in a number of opiine braconid parasitoids of fruit flies. In two species, *Diachasmimorpha longicaudata* and *Psyttalia incisi*, small males were found to have normal fertilizing potential when mated with small females, but were significantly less successful when paired with larger conspecifics. It seems likely that this reduced fertilization could result from mechanical factors – perhaps the sperm not being delivered to the correct part of the female reproductive tract. It would not be surprising if such effects were quite common, in which case one could anticipate strong selection in favour of allometric variation in male genitalia size, with small males having relatively larger external genitalia. This would be quite simple to test, though no one seems to have done so yet.

The musculature associated with the male external genitalia has been described in detail for a number of taxa, including several ichneumonids (Peck, 1937), the large Indo-Australian braconid, *Stenobracon deesae*

Figure 6.2 (see facing page) Male accessory gland of (a) the tryphonine ichneumonid, *Cosmoconus* sp., and (b)–(d) the braconid, *Charmon cruentatus*. (a) Toluidine blue-stained transverse 5 μm section showing stellate arrangement and congealed secretory product and, next to the accessory gland, the vas deferens (arrows). (b) TEM showing cells with various secretory vesicles and rough endoplasmic reticulum (white arrow). (c) Two cells with large nuclei surrounded by highly organized rough endoplasmic reticulum (white arrow) and nucleus (N). (d) Detail of cells with large electron-dense vesicles.

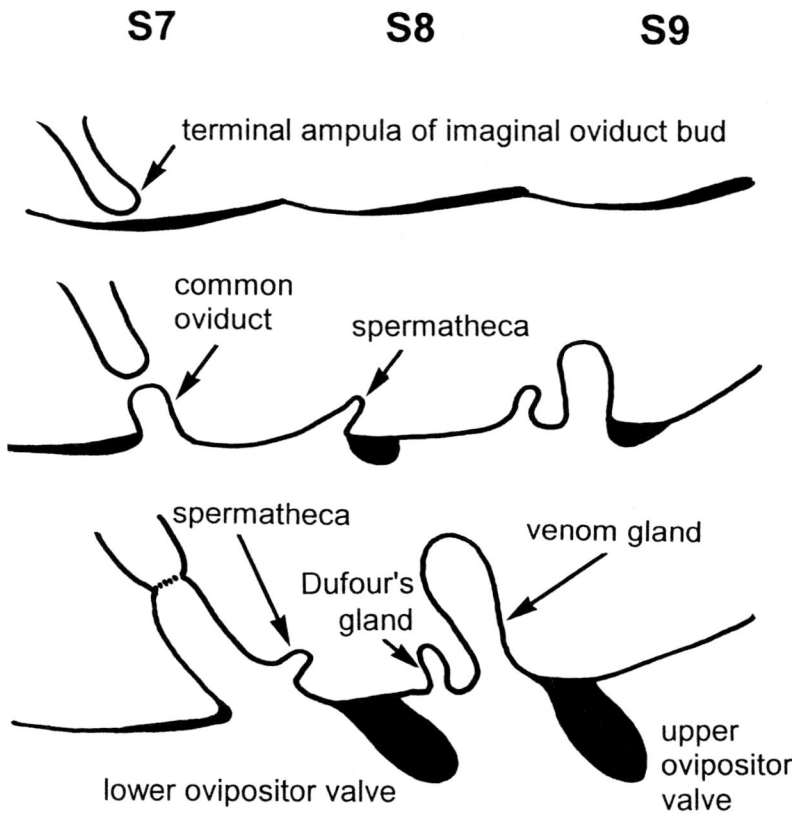

Figure 6.3 Origins of the female genitalia during metamorphosis. (Redrawn after D'Rozario, 1942.)

(Mashhood Alam, 1952, 1960), and the widely studied pteromalid, *Nasonia vitripennis* (Sanger and King, 1971).

6.6 FEMALE GENITAL SYSTEM

The female reproductive system comprises the ovipositor and its sheaths, their associated sclerites and musculature and the ovaries, vagina, spermatheca and associated glands, the venom gland (acid gland) with its associated reservoir, and the Dufour's gland (alkaline gland). In addition there may be other glandular areas, the calyx gland at the posterior of the lateral oviducts and the colleterial glands at the anterior end of the common oviduct. In scelionids, the venom gland appears to be lost and its function replaced by the common oviduct (Strand *et al.*, 1980; Rosi *et al.*, 1995)

The development of both male and female genital systems has been studied in a range of Hymenoptera, including several parasitic species by D'Rozario (1942), and in the eulophid, *Tetrastichus pyrillae*, by Rakshpal (1945). The main features of female parasitic wasp reproductive system development are summarized in Figure 6.3a–c. The ovarial imaginal bud gives rise to the lateral oviducts, but the common oviduct, its associated glands, spermatheca and the Dufour's and venom glands originate as invaginations of the ectoderm.

6.7 OVARIES AND OVIDUCTS

Parasitic wasps have one or more (usually several) pairs of meroistic ovarioles. In these, the developing oöcyte is attached to a group of trophocytes (nurse cells) which are derived from the same oögonial cell as the oöcyte itself. The trophocytes serve to supply materials to the enlarging egg, and in particular are involved with the deposition of yolk material. In the Hymenoptera with meroistic ovarioles, these are of the polytrophic form; that is, the trophocytes form a cluster at the anterior end of the oöcyte and accompany it posteriorly along the ovariole for the first part of its journey. The trophocytes are interconnected in a syncytial fashion by cytoplasmic bridges, and some of them are similarly connected to the oöcyte (Chapter 4). Dissection of wasps clearly shows different stages in egg development (Figures 6.4c,d).

The anterior end of each ovary is attached to a terminal or suspensory ligament, and these unite toward the anterior end of the metasoma (Figure 6.1g), dorsal to the gut. At the anterior end of each ovariole is the germarium, the region where division of the oögonia and prefollicular cells takes place. Cell division often continues here into adult life, though in some pro-ovigenic species there may be no cells left in the germarium by the time of emergence. In many species, the distal parts of each ovariole are swollen and act as a store for mature or nearly mature eggs (Figure 6.4d), and similarly, the distal parts of each ovariole on each side unite to form the lateral oviducts, which may also be swollen to form an egg store. These egg stores are sometimes referred to as the uteri. The walls of the lateral oviducts are sometimes conspicuously glandular – referred to in ichneumonoids as the calyx glands, which may be important in host regulation (see below). The posterior ends of the lateral oviducts unite to form the common oviduct, sometimes referred to as the vagina. The primary spermathecal duct inserts on the dorsal side of the anterior part of the vagina. The antero-ventral part of the vagina is closely apposed to the terminal abdominal ganglion (King and Ratcliffe, 1969), and this may be

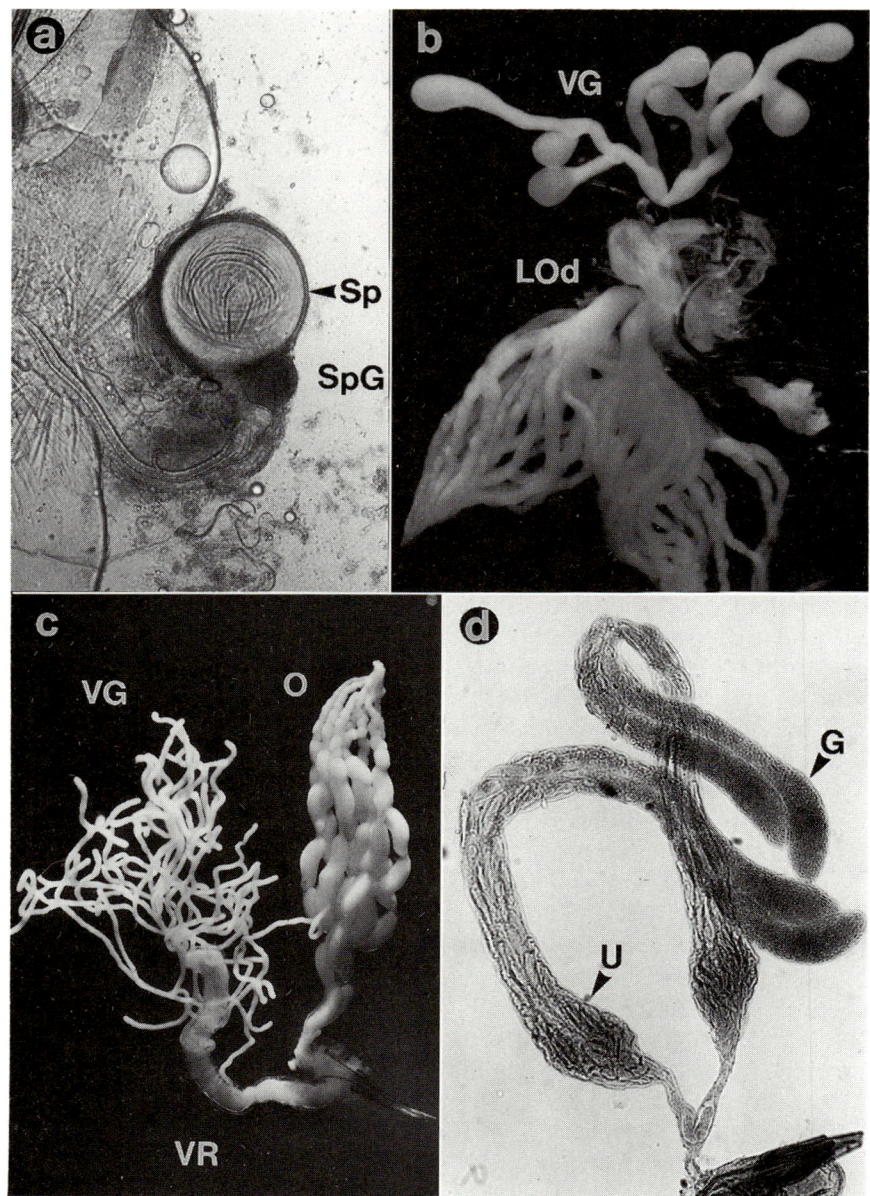

responsible for the very precise control female wasps have over the release of sperm and eggs to enable accurate sex determination.

The smallest number of ovarioles is just one pair, a condition that appears to be limited to a group of genera of aphidiine braconids related to *Aphidius*. The largest are found in the eucerotine ichneumonids, *Euceros* species, which can have up to 200 per ovary with some 3000–5000 mature eggs (Iwata, 1960a); and in trigonalyids, which have even greater numbers, with up to 700 ovarioles per ovary (Iwata, 1960b). However, most taxa have between two and 20 pairs. Within the Ichneumonoidea there is a tendency for koinobiont taxa to have larger numbers of ovarioles than idiobiont ones. The number of pairs of ovarioles shows little natural individual variation in taxa with smaller numbers of ovarioles, and is presumably genetically fixed. In taxa with larger numbers of ovarioles, considerable intraspecific variation can be found and there may even be considerable differences between the left and right ovaries of the same individual (Iwata, 1960a,b). Cold shock during the 4th larval instar has been shown to affect ovariole number in the braconid, *Habrobracon*, which normally has only two ovarioles per ovary; treated individuals have up to 11 in total (Grosch *et al.*, 1977).

The common oviduct in most parasitic wasps is generally only moderately muscularized (King and Copland, 1969; Copland and King, 1971a) and so presumably it plays only a small part in forcing the egg down the ovipositor, but rhythmic contractions are seen in some chalcidoids (section 6.15). The lumen wall of the common oviduct is furnished with cuticular spines which, as in many other insects, are believed to help move the egg into the proximal part of the ovipositor (Wilkes, 1965; Austin and Browning, 1981). At least in the Chalcidoidea the common oviduct is dorso-ventrally asymmetrical with glandular cells present on the dorsal wall. In some taxa that display egg resorption, various enzymes have been detected with activity levels being correlated to the time since the last oviposition (King and Richards, 1968), and in *Nasonia vitripennis* the

Figure 6.4 (see facing page) Dissections of female reproductive systems. (a) Spermatheca of the encyrtid, *Apoanagyrus* (= *Epidinocarsis*) *lopezi*, showing sperm; Sp, spermatheca; SpG, spermathecal gland. Ovaries (O) and venom gland of the ichneumonids (b) *Collyria coxator* and (c) a *Tryphon* sp. (left ovary removed), rendered opaque by alcohol – (b) shows absence of a venom reservoir (VR) in this egg–larval parasitoid; LOd, lateral oviduct; O, ovary; VG, venom gland. (d) Ovaries of the aphidiine braconid, *Ephedrus cerasicola*, showing germarium (G) and swollen uterine (U) regions of the lateral oviducts. (a,d: Courtesy of Anne Le Ralec.)

oviduct is seen to swell at the same time, probably as a result of increased glandular activity. The common oviduct in some scelionids is remarkably developed; for example, in *Trissolcus basalis* the walls are thick and glandular, being composed of a single layer of columnar secretory cells. These cells have a very extensively developed rough endoplasmic reticulum, indicating that they are involved in production of a proteinaceous secretion (Rosi *et al.*, 1995). In one scelionid, *Telenomus heliothidis*, the oviduct's secretion has been identified as the 'venom' component responsible for stopping host development (Strand *et al.*, 1980). Given that the oviduct is similarly modified in the other species and that both have no venom gland (or perhaps only a small remnant of one), this is likely to be a more general feature.

6.8 CALYX GLAND

The calyx gland, when present, is a swollen part of the lateral oviduct, just posterior to the egg store or uterus (Vinson, 1969), and is particularly well developed in two groups of ichneumonoids where they are the site of production of 'symbiotic' polydnavirus particles which play an important role in regulating host immune response and general physiology (Chapter 7). In some cases the fluids they secrete may also contain proteinaceous compounds involved in protecting the eggs from host defences (Stoltz, 1993; Asgari and Schmidt, 1994). The presence of polydnavirus in calyx glands can often be determined by simple visual examination: in virus-producing taxa, the calyx gland and its lumen appear distinctly bluish-white as a result of Tindall scattering. Various chalcidoids also possess variously developed calyx glands; for example, the glandular nature of the lateral oviduct in eulophids was noted by Copland and King (1971a), and Wilkes (1965) described the presence of a ring of glandular tissue around each of the lateral oviducts in *Dahlbominus fuscipennis*. However, chalcidoid calyx glands have not attracted anywhere near as much attention as those of ichneumonoids and virtually nothing seems to be known about their function.

6.9 COLLETERIAL, ACCESSORY OR UTERUS GLANDS

So-called colleterial glands are located at the anterior end of the common oviduct (vagina). In many chalcidoids the colleterial glands are conspicuous and comprise a pair, or two pairs, of pouches at the anterior end of the oviduct (Figure 6.1g) (King and Ratcliffe, 1969). These glands are also well developed in sawflies (Togashi, 1970); therefore they are probably a symplesiomorphy for the Hymenoptera, and their absence in various groups, such as all braconids and many ichneumonids, represents secondary losses. Flanders (1934) showed that in chalcidoids the large pouch-like colleterial glands at the anterior of the vagina are the source of the secretion used to form the feeding tube; the venom gland has a different function and the Dufour's gland in these is believed to be too small (King and Ratcliffe, 1969). The function of the colleterial glands in taxa that are not known to make feeding tubes can only be speculated on. For example, Flanders (1934) and Copland and King (1971a) suggest that in the Eulophidae and some other chalcidoids their secretion may serve to coat the eggs perhaps with an adhesive.

Some ichneumonids (but not braconids) possess either a single or a pair of pouch-like glands at the junction between the lateral and common oviducts. These have traditionally been termed uterus glands, despite their similarity to the colleterial glands of chalcidoids. In orussids, glands in the same place have been termed vaginal or lateral accessory glands (Cooper, 1953). Uterus gland ultrastructure appears to have only been investigated in one species, the pimpline ichneumonid, *Pimpla turionellae* (Blass and Ruthmann, 1989). Following hatching, these glands seem to take several weeks to become fully mature, by which time four distinct cell types can be discerned – one of which, from ultrastructural observation, has been postulated as playing a role in osmoregulation. The uterus glands have an attached muscle and this may be involved in expulsion of secretion, which according to Osman and Führer (1979) mostly comprises lipoprotein and hyaluronic acid. The function of these secretions remains unknown.

6.10 SPERMATHECAL STRUCTURE

An important feature of the reproductive system of Hymenoptera is that a fertilized female can select whether or not to fertilize any particular egg. It seems likely that this control will be mediated through the spermatheca, and several investigators have attempted to understand spermathecal function. The spermatheca has a relatively consistent structure throughout the parasitic Apocrita, and indeed, through most of the Hymenoptera. The spermathecal complex consists of a small, nearly spherical reservoir (capsule) which opens into a narrow primary duct which in turn opens into the anterior end of the combined oviduct (vagina) (Figures 6.1e,f,h and 6.4a). Loan (1967) provided some indication that in some euphorine braconids the spermatheca may be more elongate, but this needs verification. In addition there is usually a clearly differentiated spermathecal gland, or pair of glands. Three different arrangements for these have been described so far. In most groups, the glands connect to a single duct which unites with the primary spermathecal duct close to the reservoir (King, 1962b; Wilkes, 1965; King and Ratcliffe, 1969; Gerling and Rotary, 1974). In the case of the Aculeata, the gland emerges directly from the reservoir (Dallai, 1975). However, in the Scelionidae, the spermathecal duct is secretory and there is no separate glandular mass (Figure 6.1h; Rosi *et al.*, 1995). There are circular muscles around the spermathecal reservoir in all parasitic wasp families studied so far, but there are differences in the positions and development of other associated muscle bands and in the form of the primary duct that are probably significant in terms of function. No details of spermathecal structure or function are available for many families or superfamilies, and it is quite possible that other variants remain to be discovered.

The chalcidoid-type spermatheca is characterized by having a single glandular mass and an S-shaped primary duct (Figure 6.1e). A similar arrangement is found in Scelionidae (Figure 6.1h) but in the Ichneumonoidea the gland is paired and the primary duct is less obviously kinked (Figure 6.1f). In the Chalcidoidea a strong longitudinal band of muscle runs along one side of the kinked primary duct, and the chitinous duct walls are often spirally or annularly sculptured in this region (King, 1962b; Wilkes, 1965). It has been postulated that the S-shaped spermathecal duct and associated muscle band act as a pump to draw sperm into the spermathecal reservoir after copulation. However, the main function of this band is probably to straighten and thereby unblock the duct so that a sperm can escape and fertilize a passing egg (Wilkes, 1965). Observations of the reproductive system of the pteromalid, *Nasonia vitripennis*, dissected in Ringer's solution by King (1962b) suggested that the opening of the spermathecal duct into the combined oviduct might be protruded to facilitate fertilization. Such observations made on traumatized preparations should be treated with some caution.

Spermathecal structure and function in the Ichneumonoidea have been described by Bender (1943), Chumakova (1968) and Gerling and Rotary (1974), who all investigated *Habrobracon* (as *Bracon*) *hebetor*; Aubert (1959) for *Pimpla hypochondriaca* (as *instigator*); and Rojas-Rousse and Palevody (1981), who studied the ichneumonine, *Diadromus pulchellus* (Figure 6.1f). The spermathecae in these species are typical of those of other braconids and ichneumonids in gross structure and probably also in function. Gerling and Rotary note that the primary duct has a thick wall and is crescent-shaped near the reservoir forming the so-called horn. The spermathecal musculature in *Diadromus* was described in more detail than in the other cases. The muscles run from the spermathecal reservoir to and around the primary duct and the contiguous part of the duct running from the spermathecal gland. Rojas-Rousse and Palevody note that the cells of the vagina surrounding the opening of the spermathecal duct are modified, and they suggested that these are probably important at the time of insemination.

In the ichneumonoids, each of the two spermathecal glands has its own reservoir, which is full in virgin females. Gerling and Rotary (1974) noted that these reservoirs were almost completely emptied immediately after copulation, and therefore they suggested that the secretions may be involved in passage of the sperm into the spermathecal reservoir. It is also possible that they provide necessary nutrients for sperm survival over what may be a long period of storage. In aculeates in which the glands open into the reservoir, it has been suggested that sperm release is mediated through glandular secretion, the sperm being washed out as required (Flanders, 1939). There appears to be no direct evidence that secretions of the spermathecal glands serve any such function in parasitic species.

Within the Hymenoptera, there is considerable variation in the colour of the spermathecal reservoir, some taxa having a white or even colourless one, whilst in others it may be yellow, dark red or virtually black (melanized). It has been postulated that dark-pigmented spermathecae are an adaptation to protect sperm from the damaging effects of ultraviolet radiation (Jervis and Copland, 1995). However, consideration of the distribution of spermathecal colour types within the Ichneumonidae and Braconidae makes this seem unlikely. In both families, spermathecal colour is strongly correlated with phylogeny. In ichneumonids, black spermathecae occur in the Pimplinae and many related subfamilies (a group of relatively less derived, predominantly idiobiont and ectoparasitic subfamilies). In the Braconidae, black spermathecae are characteristic of most of the non-cyclostome, koinobiont endoparasitoid lineages such as the Microgastrinae and Helconinae. Very little intra-subfamilial variation occurs, but within the Aphidiinae it co-varies with the tribes. Given this distribution, spermathecal colour does not seem to be correlated with biology or any feature that would seem to make those with black spermathecae more prone to UV, i.e. it is independent of nocturnal/diurnal behaviour.

6.11 SPERMATHECAL FUNCTION

At least in some species, sperm appear to reach the spermatheca very soon after mating in both the Chalcidoidea and Ichneumonoidea, as demonstrated by killing and dissecting females during or immediately after copulation (King, 1960; Wilkes, 1965; Gerling and Rotary, 1974). Immediately following mating (which generally is very brief in the parasitic Hymenoptera) the ovaries of the eulophid, *Dahlbominus fuscipennis*, contract anteriorly, pulling on the spermathecal duct and causing it to be aligned with the combined oviduct (vagina) apparently to facilitate entry of the sperm (Wilkes, 1965). Females of the mutillid, *Pseudomethoca frigida*, repeatedly partly extrude and withdraw the sting after mating, leading Brothers (1972) to suggest that this might be associated with pumping sperm into the spermatheca.

Quite a few parasitic wasps refuse matings until a few days have elapsed following emergence. This might be associated with the need to mature eggs, but it is interesting to note that in the honey bee some cyto-logical evidence suggests that the spermatheca itself may not be fully functional reproductively in newly emerged queens, and therefore may require a period of further maturation before it is ready to receive sperm (Poole, 1970). Surprisingly, van den Assem and Feuth-de Bruijn (1977) showed that fertilization was temporarily inhibited when females of the pteromalid, *Nasonia vitripennis*, mated. This discovery emerged from experiments involving double matings of genetically marked (red-eye) females with both wild-type and red-eye males. After females had mated with the first male and had started laying fertilized eggs, second matings caused a cessation of fertilization until, after a while, the females resumed laying female eggs, now fertilized by a mixture of first and second male sperm. It is not known how widespread this phenomenon is, or what its functional significance or causation might be, though van den Assem and Feuth-de Bruijn tentatively suggested that there may be some sort of blocking action in *Nasonia*.

There has been some debate as to whether sperm are normally active in the spermatheca. Indeed, there is conflicting evidence about the motility of sperm whilst still in the male tract; certainly in males, when some taxa are dissected, the sperm remain virtually motionless whilst in others they seem to be swimming very actively. King (1962b) reported that in 50 dissections of fertilized females of the pteromalid, *Nasonia vitripennis*, sperm were active in every case, making it seem unlikely that their motion had been caused by some sort of trauma. Similarly, Gerling and Rotary (1974) dis-sected 40 *Habrobracon hebetor*, and in all cases they saw sperm swimming actively around the spermathecal reservoir in a circular pattern. The results of these workers is therefore at odds with those of Chumakova (1968), who reported that spermathecal sperm were inactive in the 41 parasitic wasps species she studied. In a number of Hymenoptera, the sperm have been seen to align themselves with their heads close to the opening of the duct (King 1962b), but nevertheless they appear to be unable to exit from the reservoir unless the walls of the duct in the loop region are caused to sep-arate by contraction of the surrounding muscle. However, in a dissected preparation of *Habrobracon hebetor*, Gerling and Rotary observed single sperm escaping from the spermathecal horn without any obvious major movements. In other taxa such as the pteromalid, *Pachycrepoideus vindemi-ae*, the sperm align themselves around the periphery of the spermatheca, and the thickness of the zone of sperm can be used to estimate fertilization potential (Nadel and Luck, 1985). Sperm release for fertilization in *D. fuscipennis* appears to be stimulated by movement of an egg into the oviduct, and involves straightening of the kinked spermathecal duct.

In many parasitic wasps it has been noted that females lay progressive-ly higher proportions of males as they get older (for a review, see King, 1987). There are probably both proximate and secondary reasons for this.

Of course, it is possible in some very fecund species that a female may run out of sperm, but this seems unlikely to reflect the general situation. Wilkes (1965) reported that males of the eulophid, *Dahlbominus fuscipennis*, transfer about 150 sperm to the female and that a female will produce about as many female offspring, suggesting a very frugal usage of sperm for fertilization which is consistent with that found for aculeates (Page, 1986). From the point of view of spermathecal function, Gerling and Rotary (1974) noted that the walls of the spermathecal reservoir thicken as females grow older. The possibility therefore exists that sperm release may become less likely in older females because of changes in the structure and function of the spermatheca, even if there are still ample active sperm present therein.

6.12 VENOM APPARATUS

Venom glands, sometimes referred to as poison glands or acid glands on account of their histochemical properties, and their associated reservoir and ducts are virtually ubiquitous in the Hymenoptera, and certainly are not restricted to the Apocrita or to parasitic and aculeate taxa. Indeed, homologous accessory glands are usually well developed in the phytophagous sawflies (Robertson, 1968; Togashi, 1970), and in these they presumably serve other functions, perhaps largely connected with lubricating the egg canal and perhaps in helping to force the eggs out. The venom glands may insert either on to the venom reservoir or on to the primary duct which runs from the reservoir to open either into the distal part of the vagina or into the egg canal of the ovipositor via the dorsal ovipositor valve. The whole apparatus is of ectodermal origin and therefore lined with chitin (Edson and Vinson, 1979). In many taxa it appears that secretory activity is not restricted to the gland proper, and in many cases gland cells are distributed at low densities over the reservoir and primary duct (King and Ratcliffe, 1969; Ratcliffe and King, 1969; Edson *et al.*, 1982). For example, in the braconid subfamilies, Braconinae and Doryctinae, the primary duct is highly modified with many ampulla-like outgrowths, each of which is invested with one or more large secretory cells (Figure 6.5). It would be interesting to know whether the different components of the venoms of these wasps (Chapter 7) are secreted by cells in different venom apparatus regions, and whether the mixing of these different components might bring about activation of the venom, as suggested by Copland and King (1971a).

The mode of action of the venom reservoir has been described by Génieys (1925) and Beard (1971) for the braconids, *Habrobracon brevicornis* and *H. hebetor*, respectively. In these and related braconids, the venom reservoir is surrounded by a thick, innervated layer of muscles (Edson *et al.*, 1982) and the chitinous intima is strong spirally ridged (Figure 6.5a).

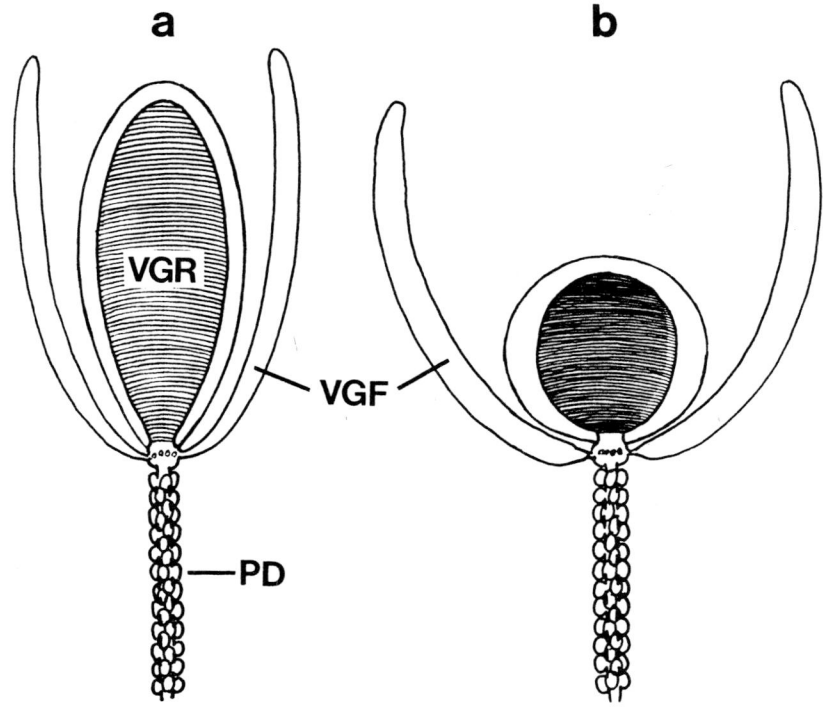

Figure 6.5 Venom apparatus of the braconid, *Habrobracon hebetor*, showing action of muscular sheath of reservoir in venom ejection: (a) at rest; (b) muscles contracted. PD, primary duct; VGF, venom gland filament; VGR, venom gland reservoir. (Redrawn and modified after Beard, 1971.)

Contraction of the muscle sheath causes shortening of the reservoir in a concertina-like fashion with expulsion of venom (Figure 6.5b); upon relaxation the spiral thickenings of the intima cause the reservoir to 'spring' back to its original volume, presumably simultaneously sucking into it more venom from the glandular filaments. However, this system is not typical of parasitic Hymenoptera in general; even in other braconids the intima is not spirally sculptured and the muscle layer is both far less well developed and lacks innervation (Edson *et al.*, 1982). A few workers have suggested that there is a valve between the reservoir and primary duct in some braconids (e.g. Venkatraman and Subba Rao, 1954; Stavraki-Paulopoulou, 1966), but my own observations failed to reveal any such mechanisms, though valves are present in some aculeates. The presence of a distinct venom reservoir is almost universal within the parasitic

Hymenoptera, but two ovo-pupal parasitoids – the ichneumonid, *Collyria* (Figure 6.4b), which attacks cephid sawflies, and a species of *Coelinidea* (Braconidae) which attacks Diptera – have only a venom gland (Pampel, 1913; Quicke *et al.*, in press).

The ultrastructure of the gland cells has been studied for a number of groups including the Braconidae (van Marle, 1977; van Marle and Piek, 1986; Edson *et al.*, 1982), the Ichneumonidae (Blass and Ruthmann, 1989) and the Chalcidoidea (King and Ratcliffe, 1969; Ratcliffe and King, 1969). They are typical class III gland cells (Noirot and Quennedey, 1974, 1991), each gland cell having a chitinous ductule which passes through an epithelial cell to open into a larger gland duct. The ontogeny of these cells has not been examined in any parasitic Hymenoptera, but it is likely to be essentially the same as that of similar secretory cells in other insect groups in which a ciliary process is first formed along the line where the duct will be, and later degenerates after the duct and end apparatus have been formed (e.g. Barbier, 1975). In the paralysing ectoparasitic braconid, *Habrobracon* (as *Microbracon*) *hebetor*, van Marle (1977) reported that light microscopy suggests the presence of two types of gland filaments based on the appearance of their secretory products but that no difference could be discerned with transmission electron microscope (TEM). However, TEM does reveal two types of cell within a single filament through differences in the electron density of their cytoplasm. In the Opiinae, different cells can be distinguished by the state of their secretory organelles which suggests that the cells may pass through a number of stages through the wasp's adult life (Carmen Gimeno, personal communication). Venom glands have been reported to contain viruses or virus-like particles in a range of ichneumonoids (Edson *et al.*, 1982; Lawrence and Akin, 1990; Rabouille *et al.*, 1994) and eucoilids (Rizki and Rizki, 1990a). These are discussed further in Chapter 7.

6.13 DUFOUR'S GLAND

The Dufour's or alkaline gland occurs throughout the whole of the Hymenoptera as a single, median, unbranched tubular gland of ectodermal origin that opens directly into the distal part of the common oviduct. Little is known for certain about its function in sawflies or parasitic wasps, though in the Aculeata it has been intensively investigated since in these it is a source of many pheromones. In the parasitic Hymenoptera, its functions have only slowly been elucidated, and even then in only a handful of species. In his detailed anatomical description of the braconid, *Habrobracon*, Bender (1943) had referred to it as the 'lubricating gland' on the assumption that it aided passage of the egg along the ovipositor (as had initially been proposed by Dufour), and Pampel (1914), Robertson (1968) and Copland and King (1971a) also made the same suggestion;

however, these views are purely surmise, and more recent work has indicated that it is primarily a source of pheromones of various sorts. Host Lepidoptera larvae treated with an acetone extract from the ichneumonid, *Campoletis perdistinctus*, are rarely parasitized (Guillot and Vinson, 1972). In two braconids, *Cardiochiles nigriceps* and *Microplitis croceipes*, a substance with the same function (i.e. permitting intraspecific host discrimination) has been shown to originate from the Dufour's gland (Vinson and Guillot, 1972). More detailed studies of Dufour's gland products have recently been carried out with the ichneumonid, *Venturia canescens*. Hubbard *et al.* (1987) demonstrated that it was the source of host-marking pheromones involved in superparasitism avoidance. As in other parasitic wasps studied to date, the gland produces a range of volatile lipophilic hydrocarbons (Mudd *et al.*, 1982). Comparisons of gas chromatograph traces of Dufour's gland extracts from different thelytokous cultures that were able to discriminate each other's parasitized hosts revealed quantitative inter-strain compositional differences (Marris *et al.*, 1996). As pointed out by the latter authors, the actual compounds used for inter-strain host discrimination may be present in such low quantities that they are currently undetectable, but the consistent differences between the detectable components indicates that the Dufour's glands of this parasitic wasp can produce pheromonal cocktails characteristic of particular genetic lines.

Weseloh (1976a) suggested that the Dufour's gland was the source of the sex pheromone in the microgastrine braconids, *Apanteles melanoscelus* and *A. liparidis*, but more recent studies have shown fairly conclusively that this is not the case in microgastrines (Weseloh, 1980; Tagawa, 1983). However, in another braconid, *Cardiochiles nigriceps*, courtship and mounting is stimulated by Dufour's gland secretions (along with cuticular components) (Syvertsen *et al.*, 1995).

The ultrastructure of Dufour's gland cells has been described for the ichneumonid, *Pimpla turionellae*, by Blass and Ruthmann (1989). As with many insect exocrine glands, two secretory cell types were distinguished on the basis of the electron density of the cytoplasm. That these two types may represent different developmental stages of the same cell type was suggested by the fact that the more electron dense type was most prevalent in glands from recently emerged wasps. In the ichneumonid, *Pimpla turionellae*, Osman and Führer (1979) and Blass and Ruthmann (1989) reported that glands produce fatty material, identified by the former workers as lecithin and cholesterol esters. The function of these remains a mystery, and in the light of the gland's pheromonal role it is possible that these could be carriers rather than active in their own right.

6.14 OVIPOSITOR STRUCTURE

The ovipositor systems of most insects can be broadly divided into two groups. In type II insects, the primitive ovipositor has been totally lost and is instead replaced by a simple, tubular elongation of the abdominal segment around the genital opening. In contrast, the Hymenoptera, along with many Hemimetabola but few Holometabola, have type I ovipositors. These are composed of a number of separate sclerotized, articulated processes, and are essentially homologous throughout the Insecta. The Hymenopteran ovipositor system is often referred to as being of the lepismatid type, since they are the only order of the Holometabola that have retained an almost entire primitive ovipositor, such as found in the primitive apterygote, *Lepisma* (Scudder, 1961a; Smith, 1969; Gauld and Bolton, 1988). Its general features are illustrated in Figures 6.6–6.10, and more or less detailed general descriptions have provided by Ross (1945), Scudder (1961a,b, 1971) and Oeser (1961). Smith (1968, 1969) provided a revised and more detailed terminology for sawflies, most of which applies equally to the parasitic wasps. Detailed and quite accurate descriptions of the ovipositor system, including associated metasomal musculature, have been provided for various specific groups, notably the Braconidae (Bender, 1943; Venkatraman and Subba Rao, 1954), Chalcidoidea (Fulton, 1933; Copland and King, 1971a,b, 1972a–c; Copland *et al.*, 1973; King and Copland, 1969; King and Ratcliffe, 1969; Copland, 1976), Cynipidae (Frühauf, 1924; Bronner, 1985; van Veen and van Wijk, 1985; Fergusson, 1988, 1990), Ichneumonidae (Baumann, 1924; Abbott, 1934), Trigonalyoidea (Oeser, 1962) and Platygastroidea (Austin, 1983; Field and Austin, 1994).

In all Hymenoptera, the ovipositor proper (and similarly, the homologous sting apparatus in aculeates) comprises three parts: a pair of lower valves and a single upper valve which is interlocked with each of the lower ones (Figure 6.6d–f). These typically form a tube encircling the egg canal down which eggs, venom and sometimes other secretions pass. When not in use the ovipositor is usually protected by a pair of ovipositor sheaths. The upper and lower ovipositor valves and the ovipositor sheaths are given a number of different technical names in many works, some of which reflect their anatomical origins, but for simplicity here, they will be referred to by their common names. Some synonyms are given in Table 6.1 (p. 148). The terminology used here refers to the relative positions when the ovipositor is directed posteriorly; in some taxa in which the ovipositor is normally directed anteriorly (e.g. Vanhorniidae, Leucospidae), the upper valves will obviously be below the lower ones. Each lower valve is connected to the upper valve along most of its length by a longitudinal T-section tongue-and-groove joint termed the olistheter mechanism (Figures 6.6c, 6.7). The groove part of the olistheter mechanism, referred to as the aulax, runs along the lower valve; the tongue, or rhachis, runs along either side of the upper valve. The olistheter enables the lower valves to slide relative to the upper one and each other – actions that play an important part in both substrate penetration and in oviposition itself. A change in the basal articulation of the ovipositor valves in all

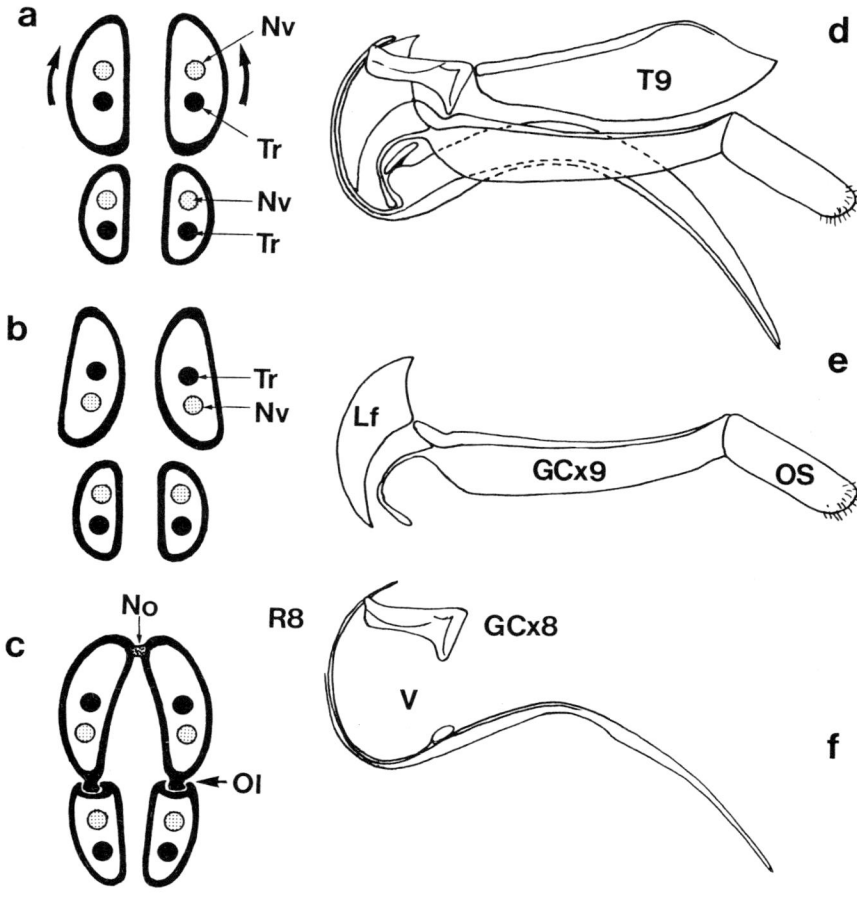

Figure 6.6 Structure and evolution of the Hymenopteran ovipositor/sting. (a,b,c) Evolutionary development of the hymenopteran ovipositor from two pairs of abdominal appendages, in which the more posterior pair that will form the upper ovipositor valve (gonapophysis 9) undergoes a 180° rotation prior to fusing medio-dorsally, and the development of the olistheter mechanism that will interlock them with the lower valves (gonapophysis 8). (Redrawn and modified after Smith, 1969.) Sting (modified ovipositor) and associated sclerites of the bethylid, *Cephalonomia* sp., showing (d) intact situation, and (e,f) isolated ovipositor sheath and lower valve complexes. (Redrawn after Oeser, 1961.) GCx, gonocoxite; Lf, gonangulum; No, notum; Nv, nerve; Ol, olistheter; OS, ovipostor sheath; R8, ramus of lower valve; T9, 9th abdominal tergite; Tr, trachea; V, valvillus.

Hymenoptera relative to other groups affords them an increased degree of movement of the lower valves relative to the fixed upper one (Smith, 1969). The rhachis and edges of the aulax have scale-like sculpture (Figure 6.8). These are large and very densely spaced in most phytophagous and xylophagous sawflies, including Xiphydriidae and Siricidae, and also in stephanids (Quicke *et al.*, 1994), but the scales are far more widely spaced and are less prominent in cephids and orussids and all other parasitic and aculeate Hymenoptera. The function of this scaling can only be guessed at, but it seems likely that it serves to reduce friction, since if it were absent one might imagine that capillary attraction would make it difficult or impossible for the valves to slide relative to one another.

That the ovipositor is derived from paired abdominal appendages and is thus a primitively segmented structure was demonstrated conclusively by Smith (1969), though other propositions have been made (e.g. Matsuda, 1976). In the Hymenoptera, the upper and lower valves each comprise an unsegmented (single segment) basal part referred to as the radix, and a distal, segmented part called the lamnium. Each valve has a lumen within which there are axons, sometimes fat cells and, in many (especially larger) taxa, haemolymph vessels and tracheae (Smith, 1969; Hawke *et al.*, 1973; LeRalec, 1991). The arrangement of nerve and tracheae shows that, as in other insects, the upper valve results from a 180° rotation of two primitively separate appendages (9th gonapophyses) and their subsequent dorsal fusion (Figure 6.6a cf. 6.6b). Smith (1972) showed that the nerves innervating the ovipositors in sawflies, are swollen to form 'ganglia' or plexuses in each segment of the lamnium, but such swellings do not appear to have been noted in any parasitic species. Hymenopteran ovipositors are devoid of intrinsic musculature and their operation depends on muscular attachments at the base of the ovipositor components within the metasoma.

The precise mechanism by which the upper and lower ovipositor valves are moved to and fro relative to one another has only been studied in detail for a few hymenopterans, e.g. Snodgrass (1933) for the honey bee and Smith (1972) for some sawflies. The mechanisms in these groups differ somewhat as a result of whether or not abdominal tergite 9 is divided, and therefore which sclerite acts as a fulcrum. The upper ovipositor valve is swollen at its base where it articulates with gonocoxite 9 (Figure 6.6d; Fergusson, 1988). The lower valves are each attached to gonocoxite 8 via a long ramnus (Figure 6.6f). Relative movement of the upper and lower valves is then achieved by the pivoting of the 8th and 9th gonocoxites. Smith noted that in sawflies and some 'lower Apocrita' the tractor muscles pull on each side of the 9th gonocoxite, and that this rotates the 8th gonocoxite against its fulcrum, so alternately thrusting the attached lower valves posteriorly.

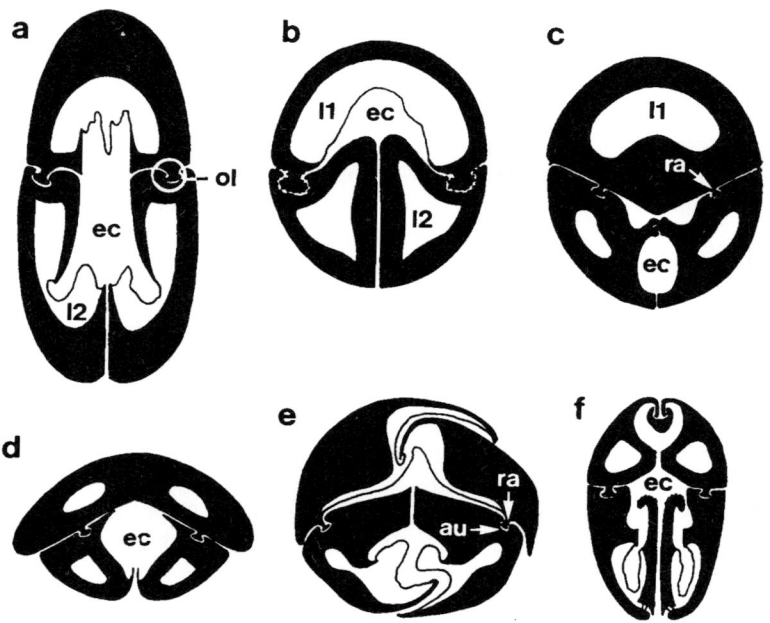

Figure 6.7 Transverse sections through medial regions (presumed radix) of ovipositors of various Hymenoptera: (a) *Xiphydria* ('Symphyta': Xiphydriidae); (b) *Sirex* ('Symphyta': Siricidae); (c) *Megalyra* (Megalyridae); (d) *Anthobosca* (Tiphiidae); (e) *Leucospis* (Leucospidae); (f) unidentified genus of Tersilochinae (Ichneumonidae). au, aulax; ec, egg canal; l1, lumen of upper valve; l2, lumen of lower valve; ol, olistheter; ra, ramus. (Redrawn from Quicke *et al.*, 1994.)

With the evolution of parasitism, several changes in the structure of the ovipositor took place. The typical sawfly ovipositor, as possessed by xyelids and tenthredinids, is rather strongly laterally compressed, the distal serrate and effectively segmented lamnium occupying most of its length. The tip of the upper valve is divided – illustrating the origin of the upper valve as a paired appendage – although the two halves are usually closely apposed and the division may not be obvious except under the scanning electron microscope or in section (Quicke *et al.*, 1994; Vincent and King, 1996). In the horntails and woodwasps (Siricidae and Xiphydriidae) which are believed by many workers to be close to the origin of the parasitic wasps, the tip is still divided but the ovipositor has a more circular cross-section (Figure 6.7), and the serrated part is more or less restricted to the apex (Figures 6.10a–c).

The parasitic sawflies, Orussidae, have a very long, thin ovipositor (which at rest is looped within the abdomen). In these, the tip of the upper valve is no longer divided, and so resembles that of all the other parasitic and aculeate Hymenoptera. The presence of repeated sensilla along the length of the toothless shaft suggests that, in these, the bulk of the ovipositor shaft is still largely the segmented lamnium, despite its lack of serrations. In parasitic wasps the upper valve tip is not divided; in fact, in those taxa that have the lumen of the upper valve divided by a mid-longitudinal septum (mostly members of the Ichneumonoidea) the lumen actually becomes fused at the tip. In most parasitic wasps, the upper valve is furnished with teeth or serrations near the tip whereas the lower ones are either simple or possess only one or two tiny teeth at the extreme tip, or a single pre-apical expansion that probably helps to lock with the substrate whilst the upper valve pushes in further. In the Ichneumonoidea, the situation is reversed, although teeth may sometimes be present on both valves. In the chalcidoids, the upper valve teeth are usually asymmetrical (except those at the extreme tip), alternating on left and right sides (e.g. Phillips and Emery, 1919; Delanoue and Arambourg, 1967; Copland and King, 1971a; Le Ralec, 1991), rather like the offset of teeth on a saw, and this may have a similar role. All parasitic wasp ovipositor teeth are scarped such that they cut on the upstroke (valve in tension), but in siricid sawflies (at least *Sirex* and *Tremex*), although the distal teeth cut in tension, those along the rest of the ovipositor shaft have a reverse scarp and so cut on the downstroke (valve in compression). The relative narrowness of the boring ovipositors of most parasitic wasps makes it impossible for cutting to take place on the downstroke, because the extra resistance would cause the ovipositor to buckle (Vincent and King, 1996). As with other insect structures that have to be especially hard (e.g. teeth of grain beetles: Robertson *et al.*, 1984) the ovipositor teeth, at least in the woodwasp, *Sirex noctilio*, and the ichneumonid, *Megarhyssa nortoni*, contain high levels of either zinc (*Sirex*) or manganese (*Megarhyssa*) (Vincent and King, 1996). A much more detailed survey is needed to see whether this is a general feature, but preliminary results suggest that the switch between Zn and Mn may have both functional and phylogenetic significance (Wyeth, Quicke and Vincent, in preparation). In some of these systems, the availability of the relevant metals may be limiting, and several parasitic wasps have been shown to have efficient uptake systems for a range of unusual and pollution-related transition metals, including zinc (Ortel, 1995).

Blackwell and Weih (1980) used the long ovipositor of the ichneumonid, *Megarhyssa*, to study the arrangement of chitin fibres and associated proteins in the cuticle by means of X-ray diffraction. They chose this taxon because its ovipositor has highly ordered, longitudinally oriented chitin fibres. However, studies of semithin sections of ovipositors stained with toluidine blue shows that the cuticular walls of many taxa are not uniform but rather are highly heterogeneous, with different parts having different fibrillar organization and staining properties (Smith, 1972; Quicke *et al.*, 1994). There have been no detailed chemical studies of ovipositor cuticle in parasitic wasps, but work on sawflies (Smith, 1972) and on the honey bee (Hermann and Willer, 1986) using fluorescence under UV illumination have both shown that particular regions may be rich in phenolics. It is suggested that these may correlate with highly resilin rich zones and it is highly likely that this is also the case in parasitic wasps. It should be noted, however, that UV fluorescence alone does not provide conclusive evidence for resilin, and that other tests are also needed.

A number of past workers, whose observations were limited to light microscopy, have mistaken sensilla on the ovipositor valves for pores, interpreted as the route by which venom was passed into the host. Subsequently, SEM showed the true sensory nature of these structures (see below). However, Nénon *et al.* (1995b) have resurrected the earlier pore notion for some structures that they observed in ichneumonids and sawflies. They further proposed, on the basis of their taxonomic distribution, that these 'pores' evolved into sensilla in higher Hymenoptera. Whilst it is not beyond the realms of possibility that there could be secretory elements in some hymenopteran ovipositors, the evidence these authors present is far from conclusive and should be treated with caution until a more thorough TEM investigation has been carried out. The same may also apply to another of Nénon *et al.*'s suggestions, namely that the large tracheae in each ovipositor valve may be part of a pneumatic system for expulsion of the egg.

6.15 MECHANICS OF OVIPOSITION

Oviposition usually requires eggs to move from the lateral oviduct or uterus-like expansion thereof, through the common oviduct and thence into and along the egg canal of the ovipositor. In the light of their electron microscopic study of egg structure in the pteromalid, *Nasonia*, King *et al.* (1969b) suggested that the first stage of oviposition involves spines (stylets) within the female reproductive tract dragging the chorion of the egg into the basal part of the ovipositor, with the oöcyte and associated vitelline membrane remaining undistorted until it finally flows through the narrow tube formed by the stretched chorion. Austin and Browning (1981) studied the internal sculpture of the egg canal in members of several insect orders and found them all to possess patterns of micro-scaling (ctenidia) directed posteriorly (Figure 6.8a–c), and they concluded that this sculpture serves to assist in oviposition either by resisting retrograde egg movements or through a ratchet-like action against the surface of the egg being extruded, or possibly both. In many parasitic Hymenoptera, the oviduct is only thin-walled and weakly muscularized and so is likely to

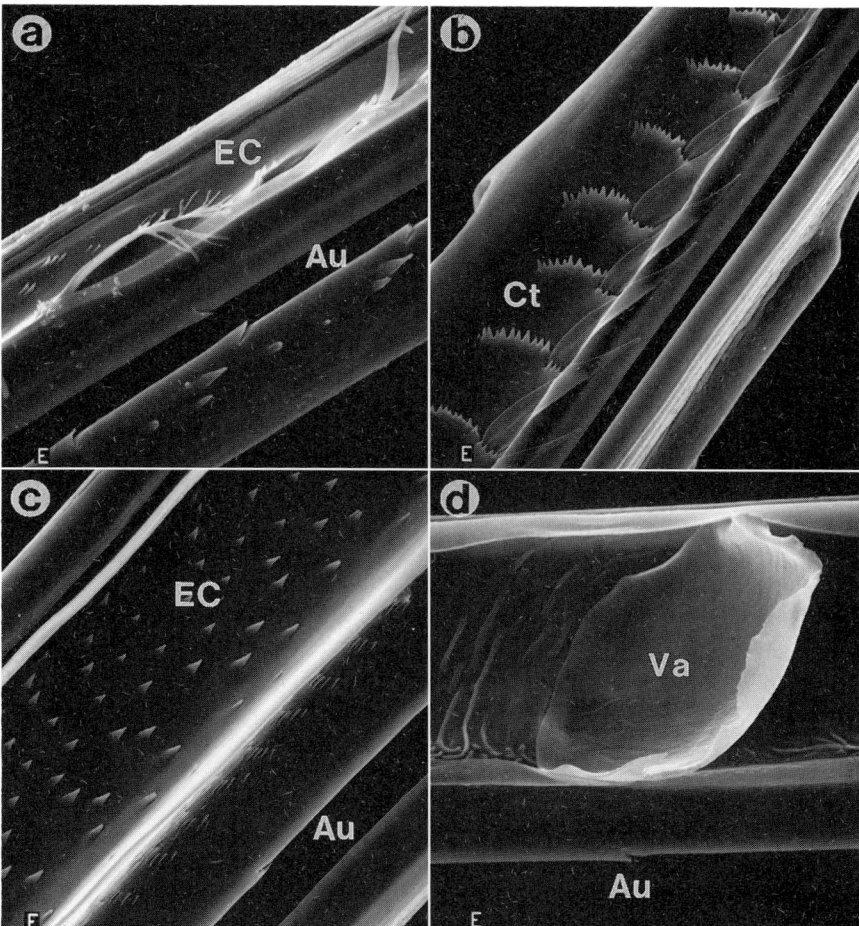

Figure 6.8 Scanning electron micrographs of the inner surface of isolated lower ovipositor valves of doryctine braconids showing a range of spiniform, scale-like and ctenidial sculpture (Ct), various possible sensilla, and in (d) a valvillus (Va). Au, aulax; EC, egg canal. (Courtesy of Habibur Rahman.)

play only a minor role in oviposition, though King and Ratcliffe (1969) note that the oviduct muscles of *Nasonia vitipennis* are well developed and presumably help to force the egg out during oviposition. Wilkes (1965) described pulsing contractions of the ovaries and lateral oviducts during oviposition in another chalcidoid, the eulophid, *Dahlbominus fuscipennis* – observed by removal of a small piece of abdominal wall! Probably much of the force needed to expel an egg down the often far narrower ovipositor channel is created by a general contraction of metasomal musculature. However, in many cases, to-and-fro movements of the lower ovipositor valves are an essential part of oviposition (Cole, 1981; Bronner, 1985). Le Ralec *et al.* (1986) noted that aphidiine braconids, despite having well-developed ctenidia, have a comparatively large egg canal relative to the small hydropic eggs, and so in these oviposition may be accomplished very rapidly. It is more difficult to envisage how these wasps routinely manage to pass only a single egg into the host in an action that may take much less than a second.

Although the egg normally passes down the egg canal completely concealed until it emerges near the ovipositor tip, among the ichneumonid subfamiliy Tryphoninae the larger part of the egg passes along the ovipositor externally, only the anchor staying within the egg canal until emerging at the tip (see Section 4.14 and Figure 4.7d–f). Observations of several members of the Chalcidoidea show that, even in some species that have more or less normally shaped eggs, the egg passes largely externally to the ovipositor – for example, in the eulophid, *Tetrastichus flavigaster*, which parasitizes psylid nymphs (Moran *et al.*, 1969) and *Mellitobia chalybii* (Hobbs and Krunic, 1971). Only in the eulophid, *Colpoclypeus florus*, has a mechanism been postulated to explain how a small part of the egg can be retained within the ovipositor while being moved from base to apex (van Veen and van Wijk, 1985). In *C. florus*, the distal part of the ovipositor has a ridge running along the egg canal wall of each lower valve, which these authors referred to as the egg-conducting ledge, and it seems likely that this helps to grip that part of the egg that passes within the ovipositor. *C. florus* is interesting in that, like a few other parasitic wasps, it oviposits near but not on its host (Dijkstra, (1986)/1987). A similar situation is found in certain other eulophids, such as *Neoplectrus* species that are koinobiont ectoparasitoids of slug caterpillars. Again the eggs of these are too inflexible and large to pass down the ovipositor completely within the egg canal; instead, they slide along the ovipositor, being held partly inside by egg guides. It is interesting that the ectoparasitic koinobiont strategies of these eulophids and of the tryphonine ichneumonids have led to a similar change in oviposition strategy, the egg no longer having to be compressible, and possession of a tough, non-plastic chorion no doubt being a considerable advantage against physical damage and desiccation (Kasparyan, 1980). Kasparyan suggests that the oviposition adaptations of these taxa may represent the sort of intermediate situation that must presumably have occurred during the origin of the Aculeata.

M.R. Shaw (1995) has observed oviposition in the enigmatic ichneumonid subfamily Adelognathinae and found that the egg is not passed down the lumen of the ovipositor at all, but rather emerges from near its base and is glued to the host's cuticle. The ovipositor thus serves only for stinging the host, a behavioural system that very closely resembles that of many less derived aculeates. Shaw also made observations on the apparently primitive braconid *Histeromerus*, and whilst oviposition itself was not observed directly, circumstantial evidence suggested that in this wasp, too, the egg may not pass all the way along the egg canal of the apically very fine ovipositor.

Oviposition can take from a fraction of a second to several hours in the case of some ichneumonids. One of the fastest observed oviposition events has been observed in the braconid, *Neoneurus*, and presumably occurs in other neoneurines, which oviposit into active adult ants (S.R. Shaw, 1993). In these the female *Neoneurus mantis* darts at its host ant, *Formica podzolica*, alights on its abdomen, oviposits and departs rapidly to a safe distance, all in under a second. At the other extreme come various taxa which use their very long ovipositors to bore through wood to reach concealed hosts (see below). For example, in rhyssine ichneumonids which attack horntail larvae (Siricidae), oviposition takes from 7 to 333 minutes as the wasps drill through the intervening centimetres of substrate (Heatwole *et al.*, 1962; Eggleton, 1989). Some aphidiine braconids oviposit very rapidly; others retain their ovipositor inside the host for a minute or so before oviposition, and presumably this extra time is spent in testing the suitability of the particular host.

In the monotypic family Pelecinidae, the ovipositor itself is very small and normally concealed, though it can be protruded directly backwards. However, the metasoma of the female is very long, slender and highly manoeuvrable and is used to reach its host scarabeid beetle larvae by digging into the soil. A detailed morphological study of *Pelecinus polyturator* by Mason (1984) showed that, as well as being capable of bending in a vertical plane, the connections between the sternites of abdominal segments 3 and 4, 4 and 5, and 5 and 6 each allow rotary movement of up to 135°.

6.16 LONG OVIPOSITORS

Most parasitic wasps have ovipositors less than 1.3 times longer than the body. This ratio is biologically significant since for most wasps the maximum length of the ovipositor is limited by the wasp's ability to align it more or less perpendicularly against the substrate. This is achieved by the wasp assuming a sort of tip-toe posture with the apex of the metasoma held aloft and the apex of the ovipositor positioned more or less between the forelegs (Gardiner, 1966; Ramírez, 1986). Townes (1975) tabulated some wasps with rather longer ovipositors, ranging from 1.6 to 8.1 times body length, and other taxa with long ovipositors were mentioned by Compton and Nefdt (1988): in Townes' table the record of an *Iphiaulax* species with an ovipositor 14 times longer than the body actually refers to *Pheloura*, another braconid with an enormously long pseudo-ovipositor but a normal, short proper ovipositor (Berland, 1951; van Achterberg,

1989). Three of the taxa listed by Townes have special ovipositor coiling mechanisms that enable them still to oviposit while in the standard tip-toe posture. These wasps are members of the Orussidae and Ibaliidae, both of which have the ovipositor coiled within the metasoma when it is not in use, and the ichneumonid genus *Megarhyssa*, which can 'retract' the ovipositor in a loop within an extensible membranous sack at the apex of the metasoma (between metasomal tergites 6 and 7) which takes up the excess length again as the wasp assumes its upright posture (Baumann, 1924; Abbott, 1934). The mechanisms in other taxa with long ovipositors lacking these morphological specializations were not known to Townes, and one of the first observations of oviposition in such wasps was by van Achterberg (1986), who described the actions of an Indo-Australian *Gronaulax* sp. (Braconidae). In this species the wasp first located its host insect larva, which lives in branches, and then walked forward, stopped and backed, slightly raising its metasoma until the ovipositor formed an angle of approximately 120° whereupon it was slowly inserted into the substrate. Van Achterberg noted that this posture is likely to deliver far less force to the ovipositor than the better known vertical stance, and so he concluded that for these very long-ovipositored wasps, the ovipositor probably mostly follows existing cracks and tunnels (but see below). More observations will be required before the general validity of this assumption can be assessed.

Several groups of parasitic wasps which oviposit into or near host eggs hidden deep within wood make use of the hole drilled by their host's ovipositor. This is true, for example, of *Ibalia* (Spradbery, 1970a), which is endoparasitic on larvae of various siricid wood wasps. Such species typically have narrower ovipositors than their hosts. Members of the ichneumonid genus *Pseudorhyssa* also make use the hole drilled by another ichneumonid parasitoid, *Rhyssa persuasoria*, which oviposits on the larva of the alder sawfly, *Xiphydria* species (Couturier, 1949). *Pseudorhyssa* is a kleptoparasitoid in that its larva hatches quickly and promptly consumes the egg of the *Rhyssa* before starting to consume the paralysed sawfly larva. The ovipositor of *Pseudorhyssa* is narrower than that of its host but its egg is considerably larger (see Section 4.11), no doubt in order to ensure that it has sufficient advantage over its victim, and the 1st instar larva is furnished with a massive, heavily sclerotized head.

6.17 OVIPOSITOR GUIDES AND OVIPOSITOR BUCKLING

Parasitic wasps that have to manipulate long ovipositors no doubt face many problems in controlling them, especially during the early stages of substrate penetration when most of the length of the ovipositor will be without support. If one looks, for example, at the braconid, *Virgulibracon endoxylophila*, 'drilling' into ironwood, it is easy to imagine the shaft of the ovipositor buckling and that any such distortion would greatly reduce its efficiency. Ovipositor guides have been described in a number of taxa with long ovipositors. Aulacidae have notches or grooves on the inner surfaces of their hind coxae and a similar adaptation is found in the ichneumonid genus, *Certonotus*. However, at least in the case of the European aulacid, *Aulacus striatus*, the female does not drill a hole into the wood herself; instead, she threads her ovipositor down the boring made by her host wood wasp's ovipositor, as is also the case with the ichneumonid, *Pseudorhyssa* (see above). In some rhyssine ichneumonid parasitoids of wood-boring hosts, the anterior metasomal sternites bear paired tubercles and a mid-longitudinal groove which help to stop the ovipositor slipping sideways during boring (Gardiner, 1966; Eggleton, 1989; Gauld, 1991; Vincent and King, 1996). Several parasitic wasps with short ovipositors also have modifications that may be involved in stabilization, such as medial emarginations of their terminal metasomal tergites, or occasionally their sternites (Kfir and Rosen, 1981; Quicke, 1987). There are very few observations of these in use, but it seems likely that they serve to prevent the ovipositor from slipping laterally, perhaps when the host substrate is particularly resilient.

Vincent and King (1996) have considered the mechanical forces that the ovipositors of the wood wasp, *Sirex noctilio*, and its ichneumonid parasitoid, *Megarhyssa nortoni*, develop when drilling into wood. Whilst *Sirex* has no problem with its very robust and relatively short ovipositor, *Megarhyssa* has a much longer and narrower ovipositor which has a far lower Euler buckling load. In order for *Megarhyssa* to drill successfully, its buckling load has to be reduced and this is achieved by a number of mechanical tricks. Firstly, the effective free length of its ovipositor is almost halved by its proximal part being engaged with the sternal ovipositor guides (see above); secondly, the end conditions are constrained both by the ovipositor guides and by drilling into a tight hole; and thirdly, by having one of the lower ovipositor valves engaged with wood fibres and held under tension. Vincent and King's modelling provides an alternative explanation for why rhyssines such as *Megarhyssa* abandon about 60% of their drillings before getting very far (Spradbery, 1970b): rather than the wasp discovering that it is drilling in the wrong place relative to its host, it may be drilling at a place where the softness of the wood leads to too loose a hole, which does not grip the ovipositor sufficiently to provide the end constraint necessary to increase the buckling load.

6.18 OVIPOSITOR STEERING MECHANISMS

Despite the fact that wasp ovipositors are without internal musculature, several observations have shown that members of various groups of parasitic Hymenoptera have the ability to bend their ovipositor tip actively –

for example, Delanoue and Arambourg (1965, 1967) for the eupelmid, *Eupelmus urozonus*, and the eulophid, *Pnigalio mediterraneus*, respectively, and Compton and Nefdt (1988) for the Agaonidae – though this ability is by no means universal and has undoubtedly evolved on several different occasions. Where these mechanisms have evolved, they appear to increase the chance of the wasp making contact with a host within a concealed cavity. To date some eight different morphological adaptations for ovipositor steering have been discovered (Quicke, 1991; Quicke and Fitton, 1995; Quicke *et al.*, 1995) and these are collectively distributed the Agaonidae, Aulacidae, Gasteruptiidae, Ichneumonidae and Braconidae.

Perhaps the most extreme and conspicuous modifications are to be found in the braconine braconids *Zaglyptogastra* spp. (Figure 6.9a), *Undabracon* spp., *Cedilla* spp. (Figure 6.9b), *Digonogastra zaglyptogastra*, and the doryctine braconid, *Heterospilus falcatus*, and also in various ichneumonids, particularly in the subfamilies Cremastinae and Tersilochinae, e.g. *Pristomerus* (Quicke, 1991; Quicke and Marsh, 1992). In these the apex of the ovipositor is formed into one or more swollen arched regions, or in the case of *H. falcatus*, an inverted arch, separated by short thinner and more flexible regions. Since both upper and lower valves are similarly modified, relative movement of these causes the thicker part of one valve to become aligned with the thinner part of the other, and since the upper and lower valves are held interlocked for their entire length by the olistheter mechanism, the ovipositor is forced to bend dorso-ventrally in order to keep the pieces connected. At least in the cases of *Zaglyptogastra* and *Pristomerus*, whose hosts are known, it has been hypothesized that the bending mechanisms enable the ovipositor tip to be manipulated within the host's hideaway, along a twig-boring or within a leaf-roll, respectively, and so increase the wasp's chance of successful parasitism. In two other ichneumonoid wasps – an undescribed genus of cosmophorine braconid, *Sinuophorus*, and the rhyssine ichneumonid, *Myllenyxis* – the apical part of the ovipositor is formed into a series of arches but the whole ovipositor is also very strongly compressed laterally, making dorso-ventral flexion seem an unlikely option. Indeed, if this lateral compression does render the ovipositor incapable of dorso-ventral flexion, it is hard to imagine how the dorsal and ventral ovipositor valves can move with respect to one another.

A wide variety of other but less conspicuous ovipositor steering mechanisms have been found in members of several other subfamilies of Ichneumonidae (Quicke *et al.*, 1994) and Braconidae (Quicke *et al.*, 1995), and also in the Aulacidae, Gasteruptiidae (Quicke and Fitton, 1995) and Agaonidae (Quicke and Fitton, unpublished observations). Some of these have been revealed by preparing semithin sections through ovipositors; others have only become apparent after dissociating the upper and lower valves of the ovipositor and examining their apices under an electron

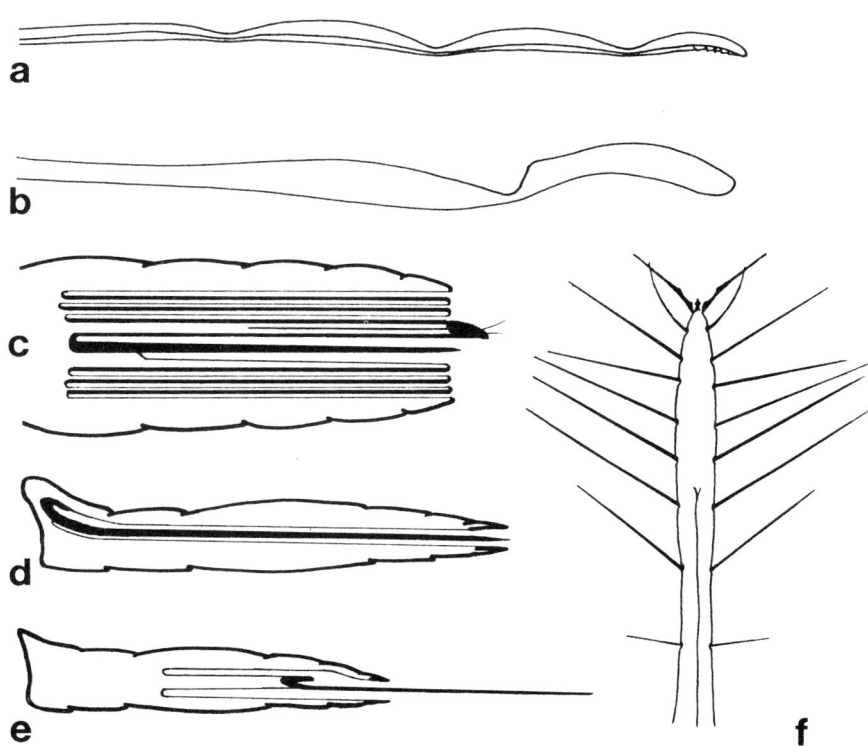

Figure 6.9 Various ovipositor adaptations. (a) *Zaglyptogastra* and (b) *Cedilla* sp. (both Braconidae) with arched ovipositor tips that are associated with bending and steering mechanisms. Telescopic ovipositor systems of two scelionids: (c) the *Scelio*-type system comprising three invaginated metasomal segments; (d,e) retracted and partially protracted *Ceratobaeus*-type system. (f) Dorsal aspect of the ovipositor of the ectoparasitic eulophid, *Colpoclypeus florus*, with forked apical structures and long 'setae'. (c,d,e: Redrawn after Field and Austin, 1994; f: redrawn after van Veen and van Wijk, 1985.)

microscope. With the exception of the mechanism found in the Agaonidae, they utilize the same general principle: near the apex of the valves there is a structure that restricts the relative movement of one part of the ovipositor with respect to the other. Thus, in the Gasteruptiidae there is a pair of abutting bosses on the apico-lateral part of the dorsal and ventral valves (Figure 6.10a,b). These bosses allow the ventral valve to be retracted relative to the dorsal one, but they prohibit it from being extended beyond the tip of the dorsal valve. If the wasp attempts to force the ventral ovipositor valve posteriorly, the whole ovipositor will be bent upwards in much the same fashion as the differential thermal expansions of the metals that comprise the bimetal strip of a thermostat causes the strip to bend and activate the switch. In the case of the Aulacidae, a family that has long been associated with the Gasteruptiidae and sometimes even regarded as a synonym of it, the ovipositor can likewise only be bent upwards but this is achieved by a somewhat different modification. In these there is a longitudinal ridge just lateral to the aulax on each lower valve, and a corresponding groove in which it runs just lateral to the rachis on each side of the upper valve. Both ridge and groove come to an abrupt end just in front of the ovipositor tip such that when the ovipositor is at rest, directed straight out posteriorly, the end of the ridge just about abuts the end of the groove. Then, as with the Gasteruptiidae, attempts by an aulacid to push the lower valves posteriorly will cause the ovipositor to curve upwards.

In the ichneumonid subfamilies Cremastinae, Lycorinae, Phrudinae and Stilbopinae, and in many banchines and members of the tryphonine tribe Phytodietini, the ovipositor can be actively bent in all directions (Quicke *et al.*, 1994). In their external appearance the ovipositors of these wasps do not seem in any way remarkable. However, transverse sections through their mid-regions and the anatomy of their basal regions show a highly specialized structure. Apart from at the very apex, the dorsal valve is longitudinally divided into two halves that are interlocked mediodorsally with an entirely separate C-section structure termed the aulaciform rod (Figure 6.7f). As observations of living wasps of these groups have shown, this structure affords them a great deal of manipulative control over their ovipositor. Since the three components that go to make up the dorsal valve are fused at the ovipositor tip, the ovipositor will bend left if the wasp differentially retracts the left piece of the valve, and right if the right side is retracted. The basal region of the dorsal valve comprises a transverse plate on which the two halves of the valve articulate (Oeser, 1961). It is presumed that relative retraction of the dorsal valve halves is brought about by rotation of the basal plate in a horizontal plane.

6.19 TELESCOPIC AND CONCEALED OVIPOSITOR MECHANISMS

Telescopic ovipositor systems occur in some Scelionidae (Austin, 1983; Galloway *et al.*, 1992; Field and Austin, 1994), in Platygastridae, in at least some Diapriidae (Huggert, 1979), in Agaonidae (Naumann, 1991) and in the aculeate family Chrysididae, especially in the Chrysidinae (Kimsey, 1992). In most of these the ovipositor is comparatively short and normally concealed within the apex of the metasoma. In the Scelionidae, two distinct types of telescopic ovipositor are known. In the *Scelio*-type, at least metasomal segments 7 and 8 are tubular and telescoped into one another except during oviposition, when the whole complex is extruded (Figure

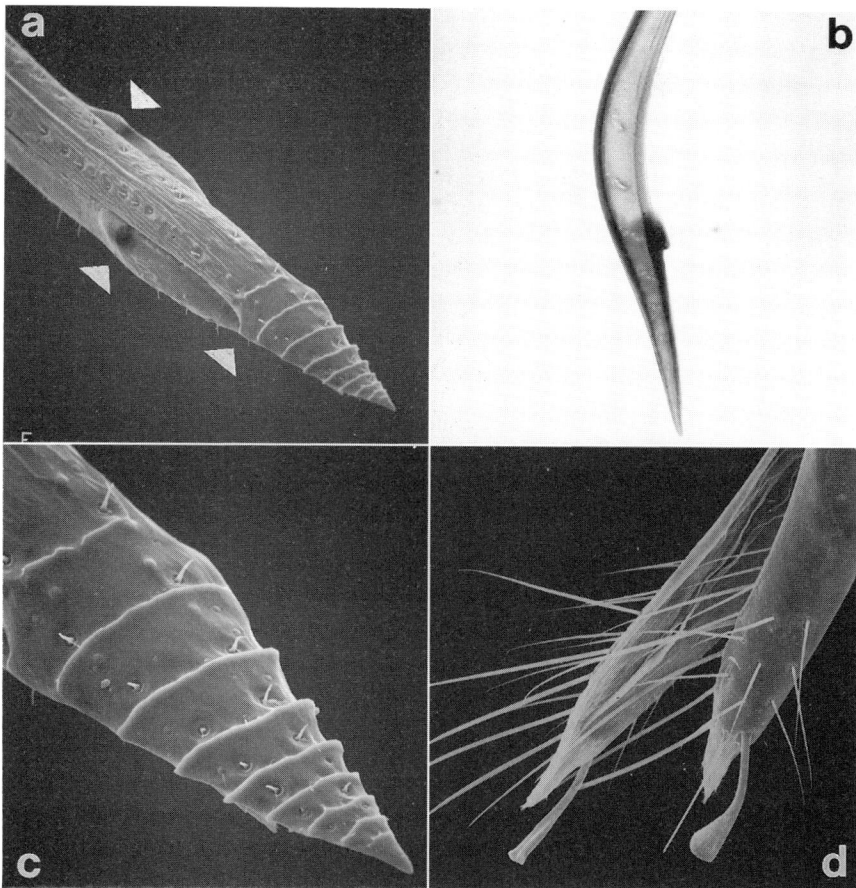

Figure 6.10 Ovipositor steering mechanisms and sensilla of *Gasteruption* sp. (Gasteruptiidae): (a) expansions of the upper and lower valves (white arrows) that are involved in the steering mechanism; (b) light microscope preparation of isolated lower valve showing increased melanization and sclerotization at tip; (c) detail of upper valve tip showing various sensilla types. (d) Ovipositor sheath sensillae of *Buluka achterbergi* (Braconidae: Microgastrinae). (Courtesy of A.D. Austin.)

6.9c). In some taxa, the last four metasomal segments are telescoped into one another. In the *Ceratobaeus*-type, only metasomal segment 8 is involved, and this, which is largely membranous, is invaginated at rest rather than being telescopically retracted (Figure 6.9d,e). In those platygastrids (e.g. *Inostemma* spp.) and gryonine scelionids in which the ovipositor is longer than the metasoma, the 1st tergite of the latter is formed into a horn-like anterior projection that overhangs the mesosoma and accommodates the extra ovipositor length.

Within the Chrysididae – and in particular the Chrysidinae, which are parasites of wasps and bees within their nests – metasomal structure is a compromise between having a heavily sclerotized anterior part, which is important when the wasp rolls itself into a protective ball, and a soft, telescoped ovipositor needed for laying in host larvae (Kimsey, 1992). The telescopic nature of the ovipositor in these (with the true ovipositor very reduced) may further be an adaptation to protect the chrysidine if it is attacked by an irate adult host. Perhaps this modification is an alternative to the ovipositor steering mechanisms described above, in that the directional control is achieved by the intrinsic musculature of the posterior metasomal segments.

Many parasitic wasps with long ovipositors have these permanently protruding from their body, like tails, but some have evolved to hold the ovipositor 'internally' while it is not in use. Of course, the ovipositor is always, strictly-speaking, external and is simply withdrawn into a pouch within the metasoma or, rarely, also within the mesosoma. In all the Cynipoidea there is a tendency to internalization of the ovipositor and this reaches its extreme in the Ibaliidae, in which the ovipositor is coiled internally through more than 300° within the blade-like metasoma (Fergusson, 1988). The ovipositor mechanism of the cecidogenic cynipids, *Biorhiza aptera* and *Diplolepis rosae*, has been described in detail by Frühauf (1924) and Bronner (1985), respectively. A very similar arrangement to the cynipoids has evolved independently in the Orussidae (Cooper, 1953). Rasnitsyn (1980) notes that the ovipositor is short in the Paroryssidae, an extinct family of Hymenoptera believed to be ancestral to the orussids, and that these wasps were probably limited to ovipositing in hosts just below the bark. The internal looping of the ovipositor in Orussidae may be seen as an adaptation enabling them to reach more deeply concealed hosts than they would otherwise reach, especially given that they lack the flexible wasp-waist of the Apocrita.

6.20 VALVILLI

Most braconids and ichneumonids have one or more articulated, flap-like processes protruding into the egg canal (Figure 6.8d) which have come to be known as valvilli (Oeser, 1961; Quicke *et al.*, 1992a). Similar and probably homologous structures are found in the stings of many aculeate families, where they function as part of the venom pumping system and are often referred to in these taxa as valves. There is considerable variation in the number, relative sizes and positions of the valvilli within the ovipositor; for example, in the ctenopelmatine ichneumonid genus, *Pion*, there are seven pairs of valvilli which diminish in size towards the ovipositor base, whilst in many others there may be only one on each valve. In a few taxa the valvilli are consistently asymmetrical, with different numbers on the left and right valves (van Veen, 1981; Quicke *et al.*, 1992a). This variation suggests that they may serve different functions in different species.

The function of valvilli in the parasitic ichneumonoids has hardly been investigated. The only functional interpretations to date are those proposed by Rogers (1972), who suggested that the single pre-apical pair of valvilli in *Venturia canescens* serve to hold the egg in place near the end of the ovipositor until the exact moment that it needs to be released, and van Veen (1981), who suggested that they serve to limit the relative to-and-fro motions of the lower valves during oviposition. Naturally it is extremely difficult to make observations of eggs emerging from ovipositors, especially in endoparasitic species. However, in a clever set-up devised by Tersac and Guerdoux (1981), the emergence of eggs from the ovipositor of *Pimpla hypochondriaca* (as *instigator*) was observed using artificial lures filled with aqueous host chrysalis extract. Their photographs show the egg emerging pre-apically from the ovipositor, very close to the position of the 1st valvillus. It is tempting to speculate, therefore, that the valvillus in this ichneumonid serves to deflect the egg and so to force it out of the ovipositor. In some chalcidoids, there is a sharp spur-like protuberance into the egg canal just before the ovipositor tip, termed the sperone by Zinna (1960). The function of this structure is not yet known but it is possible that it has a similar role to the valvillus.

6.21 MISCELLANEOUS OVIPOSITOR SPECIALIZATIONS

A very strange ovipositor with quite abnormal function has been described for the eulophid *Colpoclypeus florus* by van Veen and van Wijk (1985). *C. florus* is one of that comparatively unusual set that is a non-paralysing ectoparasitoid, its hosts being weakly concealed late instar larvae of various tortricid moths. Prior to oviposition, the wasp stings the host larva many times in the head capsule, but this process only involves penetration of the lower valves. Indeed, it would be impossible for the upper valve to do so as it is furnished with 32 pairs of long 'setae' along its length, culminating in an apical pair of articulated spines (Figure 6.9f).

Trigonalyid females, which lay several thousand eggs within a short space of time, have a specially modified 7th sternite which they use to cut a notch in the underside of the edge of a leaf into which the reduced ovipositor, with a membranous tip, inserts its eggs (Oeser, 1962). Several eucharitids also have modified ovipositors for laying in leaf tissue (Johnson *et al.*, 1986).

6.22 SENSORY SYSTEMS

Most work on parasitic wasp senses have concentrated on the antennae and the ovipositor, both of which are furnished with various types of sensory receptors. Based on general morphology, sensilla have been classified into a number of types that generally correspond quite well to what is known or can be inferred about function. Parasitic wasps as a whole display a wide range of different sensilla – in some species as many as six different sensilla morphotypes have been found on the antenna alone. Several of the types found in parasitic wasps are illustrated in Figure 6.11.

6.23 ANTENNAE

Antennae comprise an enlarged basal segment, the scape, and a small 2nd segment, the pedicellus, followed by a number of other segments that make up the flagellum. In most parasitic wasps the antennae are comparatively simple – either filamentous or with a distinct club – though more elaborate forms are found in a few taxa, especially in males in which it is presumed that they are modified for locating females via their sex pheromones. Thus males of some encyrtids, eulophids, tanaostigmatids and bethylids often have moderately to highly branched (pectinate) antennae, and male antennae are often longer than those of females. The antennae play an important role in host location in virtually all parasitic wasps, as has been demonstrated on many occasions by various partial and total excision experiments (e.g. Hays and Vinson, 1971; Weseloh, 1972; Isidoro *et al.*, in press). They are always furnished with a diverse array of sensilla, which have been subject to many studies – morphological, behavioural and electrophysiological – but there is still a very great deal that we do not know about the functions of particular sensilla types.

The roles of some sensilla have been inferred in taxa where they are differentially distributed on different antennomeres, or are positioned on particular antennomeres, and therefore one type of receptor may be invoked or eliminated in certain responses after selective partial antennectomy (Barrass, 1960; Isidoro *et al.*, 1996). For example, in the scelionid, *Trissolcus basalis*, progressive removal of the club segments leads to a great increase in the time taken for females to recognize/accept conspecific males, until recognition fails entirely when the club is completely removed (Bin *et al.*, 1989). As the club segments are the only ones to possess elongate placodiform sensilla, it is tempting to speculate that these sensilla are involved in the detection of the secretions of the male *Trissolcus* antennal sex gland (see below). However, removal of club also results in removal of virtually all of the hair-sensilla and most of the sickle-shaped sensilla, so further investigations will be needed before the exact receptors are identified with absolute certainty.

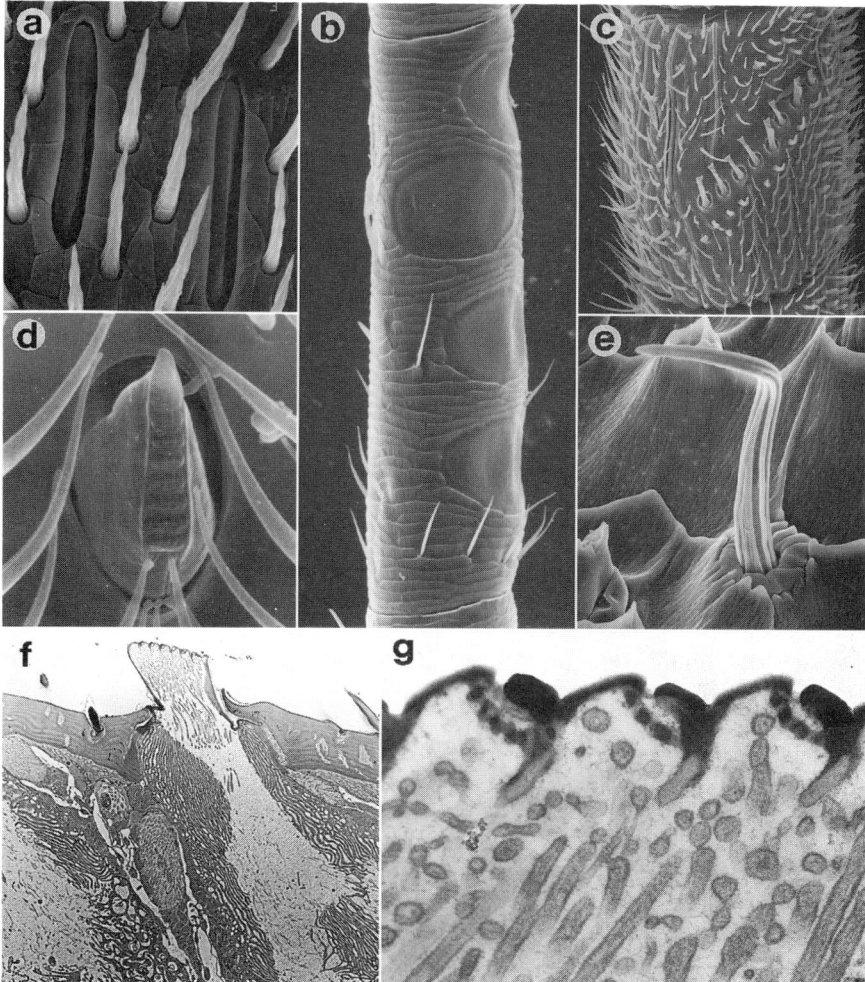

Figure 6.11 Antennal sensilla: (a) *Evania* (Evaniidae); (b) gen.sp. (Stephaniidae); (c) *Trissolcus basalys* (Scelionidae), 10th segment of female; (d) *Buluka achterbergi* (Braconidae); (e) *T. basalys*, 7th segment; (f) and (g) (detail) *T. basalys*, crypto-multiporous area, longitudinal section. (a,b: Courtesy of H. Basibuyuk; c,e,f: courtesy of F. Bin; d: courtesy A.D. Austin.)

In the case of scelionids, a group composed entirely of egg parasitoids, host acceptance probably depends on basiconic sensilla located on the clava of the female antenna (Bin, 1981). Cave *et al.* (1987) showed that in one species, *Telenomus deserti*, removal of the clava of one antenna did not prevent host egg acceptance, but that removal of the clava from both antennae completely prevented acceptance.

Electrophysiological recordings have almost always involved the responses of populations of sensilla rather than individuals. Perhaps surprisingly, the magnitudes of the responses seen in electro-antennogram recordings often seem to be strongly correlated with the wasp's positive response to various odours, such as those emanating from hosts or host foodplants as well as conspecific females (Lecomte and Pouzat, 1985; Ramachandran and Norris, 1991; Hidoh *et al.*, 1992; Rotundo and Tremblay, 1993; Vaughn *et al.*, 1996). Indeed, antennogram responses are often used as a way of assessing the importance of various odours for host location, as an alternative to whole-animal behaviour experiments using various olfactometers.

The ichneumonid, *Pimpla hypochondriaca* (as *instigator*), has a number of vesiculate structures at the tip of the antenna which apparently serve to transmit shock waves, produced in the female body, to the substrate (Henaut, 1990; illustrated in Gauld, 1991; Chapter 8). The origins and ultrastructure of these is unclear at present, but it seems likely that they are at least derived from sensilla.

6.24 ANTENNAL SENSILLA

The antennae are specialized sensory structures and are well developed in all parasitic Hymenoptera. They possess multiple types of sensilla which fall into a number of classes based on gross external morphology, nowadays usually following examination by scanning electron microscopy (e.g. Navasero and Elzen, 1991). These classes include *sensilla trichodea, basiconica, chaetica, campaniformia, coeloconica* and *placodea* as well as some special forms, but these simple definitions, based in large part on gross external appearance, are of little use and have little functional significance. Many are similar to those of other insect orders but true homology is difficult to ascertain. Although some sensilla types have also been examined using transmission electron microscopy, very little definite information exists on the functions of the different classes. Placoid sensilla are probably predominantly chemosensory; trichoid sensilla are mainly mechanosensory but some are also chemosensory, and so on. High-resolution scanning and transmission electron microscopy have shown that all placodiform sensilla, as well a number of other sensilla types (notably various ones traditionally classified as basiconic), are multiporous; that is, each sensillum has numerous sensory cells, each connected to the exterior by its own pore. These pores are often difficult or impossible to observe in untreated specimens since they are very small and frequently are concealed by proteinaceous and other secretions. For fully successful scanning microscopy these need to removed by protease digestion (Bin *et al.*,

1986; Isidoro *et al.*, 1996) or specimens can be fixed and viewed at low temperature (Olson and Andow, 1993). A far more informative classification has been proposed by Frazier (1985), who preferred to distinguish between fundamental sensilla types on the basis of three criteria: whether they have no, one or multiple pores; whether they are single or double-walled; and whether or not they are in a flexible socket. In practice this gives six combinations.

Detailed ultrastructural investigations and surveys include: Slifer (1969: Pteromalidae), Richerson *et al.* (1972: Braconidae), Norton and Vinson (1974a,b: Braconidae and Ichneumonidae), Borden *et al.* (1978: Braconidae), Barlin and Vinson (1981a: Eulophidae), Dahms (1984b: Eulophidae), Olson and Andow (1993: Trichogrammatidae) and Isidoro *et al.* (1996: various taxa). The ontogeny of antennal sensilla development has been studied in the Aulacidae by Schmidt and Kuhbandner (1983) and in the Ichneumonidae by Stepper *et al.* (1983) and Rojas-Rousse and Palevody (1983). The ultrastructure of the *sensillum coeloconicum* in the scelionid, *Trissolcus basalis*, is similar to those of other insects that are known to be thermo-hygroreceptors (Isidoro, 1992).

In many parasitic wasps there are marked sex differences in the sensorial complement. For example, male aphelinids such as *Encarsia* spp. have complex arrays of sensilla on their funicle segments which appear to be important in courtship (Viggiani and Mazzone, 1982).

6.25 PLACOID OR MULTIPOROUS PLATE SENSILLA

One of the most conspicuous antennal sensilla of parasitic and many aculeate Hymenoptera is the placodiform sensillum (Figure 6.11a–c), a group that has received by far the greatest amount of attention in terms of both functional and morphological studies. In Frazier's (1985) classification, these are examples of single-walled, multiporous and usually socketed sensilla, which are innervated by multiple neurones, and they appear to be entirely chemosensory in function. They have been most intensively investigated in the Chalcidoidea (Barlin and Vinson, 1981a; Ware and Compton, 1992) and Ichneumonoidea (Borden *et al.*, 1978; Barlin and Vinson, 1981b; Stepper *et al.*, 1983), though they are also found in the xyelid (subfamily Xyelinae only) and orussid sawflies, cynipoids, pelecinids, evanioids (Schmidt and Kuhbandner, 1983; Basibuyuk, unpublished observations) and many aculeates. There appear to be at least two types of placodiform sensillum based on general morphological features (Norton and Vinson, 1974a,b; Gibson, 1986a), and it is not clear whether these are all involved in the same functions. In the Ichneumonoidea, they are socketed (Borden *et al.*, 1978), whereas the superficially similar sensilla in Chalcidoidea are not, and further, in chalcidoids, the apex of the sensillum is often somewhat protruding and separated from the underlying flagellomere (Slifer, 1969;

Gibson, 1986a). These features suggest that the chalcidoid-type placodi-form sensillum has evolved from a normal setum which has come to lie flat along the surface of the antennal segment, but other internal features such as the possession of a longitudinal pair of cuticular thickenings, sometimes called the pendant lamellae, which define the medial pored zone could suggest that both chalcidoid and ichneumonoid types, and perhaps also the curved setae of various diapriids, may be derived from one another rather than independently (Basibuyuk and Quicke, in preparation). Less is known about placoid sensilla in other groups of parasitic wasps, but Butterfield and Anderson (1994) have described their ultrastructure in the eucoilid, *Tribliographa rapae*, a parasitoid of the cabbage rootfly, *Delia brassicae*, and Schmidt and Kuhbandner (1983) have detailed their structure in the aulacid, *Aulacus striatus*.

Before the homologies of these and other sensilla types can be worked out, it will almost certainly be necessary to study their ontogeny as well as their cuticular components. Interestingly, Stepper *et al.* (1983) have shown that the placodiform sensilla of the ichneumonid, *Pimpla turionellae*, has two trichogen and two tormogen cells, twice the typical number of each for other insect sensilla. A similar doubling is also known in the multiporous, socketed basiconic sensilla of two sawflies, *Cephus* and *Xiphydria*. Whether this represents a homology between these is yet to be determined.

Slifer (1969) proposed that the placodiform sensilla of the chalcidoid *Nasonia vitripennis* were olfactory in function and this agrees well with their morphology. A very different proposal was made by Richerson and Borden (1972b) for those of the ichneumonoid, *Coeloides brunneri*, which attacks larvae of scolytid beetles living just beneath tree bark. Richerson and Borden's experiments showed that *C. brunneri* was capable of detecting hot spots on the tree bark that differed from the surrounding bark by as little as 0.5°C, and they proposed that the placodiform sensillum of these was in fact a sensitive infra-red detector whose operation was mediated by their elongate form (section 8.8). However, it is clear that heat perception cannot be a significant factor in host location for most ichneumonoids, yet all possess elongate placodiform receptors. Further, the fact that the sensilla of *C. brunneri* extend for the entire length of the flagellomeres, which was remarked upon by Richerson and Borden, is unlikely to be directly the result of any selection for infra-red reception as this is a feature shared by all cyclostome braconid subfamilies (Quicke, 1994). Many Stephanidae have some quite remarkably large, round placoid sensilla (Figure 6.11b).

6.26 GUSTATORY SENSILLA

Some multiporous sensilla on the antenna of various parasitic wasps such as the scelionid, *Trissolcus basalis*, are associated with glands (Figure 6.11d,f,g) (Bin *et al.*, 1989), and more recent work suggests that these may be gustatory rather than olfactory (Isidoro *et al.*, 1996). Such receptors, which are apparently widespread in the parasitic Hymenoptera, may be involved in contact host recognition/acceptance (Chapter 8) or in courtship. In these, the glandular secretions normally cover the pore plate. It seems that in life the secretions probably dissolve non-volatile kairomones from the surface of a potential host such as an insect egg, and it is these that are detected by the associated sensilla. The well-known tyloids – raised or sunken patches of distinctive sculpture on the antennae of many ichneumonids and vespoids – may fall into the same category.

6.27 OVIPOSITOR SENSE ORGANS

The tip of the ovipositor is now known to be supplied with various sensilla types, depending on the taxon, though in some earlier literature these structures were misinterpreted as pores through which venom was thought to be released. There have been surprisingly few detailed morphological or functional investigations of the sense organs on parasitic wasp ovipositors, though the simple presence of sense organs has been noted in many scanning electron microscopic studies and the fact that the ovipositor acts as a sensory structure has been widely accepted. A very superficial survey was provided by Hermann and Douglas (1976) who surveyed 31 species in eight families. Most of these were aculeates, and 10 of the 11 non-aculeate parasitic Hymenoptera included were ichneumonoids. Sensilla were found in all of these taxa. Detailed morphological studies have been presented for a few specific taxa (Gutierrez, 1970; Ganesalingam, 1972; Hawke *et al.*, 1973; Greany *et al.*, 1977; Le Ralec and Rabasse, 1988; Le Ralec, 1991; Le Ralec and Wajnberg, 1990). In some cases it has been apparent that more than one kind of sensillum is present, whilst in others only single types have been found, and these may be involved in detecting both the chemical and mechanical environmental of the ovipositor tip and probably also its internal stresses and strains. In many taxa, there are various possible sensilla on the egg-canal wall near the apex of the ovipositor (Bronner, 1985; Vincent and King, 1996). These are usually peg-like or setiform and distinct from the ctenidial sculpture of the rest of the canal wall. Unfortunately, there have been no detailed ultrastructural studies of these and so it is not yet possible to confirm their sensory role, though given their location this would seem to make perfect sense. Nénon *et al.* (1995b) suggested that some of the 'pores' on the ovipositors of certain ichneumonids might be secretory, as was sometimes thought in the past, but their findings were not supported by any transmission electron microscopy and should therefore be treated with some caution.

Le Ralec and Rabasse (1988) compared the sense organ complement of three genera of aphidiine braconids with the time taken by each to oviposit. The fewest receptors were found in *Aphidius uzbekistanicus*, which oviposits in less than a second. *Praon volucre*, which takes approximately 2 seconds to oviposit, and *Ephedrus plagiator*, which takes some 20 seconds, both have more receptors and more types of receptors. However, whilst the sensilla of the *Praon* were predominantly mechanoreceptors, those of *Ephedrus* were mostly chemoreceptors, suggesting that host discrimination may be more important for *Ephedrus* than for the other aphidiines. Le Ralec *et al.* (1996) extended this work, and used correspondence analysis to examine whether phylogeny or biology had a greater influence on several features of ovipositor structure across a wider range of taxa. Their results show that both phylogeny and biology are important in determining such ovipositor features as tip shape and sensilla complement.

Ovipositor sense organs in some taxa appear to respond to relatively simple chemical compounds. Thus amino acids and inorganic ions have been found to act as oviposition stimuli, especially in egg and egg–larval parasitoids (Nettles *et al.*, 1982; Kainoh and Brown, 1994)

6.28 OVIPOSITOR SHEATH SENSILLA

There have been at least some studies on the sense organs on the ovipositors of parasitic wasps, but those on the ovipositor sheaths have received far less attention. Nevertheless, the apices of the ovipositor sheaths, probably particularly in those species that attack exophytic hosts, are supplied with numerous and diverse sensilla. Some of these are conspicuous, as is the case with those of certain microgastrine braconids (Austin, 1989) (Figure 6.10d), though mostly they require scanning electron microscopy to reveal their presence (LeRalec, 1991). A great deal more work on the role and importance of ovipositor sheath sensilla is needed.

6.29 CERCI

Both male and female Hymenoptera possess a pair of unsegmented, non-muscular appendages on the membranous posterior edge of the 10th+11th abdominal tergite, variously referred to as cerci, pygostyles or socii. These structures are well developed in sawflies, and are furnished with many sensilla. Female sawflies use these to examine the substrate during oviposition whereas in males they are used for orientation during copulation (E. L. Smith, 1970), and it is interesting to note that they are particularly well developed in males of some pollinating fig wasps (Figure 6.16), which spend their entire lives within the dark confines of a fig syconium. In the parasitic wasps they are generally far less well developed and in most species, if not all, they are not used by the females in any obvious way.

6.30 EYES AND OCELLI

Parasitic wasps have a pair of compound eyes and three ocelli sited on the top of the head on a differentiated triangular area called the stemmaticum. Much work has been carried out on insect vision in the honey bee, but very little work specifically on parasitic wasps, though vision is important in host location and escape in many species. Eyes and ocelli of nocturnal species tend to be relatively large (see below) and there are often differences between sexes, with the compound eyes of males being significantly larger than those of females. In contrast, eyes of a few taxa such as the termitophilous ypsistocerine Braconidae and various other hypogeic taxa are very reduced, though never absent. In a few extreme cases, ocelli may be totally lost, as in the diapriid, *Platymischus dilatatus*, a parasitoid of seaweed flies. Other examples are found in the males of some agaonid fig wasps, *Mellitobia* morphs (Eulophidae), and in several parasitic aculeates (e.g. some bethylids, tiphiids and mutillids).

Little is known about the spectral sensitivities of parasitic wasp eyes. *Nasonia* females have been shown to be sensitive to wavelengths in the red end of the spectrum by their diapause response to day length (Saunders, 1975), but it is not known whether this is due to compound eye or ocellar perception, or even some other internal photosensitive system.

6.31 OTHER SENSE ORGANS

Tarsal sensilla are also strongly implicated in chemoreception in some parasitic wasps (Salt, 1937; Klomp *et al.*, 1980; Nettles *et al.*, 1982), just as they are in many better studied insect groups. Hays and Vinson (1971) showed that tarsal receptors in the braconid, *Cardiochiles nigriceps*, are important for helping the wasp to orientate on the host prior to oviposition. Possible chemosensory tarsal sensilla have been illustrated in the trichogrammatid, *Trichogramma minutum*, by Schmidt and Smith (1987) and in the braconid, *Microplitis croceipes*, by Navasero and Elzen (1991); however, the long, socketed hairs they found lacked obvious pores and so their role in chemoreception must remain speculative at present. In general, it appears that palps and tarsi have fewer receptor types than antennae. For example, Navasero and Elzen found eight distinct sensilla types on the antenna of *M. croceipes*, with just four on the tarsi and only two on the palps. Tarsal chemo- and mechanoreceptors have been described in the eulophid, *Sympiesis* (Meyhöfer *et al.*, in press). The claws possess both types of receptor, whereas the manubrium only has mechanoreceptors. The membranous aroleum has no sensilla, but by inflating and deflating it, the wasp could manipulate the degree of contact between the other tarsal sensilla and the substrate, which may be important for detecting signals from and so locating its leafminer host.

The legs of all adult insects have specialized mechanoreceptors that are responsible for detecting vibrations. These are campaniform sensilla and the subgenual organ. The subgenual organ is located at the proximal end of the tibia, which is quite often distinctly swollen at this point. The organ usually comprises a bundle of from three to 50 scolopidia, each of which is composed of a sensory cell, a scolopale cell, an attachment cell and an auxiliary cell (Menzel and Tautz, 1994). In most insects the organ is essentially 'hammock-shaped' but in the aculeate Hymenoptera investigated to date it is more spherical and there is a large extra-cellular cavity filled with a mucopolysaccharide, probably gel-like substance. Despite the fact that substrate vibrations may be particularly important for many parasitic wasps, especially those that attack endophytic larvae, there appear to have been no studies of either the gross structure, ultrastructure or function of the subgenual organ in any sawflies or parasitic Hymenoptera. Given that the structure in bees, wasps and ants is unlike that found in other insect orders, investigation in parasitic wasps may be considered long overdue.

Parasitic wasps and other insects monitor the relative positions of various appendages and other body parts by means of hairplates – dense patches of small (often 1–3 μm long) mechanosensory sensilla on a surface that closely apposes the cuticle of another structure. These have been studied in only a few parasitic wasps. Virtually the whole of the Apocrita have a hairplate on the base of the 1st metasomal tergite (van Achterberg, 1977b). This is no doubt associated with sensory feedback about the angle of the metasoma relative to the mesosoma. Hairplates of the trichogrammatid, *Trichogramma minutum*, have been described in detail by Schmidt and Smith (1985a, 1987). In this small wasp, each coxa has two or three separate hairplates, each trochanter has two small hairplates (comprising only two sensilla), and others occur on the back of the head, neck, pedicellus and scapus of the antenna, and on the thorax associated with the basal wing sclerites. In addition to postural control, hairplates in many insects, including the honey bee, have been shown to be important in gravity detection (especially those in the neck region) and it seems likely that they play a similar role in parasitic wasps. In *Trichogramma* species that assess the size of host eggs through their curvature, the angle of the antenna with respect to the head may be important (Schmidt and Smith, 1985b) and it is likely that this would be detected by the hairplates on the basal antennal segments. Hairplates have also been described on the ovipositor/sting system of various Hymenoptera (Hermann and Douglas, 1976; van Achterberg, 1977b; Le Ralec and Wajnberg, 1990), and are probably associated with detecting movement and relative position of the various components of the ovipositor system.

Insect wings are furnished with numerous sensory structures. Schmidt and Smith (1985a) provide a detailed description of sensilla associated with the wings of the parasitic wasp, *Trichogramma minuta*. In trichogrammatids and other chalcidoids, wing venation is highly reduced, but the remaining veins are well supplied with sensilla comprising campaniform sensilla and innervated hairs. These are located on and near the wing veins, within which the axons run.

6.32 ALIMENTARY CANAL

The alimentary canal of adult parasitic wasps is essentially similar to that of many other insects, comprising fore- and hindguts of ectodermal origin and a midgut of mesodermal origin. Most work has been limited to light microscopy and traditional histological techniques and the most detailed investigations to date have dealt with larger species, notably the braconids, *Stenobracon deesae* (Mashhood Alam, 1954) and *Cardiochiles nigriceps* (Vinson, 1969). The fore- and hindguts correspondingly have a thin chitinous cuticular lining, the midgut does not have a cuticle but the cells of its wall are protected from direct contact with the gut contents by a type I peritrophic membrane – a membrane consisting of a concentric set of chitinous lamellae secreted by the midgut cells (Wigglesworth, 1953). According to Wigglesworth the type I peritrophic membrane in insects is secreted by cells along the length of the midgut. In the Hymenoptera, most of the studied species are aculeates (e.g. *Apis*, *Vespa*), and Mashhood Alam (1954) appears to be the only worker to have studied this process in a parasitic wasp, the braconid, *Stenobracon deesae*, in any great detail. He found that, whilst Wigglesworth's statement is probably largely true, the midgut cells at the first folds following the stomodeal (foregut) valve are particularly active in this respect.

The highly muscular proventriculus, with its typical triangular lumen in cross-section, is a valve that limits flow from the foregut to the midgut. As in many insects, the chitinous wall of the proventriculus is quite thick and is formed into posteriorly directed processes. Mashhood Alam writes about its being used to crush food particles, but it is unlikely that many parasitic wasps, including the *Stenobracon* he worked on, eat much solid food, if any (Chapter 8). The midgut epithelium comprises typical uninuclear columnar cells with brush-borders overlying multinucleate 'regenerative' cells which Mashhood Alam believes to be responsible for epithelium cell replacement.

The adult rectum of Hymenoptera and various other insects is typically furnished with several rectal pads (papillae) which are believed to function in water resorption and mineral balance. The number of rectal pads varies greatly from taxon to taxon. The plesiomorphic number is believed to be six, based on other insects and on Togashi's (1965) study of sawfly rectal pads. Most proctotrupoids have four and most chalcidoids two but only one is found in some *Elasmus* and *Eurytoma* species (Table 6.2). At the other extreme are rhyssine ichneumonids (*Megarhyssa*) which can have up to 120 rectal pads. No parasitic wasps have yet been found that completely lack rectal pads, as is the case in some bees (*Bombus* and *Psithyrus*) and siricid sawflies (unpublished observation). It seems likely that the considerable variation in rectal pad number and size in adult parasitic Hymenoptera may be correlated with water balance requirements, as found in ants by Hood and Tschinkel (1990), but this has not yet been investigated. The histology of the parasitic wasp rectal pads has been studied with light microscopy by Palm (1949), Mashood Alam (1954), Bahadur and Reddy (1966) and Vinson (1969). Electron microscopy appears to have been applied only to the rectal pads of the pteromalid, *Nasonia vitripennis*: Davies and King (1975) showed that each pad comprises four cell types arranged around a hollow core. The presence of a large number of mitochondria indicates that they are the site of active transport.

A very unusual modification of the rectum has been found in males of some braconids (subfamily Braconinae) in which it is fused to the 9th metasomal tergum, apparently forming a sort of 'pseudogland' which would allow volatiles produced in the gut to escape to the outside (Quicke *et al.*, 1996).

6.33 DESTINATION OF DIETARY HEAVY METAL CONTAMINANTS

Concern over pollution and perhaps especially the effects of nuclear contamination led to a number of studies during the 1950s and 1960s in which radioactive isotopes of heavy metals were fed to adult parasitic wasps, or more specifically to the braconid, *Habrobracon*, and this work has been summarized by Grosch (1988). Different heavy metals, as revealed by the localization of their radioisotopes, displayed different tissue specificities. Cobalt and chemically similar nickel tended to become concentrated in the wasp's

Table 6.2 Rectal pad numbers in parasitic wasp families (mostly original data)

Superfamily/Family (number of taxa examined)	Modal number of rectal pads and range	Superfamily/Family (number of taxa examined)	Modal number of rectal pads and range
Orussidae (1)	6	Vanhorniidae (1)	4
Trigonalyidae (3)	6	Monomachidae (1)	4
Stephanidae (3)	4	Pelecinidae (1)	3
Gasteruptiidae (4)	3	Chalcidoidea (20)	2 (1–2)
Aulacidae (2)	2 and 7	Ichneumonidae (140)	6 (2–120)
Evaniidae (3)	6	Braconidae (40)	4 (2–*c*.20)
Ceraphronidae (1)	2	Bethylidae (1)	4
Megaspilidae (2)	2	Chrysididae (13)	6 (3–6)†
Ibaliidae (1)	2	Embolemidae (1)	2
Cynipidae (1)	2	Plumariidae (1)	4
Figitidae* (5)	2	Sclerogibbidae (1)	6
Diapriidae (2)	2	Mutilidae (3)	6 (3–6)
Proctotrupidae (3)	4	Tiphiidae (3)	6
Heloridae (4)	4	Sierolomorphidae (1)	6
Roproniidae (1)	4	Sapygidae (3)	6 (5–6)

*   Including Charipidae and Eucoilidae.
†   All taxa examined have 6 except for *Loboscelidea*.

venom apparatus, with some cobalt also accumulating in the fat body. Zinc tended to become associated with the gut and Malpighian tubules, with the remainder mostly in the urate cells of the fat body. Cobalt and nickel had small to negligible effects on egg production and hatchability, but zinc was found to depress egg production significantly. Details of plutonium localization are not available but, as with the radio-active zinc, radiation effects led to reduced hatchability (see also section 10.25).

6.34 PIGMENTS AND COLORATION

Virtually no work has been done on the body pigments of adult parasitic wasps, nor much on the coloration of adult social wasps (Zeigler and Harmsen, 1969). Leclercq (1951) briefly surveyed the chemistry of yellow pigments across the Hymenoptera using the Ford (1947) chemical test. Pterins were identified in a number of ichneumonids and in the Leucospidae (see section 5.14), but were apparently absent from, or only present in minute quantities in, Aulacidae, Trigonalyidae, Stephanidae, Evaniidae, Pelecinidae, Cynipidae and Chrysidoidea. Unfortunately few details were provided about the actual taxa examined. However, Leclercq's preliminary survey suggests that pigment chemistry may be of some phylogenetic significance: it would probably be worth examining it in more detail and using more sensitive and discriminatory test procedures. The yellow pterin pigment in vespid wasps is present in special epidermal cells, and the bright red pigment of many large braconids (e.g. *Iphiaulax* and *Digonogastra* spp.) is likewise situated in subcuticular cells, though their chemistry is yet to be determined. The existence of a range of eye colour mutants in some wasps, such as *Habrobracon*, has prompted a number of studies on their chemistry. The dark pigments are ommochromes and are derived from the amino acid, tryptophan. In some taxa, white urate deposits can be seen through the abdominal cuticle in underlying fat body cells (Grosch, 1988), the amount present varying in accordance with adult diet.

Brown, black and some other reds are due to melanin-group pigments, and these are deposited within the cuticle itself. As in many other insects, the extent of melanization, in at least some parasitic wasps, is strongly influenced by temperature (Génieys, 1925; Liu and Carver, 1982). In *Habrobracon*, for example, maintaining pupae at elevated temperatures can lead to almost entirely testaceous (yellowish) wasps, although the black pattern is least affected at sites of muscle attachment.

Colour polymorphism is little studied in parasitic wasps. Huddleston (1975), in revising the Afrotropical ichneumonid genus *Encardia*, decided that 10 species whose original descriptions were based largely or solely on colour differences, particularly wing colour patterns, were in fact synonyms of just one: *E. picta*, which was highly variable for wing pattern. Quicke (1989) noted that in another Afrotropical species, the braconid, *Archibracon servillei*, wing colour pattern seemed to show some geographical variation, but it remained uncertain whether these variations reflected intraspecific variation rather than the existence of a sibling species complex. No such geographical trends were observed by Huddleston for E. *picta*. Seasonal variation in coloration is not uncommon, and Askew (1961) described and discussed this for a number of chalcidoids.

Quicke (1984) considered the possible significance of white-tipped ovipositor sheaths in the Braconinae. In this braconid subfamily, many species, collectively representing a number of genera, display this coloration feature, as do members of several other wasp families – for example, the Gasteruptiidae and Stephanidae. The relative lengths of white coloration and the relative lengths of the black basal region are positively correlated with relative ovipositor length across species, and these observations are consistent both with an attack deflection hypothesis and as a visual feedback mechanism such that the white coloration will help the wasp to see better the position of its ovipositor.

6.35 CIRCULATORY SYSTEMS

Almost nothing has been published concerning the heart and other circulatory pumps of adult parasitic wasps and only a very few anatomical investigations have provided descriptions of these systems. Mashood Alam's (1954) study of the braconid, *Stenobracon deesae*, is one of the exceptions. Krenn and Pass (1995) recently provided a survey of the wing circulatory pulsatile organ, sited under the scutellum, in various Holometabola. Of the ten hymenopterans investigated, three were parasitic taxa. Considerable variation was found, and it seems likely that this structure will prove to be a valuable system for future phylogenetic work. Its development may also be related to size, as Krenn and Pass failed to find a circulatory organ in the tiny *Trichogramma* investigated.

6.36 SKELETOMUSCULAR SYSTEM

There have been numerous investigations of the comparative musculature of the parasitic Hymenoptera. These reflect interest in both functional anatomy and in phylogeny, the latter studies tending to be more or less comprehensive surveys of particular systems across a wide range of taxa.

Interpretation of the groundplan for morphological features among parasitic wasps usually involves comparison with sawflies, and in particular with members of the Xiphydriidae or Orussidae, which are considered to be particularly closely related to the Apocrita (Chapter 11). The morphology of the head of non-aculeate parasitic Hymenoptera has been considered in detail by Matsuda (1965) and for several parasitic and other aculeates by Osten (1982). Various aspects of mesosomal musculature in parasitic wasps have been studied in detail by Maki (1938), Daly (1963), Matsuda (1970), Gibson (1985, 1986b), Johnson (1988) and Heraty *et al.*

(1994); and that of the anterior of the abdomen by Short (1959), though this work deals with only one parasitic species, the ichneumonid, *Rhyssa persuasoria*. Tonapi (1958) presented a survey of spiracles, their closing mechanisms and associated muscles for 75 species of Hymenoptera, of which 18 were parasitoids. Daly's study pointed out the presence of three types of muscle in the Hymenoptera: closely packed (microfibrillar), fibrillar and tubular. The first two have the contractile element cylindrical and all the nuclei are scattered along their length just beneath the sarcolemma. In tubular muscles, the nuclei occupy an axial core and the thin, lamellae-like contractile elements radiate from these. Within the Hymenoptera, Daly found closely packed muscles in the sawfly families Cephidae and Siricidae, but as his survey covers relatively few taxa it is not possible to determine whether this represents the groundplan for the Hymenoptera or a synapomorphy for these two families. Although Daly's study focused on closely packed and fibrillar muscles, he mentions that two muscles (at least in terms of position) are tubular in one thoracic segment and closely packed or fibrillar in the other; thus they could be homologous.

There have been few detailed investigations into the structure of skeleto-muscular ultrastructure in the Hymenoptera. Isidoro and Bin (1994) presented an ultrastructural study of the preocellar pit in the scelionid genus, *Basalis*, showing that this externally visible cuticular modification was associated with elongated (and microtubule-rich) epidermal cells and was in fact a suspension structure for the brain rather than a gland, as had been supposed previously by some workers. The ultrastructure of musculo-epidermis and epidermal–cuticular connections appears to have been studied in only one parasitic wasp, the eulophid, *Pediobius foveolatus* (Chapman and Hooker, 1990). Their study of musculo-epidermal connections within the metasoma revealed many differences with apparently equivalent junctions in other insects. In particular, the *Pediobius* musculo-epidermal junctions lacked interdigitations, secondary myofilament associations at the desmosomes, and microtubule insertions into the epidermal desmosomes and hemidesmosomes, whilst the epidermal–cuticular junctions lacked indentations and filamentous projections into the cuticle. Chapman and Hooker were cautious as to whether all of these differences from the standard pattern reflected genuine differences between parasitic wasps and other insects. Certainly, more studies are needed to see how representative their findings are.

6.37 FERRITIN

Ferritin is an important iron-storage protein in vertebrates and invertebrates. It is widely known now that many insects possess intracellular ferritin deposits and it has been postulated that these are also involved in a magnetic sense. Among the Hymenoptera, ferritin has been shown to occur in honey bees and ants, and also in the eulophid, *Pediobius foveolatus*

(Chapman and Hooker, 1994). In the latter, ferritin crystals, 0.2–5 μm long by 7.8–14.7 nm wide, were found in the rough endoplasmic reticulum and vacuolar systems of the metasomal fat cells and thoracic muscles. Whether the ferritin in this eulophid, which presumably does not have to navigate in the same way as bees and ants do, has any role other than iron storage remains to be discovered.

6.38 WINGS

The Hymenoptera primitively have two pairs of wings, but in flight these act as single pair because of the order's unique wing coupling system. This system comprises a fold along part of the posterior margin of the forewing and a corresponding line of hooks, the hamuli, along the anterior margin of the hindwing (Figure 6.12a,b). These features are universal among the winged species of parasitic wasps, with the possible exception of some Mymaridae, and to an even greater extent the Mymarommatidae, in which the wings are very reduced with little remaining membrane, and in some of which the hindwing is reduced to a short stalk. The hamuli in many ichneumonids form two groups, but in braconids, the basal group is represented by simple setiform structures rather than S-shaped hooks, and so in braconids these probably serve simply to support the postero-basal margin of the forewing during the upstroke. The number of hamuli varies considerably, from two, or typically three, in most chalcidoids and proctotrupoids, up to more than 20 in some ichneumonoids and trigonalyids. The presence of just three is often associated with a small group of apposed setae (Figure 6.12a), and this may be of phylogenetic significance (Basibuyuk and Quicke, in press). In other taxa, the number of hamuli appears to be largely size-dependent (Quicke, 1981), but there have been no detailed comparative studies such as those on aculeates (Richards, 1949).

Parasitic Hymenoptera display a great deal of variation in the complexity of their wings and wing venation. Most sawflies have relatively complex venation, but this is reduced in almost all parasitic taxa, sometimes to an extreme degree wherein no veins remain, as in some platygastrids. Rasnitsyn (1980) points to the fact that the evolution of the Apocrita has been associated with a general narrowing of the wings and in particular with a reduction in the anal (posterior) part. All sawflies, with the exception of the Cephoidea, have a specialized rough patch in the anal region of the forewing which interlocks with modified patches on the metanotum – the cenchri – when the wings are folded at rest (Schrott, 1986). This structure has been lost in all parasitic Hymenoptera except for the parasitic orussid sawflies. It has been suggested that this loss allows parasitic wasps to take to flight more quickly than can the sawflies, though why this should have been particularly advantageous for parasitic taxa is not clear. Relatively complex venation is found, for example, in

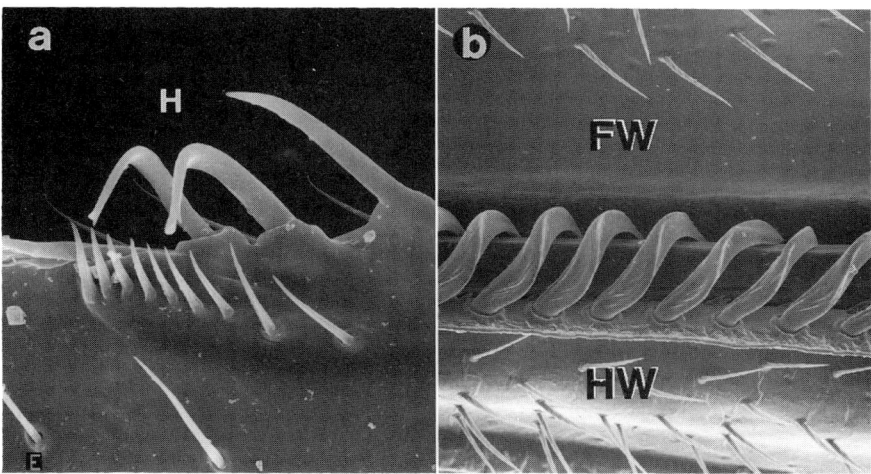

Figure 6.12 Wing coupling system: (a) hamuli (H) of left hindwing of chalcidoid showing row of apposed setae; (b) hamuli (H) interlocked with fold of posterior margin of forewing (FW) in the social aculeate, *Vespula vulgaris*. HW, hindwing. (Courtesy of H. Basibuyuk.)

Trigonalyoidea and Ichneumonoidea, but many of the 'microhymenoptera' (Chalcidoidea, Platygastroidea, Ceraphronoidea and Proctotrupoidea) have wing venation reduced to just a single main longitudinal vein along or close to the anterior wing margin. Reduction in wing venation is correlated with small size, but this is probably not the only explanation and it may also reflect phylogeny, at least to an extent. In the smallest taxa, such as mymarids and mymarommatids, wing membrane area is often very reduced; the setae on the distal and posterior wing margins become far more prominent (Figure 6.12d), and in extreme cases the hindwing may be reduced to a simple stalk, making the wasps effectively dipterous. Feathery wings are a widespread adaptation to flight in tiny insects, also being found in various small beetles, flies and thrips. However, as pointed out by Danforth (1989), these are just at the extreme end of a continuum. In contrast, in many larger Hymenoptera, the distal wing membrane has papilliform microsculpture that may be associated with retaining a boundary layer. The feathery borders of microhymenopteran wings has been suggested as serving to prevent flow separation and so allow a high angle of attack without stalling. In addition, there can be no doubt that many setae are also sensory.

Many groups show trends in wing venation such as a relative migration of veins towards the wing base and/or a reduction of the veins in the distal part of the wings. Tobias (1993) noted that throughout several groups of Hymenoptera, including the Ichneumonoidea and various aculeates, the distribution of venation was correlated with the climatic zone, with those wasps from arid habitats such as the steppe having more basally concentrated venation. Tobias concluded that this may reflect the relative windiness and exposure of these habitats in comparison with woodlands, for example. However, the distribution of venation is just one of several features of wing morphology that shows allometric variation with the overall insect size over a wide range of hymenopteran taxa (Danforth, 1989), other features being the aspect ratio, the relative position of the centroid of the wing and the relative size of the pterostigma, and so size may be a confounding factor. Interestingly, nocturnal Hymenoptera, including parasitic taxa, often show anomalous wing features for their size with relatively larger pterostigmas, low aspect ratios and distal centroids, which are all more characteristic of smaller taxa. Whether this represents an adaptive response to night flying or is a consequence of the nocturnal species having recently evolved to be larger than close diurnal relatives but not yet having evolved wing features to match their size is not clear and would benefit from further work making use of independently derived phylogenies.

In addition to veins, wasp wings usually have a series of flexion lines that are important for controlling the three-dimensional contortions of the wing in flight. Where these pass across veins, the latter are reduced, often having special striate hinge zones called bullae or fenestrae. A few taxa are capable of folding their forewings longitudinally when at rest – among the parasitic Hymenoptera this ability is restricted to the Leucospidae and Gasteruptiidae. The mechanisms involved were described by Danforth and Michener (1988). Hindwings of some ichneumonids and chalcidoids have a number of conspicuous erect setae which may be sensory or may help to maintain wing alignment when the wings are folded (Hennessey, 1981).

Wing venation plays an important role in Hymenoptera taxonomy and classification. In many taxa, intraspecific variation is rare and small venational differences can often provide reliable distinguishing features. Occasional teratological specimens occur in most groups, often possessing fewer than or more than the usual venational complement. Some authors have considered these as possible indications of ancestral venation patterns, perhaps through the re-expressions of normally masked genes (e.g. Tobias and Belokobyl'skij, 1983). However, such interpretations are far from universally accepted.

The terminology of hymenopteran wing venation has been the source of much confusion and consternation amongst novices and even experts. Several different nomenclatural systems have been widely and occasionally inconsistently used in the past. The paths of the various principal longitudinal veins have been worked out in detail by Kukalova-Peck (1991) but it should be realized that this is only one possible interpretation and is strongly influenced by her studies of fossil insect wings. Examination of more extant material may necessitate amendments (Fitton, personal communication). Other studies have considered the homologies of the veins in various other groups (Burks, 1938).

Wings are furnished with numerous sense organs, including trichoid and campaniform sensilla. The former are often especially noticeable along longitudinal veins, the latter often near vein junctions.

6.39 FLIGHT

Flight in all the Hymenoptera studied to date is asynchronous and myogenic, as it is in the Diptera and Coleoptera; that is, the power for the wing movements come from the myogenic contractions of the indirect thoracic musculature (Pringle, 1968). In asynchronous muscle, each contraction is not associated with a particular nervous impulse and the muscles may produce multiple contractions as a result of a single excitatory nerve stimulus. So in these insects, nervous control serves to turn the flight motor on or off. In antagonistic muscle pairs, as in the flight motor, contraction of one member causes extension of the other, which in turn causes an increase in its tension, and so on.

The mechanics and neurophysiology of flight have been intensively investigated in some of the larger aculeates, such as the honey bee, and it is assumed that the basics also apply to the larger parasitic taxa though little work on these has been carried out directly. Flight mechanics have been investigated particularly in some of the smaller taxa, which are among the smallest of all flying insects. In larger insects (from the size of *Drosophila* and upwards), with Reynolds numbers of 100 or more, the wings are large enough to generate turbulence and so to generate lift. However, many parasitic wasps are far smaller than a *Drosophila*, and may have Reynold's numbers of about 10 or 20. Weis-Fogh (1973) investigated hovering flight in the aphelinid, *Encarsia formosa*, and found it to involve essentially similar wing movements to those seen in larger insects. He proposed that in these small wasps lift was generated by a clap-and-fling mechanism in which the wings touch closely at the top of the up-stroke and then the fore margins of the forewings separate rapidly, causing a small vortex which provides most of the lift. The aerodynamics of flight in *E. formosa* was described in greater theoretical detail by Ellington (1975). In some small aquatic parasitic wasps, the wings are used for swimming rather than flight – for example, in the mymarid, *Caraphractus cinctus* (Jackson, 1966).

6.40 BRACHYPTERY AND APTERY

Many groups of parasitic wasps contain at least a few species that are either apterous or brachypterous at least in one sex. Non-aculeate groups in which aptery or brachyptery is known to occur are the Diapriidae,

Braconidae and Ichneumonidae, and many chalcidoid families. Aptery is particularly prevalent in the parasitic Aculeata, with females being wingless in many species of Tiphiidae, Mutillidae, Scolebythidae, Sierolomorphidae, Plumariidae, Bethylidae, Dryinidae and Embolemidae (Brothers and Carpenter, 1993). Phoretic copulation occurs in the Tiphiidae (Thynninae) and may be the main means of dispersal for these groups. Outside of the parasitic Aculeata, aptery is often associated with a hypogaeic life (Darling, 1991) or with phoresy (Naumann and Reid, 1990), or other cases in which host-searching does not require flight, such as on tree trunks (Gauld and Fitton, 1987).

Aptery is also frequently associated with particular niches that involve high levels of inbreeding, as is the case, for example, with male pollinating fig wasps (Figure 6.15). In a number of species that routinely show high levels of sib-mating, males have reduced wings or may be completely apterous – for example, the much studied pteromalid, *Nasonia vitripennis*. Other well-known examples occur among and the fig wasps, Agaonidae. In males of some fig wasp species there seems to be continuous variation in wing development; in others, males are polymorphic with some being normal-looking and fully winged and others highly modified and apterous (Figure 6.16). The winged males leave the fig to mate elsewhere; their wingless brothers are specialized for combat with other males, the victor securing matings within the fig (Hamilton, 1979).

Some species are polymorphic for wing length. This polymorphism may be either genetically inherited or cyclical as part of a complex life cycle in which some generations are winged and others wingless or apterous. Cyclical polymorphism is well known among the gall wasps (Cynipidae) but also occurs, for example, in the ichneumonid, *Sphecophaga vesparum*, a parasitoid of various social vespid wasps (Donovan, 1991). In this species brachypterous females emerge from the flimsy white cocoons within about two weeks of pupation, whereas fully winged males and females emerge either from intermediate cocoons or from robust yellow overwintering cocoons. Among chalcids, similar strict cyclical brachyptery and sexuality appears to be known only in the eurytomid, *Tetramesa maderae* (Flanders, 1950). Salt (1937, 1941) described a particularly interesting example in which males of the trichogrammatid, *Trichogramma semblidis*, hatched as fully winged individuals if they developed in an egg of the moth, but were apterous if reared in alderfly (*Sialis* sp.) eggs.

Both males and females are polymorphic for wing length in the pteromalids, *Choetospila elegans* and *Melittobia chalybii* (Howard and Liang, 1993). In *Melittobia*, the morph that develops depends entirely on the density of larvae developing on the host rather than on food quality (Freeman and Ittyeipe, 1982). Wing polymorphism has on occasion led to the same species being described twice, once from the winged and once from the brachypterous morph (Darling and Hanson, 1986).

Wing shedding occurs in a small number of parasitic wasps, notably in females of the phoretic mantid egg parasitoid, *Mantibaria manticida*. In this scelionid, once a female has located a female mantid, she secretes herself in the basal hindwing folds and removes her wings. In the pteromalid, *Bairamlia nidicola*, a parasitoid of the bird flea, *Ceratophyllus gallinae*, females often appear to bite off their own wings, and Graham (1969) suggests that this unusual behaviour may facilitate moving about within the bird nests where their hosts occur. Wing shedding also occurs in pollinating fig wasp females (Agaonidae) when they enter a new fig syconium through the small terminal ostiole. Some also lose their antennae at this stage.

Brachypterous and apterous species are often difficult to identify, not only because they lack wing characters but also because of the very considerable reduction in the size and complexity of the thorax that comes with a loss of the wing musculature and associated endoskeleton, and perhaps a real increase in leg musculature, resulting in marked structural convergence between even distantly related taxa (Reid, 1941).

6.41 GROOMING STRUCTURES AND BEHAVIOUR

All Hymenoptera use the forelegs to groom themselves, especially their antennae, and the fore tibial spur and basitarsus are usually especially modified for this purpose, forming a structure referred to as the antenna cleaner or strigil. The inner apical margin of the hind tibia is also frequently furnished with a dense patch of fine setae or a comb-like structure, also undoubtedly involved in grooming behaviour. The structure of the antenna grooming apparatus was surveyed across the Hymenoptera by Gennerich (1922), for mutillids and tiphiids (as well as for ants) by Schönitzer and Lawitzky (1987), and more recently by Basibuyuk and Quicke (1994, 1995). Considerable modifications were found in various groups, some of which are illustrated in Figure 6.13. In all Apocrita, the antenna cleaner consists of a comb of erect, closely-spaced setae situated in a more or less concave depression at the base of the fore basitarsus, and a typically curved fore tibial spur, or calcar, that may bear a similar comb, or a lamellum or a combination of these. The same structures are found in orussids, but in other sawflies the basitarsus is not notched and it lacks the comb, though in cephids and anaxyelids there is a row of more widely spaced, short erect setae which may represent an early stage in the development of the apocritan-type antenna cleaner. A highly specialized group of scale-like setae at the apex of the fore tibia in some trichogrammatids has been described by Hung (1990), who suggested that they may be involved in spreading secretions from metasomal glands, perhaps on to the wasp's wings, but there is no direct evidence to support this.

Grooming behaviours in the Hymenoptera appear to be archetypal fixed action patterns. Thelen and Farish (1977) investigated grooming behaviour

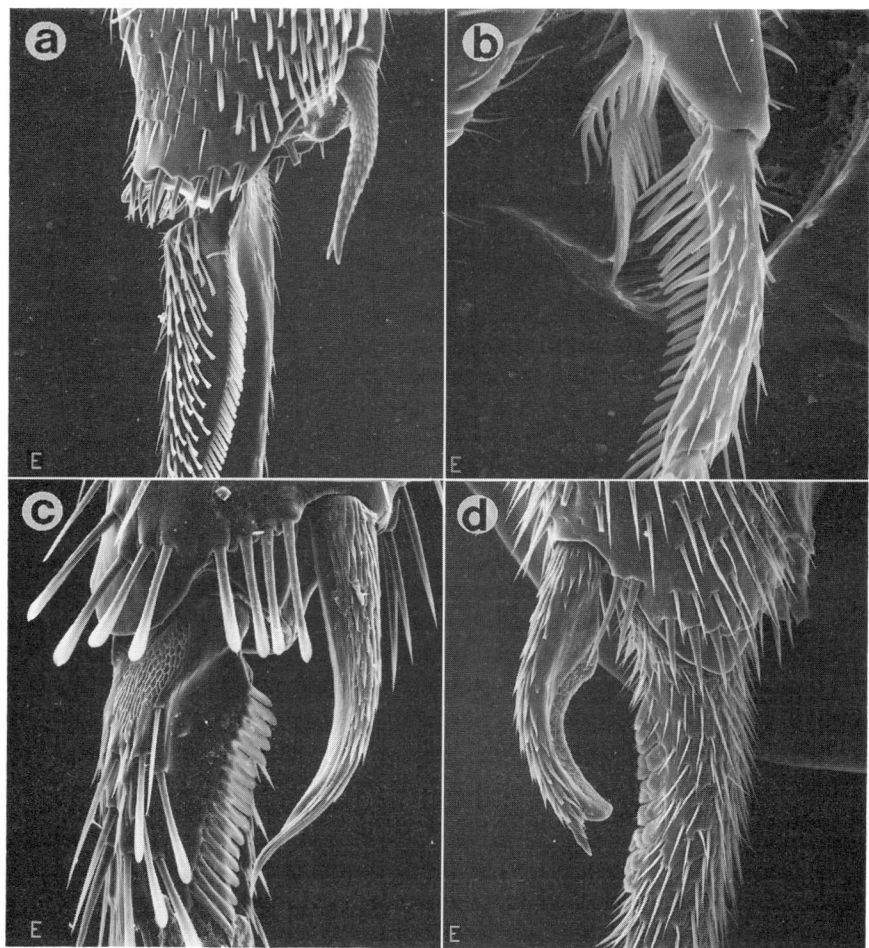

Figure 6.13 Antenna cleaner (modified fore tibial spur and basitarsus): (a) *Orussus* sp. ('Symphyta': Orussidae); (b) *Ceraphron* sp. (Ceraphronidae); (c) *Blepharotelloides* (Eurytomidae); (d) *Poecilogonalos* (Trigonalyidae). (Courtesy of H. Basibuyuk.)

in a number of mutant strains of the braconid, *Habrobracon hebetor*, and found no differences compared with wild-type stock, again illustrating the deeply fixed nature of the underlying neural circuitry. However, they did find that females and males have different grooming 'programmes', with females spending more time grooming the head and antenna, and males concentrating more on their wings. Removal of the part being groomed does not seem to affect the behaviour and sensory feedback from the groomed part does not appear to be important in controlling the fundamental movements. For example, Thelen and Farish found that *Habrobracon* individuals would make essentially the same antenna-cleaning movements even if they have been antennectomized, and vestigial-winged mutants still perform normal-looking wing grooming actions.

Farish (1972) provided the first survey of grooming behaviour across the order, and showed that a number of motor patterns appeared to be characteristic of various superfamilies. However, his data were not taken up by subsequent phylogenetic studies, perhaps partly at least because of the view that behavioural characters are less reliable indicators of relationships than are morphological ones. Further, Farish's survey missed a considerable number of major taxa, making it difficult to judge fully the phylogenetic significance of grooming behaviour. Basibuyuk and Quicke (in preparation) have extended Farish's study by investigating grooming behaviour in many of the taxa omitted from the original study. Chalcidoids, including members of the Mymaridae, together with the Scelionidae, Megaspilidae and Diapriidae, frequently groom a single antenna using both the ipsilateral and contralateral antenna cleaner simultaneously, and thus may represent a synapomorphy for the 'microhymenoptera'.

6.42 EXOCRINE SCENT GLANDS

A wide variety of exocrine glands has been discovered in parasitic Hymenoptera, but only in a very few cases is there any clear evidence concerning their functions or chemistry. Metasomal exocrine glands have been most thoroughly investigated in larger species, especially in the Braconidae, in which they are usually present only in males (Williams *et al.*, 1988; Buckingham and Sharkey, 1988; Quicke, 1991). The majority of these glands seem to be of a fairly general type referred to as Type III in the classification system of Noirot and Quennedey (1974; 1981) in which the secretory cell is separated from the cuticle by an epithelial cell (Figure 6.14). The intracellular apparatus often snakes about within the secretory cell and comprises a chitinous 'duct' structure surrounded by microvilli – the 'end apparatus'. At its proximal end, the intracellular duct comprises a porous, fibrillar rod without a lumen (Figure 6.14c, left). As it progresses distally, a lumen develops and an incomplete outer solid layer becomes apparent (Figure 6.14b). Further on, the ductule's wall becomes continuous and the fibrillar structure is lost (Figure 6.14c, right). Several of these are dealt with in detail elsewhere, notably in sections 8.27 and 6.44.

All Hymenoptera appear to have glands associated with their mandibles. These have been investigated in detail in many aculeates but their existence

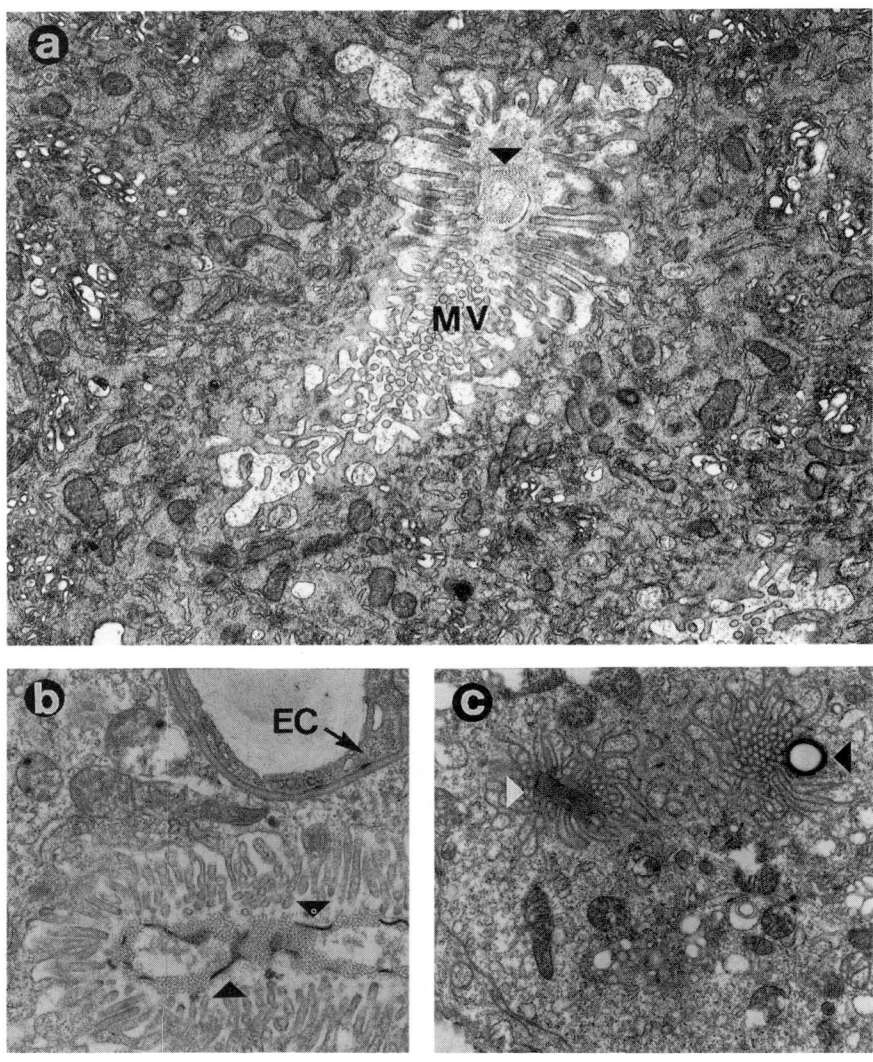

Figure 6.14 Transmission electron micrographs of Type III exocrine scent gland of the braconid, *Atanycolus ulmicola*, showing the chitinous end apparatus: (a) numerous microvilli (MV) surrounding fibrillar 'ductule' (arrow); (b) development of ductule lumen and network of more electron-dense supports (arrows), and the epithelial cell (EC) separating secretory cell from cuticle; (c) two sections through same ductule showing on left the proximal end before lumen development (white arrow), and on right the completely solid wall of the more distal part (black arrow). (Courtesy of R.A. Wharton and H. Sittertz-Bakhtar.)

and functions in the parasitic taxa have barely been touched upon (Davies and Madden, 1985; Völkl *et al.*, 1994; Jones, 1996; see below). Matthews *et al.* (1979) have described a probable metasomal gland and brush-like applicator in males of the ichneumonid *Megarhyssa*. In these, males apply their secretions to areas where other *Megarhyssa* are emerging, apparently irrespective of the sex of the emerging wasp. Because of the apparent lack of sexual discrimination, Matthews *et al.* tentatively ruled out territory marking as a possible explanation, but the precise role is uncertain. Similar metasomal glands, some with brush-like evaporatory surfaces, have been described in many braconids (Buckingham and Sharkey, 1988).

6.43 ANTENNAL GLANDS

Antennal intertwining is a common part of the mating behaviour in many parasitic wasps, and it has also been known for some time that males of several groups of parasitic wasps have specialized antennal glands. Evidence is now accumulating that these play an important role in courtship and mating. Glands are known in two chalcidoid families – Eulophidae (Dahms, 1984b) and Aphelinidae (Pedata *et al.*, (1993)/1995) – and in both families of Platygastroidea, the Platygastridae and Scelionidae (Bin and Vinson, 1986; Isidoro and Bin, 1995).

The antennae of male and some female scelionids and also those of some chalcidoids have glands that are associated with courtship and mating (Isidoro and Bin, 1995; Isidoro *et al.*, 1996), and it is likely that these will be found to be more widespread. The positions of these glands on the antennae of males and females are probably correlated with the ways that the courting wasps touch or intertwine them (see also section 6.24).

6.44 DEFENSIVE ODOURS

In addition to glands that are directly associated with mate location, courtship and mating, a number of parasitic wasps are known to emit more or less pungent odours that are derived either from special discrete glands or in some cases may be produced by numerous scattered gland units. Townes (1939) reported that several ichneumonids give off strong odours when caught and handled. Most odoriferous species, including members of the genera *Alexeter*, *Apechthis*, *Banchus*, *Ephialtes*, *Exochus*, *Phaeogenes*, *Pimpla*, *Mesoleius* and *Megarhyssa*, produce a very similar smell that reminded Townes of burnt machine oil or the smell of some carabid beetles. These belong to several subfamilies, indicating multiple origins of the feature. Townes proposed that some were models in mimicry systems and others may belong to Müllerian complexes. One genus, *Chlorolycorina*, apparently smells like lemon verbena. To date the sources of these aromas have not been determined except for some preliminary data on

Megarhyssa and its relative, *Rhyssa*. Matthews *et al.* (1979) noted that the strong smell appeared to originate from the head, and Davies and Madden (1985) have shown that the odours are produced by the mandibular glands and, in these genera, may be involved in mate location and aggregation as well as defence. Some of the chemicals involved have been identified – one of them, 6-methylhept-5-en-2-one, was previously known to be involved in the defence and alarm systems of some ants. Davies and Madden proposed that the secretory products that are specific to *M. nortoni* may help to repel *R. persuasoria* from male aggregations at potential female emergence sites, but why any interspecific repellent should be needed was not explained. Townes interpreted the smells of these ichneumonids as probably having a defensive role and suggested that they may be important in protecting the ichneumonids from attack by vespid wasps. Townes further suggested that in some cases these odours may have caused their emitters to become the models in either Batesian or Müllerian mimicry systems. In some stephanids the head is also the source of an odour that may be defensive (Davies and Madden, 1985), but the nature of it is unknown. The eucoilid, *Alloxysta brevis*, a hyperparasitoid of aphids, produces chemicals that are repellent to the ants that attend colonies of their aphid secondary hosts (Völkl *et al.*, 1994). The secretion originates in mandibular glands and also contains 6-methyl hept-5-en-2-one, as well as actinidin, and several unidentified iridoids. This defensive secretion enables *Alloxysta* females to forage for hosts among ant-attended colonies with impunity.

In the Braconidae, a few taxa produce odours that are noticeable to humans. The small North American braconid, *Paradelius rubra*, produces a strong, choking, formic acid-like odour (Whitfield, 1988) which seems almost certainly to be defensive. In contrast, males of several Opiinae produce a variety of sweet-smelling secretions from tergal glands that have become known as Hagen's glands after their discoverer (Buckingham and Sharkey, 1988). Apparently homologous tergal glands are found in the closely related Alysiinae, whilst a number of other subfamilies have other types of tergal and sternal glands. The chemical secretions of these glands in three opiine species was investigated by Williams *et al.* (1988), who showed them to be the source of several different species-specific compounds. Unfortunately, the functions of these remain largely unknown, though they have been intimated either to have a defensive role (Buckingham, 1975) or to function in courtship (Williams *et al.*, 1988). Members of both sexes of virtually all members of the Braconinae have a pair of eversible glands located between the terga and sterna of the first two metasomal segments (Quicke *et al.* in preparation). These appear to be the source of a distinctive woody but not unpleasant odour. The function of this odour is again uncertain; it does not appear to make the wasps unpalatable to lizards.

Defensive odours are also produced by some parasitic aculeates. The mutillid, *Dasymutilla occidentalis*, produces a defensive secretion whose principal constituent is 4-methyl-3-heptanone (Schmidt and Blum, 1977). Experiments with potential vertebrate and invertebrate predators of the velvet ant showed this compound to have a marked, though not ubiquitous, deterrent effect.

An interesting phenomenon that has been attributed to secretions from exocrine glands in some parasitic Hymenoptera is that of autonarcosis. The term was introduced by Doutt (1951) to cover his observations of the encyrtid, *Tropidophyrne melvillei*, which became immobilized after being kept at a fairly high density in a small tube. Within the tube they produced a coconut-like odour, and *Drosophila* introduced into the same tube were similarly affected. Price (1970) noted an apparently similar effect in some ichneumonids. It is possible the wasps were stimulated enough by each others' agitated movements that they all released their normal repellent/defensive compounds, and that in the confined space of the specimen tube the concentrations of these built up to toxic levels.

6.45 CUTICULAR HYDROCARBONS AND CHEMICAL MIMICRY

Cuticular hydrocarbons (CHCs) have been systematically investigated in only a few parasitic wasps (Howard and Liang, 1993), although they have been employed for species differentiation in rather more cases (e.g. Espelie *et al.*, 1990). In the pteromalid, *Chaetospila elegans*, Howard and Liang compared CHC profiles between males and females of different wing length morphs, and between individuals reared on different host species. The same major components were present in all wasp groups though some differences existed between the sexes and between wing length morphotypes. In the latter case, the authors were unable to tell whether the differences reflected the amount of wing cuticle present or whole body differences. The main CHC components were also the same as occurred in the cuticle of the host beetle larvae, though minor components were markedly different. Virtually no correspondence was found when Howard (1992) compared the CHCs of two bethylids, *Laelius utilis* and *Cephalonomia gallicola*, with those of their dermestid beetle hosts. Indeed, considerable intraspecific differences existed between the various developmental stages as well as between adult sexes. These results showed that these wasps do not simply transfer CHCs from their hosts to their own bodies (section 5.12).

Interesting mimetic adaptations have evolved in some braconids and eucharitids that are associated with ants. In the case of the aphidiine braconids whose hosts are aphids, it is the ants that protect the aphids that cause problems for the wasps. Ants rely heavily on their chemical senses for recognition of nest mates and food. Thus, for a parasitic wasp to be able to infiltrate the ant-protected aphid colony it must smell like its aphid hosts. This mimicry probably acts as a strong restraining factor in the host range of these aphidiines. In one case, the aphidiine appears to be olfactorily neutral as far as the ants are concerned. Thus, *Lysiphlebus cardui* are ignored by ants (*Lasius niger*) tending its host (*Aphis fabae*), whereas another aphidiine, *Trioxys angelicae*, is attacked. Hexane-washed *T. angelicae* are ignored by the ants, but when the washings are applied to the previously immune *L. cardui* the latter become targets for ant attacks (Liepert and Dettner, 1993). Thus the *Trioxys* possess cuticular hydrocarbons that stimulate an aggressive response, at least in this ant species. In contrast to the aphidiines, most eucharatids are parasitoids of ants themselves, but the wasp does not sting them directly, relying instead on their planidial larvae to find a suitable ant nest. Thus it is the juvenile stages of the wasp that face the problem of recognition and destruction by the ants. Vander Meer *et al.* (1989) investigated the cuticular hydrocarbons of an *Orasaema* species that parasitizes the fire ant, *Solenopsis invicta*. All stages of the wasp contained some of the major ant cuticular hydrocarbons, but only the vulnerable wasp pupa had them in both large quantities and similar proportions to the ants. Vander Meer *et al.* suggested that the compounds were most probably acquired from the ant hosts rather than being synthesized *de novo*.

6.46 ADULT DEFENSIVE STRUCTURES

Relatively little has been written about defensive strategies and structures of adult parasitic wasps. Gauld and Gaston (1994) drew attention to some conspicuous developments found in a number of relatively large tropical species of Ichneumonoidea that probably serve to provide protection against both vertebrate predators such as birds and lizards and invertebrate predators such as ants. Thus the large spines on the meso- and metasomas of some braconids (e.g. *Batotheca*, *Spinaria*, *Trispinaria*, *Physaraia*) probably render them difficult for birds to handle – it is notable that these wasps are heavily sclerotized compared with many braconids of similar size. Likewise the poison claws of some *Neotheronia* and *Xanthopimpla* species and the putative urticating hairs of some other pimpline ichneumonids can be quite painful (Hanson and Gauld, 1995); and in *Neotheronia*, at least, the claws seem to have a tendency to break off, which may give more time for poison to get into the victim. Because no detailed studies have been made of the poison gland in these species, it is not known whether there is any special muscular pumping mechanism involved. In contrast, the pronotal flanges of *Hymenoepimecis*, a pimpline parasitoid of adult spiders, may help to protect it from its undoubtedly dangerous hosts as well as ants.

It is well known that females of many of the larger parasitic wasps, such as ophionine ichneumonids, are capable of stinging people who handle them. These stings, whilst only occasionally being really painful, are nevertheless often very irritating and the effects, including erythemia, can persist for a considerable time after the initial sting (personal observations). It is probably this ability, perhaps combined with a degree of distastefulness in some species, that has led to many of the tropical parasitic wasps, particularly the larger braconids and ichneumonids, being involved in mimicry complexes. The dominant colour patterns that are found in the tropics are usually distinctive of particular zoogeographical regions (Mason, 1964). That the wasps themselves are actually the models in some cases is convincingly demonstrated by several of their mimics which have evolved ways of imitating the typical parasitic wasp ovipositor. Preston-Mafham and Preston-Mafham (1993) have provided a beautiful illustration of a South American bug which holds its hindlegs together straight out behind its abdomen, and some moths have evolved a pseudo-ovipositor.

Stinging is an important part of the defence mechanism of various parasitic aculeates, including some very small ones such as the African bethylid, *Scleroderma wollastoni* (Walton, 1948). Schmidt and Blum (1977) have detailed the defensive repertoire of a typical large mutillid, *Dasymutilla occidentalis*. This, like many other velvet ants, is aposematically coloured and can deliver an extremely painful sting – the sting can be protracted more than 1 cm. In addition, it runs fast, has a very tough cuticle, stridulates when disturbed and produces a defensive odour (see below). Very few ants, spiders or vertebrates are able to cause injury to the velvet ant before giving up. The combination of resilience and a powerful sting together with distinctive odours and sound production is suggestive of aide memoire 'mimicry' (Rothschild, 1984) in which additional signals (for example, memorable sounds, smells and behaviours) serve to stimulate the memory response of a potential predator so that it is reminded of a past unpleasant encounter before it has an opportunity to cause the prey any harm.

Males of some aculeate and non-aculeate Hymenoptera often show pseudo-stinging behaviour in mimicry of the female, which can often sting (Evans and Eberhard, 1970). The effect is enhanced in some by the presence in the males of a pseudo-sting, a pronged structure which resembles the true sting of a female. This may be derived from the last sternite or from the parameres of the genitalia. Such structures are largely restricted to the Aculeata, and they are known in the Mutillidae, Scoliidae and Tiphiidae (all members of the Vespoidea) as well as in various non-parasitic groups. In conjunction with the pseudo-stinging metasomal probing, the effect can be very convincing.

Many chrysidids are well known for their ability to roll up into a near-spherical, heavily sclerotized ball when threatened. This is facilitated by the ventrally concave, carapace-like metasoma.

6.47 MODIFICATIONS OF THE MOUTHPARTS

Several taxa possess specialized mouthparts for reaching floral nectaries, particular those of Compositae. For example, in many agathidine braconids the lower part of the face is elongate; in other agathidines and some braconines, it is the labio-maxilliary complex that is very elongate; in some taxa both the face and the labiomaxilliary complex are produced. A different modification for feeding from flowers involves the maxillary palps, which can be extremely elongate and have their medial faces concave such that they form a long feeding tube like a drinking-straw when they are apposed. Maxillary drinking tubes are found in several braconids belonging to at least three separate subfamilies, namely a few *Bracon* species (Braconinae): for example, *B. anthracinus*, some *Cardiochiles* species (Cardiochilinae), and the agathidine, *Aenigmostomus* (Sharkey and Mason, 1986). Collectively these features, whilst they are widespread, seem to be particularly frequent among parasitoids that inhabit arid or semi-arid regions, where Compositae are probably an important nectar source.

Perhaps the most important role of mandibles in many parasitic wasps is in helping the wasp to escape from its pupation site (see below). Frequently this is within a cocoon made by the wasp itself, but sometimes the wasp may then have to excavate a way out from a burrow deep in wood, or in some cases from within the hardened puparium of its host fly. The cuticle of the cutting edge of the mandibles of most such wasps is hardened by incorporation of high concentrations (up to 10% wt/wt) of zinc (Quicke *et al.*, in preparation).

Some groups have highly enlarged mandibles – for example, the braconid genera *Cosmophorus*, *Proclithrophorus*, *Pseudodicrogenium* and *Gnathobracon* (Quicke and Huddleston, 1991). The mandibles of female pollinating fig wasps (Agaonidae) are highly modified, bearing either an articulated lamellate appendage (Blastophaginae) or a denticulate one (Agaoninae) (Wiebes, 1982; Ramírez, 1991). These are believed to be involved in the female gaining entry into a fig through the syconium.

6.48 ADAPTATIONS FOR GRASPING AND MANIPULATING HOSTS

A lot of parasitic wasps use their normal, unmodified legs to hold and restrain small hosts during oviposition (e.g. Allen, 1990) but in a number of diverse cases particular morphological host-grasping devices have evolved. In these cases, the hosts are often adult insects or especially mobile nymphs that pose particular handling problems. These grasping mechanisms are sometimes employed in conjunction with paralysing venoms but appear in other cases to provide an alternative. One of the most conspicuous examples occurs in dryinine dryinids, whose females are furnished with chelate fore tarsi – the telotarsus (claw-bearing segment) being highly produced and toothed, and opposing an articulated, hypertrophied claw (Clausen, 1940). These chelae are used to grab the host in a variety of ways, and in some species are used to lift the plant hopper host off the substrate during oviposition (Waloff, 1974). In aphelopine dryinids that lack this foreleg modification, the mid- and hindlegs are used for the same purpose (Gauld and Bolton, 1988). Members of the euphorine braconid genus *Streblocera* have highly modified, raptorial antennae that are presumed to be involved in holding the host during oviposition but no observations on their use have ever been made (Shaw, 1985). Apparently raptorial forelegs are found in various neoneurine braconids which are endoparasitoids of adult ants, but limited field observations of one such species provided no support for this assumption (Shaw, 1993).

The putative euphorine braconid genus, *Cosmophorus*, is an endoparasitoid of adult bark beetles (Scolytidae). It has a cyclostome mouth with large mandibles, and it has been seen to use these to grasp the host during oviposition (Seitner and Nötzl, 1925; Shaw, 1985). Mandibles are also used for host-grasping by some bethylids (e.g. *Goniozus*: Gordh, 1976) though in these the mandibles are not especially modified. Many bethylids show a high degree of parental investment and may parallel or surpass some non-parasitoid aculeates in the sophistication of their host handling. Some other bethylids are known to drag their host moth or beetle larvae into concealed situations (Finnamore and Gauld, 1995). At least one, *Laelius pedatus*, uses its mandibles to shave the long setae of its dermestid beetle larval host so as to facilitate the feeding of its larva, and another African species 'slings' its host beetle larva over its back to carry it to the oviposition site (Mick Day, personal communication).

In several disparate groups the hindlegs are strongly modified. This is particularly noticeable in the Chalcididae, members of which often have greatly enlarged hind femora that are commonly furnished with spines. Cowan (1979) studied the use of the hindlegs in ovipositing females of *Chalcis canadensis* and showed that they were used both for support of the wasp whilst it manipulated host fly eggs with its other legs and for defending batches of host eggs against parasitization by other *Chalcis* females. In the latter case the hindlegs may be used to grasp an opponent or held outstretched in an apparently aggressive posture. In other chalcidids the legs are involved in grasping prey such as maggots. In *Lasiochalcidia igiliensis* they are used to hold the jaws of the host ant lion larva apart while the wasp oviposits into it (Steffan, 1961).

Within the braconid subfamily Aphidiinae, a small group of genera (for example, *Trioxys* and *Binodoxys*) have a highly modified apex to the metasoma in females which helps the wasps to grasp their aphid hosts. In *Trioxys*, which typically attacks relatively early nymphal stages, the metasomal prongs serve to facilitate oviposition by holding the hosts (Schlinger

Adult morphology and adaptations

and Hall, 1961). Members of another aphidiine genus, *Monoctonus*, use their forelegs to hold their hosts still, without any apparent structural modification (Griffiths, 1960).

The parasitic aculeates do not (with perhaps very few exceptions) use the ovipositor proper to 'drill' through substrate to reach the host. Those whose hosts are protected by cocoons or in cells constructed by wasps, like many chrysidoids, use their mandibles to make a hole through which the telescopic ovipositor can be inserted. Yamada (1991) noted a correlation within the Chrysididae between whether the posterior margin of the 3rd (terminal) segment of the metasomal carapace is dentate and the nature of the closure of the host larva's cell. Those whose host bees and wasps closed their cells with a non-sticky material, such as sand, lacked the metasomal teeth, whereas those whose host cells are sealed with resin, mud or silk do have teeth. Yamada then experimentally removed the metasomal teeth of *Praestochrysis shanghaiensis* and found that it greatly increased the time required for host-cell inspection, perforation and oviposition. The metasomal teeth were thus shown to be important for helping the chrysidid to gain purchase so that she can manipulate the host enclosure efficiently.

6.49 ADAPTATIONS ASSOCIATED WITH EMERGENCE

Most parasitoids have to escape from their cocoons upon emergence, and in many instances they may then have to find an exit from the chamber where they had been cocooned – for example, from within a host burrow or from a host pupal case or puparium. Several adaptations have evolved to help facilitate this escape. Perhaps the commonest involve escape from the cocoon and this usually entails biting through it with the mandibles. Some parasites that pupate within dipteran puparia sometimes make use of the dehiscent cap to break out, but others chew their way through. Rotheray (1981b) showed that the ichneumonid, *Diplazon pectoratorius*, uses its mandibles to cut a complex series of semicircular strips from the host hoverfly's puparium until it has made a sufficiently large opening to escape through. The mandibles of Diplazontinae are more or less tridentate and so specialized with respect to those of most other ichneumonids. This may be associated with their particular mode of egress.

One metopiine ichneumonid is believed to escape from its host lepidopteran cocoon by means of an oral secretion which softens or dissolves the fibres (M.G. Fitton, personal communication). It is quite likely that other koinobionts that emerge within tough host cocoons may use a similar secretion to facilitate egress, but there appear to be no published observations on this.

An unusual condition which appears to have evolved on only a handful of occasions is the possession of exodont mandibles. By definition these do not meet when they are closed and in operation they are direct-ed outwards. The condition is most frequently associated with members of the braconid wasp subfamily, Alysiinae, in which it is ubiquitous, but it occurs sporadically in other groups of Hymenoptera, notably in the ich-neumonid genus, *Idiogramma*, and in the Vanhorniidae, and in at least two genera of eulophids, *Exastichus odontus* and some *Omphale* species and is also found in a few opiine (genus *Exodontiella*) and ichneutine (genus *Anaprixia*) braconids. In the case of the Alysiinae and *Exodontiella*, the exodont condition is associated with their pupation site, which is within the puparium of their host fly. Their mandibles are used rather like saws to cut their way out of the puparium.

6.50 ADAPTATIONS TO PARASITISM OF HOSTS WITHIN WOOD

Fergusson (1990) and Eggleton (1989) have examined in some detail the suit of adult characters associated with parasitization of hosts dwelling deep within wood. Some of these are concerned with oviposition; others are involved in escape from the host substratum by the newly emerged host. The adaptations associated with attacking hosts deep in wood can be divided into two quite distinct sets. Firstly there is the problem of reach-ing the host for oviposition, and secondly the perhaps rather more diffi-cult problem of egressing from the host's burrow once development has been completed. These adaptations are summarized in Table 6.3.

The mesonotum is frequently furnished with strong transverse ridges which are believed to enable the wasp to gain purchase on the sides of its burrow and hence to escape from its emergence site. Such ridges are found in, among others, orussids, ibaliids, liopterids, stephanids, rhyssine and some labenine ichneumonids, doryctine braconids and pteromalids (Figure 6.15b–d) (Eggleton, 1989). The presence of this suite of characters has in the past led to quite unrelated wasps being classified together; for example, the large Neotropical pteromalid, *Leptofoenus*, was originally described in the Evanioidea (LaSalle and Stage, 1985). Parasitoids of wood-boring hosts do not always manage to get out: Beaver (1966), for example, noted that if the host of the eulophid, *Entedon leucogramma*, an ovo-larval parasitoid of scolytid beetles, pupates below the outer bark layer, the wasp cannot egress.

Some small parasitoids of wood-boring insects reach their hosts by bur-rowing through the substrate themselves. This is particularly evident in the case of the small and peculiar braconid, *Histeromerus*, studied by Shaw (1995). In this, the fore tibia is furnished with a dense cluster of spines and observations of living wasps showed that they were able to use these and their mouths to bore through solid wood and, annoyingly, through the corks of tubes meant to house them.

Table 6.3 Distribution of adaptions of adult parasitic Hymenoptera to living on endoxylous hosts and egression (modified after Eggleton, 1989).

Adaptation	Distribution (not comprehensive)
Head with spines or rough bosses	Stephanidae, Orussidae, Braconidae
Mesoscutum with strong sculpture, often comprising transverse ridges	Ichneumonidae, Stephanidae, Pteromalidae, Ibaliidae, Liopteridae
Hind femur with ventral tooth	Ichneumonidae, Braconidae, Stephanidae, Liopteridae
Long thin legs	Ichneumonidae, Braconidae, Stephanidae, Pteromalidae, Ibaliidae, Liopteridae
Pegs (chaetobothria) on legs	Ichneumonidae, Braconidae
Postgenal bridge	Ichneumonidae, Braconidae, Stephanidae, Ibaliidae, Liopteridae
Flattened thorax	Ichneumonidae, Braconidae, Stephanidae, Ibaliidae, Liopteridae
Expanded, wide head	Ichneumonidae, Braconidae, Stephanidae
Small clypeus	Ichneumonidae, Braconidae, Stephanidae, Pteromalidae, Ibaliidae, Liopteridae
Clypeal tooth	Ichneumonidae, Stephanidae
Short, robust, chisel-like mandibles	Ichneumonidae, Braconidae
Extended tergites	Ichneumonidae, Braconidae, Stephanidae, Pteromanidae, Ibaliidae, Liopteridae

6.51 ADAPTATIONS FOR PARASITOIDISM OF ADULT INSECTS

Few groups of parasitic Hymenoptera have succeeded in utilizing adult insects as hosts. Three examples occur in the Braconidae. Members of the Aphidiinae oviposit in either adult or nymphal aphids, neoneurines oviposit into adult worker ants (Shaw, 1993), and most euphorines (excluding the Meteorini) are parasitoids of adult insects or nymphs of hemimetabolous ones. Among the Chalcidoidea, parasitism of adult insects occurs in the pteromalid genera, *Tomicobia, Dirhinus* and possibly *Gastracanthus* (Graham, 1969), and also in the eulophids, *Phymastichus*. All the pteromalids attack adult bark beetles, especially members of the genus *Ips* and various weevils. In the case of the Euphorinae, Shaw (1988) has proposed that a critical stage in the shift to adult hosts was parasitism of chrysomelid beetle larvae, since these frequently occupy the same micro-habitat as their pupae and adult stages. The ovipositors of the euphorine braconid tribe, Centistini, are particularly wide and knife-like, and so adapted for stabbing between the cuticular plates of their host adult beetles. Polysphinctine ichneumonids are ectoparasitoids of immature and adult spiders and thus, as with the braconids that parasitize adult insects, the transition from attacking immature hosts to attacking adult ones may have been associated with the fact that immature and mature spiders usually live in the same sorts of places.

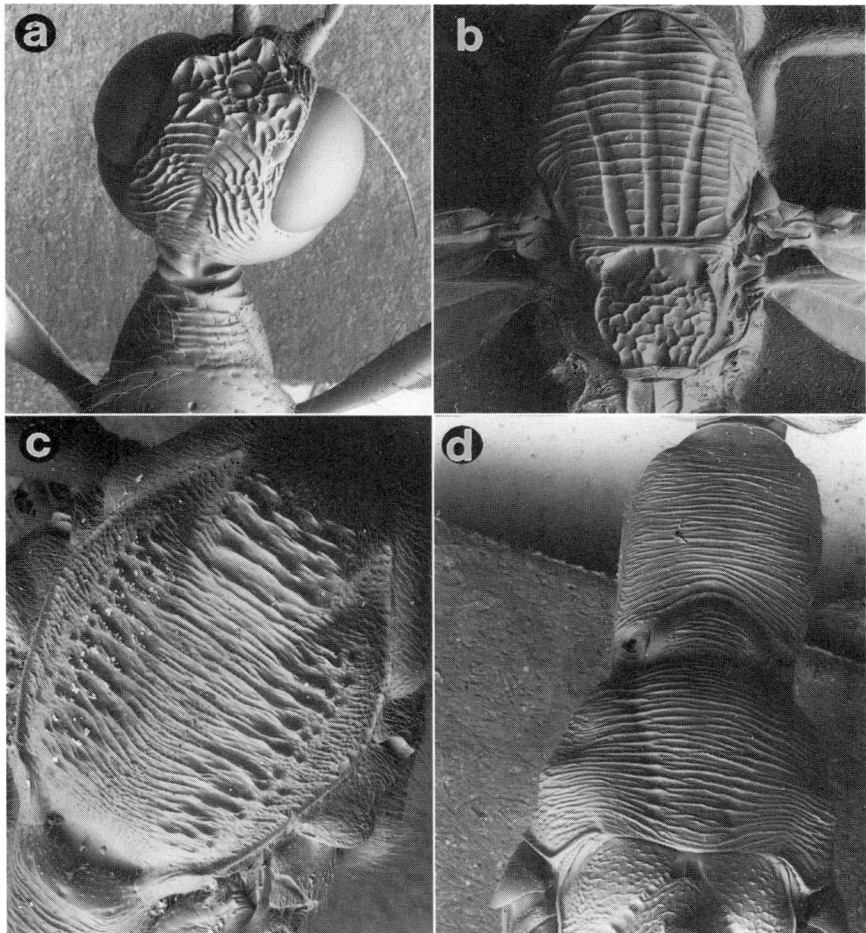

Figure 6.15 Features of parasitic wasps associated with egress from within wood: (a) head of stephanid; (b) mesoscutum of *Ibalia* (Cynipoidea; Ibaliidae); (c) mesoscutum of *Leptofoenus* sp. (Chalcidoidea: Pteromalidae); (d) mesoscutum of *Rhyssa persuasoria* (Ichneumonoidea: Ichneumonidae).

6.52 ADAPTATIONS TO NOCTURNAL ACTIVITY

A number of parasitic wasps are nocturnal, being frequently collected at ultraviolet moth traps and other lights. Some of these are typical of other members of their taxonomic groups, but many appear superficially remarkably similar, usually being a more or less uniformly yellow-brown colour. Their similarity does not stop at their coloration, but includes a number of other morphological and biological features that have been termed the 'ophionoid-facies', named after the ichneumonid subfamily, Ophioninae, many members of which are nocturnal and display the same distinctive features (Gauld and Huddleston, 1976; Huddleston and Gauld, 1988). The ophionoid-facies comprises the following features, though not all species necessarily display the entire set:

- body more or less uniformly yellowish to yellow-brown
- large ocelli
- large eyes
- long antennae
- long legs
- koinobiont
- endoparasitic
- hosts are only accessible nocturnally.

Wasps displaying the ophionoid facies are not restricted to the 'Parasitica' but are also known among the parasitic and non-parasitic Aculeata (e.g. many male Embolemidae, Mutillidae). The reason for the predominantly yellow-brown coloration of most nocturnal species is unclear but perhaps it is a physiological default, rather as many deep-sea animals are vivid red. The long legs and antennae are no doubt adaptations for host and, perhaps, predator detection. The last three features in the list tend to go together – hosts that are only available at night, such as many lepidopteran larvae that hide during the day and emerge to feed at night, would be unsuitable for attack by idiobionts while so exposed (but see section 3.12). Similarly, most koinobionts are endoparasitic, though members of the tryphonine ichneumonid genus *Netelia* are ectoparasitic koinobionts and they are a conspicuous component of the nocturnal parasitic wasp fauna throughout much of the world. Danforth (1989) noted that across the Hymenoptera as a whole, nocturnal species are larger than closely related diurnal taxa, but it is not clear whether this also applies within parasitic taxa (section 6.38).

6.53 AQUATIC PARASITIC WASPS

It has been known for some time that some parasitic wasps are partially aquatic as adults and attack the submerged eggs of dragonflies and aquatic bugs such as those of the genera *Noterus*, *Ranatra* and *Norcoris* (Rimsky Korsakov, 1917, 1933; Henricksen, 1922; Fursov and Kostyukov, 1987). A few parasitic wasps could be described as semi-aquatic; for example, the diapriid, *Trichopria popei*, parasitizes floating sciomyzid pupae, and thrusts the end of its abdomen through the surface film to lay eggs into the submerged anterior part of the pupa (O'Neill, 1973). In some more completely aquatic taxa, such as the mymarid, *Caraphractus*, and the eulophid, *Aprostocetus* (*Tetrastichus*) *nadens*, the wasps use their wings literally to fly under water (Jackson, 1958; LaSalle and Schauff, in Hanson and Gauld, 1995). Alternatively, as in the trichogrammatid, *Prestwichia*, swimming is carried out by the legs. The life history of the aquatic mymarid, *Caraphractus cinctus*, was described in considerable detail by Jackson (1958, 1966). This small wasp is a gregarious parasitoid of dytiscid beetle eggs of the genus *Agabus*, which it attacks under water – where mating also occurs, though it can also mate in the air. The wasp can break through the surface film both to enter and leave the water. The wasps seem to have no special morphological adaptations to aquatic respiration, and therefore it is presumed that gas exchange takes place across the body surface, something permitted by the wasp's very small size.

Some cryptine ichneumonids have been reared from cocoons of aquatic Lepidoptera or gyrinid beetles. The probably related ichneumonid subfamily, Agriotypinae, which comprises only the genus *Agriotypus*, is a specialist idiobiont ectoparasitoid of caddis prepupae and pupae living in fast-flowing streams. Whilst most aquatic parasitic wasps are small, since they attack host eggs, *Agriotypus* provides the exception since they are large, reaching up to 1 cm. The adult wasps themselves have slender legs and sharp claws which are believed to help them to crawl over submerged rocks in search of hosts. The larvae live inside their host caddis cases surrounded by water, so presumably respiration takes place through their thin cuticle (Fisher, 1932). However, at the prepupal stage small bubbles of gas appear that Fisher suggested could be carbon dioxide. As the gas bubble grows to fill the cocoon it presumably acts as a gas exchange organ, the surface area of which is greatly increased by the silk ribbon made by the larva before pupation. The importance of the ribbon was demonstrated by Fisher, who showed that its removal led to death of the *Agriotypus* pupa if the caddis case was submerged.

6.54 ADAPTATIONS OF FIG WASPS

Although not parasitoids, the agaonids are frequently considered together with their parasitoid relatives because many of their life history features provide interesting parallels. This group of wasps displays a remarkable set of special adaptations associated with their mutualism with figs. Females of many species have specialized pockets, called corbiculae in analogy to the functionally equivalent structures of bees. These are used for carrying pollen from the fig in which they emerge to a new one in which they will oviposit. Ramirez (1969) showed that in most *Blastophaga* spp. there are corbiculae on each fore coxa and a pair on the mesosternum, whereas only sternal ones occur in various other taxa. Corbiculae appear to be absent from some agaonids such as *B. psenes* and *Tetrapus* spp., and

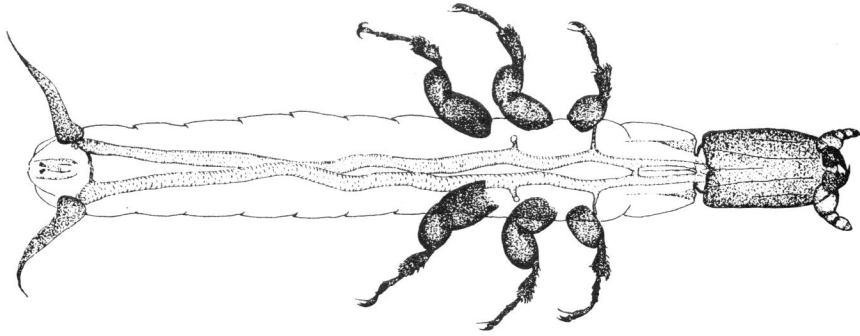

Figure 6.16 Male fig wasp, *Sycophaga cyclostigma* (Agaonidae). (Courtesy of S. Compton.)

instead the fig pollen may be carried in the digestive tract and perhaps even regurgitated.

Male pollinating fig wasps are often highly modified (Figure 6.16). They all mate with females within the fig syconium and have not only lost the power of flight but have also become highly adapted for combat with other males in competition for mates, and some authors have described their features as being paedomorphic. In some figs, males may hatch while the syconium is still filled with liquid, and so the wasps face the problem of how to respire and prevent their tracheal system from becoming waterlogged. This is achieved by means of highly developed hydrofugal setae on the peritremes (Compton and McLaren, 1989). A remarkable instance of parallel evolution has recently come to light in South American figs, in that some highly specialized males, closely resembling those of pollinating fig wasps, have now been shown to be doryctine braconids of the genus, *Psenobolus* (Shaw, in Hanson and Gauld, 1995; Ramirez and Marsh, 1996).

7

Physiological interactions of parasitic wasps and their hosts

7.1 INTRODUCTION

Parasitic Hymenoptera can cause any of a number of changes in the host insect, particularly in its development, sometimes delaying and sometimes accelerating it (e.g. Varley and Butler, 1933). Of course, many of the interactions are concerned with overcoming host defences, such as by blocking a host's encapsulation response, or by paralysing it either temporarily or permanently (Vinson, 1990). Parasitism of some hosts leads to a general reduction in their immune response against foreign objects such as parasitoid eggs, Sephadex beads or yeast particles (Streams and Greenburg, 1969; Stoltz and Guzo, 1986), and can increase susceptibility to virus infections (Kyeipoku and Kunimi, 1996), whereas in others the parasitoid's defence is much more specific. Other effects include inducing precocious development and moulting, increasing or reducing host growth rate, terminating host growth, and retarding host development. Some parasitoids may even alter a host's behaviour in ways that will advantage the wasp. These changes result from chemical, often hormonal, interactions between the parasitoid and host. The chemicals involved may originate from the female parasitoid at the time of oviposition in the form of venoms and other secretions, or they may be derived from the parasitoid's egg, embryo or larva. All of these effects are the subject of the present chapter.

7.2 VENOMS

In this chapter the term venom is used specifically to refer to the products of the wasp's venom glands (acid glands; Chapter 6) rather than the whole mix of substances that a wasp injects into its host which may additionally include secretions of the oviducts, colleterial glands and perhaps the Dufour's glands. For the most part, the actions of these other glandular tissues are uncertain, though in the case of some Scelionidae the highly developed common oviduct gland appears to be the site of production of a 'venom' that prevents the host egg from developing any further (Strand *et al.*, 1980). A range of actions that have been attributed to parasitic wasp venoms is summarized in Table 7.1.

Venoms serve very different functions in endoparasitoids and ectoparasitoids. In most (if not all) endoparasitoids, the female wasp injects a cocktail into the host that will help to prevent it from harming the egg. Most ectoparasitoids – that is, ectoparasitic idiobionts – paralyse their hosts and so inject various paralysing agents. Venoms have been most thoroughly investigated in aculeates and only a very few studies have been carried out on venoms of non-aculeate parasitoids (Piek, 1986). Of the idiobiont ectoparasitoids, only for *Habrobracon hebetor* has any signifi-

Table 7.1 Summary of effects of parasitic wasp venoms

Gross venom effect	References
Death	Beard, 1964; Rivers *et al.*, 1993
Permanent paralysis	Tamashiro, 1971; Piek *et al.*, 1974; Skinner *et al.*, 1990
Temporary paralysis	Gurney, 1953; Olmi, 1994
Cessation of development	Bocchino and Sullivan, 1981; van Veen and van Wijk, 1987; Coudron and Brandt, in press
Reduced growth rate	Morales-Ramos *et al.*, 1995
Reduction in respiration and reduced Q_{10}	Payne, 1937; Waller, 1965; Piek and Spanjer, 1986
Blocking host ecdysis	Uematsu and Sakanoshita, 1987
Specific prevention of encapsulation	Kitano, 1982
Suppression of host thoracic gland activity	Tanaka and Vinson, 1991
In vitro uncoating and *in vivo* persistence of polydnavirus DNA	Stolz *et al.*, 1988a
Increased haemolymph free amino acids	Guerra *et al.*, 1993; Morales-Ramos *et al.*, 1995
Reduced cholinesterase activity	El-Sawaf and Zohdy, 1976
Inhibition of phenol oxidase	Willers and Lehmann-Danzinger, 1984
Change in haemocyte morphology	Rizki and Rizki, 1990b, 1991
Changes in host lipid metabolism	Rivers and Denlinger, 1995
Selective destruction of haemocytes	Rizki and Rizki, 1991
Inducing cuticular encystment	Arthur and Ewen, 1975; Ewen and Arthur, 1976

cant amount of work been carried out on the nature and mode of action of its paralysing venom (section 7.4).

Few chemical constituents of parasitic wasp venoms have been identified and even fewer have been associated with particular functions, though a few have been shown to have neurophysiological effects (section 7.5). Leluk *et al.* (1989) have carried out a general survey of venom constituents in several hymenopterans. They found that venoms of all aculeates examined (social wasps, bees and ants) contained polypeptides and low molecular weight proteins but that these were absent in the six ichneumonids investigated. The work of Skinner *et al.* (1990) on bethylid venom further supports this assertion. Leluk *et al.* also showed immunological similarities between some venom proteins of a braconid wasp (*Chelonus* sp.) and those of various aculeates, suggesting that despite their different biologies, there may be some evolutionary continuity of venom composition and perhaps functionality. It would be interesting to know whether this similarity extends to other braconids and ichneumonids since collectively these form the sister group of the Aculeata.

7.3 SPECIFICITY OF VENOM ACTION

Most paralysing venoms have effects on the host central nervous or neuromuscular systems. Because of the relative pharmacological conservancy of these systems among insects, most paralysing venoms are active against a wide range of target species. Nevertheless, several studies have revealed differences in susceptibility of different hosts that appear to reflect differences in pharmacology rather than simple size/dose effects (Beard, 1972, 1978; Drenth, 1974; Temerak, 1983). Klein and Beckage (1990) showed that two congeneric dermestid hosts of the bethylid *Laelius pedatus* were not equally susceptible to the parasitoid venom's paralysing action. Whilst 100% of *Trogoderma variabile* host larvae remained paralysed and parasitized until death, 13–39% of *T. glabrum* recovered from envenomation and completed development, although more than half of them subsequently failed to reproduce. Beard (1952) investigated whether the failure of *Habrobracon* (Braconidae) venom to paralyse unnatural hosts such as the European cornborer, *Pyrausta nubilalis*, was due to some neutralizing action of the *Pyrausta* haemolymph. However, this was not the case as haemolymph from envenomated (but unaffected) *Pyrausta* could still cause permanent paralysis of the natural host, *Galleria*.

Non-paralysing venoms have also been shown to have a range of specificities. Rivers *et al.* (1993) showed that the venom of the pteromalid, *Nasonia vitripennis*, an idiobiont endoparasitoid of Diptera puparia (though strictly ectophagous on the pupa or pharate adult), has differential toxicity (LD_{50}) to different test Diptera. The highest toxicity was found against the wasp's natural host, *Sarcophaga bullata*, with larvae, pupae and adults being equally susceptible. Non-hosts displayed various lesser degrees of susceptibility, indicating that the venom was specialized to act against the natural host. However, *Musca domestica* died too quickly, making it unsuitable for parasitization by *Nasonia*. *Nasonia* was found to be immune to its own venom but, since honey bees were susceptible, this did not reflect a general lack of effect against Hymenoptera. As might be expected of a pupal endoparasitoid, *Nasonia* venom had no paralysing activity, but it is rather surprising in this case that *Nasonia* venom has any lethal properties at all.

A wider range of action was found by Coudron and Puttler (1988) who artificially exposed a number of host and non-host species to parasitism by *Euplectrus plathypenne*, a gregarious eulophid ectoparasitoid of Lepidoptera (Noctuidae and Sphingidae) larvae. In addition to parasitism, the authors injected various hosts and non-hosts with a crude extract of the female *Euplectrus* lower reproductive tract. In natural hosts, parasitization leads to arrested development (inhibition of moulting) and both their sets of experiments showed that this was a general response of lepidopteran larvae to its venom, including both natural and factitious host species. However, very few of the tested non-lepidopteran species showed any delayed or arrested development, indicating that the *E. plathypenne* 'venom' is more or less specific for Lepidoptera larvae.

7.4 PARALYSING VENOMS

Paralysing venoms, probably because of their rapid effects, have been studied in depth in many organisms, particularly with an eye to their potential in pharmacological or pesticide science. Those of parasitic wasps, despite their high potency, have been comparatively little studied either pharmacologically or chemically compared with venoms from aculeates, let alone those of spiders, scorpions and other venomous arthropods. This probably reflects the relatively small size of most parasitic wasps and the difficulties of keeping them in culture, as well as taxonomic problems. Piek and Spanjer (1986) summarized the literature on paralysing actions of venoms for a large number of parasitic and solitary wasps.

Most idiobiont ectoparasitoids of larval stages inject a venom into the host that induces permanent paralysis. The adaptive advantage of this is obvious in that a still-mobile host potentially poses a considerable threat to survival of the developing parasitoid. Although koinobiont parasitoids allow their hosts to continue to move and develop, many inject a venom which induces a temporary paralysis and so allows the wasp better to place her egg(s) and sometimes also to host-feed.

Relatively little is known of the effects of venoms of many parasitic aculeates. Most of those that attack larval hosts are idiobionts and cause long-term paralysis; others have at most a temporary effect, and subsequently the parasitoid larva develops as a koinobiont. Permanent though sometimes incomplete paralysis is induced by various parasitic aculeates, including bethylids (Skinner *et al.*, 1990), and in secondarily parasitoid pompilids (Piek and Spanjer, 1986), not to mention the truly predatory prey-transporting pompilids, and sphecids. Gurney (1953) showed that crickets parasitized by rhopalosommatids are torpid but not totally immobile. Temporary paralysing effects are also widespread, being produced by many koinobionts and some idiobionts. Among the parasitic aculeates they have been recorded in tiphiids (Clausen *et al.*, 1927), pompilids (Grout and Brothers, 1982) and gonatopodine dryinids (Olmi, 1994).

Paralysing venoms are not restricted to primary parasitoids but are also present in some hyperparasitoids. For example, Bocchino and Sullivan (1981) describe the paralysing action of the venom of *Asaphes lucens*, a pteromalid ectoparasitoid of *Aphidius smithi*, a primary aphid endoparasitoid. Not surprisingly, observations of venom effects on endoparasitic primary hosts are seldom made.

Dowry *et al.* (1995) demonstrated that paralysing 'venoms' are not only produced by adult female wasps, and that in at least one instance larvae also produce a paralysing salivary secretion. In the case they studied, 1st instar larvae of the chalcidoid, *Eupelmus orientalis*, an ectoparasitoid of larvae of the beetle *Callosobruchus maculatus*, caused paralysis of their host in the absence of the female injecting venom. Only 1st instar larvae produced the paralysing saliva, and its effect was limited to paralysis.

7.5 NEUROBIOLOGICAL ACTIONS AND CHEMISTRY OF PARALYSING VENOMS

There has been considerable work on the neurophysiological effects of venoms of aculeate Hymenoptera, but only scant work on those of the 'Parasitica'. This is no doubt due to the small size of the majority of parasitic wasps, which makes them unattractive for obtaining venom in sufficient quantity for neurophysiological investigation, and to the consequent requirement for considerable numbers of individuals from which to extract venom. An ancillary factor is that while many aculeates can be persuaded to sting through an appropriate artificial membrane, thus allowing the collection of pure venom, such techniques are inappropriate for parasitic wasps whose venoms are not employed primarily for defence. Instead most studies have made use of either whole wasp homogenate, or homogenized whole venom apparatus.

The most thoroughly investigated venom of a parasitic wasp is that of the small braconid, *Habrobracon hebetor* (usually referred to under the generic names *Bracon* or *Microbracon* in this literature, and also often referred to as *H. juglandis*; but see also Appendix A). Beard (1952) noted that in hosts paralysed by *Habrobracon* venom, the heart and gut continue to function, showing that the venom is specifically effective against the skeletomuscular system. Beard tested several dilutions of *Habrobracon* venom and found that it remained potent at very high dilutions – he estimated that permanent paralysis could be achieved by a dilution of $1 : 2 \times 10^8$ of host haemolymph. Using a standard neurophysiological preparation, the locust leg, Walther and Rathmayer (1974) showed that the venom blocks neuromuscular transmission by acting presynaptically at the excitatory glutamatergic synapses. El-Sawaf and Zohdy (1976) showed that *Habrobracon* venom causes a reduction in acetylcholineesterase activity in *Corcyra* larvae, and they interpreted this in terms of interference with synaptic transmission and possibly neuromuscular block. The latter is highly unlikely, as excitatory neuromuscular transmission in insects is mediated by L-glutamate (and inhibitory by GABA and L-glutamate) – acetylcholine only being present in the CNS. Ligation experiments by Walther and Rathmayer provided no evidence of effects on either central nervous system neurotransmission or the target insect's sensory system. Subsequent ultrastructural investigation showed that the venom causes an accumulation of presynaptic neurotransmitter vesicles in the nerve terminal, and thus probably blocks neurotransmitter release by interfering with the fusion of the vesicles with the synaptic membrane (Walther and Rathmayer, 1983). This conclusion was supported by the electrophysiological study of Piek (1982) using the locust leg preparation which showed that *Habrobracon* venom reduces the frequency but not the amplitude of the miniature excitatory post-synaptic potentials which result from the spontaneous release of single L-glutamate neurotransmitter vesicles into the synaptic cleft. Inhibitory neurotransmission, mediated either by γ-amino-butyric acid, or by a different class of L-glutamate receptors, was not affected, indicating that the action of the *Habrobracon* toxins is specific to the vesicles in the excitatory glutamatergic system.

The relative lability of the *Habrobracon* toxins has made chemical investigation difficult (Piek and Spanjer, 1986). However, Spanjer *et al.* (1977) and Visser *et al.* (1983) have partially characterized the active principles, which appear to be two large molecular weight proteins, with molecular weights of 43.7 kDa (A-MTX) and 56.7 kDa (B-MTX). Both proteins appear to have the same general activity as the whole venom but differ in dose–response relationships. Also, in *Habrobracon hebetor* venom, Fukushima *et al.* (1990) demonstrated the presence of a novel compound – a non-cyclic homo-diterpene – but the function of this is not known.

Among the parasitic aculeates, Skinner *et al.* (1990) have partially characterized the venom components of the bethylid, *Goniozus legneri*, which causes permanent partial paralysis of various lepidopterous hosts. It was found to contain the polyamine, putrescine, the amino acid, proline, and small amount of dopamine, but the bulk of the active polar fractions could not be identified because of difficulties with purification. The presence of a polyamine is of interest because polyamine components are present also in the paralysing venoms of various spiders (Quicke, 1988; Quicke and Usherwood, 1990), and the venom of the sphecid, *Philanthus triangulum*, and in both of these they are believed to act by blocking cation channels in nerve or muscle plasma membranes. Proline, too, is neurotoxic and is also known to occur in some paralysing spider venoms. Dopamine is a major constituent of the venoms of various stinging Hymenoptera and is believed to play an important role in causing pain (Piek, 1986). While the neurophysiological actions of only a few parasitic wasp venoms have been investigated experimentally, it is possible to deduce quite a lot from observation of their actions on their hosts.

7.6 EFFECTS OF NON-PARALYSING VENOMS ON HOST DEVELOPMENT AND PHYSIOLOGY

Depending on the particular host–parasitoid complex under consideration, parasitization can lead to delays, inhibition or acceleration of host development. Delayed development with suppression of metamorphosis may be brought about either by the secretion into or on to the host of juvenile hormone (JH) (Beckage and Riddiford, 1982; Hegazi *et al.*, 1988; Beckage, 1991; Höller *et al.*, 1994a), or in some cases the parasitoid may produce an inhibitor of the enzyme, juvenile hormone esterase, which would normally reduce the levels of JH in the host, permitting continued development. The exact source of parasitic wasp JH secretions may vary from system to system, and in some it appears that the teratocytes may be the principal source (Joiner *et al.*, 1973; Zhang and Dahlman, 1989; Grossniklaus-Buergin and Lanzrein, 1990). Höller *et al.* (1994a) provided some evidence to suggest that *Dendrocerus carpenteri*, a megaspilid hyperparasitic on aphids via aphidiine braconids, uses juvenile hormone to prevent their primary host from developing beyond the onset of pupation. The megaspilid is an external parasitoid of the aphidiine, which is attacked either in its final larval instar or in its prepupal stage. Following oviposition, the host aphidiine is not immediately killed, but the venom injected by the megaspilid prevents the completion of pupation and hence prevents the aphidiine from developing a sclerotized pupal cuticle, allowing the megaspilid larva to feed unhindered.

Coudron and Puttler (1988) investigated the generality of the effects of stings from and injections of female reproductive tract extracts into natural and a wide range of factitious hosts of the ectoparasitic eulophid, *Euplectrus plathypenae*. In all natural caterpillar hosts, stings and injected extracts caused an arrest in the host's development, and had a similar effect in most of the factitious hosts. Venom often plays an important part in preventing encapsulation – sometimes in conjunction with polydnaviruses (section 7.12). Kitano (1982) showed that if the venom gland was experimentally removed from the braconid, *Cotesia glomerata*, the wasp's eggs were encapsulated by the host, *Pieris rapae*. However, encapsulation of DEAE-Sephadex A-50 particles injected into envenomated *Pieris* larvae shows that the *Cotesia* venom is not simply knocking out the caterpillar's encapsulation ability, but rather is involved in a very specific interaction between the host and parasitoid.

The venoms of many parasitoids appear to alter host metabolism, probably in ways that usually render the host a better food source (Rojas *et al.*, 1996). *Nasonia vitripennis* venom causes an increase in fat-body lipid content in its preferred host, *Sarcophaga bullata*, and this change appears to enable the host to support more developing *N. vitipennis* larvae (Rivers and Denlinger, 1995). Other, unnatural hosts significantly did not show this response, indicating a considerable degree of coevolution.

Strand (1986) has studied the pathologies caused by egg parasitoid venoms on their hosts. His work concerned two species: the scelionid, *Telenomus heliothidis*, and the trichogrammatid, *Trichogramma pretiosum*, both of which attack *Heliothis virescens*. In order to separate the effects of the parasitoid's venom from that of possible larval secretions or teratocytes (in the former case), the female wasp's oviposition sequence was interrupted after venom had been injected but before an egg had been laid – a sort of artificial pseudoparasitism. In the case of *Telenomus*, the 'venom' appears to originate from an enlarged gland on the common oviduct (Figure 6.1h), rather than the true venom gland as in most other Hymenoptera. Its action could account for only part of the observed pathology. Pseudoparasitized eggs failed to hatch but the tissue did not become necrotic in the way it does in parasitized ones, and Strand was able to provide evidence that necrotization was due to teratocyte secretions. With *Trichogramma*, on the other hand, pseudoparasitized hosts exhibited the same pathology as was found in parasitized ones, indicating that maternal factors were solely responsible for the observed degeneration of the host's tissues.

7.7 CHEMISTRY OF NON-PARALYSING VENOMS

Compared with paralysing venoms, even less work has been carried out on those of non-paralysing species. In general, cytological and chemical analyses have shown them to contain various proteins. Leluk and Jones (1989) used radio-labelled antibodies raised against venom proteins of the ovo-larval braconid parasitoid, *Chelonus* sp. nr *curvimaculatus*, to investigate their persistence in the host, *Trichoplusia ni* eggs. The wasp proteins were detectable and apparently stable for two days following oviposition, but were then rapidly broken down by host proteolytic enzymes just before the host emerged. Shimizu *et al.* (1993) demonstrated the presence of the aromatic amino acids and their derivatives, tyrosine, tyramine, hydroxyphenylacetic acid, tryptophan and kynurenine, in the venom reservoir and venom of the braconid, *Apanteles kariyai*, but the function of these is uncertain. Coudron and Brandt (in press) have isolated a 40 kDa protein from the venom of the eulophid, *Euplectrus comstockii*, which is responsible for the arrest of host, *Trichophisia ni*, development (see also section 7.9), and Digilio *et al.*, (in press) provide evidence that the chemical in *Aphidius* venom that causes developmental arrest is also a protein. Krishnan *et al.* (1994) have shown that the venom of a *Chelonus* sp. (Braconidae) contains an active chitinase despite the fact that the reservoir in which it is stored has a chitin lining. As *Chelonus* spp. are ovo-larval parasitoids of Lepidoptera, the function of this chitinase remains a mystery. Chemical studies on a protein in the venom of the braconid, *Chelonus* sp. nr *curvimaculatus*, showed that it has a molecular weight of 33 kDa and that under native conditions it produced a stable dimer (Jones *et al.*, 1994).

Several studies have shown that in endoparasitic koinobionts venom components play an important part in protecting the wasp's eggs and larvae from encapsulation (Kitano, 1982). Rizki and Rizki (1984, 1990a,b, 1991) have investigated in detail the way that venom components of the eucoilid, *Leptopilina heterotoma*, prevent their host *Drosophila* larvae from encapsulating the wasp's eggs. They have shown that the wasp venom contains a compound termed lamellolysin, which causes morphological changes in the host haemocytes (lamellocytes) that would otherwise encapsulate the wasp's eggs. The venom-modified lamellocytes are no longer adhesive, and assume a bipolar and ultimately thread-like shape before fragmenting. The shape change induced by the wasp factor is dependent on microtubule formation, and it can be blocked by simultaneous application of vinblastin, an inhibitor of tubulin polymerization (Rizki and Rizki, 1990b).

7.8 HOST CASTRATION AND PSEUDOPARASITISM

Host castration is the process in which the action of a parasite or pathogen inhibits development of its host's gonads. It is probably a widespread phenomenon occurring in parasitoids of both juvenile and adult hosts, and is caused by both aculeate and non-aculeate wasps. However, the mechanisms involved and the real energetic consequences remain comparatively little studied in the parasitoid Hymenoptera. The most obvious advantage of host castration to the parasitoid is that it frees for the parasitoid's immediate use those resources that a host would normally invest in gonads, a tissue that is of no use to the parasitoid itself. Godfray (1993) argued that host castration ought to be more prevalent or obvious in female than in male hosts because ovaries will nearly always (ultimately) provide a significantly larger resource than testes. However, there are a considerable number of cases in which parasitoids do indeed seem to castrate male hosts. This may be because testis development in some hosts has been better studied than ovarian bud development, and that is apparently one reason why, for example, most publications on host castration by *Ascogaster* species refer specifically to the testes, though female hosts are usually similarly affected (Brown and Reed-Larsen, 1991). Male gonadal imaginal buds are also often larger than female ones. Another hypothesis that might be appropriate in some cases is that the parasitoid could benefit from making a male host more like a female one – many female insects are larger than their conspecific males and therefore potentially offer a larger reward. This would only succeed if there was a targetable system that could convert a genetically male host to a female one, but this is quite likely. Under these circumstances, the apparent reductions in testis size that has been observed in several parasitized male insects may reflect other feminization events, and could in itself be of little significance. Feminization is well known in plant hoppers parasitized by dryinids and the degree of change can be very considerable, affecting size, coloration and genitalic morphology (Clausen, 1940). The physiology of this merits further work. Unfortunately, there have been no detailed comparative studies of growth in pseudoparasitized and unparasitized hosts.

In parasitoids of adult insects, such as many euphorine and aphidiine braconids, the profit from host castration is direct (Smith, 1952) and can have important implications for their potential role in biocontrol. Soldán and Starý (1981) showed that parasitization of aphid nymphs and adults by aphidiine braconids caused progressive ovary degeneration and embryo resorption, largely as a result of indirect actions rather than direct consumption by the aphidiine larva, which only occurs near the end of the aphidiine's development. These authors attribute the presence of well-developed ovaries in 1st instar aphid nymphs as one of the features enabling aphidiines to attack early host aphid stages, because the ovarian tissue provides a large and readily available food supply.

Brown and Kainoh (1992) showed that the braconid egg–larval parasitoid, *Ascogaster reticulatus*, causes complete castration of its larval host, the tortricid *Adoxophyes* sp. That is, in parasitized hosts no gonadial imaginal buds, either male or female, form at all. If the parasitoid egg fails to hatch or develop, as tends to happen when older host eggs are attacked, the host *Adoxophyes* continues to develop to adulthood but fails to develop any gonadial tissue, or the tissues that develop will be abnormal. In caterpillars of *Pseudaletia separata* parasitized by the micrograstrine braconid, *Cotesia kariyai*, Tanaka *et al.* (1994) have shown that the inhibition of testis growth is partly caused by polydnavirus but that complete inhibition depends on the presence of both virus and venom (Figure 7.1). However, in the *Ascogaster* system, host castration might also potentially be brought about by secretions of the wasp larvae themselves since it has been shown that larvae excised from hosts will release immunoreactive ecdysteroids into the culture medium (Brown and Reed-Larsen, 1991).

Pseudoparasitism is a term applied to failed parasitization events such that no parasite egg is laid (or it fails to develop), but the host, even though

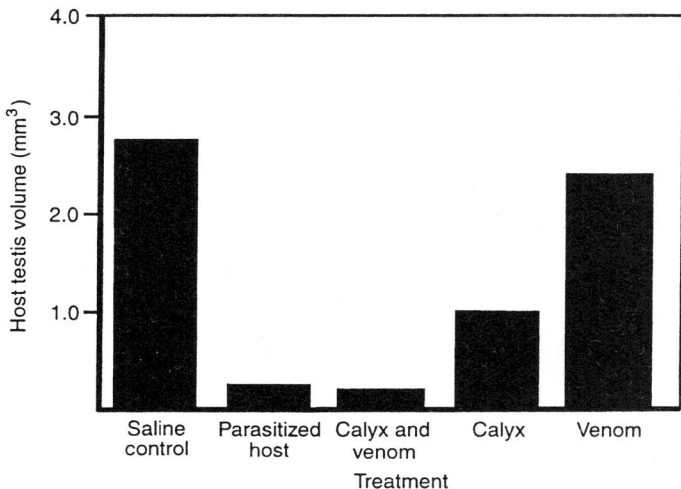

Figure 7.1 Effects of injecting various parasitoid-derived extracts from the braconid, *Cotesia kariyai*, into host moth larva, *Pseudaletia separata*, on testis development compared with development in normally parasitized hosts. (Redrawn from Tanaka *et al.*, 1994.)

it survives to adulthood, is castrated as a result of components of the parasitoid's venoms, and so cannot give rise to a subsequent generation (Jones, 1985a; Jones *et al.*, 1986a). Sometimes the term is used more generally to refer simply to situations in which a parasitoid envenomates the host but no egg is laid (Strand, 1986). Pseudoparasitism appears to be quite widely distributed among the parasitic wasps. Two well-studied cases have been described in the chelonine braconids by Reed-Larsen and Brown (1990: *Ascogaster quadridentata*) and Brown and Kainoh (1992: *A. reticulata*).

Münster-Swendsen (1994) discussed the incidence and significance of pseudoparasitism in the tortricid, *Epinota tedella*, probably due to the microgastrine braconid, *Apanteles tenellae*, in the field. Observations of reproductive potential of the host and the frequency of true parasitism showed a strong negative correlation, indicating that pseudoparasitism may be a significant factor in host population regulation. In fact when 30% of hosts were killed by parasitism, another 40% may be partially to wholly castrated through pseudoparasitism.

It is extremely difficult to see any way in which pseudoparasitism can be of selective advantage to a wasp, especially under natural conditions, and it is possible that it is simply a consequence of the normal physiological effect of the wasp's venom and represents either partially successful escape reactions by the host, or the wasp making a decision that that particular host is unsuitable for oviposition.

7.9 ENDOCRINE INTERACTIONS

Endocrine interactions are often bidirectional; that is, the hormones of the host may influence the parasitoid, and vice versa. Some parasitoids are host conformers and others are host regulators (Beckage, 1985, 1991; Lawrence, 1986, 1988b) (Table 7.2). Many instances are known in which parasitic wasp development is synchronized with that of the host. For example, parasitoid first moults are often synchronous with host moulting or metamorphosis, and parasitoid and host diapause are also commonly synchronized (Beckage, 1985). In such cases, it is presumed that the parasitic wasp larva is sensitive to hormonal changes taking place within the host. When parasitoids cause their hosts to cease development, this can be due to any of a number of actions, including inhibition of synthesis of prothoracicotrophic hormone (produced in the host insect's brain), effects on ecdysteroid production or a more general slowing of host metabolism. Experiments involving ligation of host larvae to prevent the spread of various hormones to other body parts have often shown marked effects on the development of parasitoid embryos or larvae behind the ligature (e.g. Smilowitz, 1974; Baehrecke and Strand, 1990).

Many parasitoids are known to slow the development of their hosts by secreting juvenile hormone, as has been demonstrated for the braconid, *Cotesia congregatus*, by Beckage and Riddiford (1982), and the lack of production of 20-hydroxyecdysone from ecdysone (Beckage and Templeton, 1986). Uematsu and Sakanoshita (1987) showed that small quantities of venom from the eulophid, *Euplectrus kuwanae*, were sufficient when injected into host, *Argyrogramma albostriata*, caterpillars: their subsequent moults were blocked and they died. In this parasitoid, numerous rhombic crystals could be seen in the anterior chamber of the venom reservoir, but their chemical nature was not determined. It is well known that other hymenopterous parasitoids accelerate host development, often causing premature moults or pupation and the associated behaviours such as

Table 7.2 Comparison of endocrine-conforming and endocrine-regulating host–parasitoid interactions (based on Lawrence, 1986)

Conformers	*Regulators*
• First moult obligatorily synchronized with host metamorphosis	• Moult synchrony or its absence depends on stage of host development
• Do not disrupt host development	• Disrupt host development and/or behaviour–particularly through change in juvenile hormone or ecdysteroid metabolism

pupation site-seeking and cocoon construction. There have been several suggestions as to what the causal factors in accelerated host development might be, including direct release of ecdysteroids from the parasitoid larva, release of chemicals from the parasitoid that affect the host's neuroendocrine cells, or more simply, effects of the traumatism of oviposition itself (Lawrence, 1988a). Brown *et al.* (1993) demonstrated that larvae of *Ascogaster reticulatus*, an endoparasitic koinobiont braconid, release ecdysteroids into their hosts. In another chelonine braconid, *Chelonus* sp. nr *curvimaculatus*, parasitic on *Trichplusia ni*, parasitism results in a suppression of host juvenile hormone levels and hence leads to premature pupation (Jones, 1985a, b). Jones *et al.* (1992) have shown that secretions of the female wasp in truly and pseudoparasitized hosts suppress host ecdysteroid production in the precocious prepupa – thus moulting is entirely under the control of the parasitoid.

Transplantation experiments have shown that larvae of the diplazontine ichneumonid, *Diplazon fissorius*, probably secrete something into larvae of their host hoverfly, *Epistrophe bifasciata*, so as to bring about premature ecdysis. The changes also occurred when the host larva's head was ligatured, suggesting that the compounds involved may be ecdysteroids themselves rather than messengers higher up in the endocrine pathway (Schneider, 1951). Ecdysteroid-type compounds are also apparently secreted by larvae of the braconid, *Ascogaster reticulata* (Brown and Reed-Larsen, 1991).

Lawrence (1988b) investigated the endocrine dependence of the koinobiont parasitoid, *Diachasmimorpha* (as *Biosteres*) *longicaudata*, and its tephritid fly host, *Anastrepha suspensa*. The wasp oviposits into an early instar host larva and hatches quickly, but it then remains as a 1st instar larva until the host pupates. Parasitoid larvae would hatch and remain alive in an *in vitro* system, but would not moult. However, moulting could be initiated either by transplanting into unparasitized pupal hosts or by the addition of 20-hydroxyecdysone (1 x 10^{-6} M). Lawrence's accompanying electron microscopical study showed that the failure to moult in the absence of hormone resulted at least in part from an associated inability to synthesize new cuticle.

7.10 SYMBIOTIC VIRUSES, VIRUS-LIKE PARTICLES AND FUNGI

It has been known for a long time that some parasitic wasps inject into their hosts a particulate substances derived from various parts of the female reproductive system. Some of these contain DNA and are thus referred to as viruses, while others appear to lack nucleic acids and are best referred to as virus-like particles (VLPs). The exact status of a number of such entities remains to be clarified so it is not always possible to be definite about whether or not they are true viruses. In addition to these, some parasitic wasps have yeast-like fungi associated with the female reproductive system, some of which appear to be associated with the digestion of host tissue. The parasitic sawfly, *Orussus sayi*, has numerous cylindrical particles in the well-developed vaginal (lateral accessory) gland whose nature and functions have yet to be determined (Figure 6.1g; Cooper, 1953).

The most widely studied of the parasitoid-associated viruses are found in various members of the Ichneumonoidea, and have a unique combination of biology and structure which leads them to be grouped together in the single family Polydnaviridae. These polydnaviruses will be considered in some detail in the next section. In addition to the polydnaviruses, several other types of wasp-associated virus have been described, though far less is known about their biologies and effects. In some instances, a single wasp species may be associated with two or more types of virus or virus-like particles (Stoltz and Vinson, 1979; Hamm and Styer, 1985; Stoltz *et al.*, 1988b; Hamm *et al.*, 1992).

7.11 POLYDNAVIRUSES

The importance of polydnaviruses in suppressing the host immune response was first noticed by Edson *et al.* (1981), and this opened a whole new area of physiological and molecular-genetic research. Polydnaviruses (PDV) are produced in the calyx glands, located at the posterior end of the ovaries on the lateral oviducts of the female parasitoid, and they are known to occur in two families: the Ichneumonidae and Braconidae. Within the former, they have been discovered in members of the subfamilies Campopleginae and Banchinae (Stoltz *et al.*, 1981; Cook and Stoltz, 1983; Stoltz and Whitfield, 1992), whilst in the Braconidae, they occur in the Microgastrinae, Cardiochilinae, Miracinae and Cheloninae, and possibly in some related taxa that have yet to be investigated. All the taxa of Braconidae, and most of the Ichneumonidae, that possess polydnaviruses are koinobiont endoparasitoids of Lepidoptera larvae. The four braconid subfamilies in which polydnaviruses are found belong to a putatively monophyletic group referred to as the 'microgastroid' lineage (Quicke and Achterberg, 1990), raising the possibility that their possession of polydnaviruses represents a synapomorphy indicating descent from a common ancestor. As regards the distribution of polydnaviruses in the Ichneumonidae, recent independent phylogenetic analyses (Fitton *et al.*, in preparation) show that the Campopleginae and Banchinae belong to a larger grouping, the so-called Ophioniformes, but they do not themselves appear to form a monophyletic group, though at this stage it would be unwise to rule out the possibility completely. However, in the Ichneumonidae, it certainly appears more likely that polydnavirus distribution reflects some degree of horizontal rather than longitudinal transmission. The polydnaviruses

of ichneumonids, often termed ichnoviruses, differ from those of the
braconids, or bracoviruses, in a number of ways (Table 7.3). Ichnoviruses
have two membranes, not one (Figure 7.2); they lack a cap; and the
nucleocapsids are spindle-shaped (baculiform) rather than rod-shaped
like the bracoviruses.

Despite these differences, which would normally be considered sufficient to place them in separate virus families, their biology, structure and
functions are sufficiently similar for them to be regarded as constituting a
single family of viruses, the Polydnaviridae (Stoltz *et al.*, 1984). As their
name suggests, each polydnavirus particle contains several different circular, double-stranded DNA molecules. For example, in the case of polydnaviruses of the braconid, *Cotesia congregata*, each virion contains between
15 and 20 circular DNA molecules (Beckage *et al.*, 1994), whilst different
species of *Chelonus* have between 12 and 15 size-classes of DNA molecules
(Jones *et al.*, 1986b). The number of DNA loops in each virus is variable,
even within a species, and the exact DNA types present in a single virus
vary. However, as vast numbers of polydnaviruses are injected into a host,
the inter-particle variability probably makes no significant difference to
the total effect. In addition to the DNA, each virus particle contains a number of distinct proteins.

Although polydnaviruses are produced in the cells of the calyx gland of
the female parasitic wasp, they do not express their genes within her.
Further, it is not known for certain whether the viruses themselves actually replicate in the calyx cells or whether they are each produced directly from copies of their genes that are integrated within the calyx cell
nuclear genome. Once in the host insect, the virus genes are expressed but
the viruses do not replicate in the host (Stoltz, 1993).

Table 7.3 Differences between ichnoviruses and bracoviruses

Ichnoviruses*	*Bracoviruses*
• Nucleocapsids fusiform (or spherical)	• Nucleocapsids cylindrical
• Nucleocapsids surrounded by two lipid bilayer membranes	• Nucleocapsids, individually or in groups, surrounded by a single lipid bilayer membrane†
• Activity seems to be largely independent of wasp venom	• Activity appears to depend on simultaneous presence of wasp venom

*   Currently placed in the genus *Polydnavirus*; no generic name has been proposed for
    braconid polydnaviruses (Stolz *et al.*, 1988b).
†   Stolz and Whitfield (1992) have proposed that the presence of multiple nucleocapsids
    per envelope in some Cotesiini (tribe of microgastrine braconids) compared with only a
    single one in others is a phylogenetically informative character for these braconids.

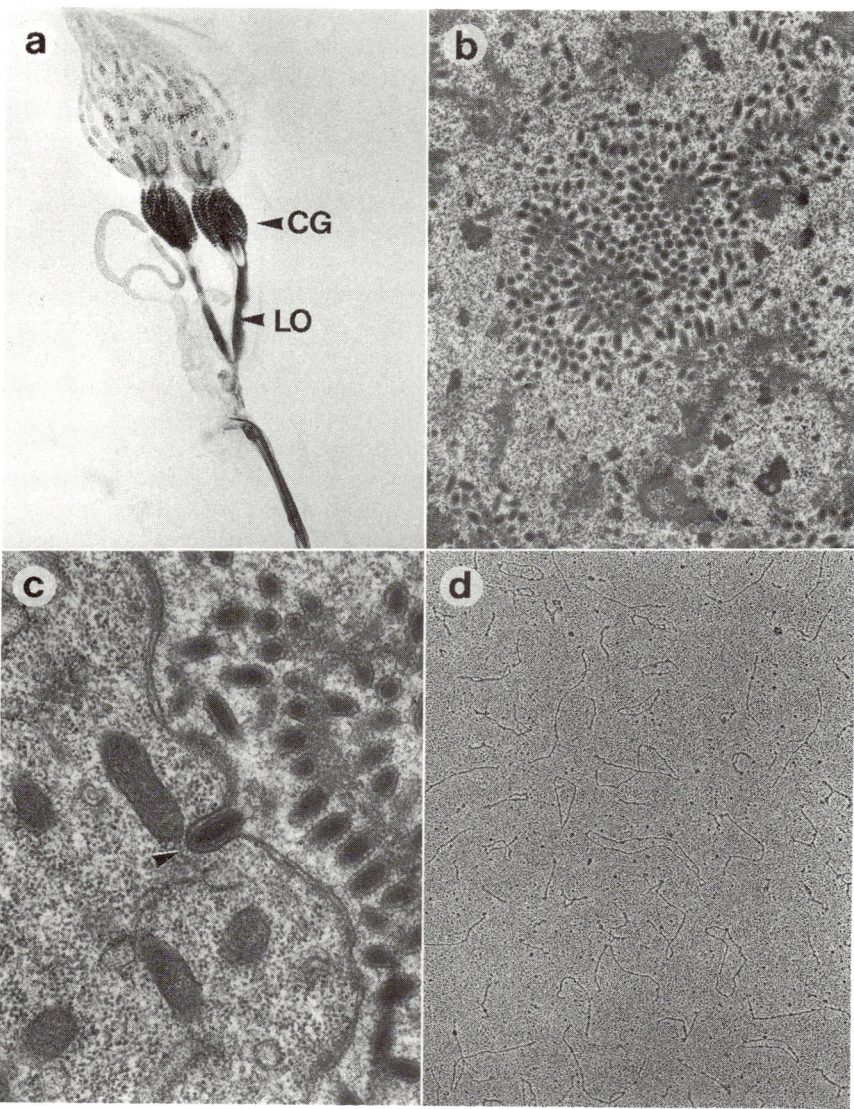

Figure 7.2 Electron micrographs of polydnaviruses in the campoplegine ichneumonid, *Hyposoter didymator*: (a) Feulgen-stained female reproductive system showing intense reaction in the calyx gland (CG) and lumen of lateral oviducts (LO) indicating polydnavirus DNA; (b) virus 'replication' in the calyx epithelial cell nucleus with *de novo* formation of enveloping membranes; (c) budding of polydnavirus particles through calyx cell nuclear membrane; (d) isolated polydnavirus DNA molecules showing variously sized circular molecules. (Courtesy of Nathalie Volkoff.)

The timing of the beginning of polydnavirus development (replication) appears to be closely related to the parasitic wasp's developmental stage (Norton and Vinson, 1983; Volkoff *et al.*, 1995b); the insect hormone, 20-hydroxyecdysone, stimulates polydnavirus replication in excised ovaries of the ichneumonid, *Campoletis sonorensis* (Webb and Summers, 1992). Thus the viral replication is probably triggered by hormone levels in the developing parasitic wasp pupa.

Federici (1991) noted that, structurally, the bracoviruses rather closely resemble baculoviruses. Most ichnoviruses resemble ascoviruses but the polydnaviruses of the ichneumonid, *Bathyplectes*, having spherical nucleo-capsids, more closely resemble entomopoxviruses. Baculoviruses and ascoviruses are common viral pathogens of lepidopterous larvae, the major hosts of the polydnavirus-bearing wasps, whilst entomopoxviruses are most frequently reported from Coleoptera which are the hosts of *Bathyplectes*. The possibility therefore exists that, as presently constituted, the Polydnaviridae may constitute a polyphyletic assemblage with different polydnaviruses having originated from interactions between wasps and particular host virus groups. Federici also noted that some of the non-polydnavirus viruses associated with the female reproductive tracts of parasitic wasps and injected into their hosts may represent intermediate stages in the capture of originally free-living viruses by parasitoids for the benefit of the latter.

The unique type of life history of polydnaviruses draws their status as true viruses into question – even more so now that it has been shown that the genes for the polydnavirus DNA are integrated within the nuclear genome of the parasitic wasp (Stoltz, 1990; Fleming and Summers, 1991). Federici (1991) showed that polydnaviruses satisfy only two of the five criteria normally used to define a virus; specifically, they do not to replicate viral nucleic acid in the host (and may not in the wasp), they do not appear to carry genes encoding all of 'viral' proteins, and they do not assemble and release progeny viruses from host cells. The question of the origin of polydnaviruses has been considered at length by Stoltz and Whitfield (1992), who suggested a total of four possible scenarios:

- They originated from host viruses.
- They originated from parasitic wasp viruses.
- They represent purely wasp genes that are copied and delivered to their expression site within the host.
- They represent a combination of wasp genes and virus genes.

Distinguishing between these is not easy. Morphologically, bracoviruses bear some resemblance to baculoviruses, but this is at best weak evidence for a viral origin. Immunological similarities have been found between polydnaviruses and venom components in the ichneumonid *Campoletis sonorensis* (Webb and Summers, 1990), suggesting that they may simply be gene vectors. However, this is apparently not the case in the braconid

Chelonus sp. nr *curvimaculatus* (Leluk and Jones, 1989; Chelliah and Jones, 1990). The fact that the polydnavirus genome is distributed across the wasp's genome may suggest that the viruses are simply a means by which the wasp encapsulates various genes to be expressed in the host, but this in itself does not mean that some of the polydnavirus genes, or at least the excision mechanisms, are not of viral origin (Stoltz, 1993). The ovaries of insects are known to harbour many different viruses, often without obvious pathology, and so they may be a relatively permissive tissue. Thus the possibility that viruses persisting in the ovary might have become genetically integrated with the host, and subsequently gained fitness by assuming a gene vector role, may be the most likely scenario for the origin of polydnaviruses. Perhaps we shall never know for certain.

Despite the current uncertainties regarding the origins and monophyly of the Polydnaviridae, they might well turn out to be one of the most diverse (speciose) groups of viruses known – given that the viruses from each species that possess them appear to differ considerably from one another and considering the numbers of wasp species involved.

Polydnavirus DNA is transcribed and expressed in the host. Strand *et al.* (1992) and Hayakawa *et al.* (1994) have shown that the viral transcripts are not produced equally in all host tissues but are most obvious in haemocytes – an observation that correlates well with the known effects of polydnaviruses. It has now been shown that *Microplitis demolitor* PDV specifically targets one morphotype of host haemocyte (Strand, 1994), in which it induces apoptosis (Strand and Pech, 1995), but it is not yet certain that other systems show the same degree of target cell specificity. Hayakawa *et al.* further showed that purified *Cotesia kariyai* (Braconidae) polydnavirus only produced transcripts when injected into its normal armyworm host, *Pseudaletia separata*, and not when injected into the related, but non-host, common cutworm, *Spodoptera litura*. They may, therefore, be important factors in the determination of host specificity.

Xu and Stoltz (1991) and Dib-Hajj *et al.* (1993) have investigated the gene sequence of polydnaviruses from the ichneumonid *Campoletis sonoriensis*. Dib-Hajj *et al.* found three members of a cystein-rich gene family, the sequence of which showed similarities to molluscan α-conotoxins that act as ion-channel ligands. Whether this is indicative of their function remains to be determined.

7.12 PHYSIOLOGICAL EFFECTS OF POLYDNAVIRUSES

The physiological effects of polydnaviruses on the host insect have been most thoroughly investigated in only a few parasitic wasps. Notable among these are the campoplegine ichneumonid, *Campoletis sonorensis*, from south-western United States, a parasitoid of *Heliothis virescens* and on various braconids, notably species of the microgastrine genera, *Cotesia* and

Microplitis, the chelonine genus, *Chelonus*, and the cardiochiline genus, *Cardiochiles*. The effects known so far are diverse and are summarized in Table 7.4.

Polydnaviruses of ichneumonids and braconids appear to differ in their need for the simultaneous presence of wasp venom components for their action, particularly when it comes to preventing encapsulation (see also section 7.8). Tanaka and Vinson (1991) showed that calyx fluid alone of *Cardiochiles nigriceps* (containing polydnavirus) was not sufficient to prevent pupation of host (*Heliothis virescens*) larvae. Instead, normal arrest of host development depended on the presence of both calyx and venom gland products. The venom component that interacts with the polydnavirus was shown to be a moderately large protein (66 kDa). In the case of the microgastrine braconid, *Cotesia karayai*, venom and polydnavirus interact to bring about host castration (Figure 7.2; Tanaka *et al.*, 1994), though calyx secretion alone caused some reduction in testis size. Noguchi *et al.* (1995) showed that *C. kariyai* polydnavirus causes an elevation in the dopamine concentration of its host's haemolymph and central nervous system. Dopamine is an important chemical in the melanization pathway in insects, and its build-up might signify a blocking of the pathways that convert dopamine ultimately to melanin. Further, this increase

Table 7.4 Summary of polydnavirus effects on hosts (modified after Stolz and Whitfield, 1992)

Effect of polydnavirus	*References*
Specific reduction in host immune response against the egg/larva of the parasitoid	–
General reduction in host immune response	Edson *et al.*, 1981; Stolz and Guzo, 1986; Davies *et al.*, 1987; Strand and Noda, 1991
Increase in host dopamine levels	Noguchi *et al.*, 1995
Inhibition of host haemolymph phenoloxidase activity	Stolz and Cook, 1983; Beckage *et al.*, 1990
Haemocyte transformation	Stolz and Guzo, 1986
Block haemocyte spreading	Strand and Noda, 1991
Reduction in haemocyte count	–
Prothoracic gland degeneration	Dover *et al.*, 1988
Reduction in phenoloxidase activity	Strand and Noda, 1991
Elevation of growth-blocking peptide	Noguchi *et al.*, 1995
Slowed or arrested development	Tanaka, 1987; Tanaka and Vinson, 1991
Reduced haemolymph viscosity	Davies *et al.*, 1987
Pigmentation changes	Beckage *et al.*, 1990
Reduced host feeding	Vinson *et al.*, 1979

seems to be related to an increase in growth-blocking peptide and this in turn represses juvenile hormone esterase activity and so delays pupation.

Guzo and Stoltz (1985) showed that a combination of venom and polydnavirus from the braconid *Cotesia melanoscela* injected into the larva of the tussock moth, *Orgyia leucostigma*, rendered the latter susceptible to parasitism by three species of the ichneumonid genus, *Hyposoter*, none of which could develop on untreated *O. leucostigma* larvae. The normal encapsulating response of *Hyposoter* eggs was inhibited by the *C. melanoscela* venom alone, but complete development of the *Hyposoter* larva depended also on the presence of active polydnavirus particles. Similarly, in the *Microplitis demolitor* (Braconidae)/*Pseudoplusia includens* (Noctuidae) system, Strand and Noda (1991) demonstrated synergism of polydnavirus action by venom.

7.13 OTHER VIRUSES

Not all symbiotic viruses in parasitic wasps are polydnaviruses. Several other types of virus have been found associated with the female reproductive tract, including in the calyx and venom glands, in various groups. Electron microscopy has revealed these other forms to be morphologically diverse. Thus in *Meteorus*, another braconid parasite of Lepidoptera, two types of virus have been observed in the venom glands (Edson *et al.*, 1982). Both Edson *et al.* and Lawrence and Akin (1990) reported particles from the venom gland of the opiine braconid *Diachasmimorpha* (as *Biosteres*) *longicaudatus*, a parasitoid of tephritid fly larvae. They are observed inside vacuoles and are believed to enter the venom. The female reproductive system of some wasps possesses more than one type of virus. In addition to polydnaviruses, the ctenopelmatine ichneumonid, *Mesoleius tenthredinis*, also possesses a baculovirus in its ovary (Stoltz, 1981), whilst Krell (1987) has shown that some campoplegines possess long, enveloped viruses. Much more work on these other types of virus needs to be carried out in order to understand their relationships with the wasps and one gets the feeling that we have only just started to scratch the surface of their distribution and function.

Because of the interest in their PDVs, members of the braconid subfamily Microgastrinae have been intensively investigated. As a result, many viruses associated with the reproductive system have been found in these wasps, but as yet very little is known of their functions, actions and modes of transmission. Several very elongate, enveloped virus-like particles (VLPs), which show similarities to members of the Baculoviridae, are found in the calyx secretions of *Cotesia congregata*, *C. hyphantriae*, *Microplitis croceipes* and *M. mediator* (Stoltz and Vinson, 1977, 1979; Tanaka, 1987; DeBuron and Beckage, 1992). Stoltz and Faulkner (1978) described a biconvex virus from the reproductive tract of female *Cotesia melanoscela* which replicates within haemocytes of its host. A filamentous virus from *C. marginiventris*, referred to as CmFV, is known to be transmitted to the noctuid host larva in whose cells it replicates (Styer *et al.*, 1987). However, while it appears that the CmFV has some effects on host cells that may be advantageous to the parasitoid larva, it has not yet been possible to distinguish these effects from those of the polydnaviruses that are transmitted into the host in far larger numbers. The *M. mediator* VLPs are produced in the calyx gland and they become attached to the outer chorionic layer, as with the *Venturia canescens* VLP discussed in the next section (Tanaka, 1987). Another *C. melanoscela* virus, referred to as CmV2, has been found only in certain strains (Stoltz *et al.*, 1988b). It is known that this virus replicates in both the wasp and its host, has a genome consisting of a single double-stranded DNA molecule (*c.* 125 kb), and is transmitted vertically along the maternal line. It occurs in the ovarian calyx and various other tissues in both males and females, but it is apparently absent from the testes. From its morphology, Stoltz *et al.* suggest that CmV2 may be related to the Ascoviridae, but it differs from typical ascoviruses in its membrane structure. CmV2 is apparently non-pathogenic to the wasp's gypsy moth host, and it is to date only known from the French and Nova Scotia strains of *C. melanoscela*; it is apparently unnecessary for successful parasitism, and it is possible that this virus represents some sort of intermediate stage as would have occurred in the evolution of the polydnaviruses.

Rabouille *et al.* (1994) discovered that both wild and laboratory populations of the ichneumonid, *Diadromus pulchellus*, are infected with a reovirus which they termed DpRV, and Hamm *et al.* (1994) described a putative reovirus-like particle in another microgastrine braconid, *Microplitis croceipes*. DpRV is present mainly in the gut of the adult wasp but also in smaller quantities in the venom gland. The virus is interesting for a number of reasons. Firstly it appears to have little affinity with any other known Reoviridae, though terminal inverted repeats suggest a possible affinity with the genus, *Orthoreovirus* (Bigot *et al.*, 1995). Secondly, it displays variation dependent upon the ploidy of the wasp. In haploid wasps, each virus genome consists of 10 segments whereas diploids, females and males, have an extra segment. It would be interesting to know if the *Microplitis* virus is related to the ichneumonid virus.

7.14 VIRUS-LIKE PARTICLES

The term virus-like particle (VLP) refers to various structures that superficially resemble true viruses but probably do not contain nucleic acids. The best-known parasitic wasp VLP is that of the campoplegine ichneumonid, *Venturia canescens* in which they were first reported by Rotherham (1967). Although it is devoid of nucleic acids, in general appearance and in its site of production (within the calyx gland cell nuclei), it closely resembles polydnaviruses. This VLP apparently coats the wasp's eggs by sticking to the 'flocculent' outer chorionic layer. Feddersen *et al.* (1986) have shown that this VLP has antigenic properties that are similar to proteins of the host, and they surmised that it protects the parasitic wasp's eggs by rendering them chemically invisible to the host – a sort of chemical camouflage. Significantly, they have shown that injection of isolated VLPs into the host does not render protection or apparently have any effect on the host's ability to encapsulate *V. canescens* eggs that have been cleaned of VLPs. A similar phenomenon has been described for the braconid, *Cotesia rubecula*, which produces polydnaviruses but whose calyx fluid also contains a non-particulate protein, related to those of its polydnavirus, which coats eggs and renders them immune from encapsulation by the host (Asgari and Schmidt, 1994). In addition to showing antigenic similarity to host proteins, Schmidt *et al.* (1990) showed that haemocytes of the host *Ephestia kuehniella* caterpillars produce a protein that is structurally similar to a *Venturia* VLP protein. When haemocytes are induced to produce larger quantities of the protein, they have a reduced tendency or ability to encapsulate or to spread. Theopold *et al.* (1994) have cloned one of the protein coding genes of the *V. canescens* VLP and shown that the 1.4 kb gene codes for a 40 kDa protein. Southern hybridization of the cloned VLP protein gene and enzymatically cut wasp DNA indicated that the gene was represented in the wasp's genome by only a single copy. Given the widespread occurrence of polydnaviruses in campoplegine ichneumonids (Stoltz and Whitfield, 1992), it seems highly likely that the *V. canescens* VLP was originally derived from a polydnavirus.

Another well-studied group of such particles is found in members of the eucoilid genus, *Leptopilina*, which are endoparasitoids of *Drosophila* larvae (Rizki and Rizki, 1990a). These particles are produced in the venom gland and appear to be modified as they pass into the venom reservoir. There has been some debate as to whether these particles contain nucleic acids and so they will be regarded as VLPs at present. The *L. heterotoma* VLPs play a very important role in overcoming the host's immune defence reaction. In *Drosophila* haemolymph there are special haemocytes, termed lamellocytes, which adhere to and encapsulate foreign particles such as parasitoid eggs. The VLPs appear to enter these lamellocytes and bring about changes in their morphology and surface properties such that the encapsulation reaction is blocked.

7.15 SYMBIOTIC YEASTS

Several groups of insects have yeast-like symbionts, and an association with fungi may have been crucial for the initial evolution of the parasitoid way of life in the Hymenoptera (Eggleton and Belshaw, 1992). Among the parasitic Hymenoptera such relationships have been recorded in the encyrtid *Comperia* (Lebeck, 1989), the ichneumonid *Pimpla turionellae* (Middeldorf and Ruthman, 1984) and the braconid *Adelurola apii* (Keilin and Tate, 1943). In *Comperia*, the yeast-like organism is injected into the host (cockroach) eggs, along with other venom components. Within the host egg, the fungi appear to do little, though some are ingested by the newly emerged parasitoid larva within whose gut they produce a large population and where pleomorphic and filamentous forms occur. In contrast, in the case of the braconid, *Adelurola*, the symbiotic fungus multiplies rapidly in the haemolymph of its celery fly host, and is consumed by the parasitoid larva. As with *Comperia* yeasts, filamentous forms of the fungus develop within the parasitoid larva's gut.

7.16 STIMULATION OF KOINOBIONT DEVELOPMENT

There has been considerable interest in what factors stimulate development in koinobiont endoparasitoids. Many species with this way of life emerge from their egg soon after it has been laid, but then remain as a 1st instar larva until their host reaches its last larval instar and pupates or, in the case of some hyperparasitoids, until the secondary host is parasitized by the primary host-to-be. Only then does the parasitoid larva commence feeding and it usually completes its development rapidly from this point onwards. Corbet (1968) proposed that in the ichneumonid, *Venturia canescens*, a parasite of *Ephestia kuhniella*, the stimulus could be the decrease in osmolarity and in free amino acids in the host's haemolymph, and not directly due to host hormones. In many cases, though, endocrinological stimuli are likely to play a significant role even if they are not the only factors involved.

7.17 TERATOCYTE FUNCTION

Teratocytes, also known as giant cells, derive from the serosal membrane of several groups of koinobiont endoparasitic wasps, and continue to develop in the host insect in parallel with the wasp larva (Chapter 5). For a long time they were assumed to form an additional means of sequestering host nutrients ready for consumption by the parasitoid larva when it completes its development. However, while ultrastuctural studies have shown them to develop a dense microvillar surface (Figure 5.1) as would be expected of cells with an important trophic function, they have also shown that they probably play a secretory role too (e.g. Tremblay and Laccarino, 1971; Strand, 1986). In the microgastrine braconid, *Cotesia kariyai*, for example, teratocytes produce a substance that inhibits encapsulation (Tanaka and Wago, 1990).

Führer and El-Sufty (1979) demonstrated that teratocytes of the braconid, *Cotesia* (as *Apanteles*) *glomerata*, produce a fungistatic compound which is released into the haemolymph of its host (larvae of the large white butterfly, *Pieris brassicae*). El-Sufty and Führer (1981) showed that parasitization by *A. glomeratus* causes a pathology of the host's cuticle that permits penetration by hyphae of the entomopathogenic fungus, *Beauveria bassiana*. Thus the teratocytes help to compensate for presumably disadvantageous side effects of the parasitization process.

Juvenile hormone production and release by teratocytes of the braconid, *Cardiochiles nigriceps*, were demonstrated by Joiner *et al.* (1973), and this discovery suggested that they may play a rather more active role in host regulation. Not only are *Cardiochiles* teratocytes involved in JH synthesis; they also play a role in ecdysteroid processing. Pennacchio *et al.* (1994c) have provided evidence that the teratocytes convert active 20-hydroxyecdysone to polar inactive forms. This has the effect of inhibiting pupation despite an overall increase in ecdysteroid titre. Teratocytes of the same species also synthesize and release a number of proteins when cultured *in vitro* (Vinson *et al.*, 1994a), a procedure that is likely to be of great help in elucidating the roles of the various compounds produced.

In the case of the scelionid, *Telenomus heliothidis*, Strand (1986) demonstrated that much of the necrosis observed in host eggs was due to substances (presumably enzymes) released by the teratocytes. Strand found that teratocyte culture medium became positive for leucine amino peptidase, esterase and phosphatase, and these are therefore implicated in host necrosis. This of course begs the question: how do the teratocytes themselves and the parasitoid larva survive in this cocktail? The parasitoid larva may be protected by its chitinous cuticle, but the teratocytes presumably must have specialized cell membranes.

7.18 PHYSIOLOGICALLY ACTIVE LARVAL SECRETIONS

Some inconclusive evidence that larval secretions also cause long-lasting developmental and physiological effects in the host was initially provided by Schneider (1951) who found that larvae of the ichneumonid, *Diplazon laetatorius*, affect development of its hoverfly host, *Epistrophe balteata*. The effects included reduction in size of parts of the host central nervous system and abnormal wing development. Schneider suggested that the effects may have been caused by secretions of the wasp larva's salivary glands, but without any clear evidence. Willers and Lehmann-Danzinger (1984) showed that not only do compounds in the venom of pimpline ichneumonids that parasitize lepidopterous pupae cause a reduction in the activity of larval phenoloxidases, so too do anal secretions of the parasitoid larvae. The exact sources of the larval toxins was not identified with total certainty, but anatomical investigation suggest that they may derive from a modified hindgut pouch with glandular cytology and/or from a pair of differentiated Malpighian tubules (Chapter 5; Figure 5.5). Whereas the true venom components have a low level of effect but a long duration of activity, the compounds secreted by the larva are more potent but have only a short duration. This is almost exactly what would be expected: as the larva is permanently present in the host, there is no need for its secretions to have long-lasting effects. Larvae of some chelonine braconids release ecdysteroids (Brown and Reed-Larsen, 1991), and larvae of *Pteromalus puparum*, an unrelated idiobiont parasitoid of Lepidoptera pupae, appear to produce cytotoxic compounds when tested against lepidopteran cell cultures (Boulétreau and Quiot, 1972).

Führer and Willers (1986) showed that larvae of the pimpline ichneumonids, *Pimpla turionellae* and *Itoplectis conquisitor*, secrete a bactericidal and fungicidal substance from the hindgut, and possibly also from two hypertrophied glands (again putatively identified as modified Malpighian tubules). These secretions may be important in limiting the detrimental effects that can result when the parasitoid damages the host's gut, releasing bacteria into the host's haemocoele (Willers, 1980). Ligation experiments with the braconid, *Cardiochiles nigriceps*, by Vinson *et al.* (1994a) showed that both the head and anal vesicle (Chapter 5) change the protein composition of media in which they are cultured, suggesting that these too may be secreting physiologically active proteins.

The larvae of several other groups of parasitic wasps have highly modified rectums, and most have a pair of hypertrophied Malpighian tubules associated with them, but only a few have been examined for probable secretory roles. The anal vesicles of various ichneumonoids have been shown to be involved in excretion of nitrogenous waste (Edson and Vinson, 1976) but the possibility that they also secrete factors that interact with host physiology cannot be excluded. Certainly, the hypertrophied rectal cells of alloxystine charipids, which are hyperparasites of aphidiine braconid primary parasitoids within aphids, are secretory (Nahif and Madel, 1990) and it seems very likely that their products are involved in interactions with either their primary or secondary hosts.

Parasitoid larval saliva may be involved in regulating and redirecting host metabolism so that the host's nutritional value to the parasitoid will be increased. In the case of the pteromalid, *Catolaccus grandis*, 81% of host boll weevil larvae are not paralysed by the female wasp, and host paralysis is not essential for larval survival. Not only can 1st instar larvae of this parasitoid survive on unenvenomated, unparalysed hosts; their saliva causes changes in the host's amino acid composition that are similar to those caused by *C. grandis* venom (Morales-Ramos *et al.*, 1995). Apparently the female wasps do not have the capacity to inject a full venom dose 'into' all hosts encountered, especially as the wasps get older. Clearly the ability of the 1st instar larva to bring about similar effects to the mother's venom is an advantage, and it would be interesting to know whether the same chemicals are involved.

7.19 ENCAPSULATION AND OTHER HOST DEFENCES

By far the commonest physiological host defence is the ability to encapsulate eggs or early instar larvae of a parasitoid. The general mechanism, which has been described in detail by Salt (1970), Norton and Vinson (1977) and Lackie (1988), involves the adhesion of host haemocytes to the surface of the parasitoid egg or larva. The deadly effect of encapsulation on parasitoid eggs, etc., appears to be due to a combination of factors including starvation from food and/or oxygen and the release of toxic compounds into the capsule. Eggs and early instar larvae of some parasitoids are particularly sensitive to reduced O_2 concentration, and obtaining sufficient oxygen within the host in the absence of a tracheal connection to the outside seems to be a general problem for endoparasitic Hymenoptera (see Chapter 5).

Once haemocytes, or probably a particular class or subset of classes of haemocytes, have adhered to a foreign body, they undergo a dramatic morphological change, involving flattening and spreading, which results in the formation of a dense capsule, composed of several haemocyte layers. Cell flattening appears to involve microfilament assembly and is blocked by cytochalasin B. Finally, the capsule cells darken, i.e. melanize, and harden. Melanization involves several biochemical steps, the first of which is conversion of the amino acid, L-tyrosine to L-DOPA (3-4 dioxyphenylalanine) by the enzyme phenoloxidase, within the host's haemolymph (phenoloxidases are also involved in melanization and sclerotization of insect cuticle). This step is usually followed in succession by conversion of L-DOPA to dopaquinone, to dopachrome and finally melanin, some of these steps also involving phenoloxidase. Details of the complicated chemistry involved and the triggering of phenoloxidase activity are described by Sugumaran and Kanost (1993). Using phenoloxidase-deficient mutants of *Drosophila melanogaster*, Rizki and Rizki (1990c) showed that phenoloxidases are not themselves involved in the recognition of foreign bodies such as parasitoids. As phenoloxidases play such a central role in melanization, and hence in defence against pathogens and parasitoids, it is not surprising that many parasitoid defence reactions involve blocking its products.

The term haemocyte encompasses a number of different types of cell and each type presumably serves particular functions (Lackie, 1988). Unfortunately, most types are defined on morphological grounds and homologies of particular types between insect orders are largely speculative. Some insects have characteristic classes of haemocyte that appear to be lacking in other orders – for example, the crystal cells of Diptera, which are packed with a precursor of phenoloxidase and are important in their defence against parasitoids. In Lepidoptera, encapsulation involves two classes of haemocyte: the plasmatocytes and the granular cells.

Drosophila lamellocytes are induced to change shape from discoidal to bipolar in individuals parasitized by the eucoilid *Leptopilina heterotoma* (Rizki and Rizki, 1990b). The change can be blocked *in vitro* by both vinblastin and vincrystin, two agents that block assembly of tubulin to form microtubules, perhaps implying that the wasp female injects compounds into the host that stimulate microtubule assembly. Rizki and Rizki (1990c) showed that a phenoloxidase-deficient mutant strain of *Drosophila melanogaster* was still able to encapsulate eggs of its eucoilid parasitoid, *Leptopilina boulardi,* but that the capsules did not subsequently melanize. So at least in the case of *Drosophila,* the formation of a capsule around a foreign body such as a parasitoid egg is a separate process from its subsequent melanization.

Encapsulation ability is by no means fixed within a species. In the case of *Drosophila melanogaster* and its parasitoids, *Leptopilina boulardi* and *Asobara tabida,* there is considerable geographical variation in the host's ability to encapsulate eggs and in the parasitoid's ability to avoid encapsulation (Boulétreau, 1986; Kraaijeveld, 1994; Kraaijeveld and van Alphen, 1994; Kraaijeveld *et al.,* 1995). The encapsulation ability of *D. melanogaster* against *A. tabida* is highest in north-west Europe and much lower in the south of France. Geographical variation in encapsulation ability is also found in many other host parasitoid systems (e.g. Rizki *et al.,* 1990). Kraaijeveld (1994) and Monconduit and Prevost (1994) showed that the ability of *Drosophila* to encapsulate eggs of *A. tabida* was inversely proportional to the 'stickiness' of the parasitoid's eggs. Monconduit and Prevost showed that whilst *D. melanogaster* from Lyon in France failed to encapsulate the eggs of local *Asobara,* the sibling species, *D. simulans,* was successful in encapsulating all the eggs. Dissections showed that in *D. melanogaster* the *Asobara* eggs were fully attached to host tissues, whereas in *D. simulans* they were always partly free. Stickiness appears to allow part of the egg to evade attack by the fly's haemocytes, thus leaving part of it free so that the parasitoid larva can both respire and emerge. Kraaijeveld categorized 'stickiness' according to a four-point scale depending on whether or not, and if so by how much, the egg was associated with host tissue. This semi-quantitative approach has enabled Kraaijeveld to correlate egg stickiness with the parasitization success, or virulence, of *Asobara* (Figure 7.3). In southern Europe, the main host of *A. tabida* is *D. melanogaster*; in north-west Europe it is *D. subobscura* (Kraaijeveld and van der Wel, 1994). Thus the regional variation in virulence of *A. tabida* against *D. melanogaster* may reflect the degree of co-adaptation with its main host. The encapsulation ability (resistance) of *Drosophila* larvae against *A. tabida* is also correlated with their total haemocyte count, both between and within species (Eslin and Prevost, 1996); it appears to be a specific response in that it was not correlated with *Drosophila*'s encapsulation ability against another endoparasitoid, the eucoilid, *Leptopilina boulardi*. In the *Asobara tabida–Drosophila* system, encapsulation ability appears to be highly temperature dependent (Kraaijeveld and van der Wel, 1994; Kraaijeveld *et al.*, 1995). When reared at a constant temperature of 15°C, survival of *A. tabida* (Woereleuse Verlaat strain) in *Drosophila melanogaster* was 65% compared with only 9% at 20°C.

The variation in encapsulation ability of different strains of *D. melanogaster* against *L. boulardi* has been used to investigate the relationship between the host–parasitoid defence reaction and the inducibility of anti-bacterial defence reactions, namely the production of cecropins and other bactericidal peptides (Coustau *et al.*, 1996). The relationship discerned is in fact rather surprising in that resistant *Drosophila* did not produce cecropins in response to parasitization, but a susceptible host strain did. This result demonstrates that the two response types are different but may share some common regulatory mechanisms.

A number of instances is known in which the incidence of encapsulation is reduced with increasing number of parasitoid eggs laid (i.e. with superparasitism) (Streams, 1971). Blumberg and Luck (1990) showed that different strains of the encyrtid, *Comperiella bifasciata*, are characterized by their different tendencies to superparasitize their diaspid scale host, *Aonidiella aurantii*. Eggs of a Californian strain of *C. bifasciata* were far more likely to be encapsulated than were those of an Israeli strain (26% as opposed to 6%). Encapsulation was reduced when members of the Californian strain of *C. bifasciata* superparasitized their hosts and it was

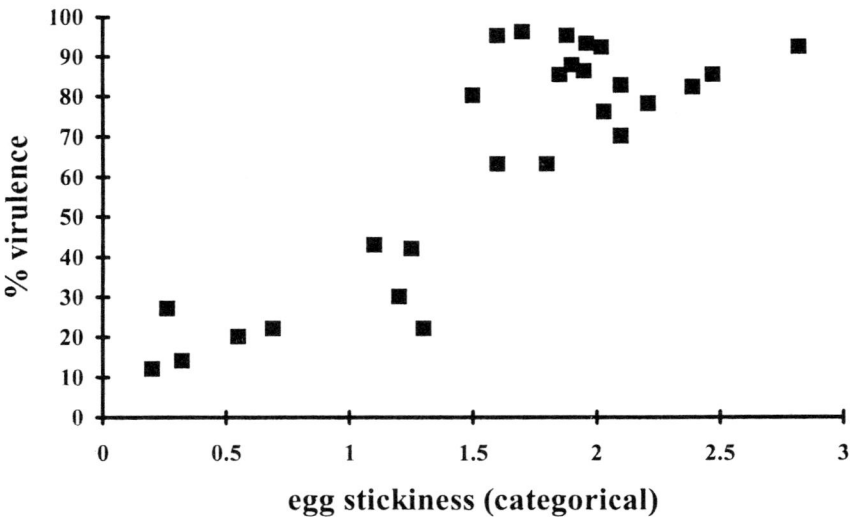

Figure 7.3 Relationship between egg stickiness and virulence in the alysiine braconid, *Asobara tabida*, a parasitoid of *Drosophila melanogaster* and *D. subobscura*. (Redrawn from data from Kraaijeveld, 1994.)

found that superparasitism was significantly more common in the Californian strain than in the Israeli one. Blumberg and Luck therefore suggested that superparasitism in Californian *C. bifasciata* is an adaptation for avoiding encapsulation. Hegazi *et al.* (1991) provided evidence that larvae of the noctuid, *Spodoptera littorale*, may produce two different types of capsule, permanent and impermanent, the former being formed in lightly parasitized hosts and the latter in heavily superparasitized ones. Failure to produce permanent ones might reflect depletion of some important component involved in capsule completion.

In vitro assays of encapsulation are difficult to devise because the haemocytes involved have a strong tendency to adhere to surfaces or to clump or undergo autolysis. Ratner and Vinson (1983) developed an *in vitro* encapsulation system with *Heliothis virescens* haemolymph, and Pech *et al.* (1995) have succeeded with those of another noctuid, *Pseudoplusia includens*. Ratner and Vinson's method was subsequently employed by Davies and Vinson (1986) to investigate the way that eggs of the braconid, *Cardiochiles nigriceps*, normally avoid encapsulation. Their results suggested that the fibrous, outer chorionic layer blocked recognition by the *H. virescens* haemocytes. *C. nigriceps* eggs are encapsulated by caterpillars of the related moth, *H. zea*, and Davies and Vinson suggested that this may be because the haemolymph of this non-host species degrades and removes the fibrous layer.

Encapsulation is not likely to be without its own costs to the host. Although a host will usually survive parasitism if it encapsulates the parasitoid, this process must almost certainly have a metabolic cost and perhaps also in terms of resource. To date little work has been carried out on this side of the equation. Carton and David (1983) showed that *Drosophila* that survived attack by the eucoilid, *Leptopilina*, experienced a reduction in fitness compared with unparasitized larvae, but whether this was due to the costs of encapsulation or to damage caused by parasitoid toxins could not be determined. Further, it should be noted that the ability to encapsulate a parasitoid may also involve costs to the host, and this might be a reason underlying the geographical variation in encapsulation ability observed in some systems; otherwise, one might expect that all hosts would be equally good at encapsulation.

7.20 NON-ENCAPSULATING HOST DEFENCES

In a few instances a parasitoid's egg may fail to develop within a given host without encapsulation occurring. For example, Henter and Via (1995) have shown that eggs of the aphidiine braconid, *Aphidius ervi*, fail to develop in some resistant pea aphid host strains (*Acyrthosiphon pisum*) and eventually disintegrate. Such failures are usually attributed to some sort of humoral defence mechanism though in no case has any details been worked out. This may be the most likely explanation but there are a few other possibilities that need to be considered. Perhaps, for example, some hosts simply lack a substance that is necessary for further parasitoid development – a sort of chemical crypsis. Consideration of the ability of another aphidiine, *Monoctonus paludum*, to develop in different aphids is of interest because its eggs are laid within the aphid's central nervous system. Griffiths (1961) followed the development of this species in its natural host, *Nasonovia ribis nigri*, and four other lettuce-infecting aphids in which the *Monoctonus* will oviposit indiscriminately, but in which it will fail to develop. In most of the non-host aphids, embryogenesis was retarded and failed before eclosion, whilst in some others the wasp embryo cells became vacuolated at an early stage. In one aphid, *Aulacorthum circumflexum*, the embryo became encapsulated within the ganglion, apparently as a result of small haemocytes migrating into the ganglion, perhaps through the oviposition hole. Death of a parasitoid without encapsulation may also occur at a later stage in its development. Evidence of such later acting humoral responses also comes from aphids and their aphidiine braconid primary parasitoids.

Arthur and Ewen (1975) described an apparently unique host defence reaction against the banchine ichneumonid, *Banchus flavescens*, by factitious noctuid hosts. The preferred host is the Bertha armyworm, *Mamestra configurata*, but the *Banchus* will also oviposit in *Trichoplusia ni*, the cabbage looper. The egg of the wasp becomes encysted between the host's epidermal cells and the endocuticle. The host's response appears to be triggered by venom gland secretions of the wasp, and the cyst which forms around the wasp egg as a result is shed at the host's next moult. The same response was also found in three other noctuids, and so it seems to be a characteristic feature of the wasp's venom (Ewen and Arthur, 1976).

7.21 USE OF IMMUNOLOGICALLY FAVOURABLE SITES OR HOST STAGES

Some endoparasitoids of larval hosts simply deposit their eggs in the host haemocoele, but there it is most vulnerable to attack by host haemocytes and subsequent encapsulation. Not surprisingly, therefore, many parasitoids have evolved to lay their eggs in places where they will be less prone to attack (Table 7.5). Such immunologically protected sites include the nervous tissue, gut wall, fat body – indeed almost anywhere that haemocytes cannot reach with ease. In a few cases, it appears that wasp factors cause a proliferation of host cells around the developing embryo or early larva, and these cells may provide a barrier separating the wasp from the host's haemolymph. Such structures form around the aphelinid, *Eretmocerus* (Gerling *et al.*, 1990), and some dryinids (Clausen, 1940). In these dryinids, the structure formed has been likened to a placenta or plant gall (Keilin and Thompson, 1915). What is especially interesting is how the wasps manage to locate the desired target site within the host so accurately. In Chapter 6, various ovipositor steering mechanisms are described which may facilitate such accurate positioning.

The metopiine ichneumonid, *Chorinaeus funebris*, inserts its ovipositor through the host's anus and deposits her egg in the rectum (Aeschlimann, 1974). The egg is thus technically laid externally, entrance into the host being made through the gut wall by the 1st instar larva. The eggs of some tersilochine and anomalonine ichneumonids are furnished with an anchor which is used to fix them to host tissues internally (Gauld, 1976), and this serves to protect part of the egg from encapsulation very much as the stickiness of some *Asobara* eggs do (see sections 4.14 and 7.19).

7.22 PHYSIOLOGICAL SUPPRESSION OF SUPERNUMERARIES

Many endoparasitic wasps have large sickle-shaped mandibles in their 1st instars which they use to eliminate supernumeraries (Salt, 1961; Hegazi *et al.*, 1991). In a few taxa there is evidence that one dominant larva secretes compounds into the host which suppress the development

Table 7.5 A few selected examples of the use of immunologically favourable sites egg laying

Parasitoid	Family	Host order	Oviposition site	References
Amblyteles	Ichneumonidae (Ichneumoninae)	Lep.	Salivary gland	Strickland, 1923; Salt, 1968
Chorinaeus funebris	Ichneumonidae (Metopiinae)	Lep.	Through anus into hindgut	Aeschlimann, 1974
Poecilostictus cothurnatus	Ichneumonidae (Ichneumoninae)	Lep.	Gut wall	van Veen, 1981
Anomalonines	Ichneumonidae (Anomaloninae)	Lep.	Through body into gut wall	Tothill, 1922
Trieces	Ichneumonidae (Metopiinae)	Lep.	Through body into gut wall	Dijkerman, 1988
Triclistus	Ichneumonidae	Lep.	Ganglia	Dijkerman, 1988
Monoctonus	Braconidae (Aphidiinae)	Hom.	Ganglia	Griffiths, 1960, 1961
Agathidines	Braconidae (Agathidinae)	Lep.	Ganglia	Quednau, 1970; Odebiyi and Oatman, 1972
Coccophagus	Aphelinidae	Hom.	Ganglia	Askew, 1968
Platygaster	Platygastridae	Dipt.	Embryonic mid gut and brain	Clausen, 1940, and references therein
Microterys	Encyrtidae	Hom.	Alimentary canal	Askew, 1968
Callaspidia	Figitidae	Dipt.	Cerebral ganglion	Rotheray, 1979

of other conspecific eggs, and Salt listed 15 cases in which such physiological suppression had been suggested. Clear demonstration of this has not been easy, though the work of Fisher (1958, 1971) provided fairly clear evidence for it in the ichneumonids, *Venturia* (as *Nemeritis*) *canesens* and *Diadegma* (as *Horogenes*) *chrysostictos*, both of which are parasitoids of *Anagasta* (= *Ephestia*); and Mackauer (1986) showed similar effects with aphidiine braconid parasitoids of aphids. Fisher (1963) also showed that lack of availability of sufficient oxygen may play a role in the suppression of supernumeraries in the *Venturia–Anagasta* system, though only the egg and 1st instar are particularly susceptible to oxygen deprivation (Chapter 5). Hegazi *et al.* (1991) noted that whilst many supernumerary larvae of the braconid, *Microplitis rufiventris*, are eliminated by physical combat, some supernumerary 1st instars are encapsulated without any sign of physical injury (the normal trigger for encapsulation), and so larvae of this braconid may employ a combination of physical and physiological suppression. Silvers and Nappi (1986) provided clear evidence of physiological suppression involving chemical secretion in the eucoilid, *Leptopilina heterotoma*, a parasitoid of *Drosophila*, using an *in vitro* system. When more than one egg is laid in a single host only one them will complete development; after 30 hours, one will appear considerably larger and at a more advanced stage of development than the other. Silvers and Nappi showed that if these 30-hour-old 'dominant' eggs were transferred to culture medium, 81% of them continued to develop and would hatch. However, if pairs of such eggs were placed together in culture medium, the development of one or both of them failed, suggesting that the eggs release toxins into the surrounding medium that impair the development of conspecific competitors.

7.23 EFFECTS ON HOST BEHAVIOUR

Many koinobiont parasitoids have been reported to induce behavioural changes in their hosts, especially towards the end of their development. The braconid *Chelonus inanitus* causes its host caterpillar (*Spodoptera littoralis*) to go into the soil at its 4th instar rather than its normal 6th instar (Rechav and Orion, 1975), no doubt having by this time reached the ideal size for the braconid to complete its development. Many similar examples are known and probably most of these reflect changes induced by modifications to the host's hormonal system, really reflecting no more than an acceleration of normal behaviour patterns. Biologists have been more excited by the apparent induction in some species of host behaviours that are not observed in unparasitized individuals. Two conflicting possibilities exist. One is that the parasitoid has managed to change the host's behaviour in some way that will increase its chance of survival, perhaps through causing the host to move to a place where it will be relatively protected from predation or hyperparasitism (Brodeur and McNeil, 1989, 1990, 1992). The other possibility is that the parasitized host changes its behaviour so as to increase the likelihood of its being eaten or attacked by a hyperparasitoid – adaptive host suicide (Shapiro, 1976; Smith Trail, 1980; McAllister and Roitberg, 1987). The first of these involves straightforward individual selection on the parasitoid, whereas the adaptive suicide scenario would require kin selection on the host. Godfray (1993) has considered these in some detail, and while both possibilities may still apply in some cases, at present it seems likely that many will show that the parasitoid has the upper hand.

A remarkable parasitoid, or rather hyperparasitoid, effect has been described by Boenisch and Jürgens (1994). These workers found that grain aphids, *Sitobium avenae*, produced significantly more offspring when in the presence of females of their hyperparasitoid, the megaspilid, *Dendrocerus carpenteri*. The effect, apparently attributable to volatiles given off by the megaspilid, is difficult to explain in evolutionary terms with the information available. Although aphid clones may experience less mortality in areas where there are high densities of primary parasitoids, this hardly seems likely to be a reason for them to produce fewer offspring under these conditions – unless, perhaps, they change their behaviour in some other compensating way, say through dispersing.

7.24 EFFECTS ON HOST FOOD CONSUMPTION

There have been numerous studies on the effects of koinobiont parasitoids on the food budgets of their hosts. Both increased and decreased (Powell, 1989) rates of food consumption have been observed and often the effects seem to vary according to the developmental stage of the parasitoid (Duodu and Antoh, 1984) (Figures 7.4 and 7.5). There appears to be a general tendency for gregarious parasitoids to cause an increase in host food consumption but for solitary ones to bring about a reduction in feeding (Slansky, 1986). Changes in host feeding may manifest themselves very soon after parasitization, and in those cases in which feeding is reduced, such effects may provide an additional benefit of parasitoid release in biocontrol programmes (Kumar and Ballal, 1992). Cloutier and Mackauer (1979) investigated the effects of parasitism of the aphidiine braconid, *Aphidius smithi*, on its host aphid, *Acyrthosiphon pisum*. Feeding, food assimilation and growth rates were reduced in comparison with unparasitized aphids whilst the parasitoid was an embryo, but increased to higher than normal levels later in development, only to fall again as the parasitoid completed its development and killed the host. Cloutier and Mackauer suggested that this complex sequence of effects might reflect different differences in nutrient requirements between the parasitoid and host. Not only do parasitoids affect host feeding behaviour; they also

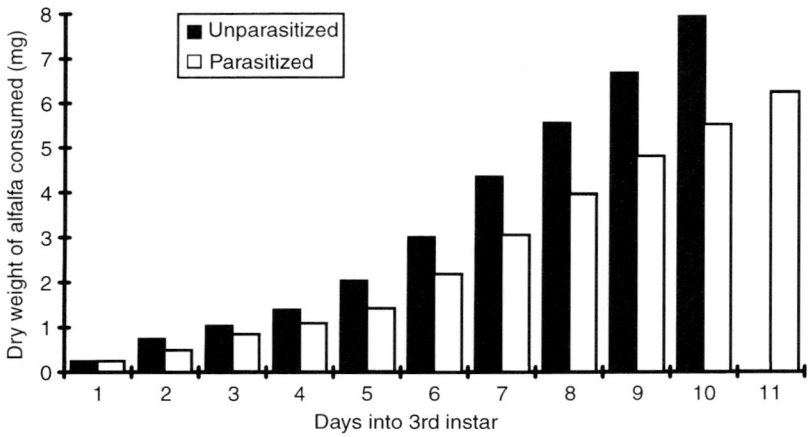

Figure 7.4 Effect of parasitization by the ichneumonid, *Bathyplectes curculionis*, on food consumption by 3rd instar larvae of the alfalfa weevil, *Hypera postica*, at 22°C. (Redrawn from Duodu and Davis, 1974.)

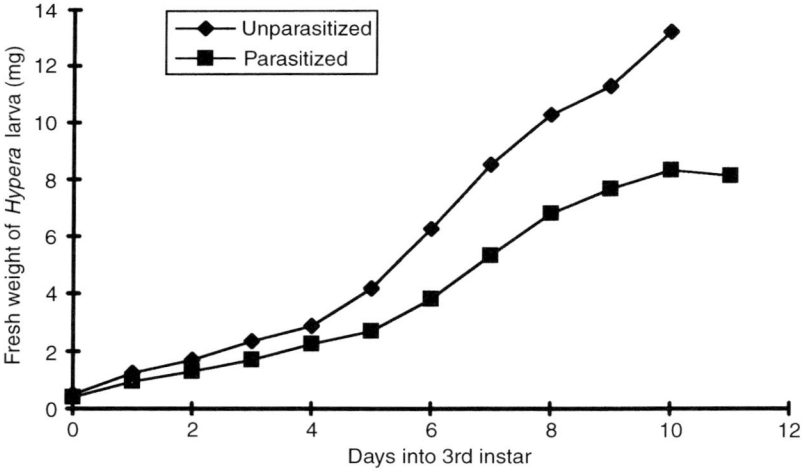

Figure 7.5 Effect of parasitization by the ichneumonid, *Bathyplectes curculionis*, on growth of 3rd instar larvae of the alfalfa weevil, *Hypera postica*, at 22°C. (Redrawn from Duodu and Davis, 1974.)

induce many changes in host metabolism, modifying concentrations of carbohydrates, lipids and nitrogenous compounds (Dahlman and Vinson, 1975; Thompson, 1982; Karnavar, 1984; Hawlitzky and Boulay, 1986; see section 5.12).

7.25 HOST RELATIONSHIPS OF PARASITIC WASPS AT FAMILY LEVEL

Some families are extremely polyphagous whilst others are largely or entirely restricted to one host order (Appendix B). There is also great variability in the range of life history styles displayed by the members of a given parasitoid family. Some taxa can be facultatively primary parasitoids or hyperparasitoids; some attack members of many host orders; some can attack hosts at different stages of development, etc. The chalcidoid, *Pachyneuron concolor*, can develop as either a hyperparasitoid of encyrtids parasitizing various scale insects, mealy bugs or even coccinellid larvae, or as a primary parasitoid of Diptera, particularly on the puparium of the chamaemyiid, *Leucopsis*, a predator of aphids (Rosen and Kfir, 1983). Rosen and Kfir suggested that, in the case of *P. concolor*, host selection may be based on locating a soft-bodied host within a hard, dry external shell, and that it does not matter what the actual host taxon is. How the *P. concolor* egg and larva manage to circumvent the immune responses of such a diverse range of hosts remains a mystery.

8

Adult behaviour

8.1 HOST LOCATION

Much has been written about the various phases involved in finding suitable hosts (Weseloh, 1981; Vinson, 1985, 1988; Vet and Dicke, 1992). Initially the parasitoid must find the right habitat, if it is not already there; it must then find appropriate locations for its hosts (for example, the right type of plant) and then it must find a host. Sometimes hosts may not be attractive at all in isolation from their natural habitat (Matthews, 1974). These stages have been categorized as host location by Vinson (1985).

Once a potential host has been located, the parasitoid must confirm its identity (is it the right species?) and its suitability for parasitism (is it big enough, at the right stage, not already parasitized, etc.?) before finally making a decision to oviposit. These are categorized as host acceptance stages.

Parasitoids often home in on volatile chemicals associated with a host or likely host site, such as chemicals emanating from host frass, symbiotic fungi (Madden, 1968; Spradbery, 1974; Vinson, 1991), or damage to plants caused by host feeding. However, unlike various Diptera (Tachinidae) that attack Orthoptera and other sound-producing insects, parasitic wasps do not make use of airborne sound in host location, and parasitic Hymenoptera appear to lack tympanic ears. Vibrations through substrates, on the other hand, are probably important for many hymenopterous parasitoids of concealed hosts. Once a likely site has been located, parasitic wasps often show responses to another, less volatile set of compounds – for example, honey dew or trails – which require closer antennation. Once located, a potential host often has to pass certain tests. Its correct identity may be determined by water-soluble, non-volatile substances that are effectively tasted by the wasp (Strand and Vinson, 1982; Vinson et al., 1986; Bekkaoui and Thibout, 1992), and such behaviours may be mediated by the gustatory antennal sensilla present in some groups (Bin et al., 1989; Chapter 6). Host-related signals are usually only of interest to female parasitoids, and numerous behavioural and electrophysiological studies have shown that chemicals that attract females to food plants or hosts often elicit no response in males – though this is not necessarily the case (e.g. Rotundo and Tremblay, 1993).

Evolution has in general led to prey organisms giving off as few signals as possible that could alert predators to their presence; thus caterpillars and their frass are generally virtually odourless. There are obvious disadvantages to most potential host insects should they emit strong signals that could attract predators or parasitoids, and it is likely that most insects are as odourless and visually inconspicuous as evolution can make them. For this reason parasitoids have had to evolve to make use of cues emanating not only directly from their hosts but also those produced as a result of the host's actions, such as feeding and defecating.

Host location is not simply a matter of homing in on chemical cues but may also involve vision, sound, touch and even heat. Host location in the ichneumonid, *Pimpla instigator*, involves the wasp detecting echoes of vibrations that it transmits to the host substrate (Henaut and Guerdoux, 1982). In their experiments, Henaut and Guerdoux used the filter tips of cigarettes as a neutral, non-odoriferous substrate that the *Pimpla* females would probe with their ovipositors (Figure 8.1). When filters were hidden within a tube made of thin paper, the wasps were able to detect their presence and would probe them with their ovipositors. When thicker paper was employed, the ability to detect the filters was blocked.

8.2 HOST FRASS

Frass is a potent source of host location kairomones, especially for parasitoids whose hosts are exophytic and feed on low-growing plants (e.g.

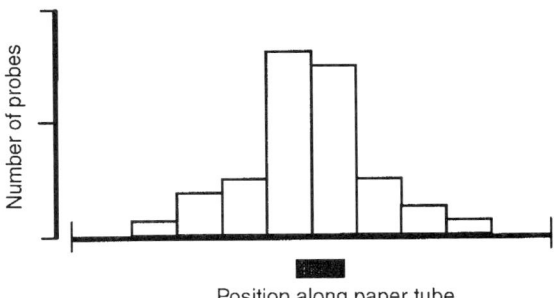

Figure 8.1 Frequency of ovipositor probings by the ichneumonid, *Pimpla instigator*, along a paper tube containing an oviposition lure (a cigarette filter, indicated by the shaded rectangle). *P. instigator* uses a form of echolocation to detect hosts. (Based on Henaut and Guerdoux, 1982.)

Lewis, 1970; van Leerdam *et al.*, 1985; Ding *et al.*, 1989; Agelopoulos and Keller, 1994), and may either cause long-range attraction or stimulate host searching (e.g. Odebiyi and Oatman, 1972; van Leerdam *et al.*, 1985). Several studies have isolated and identified the particular volatiles emanating from host frass that are attractive to parasitic wasps (e.g. Auger *et al.*, 1989; Nemoto *et al.*, 1987; Fukushima *et al.*, 1989; Ramachandran *et al.*, 1991). Ramachandran *et al.* identified several active volatiles from frass of soybean looper moth (*Pseudoplusia includens*) larvae that attract the braconid, *Microplitis demolitor*. Fukushima *et al.* identified 9,10-dihydroxy-cis-12-octadecanoic acid as the kairomone present in the frass of the almond moth, *Cadra cautella*, that is responsible for eliciting stinging behaviour in the braconid, *Habrobracon hebetor*. Interestingly, Nemoto *et al.* showed that the compound that acts as a kairomone for the ichneumonid, *Venturia canescens*, in the frass of both *C. cautella* and the Indian meal moth, *Plodia interpunctella*, belongs to a different group: 2-acylcyclohexane-1,3-diones. In the *Pseudoplusia–Microplitis* system, the frass kairomones depend on the caterpillar's diet and are not present, for example, when the *Pseudoplusia* are fed on artificial diet; the same effect was found for the braconid, *Macrocentrus grandii*, and its host, the European corn borer, *Ostrinia nubilalis* (Ding *et al.*, 1989). Thus frass kairomones may be part of a more complex tritrophic interaction. Frass can also provide information about the age of a potential host: for example, the pteromalid, *Trichomalus perfectus*, is stimulated to oviposit into rape pods in the absence of hosts if the pods are injected with 3rd instar host frass extract, but not with that of younger ones (Dmoch and Rutkowska-Ostrowska, 1978).

8.3 HONEYDEW AND OTHER APHID SECRETIONS

Honeydew, a sugary waste product of aphids and many other Homoptera, is another reliable indicator of the presence of hosts and has been shown to be an important source of kairomones for many aphid and coccid parasitoids (Bouchard and Cloutier, 1984). In addition to producing honeydew, aphids have specialized glandular defence organs which open at the prominent abdominal cornicles. The secretions produced from the cornicles of the pea aphid, *Acyrthosiphon pisum*, also act as host-searching kairomones for its aphidiine braconid parasitoid, *Aphidius ervi* (Battaglia *et al.*, 1993).

8.4 TRAIL FOLLOWING

Many insects leave chemical trails behind them when move around. Several parasitic wasps have been shown to be able to detect and follow these trails. For example, Howard and Flinn (1990) showed that experienced females, but not males or naïve females, of the bethylid,

Cephalonomia waterstoni, make use of the trails left behind by their cucujid beetle hosts, *Cryptolestes ferrugineus*. In this case, and probably in all others, the trails are transitory and in the above system they lasted for less than one week.

Klomp (1981) showed that the ichneumonid, *Poecilostictus cothurnatus*, a parasitoid of pine looper moth (*Bupalus pinarius*) larvae made use of the chemical trail left behind by larvae as they moved about. The wasp flies a short distance and walks a while; if it encounters a host trail, it follows it until either it finds a host or the trail vanishes. In the latter case, the wasp carries out area-specific searching; if that does not lead to a host or another trail, it flies to another site a short distance away. That the trails were the result of kairomones in the host's cuticle was demonstrated by dragging host caterpillar skins over a filter paper to form artificial trails. The braconid, *Microplitis mediator*, similarly follows the trail of its armyworm host, *Mamestra configurata* (Arthur and Mason, 1986). Many larvae produce mandible gland secretions as part of their normal feeding behaviour and in the case of the flour moth, *Anagasta kuehniella*, Corbet (1971) showed that this secretion causes the ichneumonid, *Venturia canescens*, to show oviposition movements. The secretion also probably functions in attracting the parasitoid or arresting its searching movements.

8.5 HOST PHEROMONES

Hiding oneself from attack by parasitoids is often incompatible with communicating with conspecifics. Several intraspecific chemical communication systems have been usurped by parasitoids in their search for suitable hosts (Vet *et al.*, 1991; Vet and Dicke, 1992; Wiskerke *et al.*, 1993; Dicke *et al.*, 1994; Leal *et al.*, 1995). The majority of these work either because the communicating adult hosts remain near their vulnerable immature stages, as in many gregarious bugs, or because the ovipositing female cannot avoid leaving chemical traces near her oviposition site. Some parasitoids have even managed to make use of host pheromones produced at a distance from suitable host life stages.

Normally, sex pheromones of hosts are not likely to be of much use to parasitoids as they are produced by adult insects, a stage for which only a few hymenopterous and other parasitoids have evolved to utilize as hosts (see section 1.10). However, when the adults themselves are potential hosts and when the adults necessarily live in close proximity with their offspring, then sex pheromones will undoubtedly carry information of interest to non-mates as well as mates. Other insects which form aggregations as the result of pheromonal communication – for example, many stink-bugs and soldier-bugs (Hemiptera: Alydidae, Lygaeidae, Pentatomidae) – leave themselves open to such attack as well (Harris and Todd, 1980; Aldrich *et al.*, 1984; Aldrich, 1995; Leal *et al.*, 1995).

That host sex pheromones may play an important role in host location by some parasitoids has become apparent from numerous studies of scale insects, aphids and other Homoptera in which pheromone traps were found to collect large numbers not only of male homopterans but also female parasitic wasps. In what appears to have been the first demonstration of a host sex pheromone attracting parasitoids, Sternlicht (1973) showed that sex pheromones of the California red scale, *Aonidiella aurantii*, attracted not only coccid males but also considerable numbers of two species of parasitic wasps, *Aphytis melinus* and *A. coheni*, both of which attack the *Aonidiella*. Rice and Jones (1982) caught specimens of the aphelinid, *Prospaltella perniciosa*, in traps using pheromones of the San Jose scale insect, *Quadraspidiotus perniciosus*, and similarly, aphid sex-pheromone traps have been found to be attractive to their specific aphidiine braconid parasitoids (Powell *et al.*, 1993; Hardy *et al.*, 1994).

Sex pheromones also appear to be important to some egg and egg–larval parasitoids of Lepidoptera because the egg masses or leaves are unavoidably contaminated with traces of the female's sex pheromones (Noldus *et al.*, 1991; Zaki, 1985). Often, moths only release their pheromones at night and the parasitoids that are attracted to them are only active by day (Noldus, 1989); the pheromone traces left can be extremely small, though still enough to elicit responses in the parasitoids. Similarly, many ovipositing Lepidoptera leave scales at or near their eggs, and the volatiles given off by the scales can also act as a stimulant for parasitoids (e.g. Chiri and Legner, 1986). In some lepidopterans, scale-shedding may be accidental, but in others it appears to be adaptive and could be involved in epideiotic behaviour. In the latter case, their scales are particularly likely to be giving off pheromone cues to parasitoids.

Yasuda and Tsurumachi (1995) showed that the scelionid, *Gryon pennsylvanicum*, a parasitoid of the eggs of the coreid bug, *Leptoglossus australis*, is attracted to the male sex pheromone of its host. Yasuda and Tsurumachi showed that egg masses of *L. australis* placed away from males were less heavily attacked by the *Gryon* than those located near male bugs. In this bug, females attracted to males appear to lay their eggs close to their mates and so the male sex pheromone acts as a reliable sign for host location. Another scelionid, *Telenomus euproctidis*, which is a phoretic species that oviposits in eggs of the tussock moth, *Euproctis taiwana*, has been found to occur most commonly on virgin host females, indicating that it is probably attracted by the female moth's sex pheromones (Arakaki *et al.*, 1996). It is not known whether any hymenopteran parasitoids of adult insects also home in on their hosts' sex pheromones, as do some tachinid flies (Mitchell and Mau, 1971).

The braconid, *Opius lectus*, makes use of the oviposition-deterring pheromone of its host fly, *Rhagoletis pomonella*, in order to locate sites where young host larvae are likely to occur (Prokopy and Webster, 1978).

8.6 THE ROLE OF SILK AS A SOURCE OF KAIROMONES

Many parasitic wasps have been shown to use host silk as a source of important host location kairomones. The wasps involved are mostly either pupal parasitoids or attack spider egg masses. For example, Weseloh (1977) showed that the braconid, *Cotesia* (as *Apanteles melanoscela*) respond-ed to silk of its gypsy moth host; Sandlan (1980) obtained similar results with *Pimpla* (as *Coccygomimus*) *turionellae*, a highly polyphagous parasitoid of lepidopteran pupae; and Frilli (1965a) reported that *Venturia* (as *Devorgilla*) *canescens* will probe through silk of its host, *Anagasta*, even if the host has been removed. Silk may also be attractive to some larval para-sitoids whose hosts make silken retreats (e.g. Ding *et al.*, 1989). The response to host cocoons by the ichneumonid, *Diadromus pulchellus*, does not depend on either the texture or appearance but rather on the presence of non-volatile and water-soluble compounds (Bekkaoui and Thibout, 1993). These compounds and the wasp's response are host-species specif-ic and independent of the host's diet.

Gauld (1988a) proposed that silk produced by hosts may be an important source of both volatiles and contact infochemicals for parasitic wasps, based on his deductions concerning a number of possible host group shifts among the Ichneumonidae. Many cryptine ichneumonids attack lepidopteran and other endopterygote pupae and prepupae that are concealed within cocoons. However, spider egg cocoons, within which the cryptine larvae are predators, are also attacked by members of several cryptine genera (Fitton *et al.*, 1987, as Phygadeuontinae), and members of one genus, *Obisiphaga*, have evolved to attack pseudoscorpion egg masses in which its larvae are also predators. A similar situation occurs in the Eurytomidae, with *Eurytoma* species that are predators within spider egg sacs being most closely related to species that are pseudohyperparasitoids of braconid cocoons (Hanson, in Hanson and Gauld, 1995). The most obvious connec-tion between these host groups is the presence, in all cases, of silk.

Austin (1985) notes that whilst some ichneumonid parasites of spider eggs are apparently closely related to other species that attack adult spi-ders, others such as various *Gelis* species (Cryptinae) and some mesostenines belong to groups that attack other hosts in cocoons, and therefore probably use silk as one of their main host location cues.

8.7 HOST-CREATED VIBRATIONS

Host-induced vibrations have long been suspected as playing an impor-tant role in location by parasitoids of concealed hosts such as wood-boring beetle or sawfly larvae whose feeding naturally produces sounds. Recent detailed work by Casas and colleagues using laser vibratometry has shown that vibrations caused by moving and feeding of the leaf-miner lar-vae, *Phyllonorycter malella*, provide a great deal of information about the position, activity and developmental stage of the potential host (Casas, 1989; Meyhofer *et al.*, 1994). The use of laser vibratometry is important because vibration patterns in the leaf can be greatly affected by applying mechanical pick-ups.

8.8 TEMPERATURE

Richerson and Borden (1971, 1972a,b) carried out a series of experiments on the host-locating mechanism of the braconid, *Coeloides vancouverensis*, a parasitoid of subcortical scolytid beetle larvae, *Dendroctonus pseudotsugata*, and Richerson and Borden (1972a) found no evidence for sound, sonar or chemical stimuli. They reported that the oviposition sites of *C. vancouverensis* were between 0.5 and 0.9°C hotter than points at one wasp-antenna's length in any direction around it. They then reported that artificial hot spots, created using small, host-sized heating coils under Douglas fir bark, were attractive to the parasitoid in the absence of hosts. However, the possibility that the heat created a substrate-related chemical cue that stimulated the wasps, rather than the heat itself being responsible, cannot be eliminated. Mills *et al.* (1991) tried to repeat Richerson and Borden's results using another scolytid–parasitoid system, this time involving the Eurasian scolytid beetle, *Ips typographicus*, and its braconid parasitoids, *Coeloides bostrychorum* and *Dendrosoter middendorffii*. Mills *et al.* failed to detect any host metabolism-induced 'hot spots' on the tree bark, but they did show that host location was significantly disrupted by a wax barrier, and that parasitoids would continue to probe bark from which hosts had been removed 24 hours previously or bark in which hosts had been killed by deep freezing. Collectively, these results strongly suggest that in this system heat plays little or no role in short-range host location, and that chemical cues most probably constitute the major set of cues. However, there is no fundamental reason why heat might not be used in host location if there are situations in which a host is associated with some sort of literal hot spot.

8.9 VISUAL CUES

Most work on host location by parasitic wasps has concentrated on chemical and other invisible cues, but many parasitoids undoubtedly also use their sight to find hosts (e.g. Sandlan, 1980; Henriquez and Spense, 1993). Host acceptance in two ichneumonid parasitoids of *Yponomeuta* moth larvae was studied by Dijkerman (1988). In both, the behaviour of the potential host was crucial for oviposition by the wasp, though the role of vision is not known for certain. In *Trieces tricarinatus*, it is important that the host is walking; in *Triclistus yponomeutae*, the host had to be performing its disturbance movements. Wasps can also be trained to associate hosts with visual targets (see section 8.17).

8.10 HOST PLANTS AND TRITROPHIC EFFECTS

It has been known for a long while that plants play a role in host location by parasitoids, and an increasing number of studies have identified specific plant volatiles that attract parasitic wasps to them (e.g. Elzen *et al.*, 1984; Takabayashi *et al.*, 1991; Vinson *et al.*, 1994b) or give rise to strong electroantennogram responses (LeComte and Pouzat, 1985; Ramachandran and Norris, 1991; Rotundo and Tremblay, 1993). In general, freshly damaged plants give off a distinctive odour, and gas chromatography shows that this is a complex mixture of volatiles – the so-called green-leaf volatiles. As shown by Ding *et al.* (1989) using the braconid, *Macrocentrus grandii*, a parasitoid of *Ostrinia nubilalis*, the European corn borer, the attractiveness of some plants to parasitoids of their herbivores is enhanced by simple physical damage of the plant, i.e. cut corn stems.

Much interest has recently been raised by the work of Turlings and colleagues which has shown that parasitoids of exophytic hosts may be attracted by compounds that plants release as a result not just of being damaged but more specifically by being damaged and at the same time exposed to saliva of the herbivorous host, as always happens when caterpillars are feeding (Turlings *et al.*, 1990, 1993a; Turlings and Tumlinson, 1991, 1992; Turlings, 1994). Unlike the normal green-leaf volatile emissions caused by simple mechanical damage, contact with insect saliva or regurgitate causes additional chemicals to be released, and in particular, specific terpenoids. The initial work dealt with corn seedlings damaged by larvae of the noctuid moth, *Spodoptera exigua*, and its braconid parasitoid, *Cotesia marginiventris*. Evidence has rapidly accumulated to suggest that the release by plants of volatiles (synomones) as a specific response to contact with caterpillar saliva is the rule rather than the exception (Geervliet *et al.*, 1994), and it has now been shown that this effect is not restricted to plants being attacked by caterpillars, but can also be elicited by aphid feeding, as was recently demonstrated by Du *et al.* (1996) for broad-bean plants infected with the pea aphid, *Acyrthosiphon pisum*. Further, at least some of the plant responses are systemic (Dicke, 1994), and the same effect can be obtained by placing cuttings in water with added caterpillar regurgitate. Potting *et al.* (1995) described an elegant series of experiments using a Y-tube olfactometer to quantify relative attractiveness of maize plants to the braconid, *Cotesia flavipes*, a parasitoid of the maize stem borer, *Chilo partellus*, after various manipulations. The results of one experiment are represented in Figure 8.2. In this experiment, the responses of *C. flavipes* to isolated maize leaves were compared for leaves from plants which had *Chilo* larvae in their stem, control plants with artificial holes bored in the stem, and plants with artificial holes into which regurgitate from *Chilo* larvae was injected. The *Cotesia* females showed a clear preference for leaves from plants treated with regurgitate over leaves from simply damaged plants. Of course, the plant volatiles released as a result of this systemic plant effect are only part of the full range of chemical interactions, and various degrees of attraction were elicited by plants alone, artificially damaged plants without regurgitate, host larvae alone, larval frass, and other host-associated cues.

The question arises as to whether or not this signalling system has evolved because plants gain benefit by attracting parasitoids. Turlings (1994) has shown that the release of caterpillar-feeding induced volatiles (terpenoids and indoles) by corn plants does not begin until several hours after the start of caterpillar feeding. The release of terpenoids lasts for approximately two days and is maximal during the daytime. Is this related to photosynthetic activity or is it specifically attuned to the host searching habits of *Cotesia*? Both would be maximal at midday. It would be interesting to know whether plants whose main pests and their parasitoids were nocturnal had a shifted response.

It should also be noted that not all plants seem to produce special volatiles in response to insect damage. In their study of the Brussels sprout–*Pieris brassicae*–*Cotesia glomerata* system, Mattiacci *et al.* (1994) showed that the plant's chemical responses to mechanical damage, caterpillar damage and caterpillar regurgitate were qualitatively similar and differed in that caterpillar-related damage produced a markedly enhanced release of normal green-leaf volatiles rather than inducing specific terpenoids or other compounds. In this system, *Cotesia*, the braconid parasitoid, seems to respond to the sprout's normal green-leaf volatile spectrum, and is attracted more by leaf treatments that produce the greatest release of these compounds.

Several questions remain to be answered about these tritrophic interactions. For example, how much energy does the release of these compounds cost the plant? And why don't the insect-induced plant volatiles attract other herbivores – or do they?

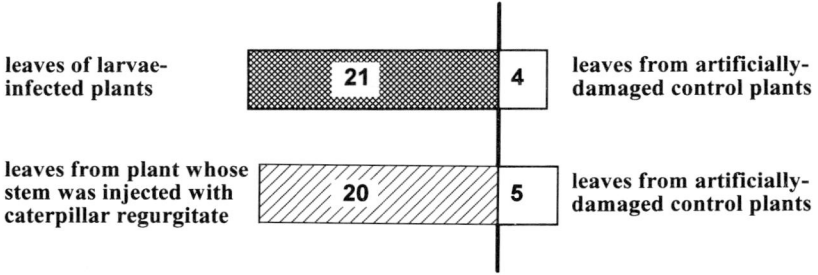

| leaves of larvae-infected plants | 21 | 4 | leaves from artificially-damaged control plants |
| leaves from plant whose stem was injected with caterpillar regurgitate | 20 | 5 | leaves from artificially-damaged control plants |

Figure 8.2 Attraction of female parasitoid, *Cotesia flavipes*, due to systemically induced volatile release in a maize plant caused by larvae of the moth, *Chilo partellus*, or larval regurgitate. (Redrawn after Potting *et al.*, 1995.)

8.11 OVIPOSITION STIMULI

Once the parasitoid has located and accepted a host, there needs to be a stimulus to initiate oviposition. In some cases, it has been shown that simple chemicals can suffice for this purpose, though in many cases, when the exact oviposition site is very precisely defined (for example, in those taxa that only oviposit in the host's ganglia), it seems likely that more sophisticated stimuli may be involved. Amino acids, but not all and not the same ones in all cases, often act as oviposition stimuli in the endoparasitoid braconids and ichneumonid, *Ascogaster reticulatus* (Kainoh *et al.*, 1989), *Chelonus* nr *curvimaculatus* (Kainoh and Brown, 1994) and *Itoplectis inquisitor* (Arthur *et al.*, 1972).

8.12 KLEPTOPARASITISM

A few parasitic wasps are facultative or obligate kleptoparasitoids of other parasitic wasps. That is, they attack individuals that have already been attacked and consume them after initially killing the primary parasitoid. The ichneumonid, *Pseudorhyssa*, has been known for a long time to be a kleptoparasitoid of members of the related genus *Rhyssa*, members of which attack several large wood-boring hosts but most commonly siricid woodwasp larvae. *Pseudorhyssa* females locate ovipositing *Rhyssa* or their borings and, after the *Rhyssa* has laid her eggs, insert their ovipositors down the boring made by *Rhyssa*'s thicker ovipositor (Couturier, 1949; Spradbery, 1969). Despite the smaller diameter of the *Pseudorhyssa* ovipositor, her egg is larger than that of *Rhyssa* and hatches quickly. The 1st instar *Pseudorhyssa* larva consumes the *Rhyssa* egg and then proceeds to consume the paralysed host of the *Rhyssa*.

Two further examples are afforded by eurytomids. *Eurytoma waachti* only attacks its *Pissodes* weevil larvae hosts after they have been paralysed by ichneumonids of the genus *Scambus* (Roques, 1976), whilst *E. pini* only parasitizes larvae of the European pine shoot moth, *Rhyaciconia buoliana*, that have been paralysed by any of various other wasps (Arthur, 1961). In the latter case, it was found that *E. pini* will oviposit in coddled hosts, so it is presumably a lack of movement on the part of the host that is important in the decision to oviposit rather than any chemical signals from the primary parasitoids. In these examples, the associations are obligate, but in some cases two parasitoids may have a more facultative kleptoparasitic association. For example, Arthur *et al.* (1964) showed that the ichneumonid, *Temelucha interruptor*, a parasitoid of the pine shoot moth, *Rhyacionia buoliana*, was not very good at finding its host, but was good at finding hosts that had already been parasitized by the braconid, *Orgilus obscurator*. The *Temelucha* could readily develop in unparasitized

Rhyacionia larvae, and when it attacked already parasitized ones (multi-parasitism) it presumably killed the existing *Orgilus* larva.

8.13 HOST ACCEPTANCE

Host acceptance is the term used to describe the decision of a wasp to oviposit in a host once it has been located. It usually follows a series of steps involving a number of different cues. The final cue is often a compound associated directly with the host; for example, the glue that attaches host eggs to a plant. Acceptance can occur at the time of first external contact or examination, or the final decision to lay may not take place until the wasp has made an attack and inserted its ovipositor (e.g. Pennacchio *et al.*, 1994a).

Several studies have looked for a relationship between host acceptance and the host's ability to encapsulate the parasitoid's eggs. In polyphagous and oligophagous species, it might be expected that selection will have led them to prefer host species with less well-developed defence reactions, all other things being equal. Kraaijeveld *et al.* (1995) found exactly this for the alysiine braconid, *Asobara tabida*, and its *Drosophila* hosts. *A. tabida* will attack several *Drosophila* species occurring in decomposing fruit, but its probability of avoiding encapsulation depends on the *Drosophila* species and strain involved. The order of preference of *A. tabida* for its potential hosts has been shown to correlate well with its survival chances (van Alphen and Janssen, 1982). The effect is also found on an individual wasp basis. Thus, when Kraaijeveld *et al.* presented *D. melanogaster* larvae to individuals of a particular *A. tabida* strain, those females that had the highest offspring survival rates were most likely to accept a larva once encountered.

8.14 DISCRIMINATION AND SUPERPARASITISM

An ability to distinguish between hosts that are suitable for oviposition and ones that are not is broadly termed host discrimination – for example, healthy hosts versus sick ones, parasitized versus unparasitized, large versus small. However, the term is most frequently used nowadays to indicate discrimination between unparasitized hosts and previously parasitized ones (van Lenteren *et al.*, 1978). Most parasitoids studied appear to be able to discriminate between parasitized and unparasitized hosts, but this does not appear to be true of all species. For example, several studies failed to find any evidence for host discrimination by *Microplitis croceipes*, a braconid parasitoid of *Heliothis* larvae (Lewis and Snow, 1971; Vinson and Guillot, 1972; Eller *et al.*, 1990).

Superparasitism, as opposed to multiparasitism, is the laying of more eggs in a host individual than is normal for the parasitoid. A female may self-superparasitize but more typically superparasitism results from two different parasitoid females finding and ovipositing into the same host. Almost by definition, superparasitism generally leads to a reduction in fitness of one or both of the parasitoids because a given host cannot normally support two broods. Superparasitism has therefore attracted considerable attention from the point of view of how wasps generally manage to avoid superparasitizing by being able to discriminate between hosts that have and those that have not already been parasitized.

The success of the second female in ovipositing in a host often depends greatly on the time interval between her own and the previous female's attacks. Eller *et al.* (1990) showed that in *Microplitis croceipes* the survival probability of a superparasitizing larva is equal to that of the first laid egg if the interval between ovipositions is less than a day. They were able to distinguish between eclosing adults from the first and the superparasitizing ovipositions by their sex, having used one virgin and one mated female for the two attacks (with appropriate controls). This technique is suitable where it can be demonstrated that there is no intrinsic difference in the developmental success between the sexes. Previous studies, using genetically marked individuals, have sometimes been problematical because the genetically marked strain may be less fit (Bakker *et al.*, 1985).

In many endoparasitic wasps, the 1st instar larva has distinctive sickle-shaped and proportionately larger mandibles than later instars. This feature is believed to be primarily an adaptation to preventing super- and multiparasitism. Interestingly, in trigonalyids it is the 3rd instar larva that has massive mandibles for eliminating competitors. The eggs of these hyperparasitoids are consumed first by the secondary host caterpillar, which may consume many such eggs, but these hatch to form quiescent 1st instars, at which stage they remain unless their secondary host is subsequently parasitized by a tachinid fly or ichneumonid wasp primary host. Only then do specialized combative forms of larva develop. There is also evidence that the development of a superparasitizing larva can be inhibited by physiological suppression induced by secretions of the first larva (Smith, 1952; Salt, 1961).

Discrimination can be broken into three categories: interspecific host discrimination, intraspecific discrimination and self-discrimination. Avoidance of intraspecific superparasitism, especially by solitary parasitoids, will be advantageous under most circumstances. There are exceptions, and these have been explored in considerable detail by a number of groups (e.g. van Alphen and Visser, 1990; Speirs *et al.*, 1991; Visser *et al.*, 1992). Basically, superparasitism will usually be disadvantageous because it will often mean wasting an egg as well as the time spent ovipositing. Therefore, if a female can detect at an early stage that a potential host is already parasitized, she can save time and her egg by moving on to try and find another, unparasitized host. However, superparasitism can also be adaptive if a number of criteria are satisfied, i.e.: unparasitized hosts are rare; superparasitized eggs have a better than zero chance of survival; the wasp is not egg limited; and more than one female is foraging for hosts within the same patch. Weisser and Houston (1993) have reviewed the evidence for both self and non-self superparasitism, and provide a dynamic programming model for investigating the criteria that may make it adaptive.

Superparasitism can also be adaptive in certain physiological situations. For example, in the encyrtid, *Comperiella bifasciata*, females of the Californian strain are poorly adapted to parasitize the California red scale, *Aonidiella aurantii*, compared with an Israeli strain of the same parasitoid species – because, when laid singly, eggs of the Californian strain suffer a relatively high level of encapsulation (Blumberg and Luck, 1990). However, encapsulation of Californian *C. bifasciata* eggs is markedly reduced when the wasp superparasitizes; and, interestingly, the Californian wasps superparasitize far more frequently than their Israeli conspecifics.

Host discrimination is presumably much easier for ectoparasitoids or in species where there is an obvious external marker. The highly specialized eggs of encyrtids with their aeroscopic plate that remains connected to the outside of the host provide just such a marker, and several workers have produced evidence that the external egg stalk is used by females in their host discrimination decisions (van Baaren and Nénon, 1994). There is considerable variation in the ability of parasitoid species to discriminate between hosts parasitized by conspecifics. Van Lenteren *et al.* (1978) provide a number of examples of both good and bad discriminators. Arakawa (1987) described an interesting case of ovicide by superparasitizing *Encarsia* in which the second female uses her ovipositor multiply to pierce and kill the egg of the first female in the whitefly host. Similar ovicidal behaviour has been reported for the ectoparasitic braconid, *Habrobracon hebetor* (Strand and Godfray, 1989; Antolin *et al.*, 1995), and superparasitizing female bethylids also actively destroy the brood of the first female (section 8.27). Interspecific ovicide occurs in some if not most inquilines of cynipid gall wasps, again the inquiline female using her ovipositor to destroy the original gall-former's egg (Shorthouse, 1980)

Van Dijken *et al.* (1992) have surveyed the evidence for self-discrimination in parasitoids and found that six out of the nine species for which relevant information was available appeared to be able to recognize eggs parasitized by themselves. Subsequently this ability has been found in a few other taxa (e.g. Danyk and Mackauer, 1993; Ueno, 1994). The ability of the ichneumonid, *Itoplectis naranyae*, to distinguish between self-parasitized and non-self-parasitized hosts depends on the time that has lapsed after the first parasitization event (Ueno, 1994). Females could detect self-parasitized hosts externally up to 30 minutes following initial parasitization, and strongly avoided self-superparasitism. However, *I. naranyae* appears unable to distinguish between hosts parasitized by full sisters, half sisters or non-kin.

Host discrimination is not only important for wasps to avoid superparasitism, it is of course also important for hyperparasitoids, which have to detect hosts that have previously been parasitized by the correct primary. Baur and Yeargan (1994a) examined host discrimination by the ichneumonid, *Mesochorus discitergus*, an endoparasitic hyperparasitoid in the primary endoparasitoid braconid, *Cotesia marginiventris*, in various larval noctuid hosts. They showed that the ichneumonid can discriminate parasitized (suitable) hosts from unparasitized ones after about one hour. Injection of crude extracts from different parts of the *C. marginiventris* reproductive system into unparasitized hosts showed that components of the calyx region, and particularly of extracts of the venom apparatus, increased handling time in a similar, though less extreme, way as caused by true parasitism. This indicates that host discrimination kairomones for the *Mesochorus* may either be contained within the venom system of the primary, or that components of the latter interact with the host's haemolymph or other tissues to produce such kairomones. Chemicals might also be released from the egg, embryo or larva of a parasitoid within a host.

Some interesting cases of discrimination occur in various heteronomous autoparasitoids (see section 2.26) – for example, *Encarsia deserti*, an aphelinid which attacks the whitefly, *Bemisia tabaci* (Gerling *et al.*, 1987). Virgin, male-producing females did not discriminate between suitable hosts (that is, ones that already contained a mature *E. deserti* larva) but mated females, which only ever produce female offspring, discriminated in that they only laid in unparasitized *Bemisia*.

Discrimination between patches rather than between individual hosts is displayed by females of the aphidiine braconid, *Aphidius uzbekistanicus*, which detects when its hyperparasitoid, *Alloxysta victrix*, is also present in a patch of hosts, and as a result leaves the patch (Höller *et al.*, 1994b). Analysis of air blown over *Alloxysta* individuals showed that they give off considerable amounts of 6-methyl-5-hepten-2-one (see also section 6.44), and this chemical was shown to be largely or entirely responsible for the patch-leaving response of the *Aphidius*.

8.15 HOST MARKING

Chemical secretions are probably involved in most instances in which a female parasitoid leaves behind a definite signal that she has already attacked a particular host or patch of hosts. However, in egg parasitoids of the Scelionidae, physical scratching of the host egg seems to be a significant marker, even if not the only one. For example, in *Telenomus sphingis*, following oviposition a female backs over her host with the tip of her ovipositor and scratches it on the host egg's surface in a sinuous fashion. The whole marking process takes about a minute and a half per host, and about 77% of host eggs are marked (Rabb and Bradley, 1970). Of the eggs that were not marked, it turned out that the *Telenomus* female had only actually oviposited in 15% of them. That the marking is an effective deterrent to other females was clearly demonstrated in another *Telenomus* species by Strand and Vinson (1983), who removed eggs from a female either after drilling but before oviposition or after oviposition but before marking, and presented them to another female. The second *T. heliothidis* accepted 96% of the first category and 94% of the second; in comparison, only 4% of second females accepted eggs that had been marked.

Guillot and Vinson (1972) provided evidence that host marking in the ichneumonid, *Campoletis perdistinctus*, involves a chemical from the Dufour's gland, and Höller *et al.* (1994a) provided evidence suggesting that one of the chemical markers used by the megaspilid hyperparasitoid, *Dendrocerus carpenteri*, is juvenile hormone, though its source remains to be determined. In the latter case, the use of JH appears to have arisen as a by-product of the application of this hormone to its aphidiine braconid primary parasitoid host, as a developmental regulator.

It might be thought that in the case of ectoparasitoids of exposed hosts, or at least of hosts that the female parasitoid can visually examine, marking would be unnecessary since a parasite egg would be quite obvious. Barrera *et al.* (1994) have provided experimental evidence that in the case of the bethylid, *Cephalonomia stephoderis*, host-marking pheromones are also employed. How widespread such pheromones are among ectoparasitoids that have complete physical access to their hosts remains to be determined.

8.16 MULTIPARASITISM

Whereas superparasitism is the use of a single host more than once by a single parasitoid species, multiparasitism is the use of a single host individual by two or more different parasitoid species. Multiparasitization events nearly always result in the death of one of the parasitoids and quite often both (Strand *et al.*, 1990a), though which survives depends both on the species and on the interval between ovipositions. Usually one of the species has a fairly consistent advantage.

It may at first seem surprising, therefore, that many species fail to discriminate between hosts that have been previously attacked by another species and still oviposit in these (Vinson, 1976; Weisser and Houston, 1993). For example, neither the braconid, *Asobara tabida*, nor the eucoilid, *Leptopilina heterotoma*, which are sympatric primary parasitoids of *Drosophila* larvae, shows any sign of interspecific host discrimination (van

Strein-van Liempt and van Alphen, 1981) though both show marked intraspecific host discrimination. The same result was obtained by Okuda and Yeargan (1988) for two scelionid parasitoids of stink bug eggs, *Telenomus podisi* and *Trissolcus euschisti*. This was partly explained by Bakker *et al.* (1985) who showed that the first species to start to discriminate will generally be at a disadvantage, but Turlings *et al.* (1985), who modelled this, showed that even under extreme conditions there will be no advantage in avoiding multiparasitism as long as females have lots of eggs. The latter criterion will not always hold, of course, and experimental evidence from Mackauer's group using competing species of aphid parasitoid suggests that selection pressure to evolve interspecific discrimination is greater in the species whose offspring are less competitive and so more likely to be killed (Chow and Mackauer, 1985, 1986; McBrien and Mackauer, 1990, 1991). A similar situation appears to apply to scelionid parasitoids of stink bug eggs (Weber *et al.*, 1996) and hyperparasitoids of aphids (Scholz and Höller, 1992).

The aphelinid, *Aphytis holoxanthus*, not only discriminates diaspid scale hosts containing eggs of the related *Pteroptrix smithi*, but actually selects in favour of them, the *Aphytis* being a superior competitor at this life history stage (Steinberg *et al.*, 1987). No distinction of hosts containing 1st instar larvae was found, perhaps because the advantage has lapsed by this time. In the case of two mymarid parasitoids of weevil eggs, *Anaphes sordidatus* and *A. n. sp.*, van Baaren *et al.* (1994) suggested that the ability to display interspecific host discrimination might reflect their recent speciation from one another, as the infochemicals involved will probably not have diverged much from when they were used in intraspecific discrimination.

Occasionally, multiparasitisms have been found to be successful with offspring of both attacking species surviving to adulthood. One such case is described by Miller (1982) and concerns the alfalfa looper, *Autographa californica*. Of its six common parasitoids, successful multiparasitism only occurred frequently between the microgastrine braconid, *Apanteles yakutatensis*, and the tachinid fly, *Madremyia saundersoni*. Browning and Oatman (1984) provided another example involving a tachinid, this time with an encyrtid. Why successful multiparasitism can occur between these two and not, or hardly ever, between any of the other possible parasitoid pairs seems to depend on several factors. Probably most important is that neither of these two parasitoids would normally consume the entire contents of the host in single parasitism development, so there would be food left over that could suffice for the second species. Secondly, the fly maggot has no mandibles suitable for attacking multiparasitizing competitors, whereas this species of *Apanteles* is gregarious and so its larvae may not have evolved to show aggression against others they may encounter in a host.

8.17 LEARNING

Learning to associate environmental stimuli with hosts, as well as food, has been an important factor in the success of the parasitic Hymenoptera and has recently been reviewed by Turlings *et al.* (1993). It is now known that such learning is widespread, if not ubiquitous, and that it can involve any of a large range of stimulus types. There is also some evidence that learning ability may be inherited and so different parasitoid genotypes may be more or less influenced by their experiences (Prevost and Lewis, 1990). Host-location stimuli can be experienced either at the time of eclosion, or later during encounters with hosts. There is also some evidence that parasitic wasps may be able to learn host-location cues prior to their emergence, and perhaps even ones that they are exposed to as larvae (Turlings *et al.*, 1993b; Cortesero and Monge, 1994), though such cues are probably of less importance than ones learned post-emergence. It seems quite likely that volatile chemicals may be detected by pharate parasitoid pupae, but it is harder to imagine how chemicals sensed by parasitoid larvae could still be recognized by the adult. One possibility is that traces of chemicals may be carried over to the adult stage, either on or within a parasitoid, and that it is these that act as host-location cues – this is referred to as the chemical legacy hypothesis (Corbet, 1985). Carrying out experimental manipulations that exclude the possibility of contamination with host or host-habitat odours is very difficult, and in any case it must be nearly impossible to control for compounds stored within the parasitoid.

In a very impressive series of experiments in a flight chamber, Wäckers and Lewis (1994) showed that the braconid, *Microplitis croceipes*, a parasitoid of *Helicoverpa zea* larvae, can learn to associate both odours and visual cues with the presence of hosts (Figure 8.3). Parasitoids were initially trained to find host larvae located on a target that was identified with an extract of host frass from hosts that had been fed on cotton leaves. The alternative non-reward target was marked with a frass extract from larvae fed on cotton petals. In the reciprocal test, the odour cues were swapped. In the tests the wasps were presented with just the two odour-marked targets, and it was found that they showed a highly significant preference for the odour which was associated with hosts during training (Figure 8.3). In a second choice experiment, Wäckers and Lewis trained wasps to find hosts on one of two differently coloured background targets – one completely orange, the other with black and white stripes. Again in tests, *M. croceipes* females demonstrated a very clear preference for the target colour on which hosts had been presented during training (Figure 8.4). The general shape of the host microhabitat can also provide learning cues. Females of the ichneumonid, *Exeristes roborator*, were trained by Wardle and Borden (1990) to locate hosts presented either in a styrofoam cylinder or in a styrofoam sphere. After training, the wasps concentrated their searching on the artificial microhabitat that they had come to associate with the presence of hosts.

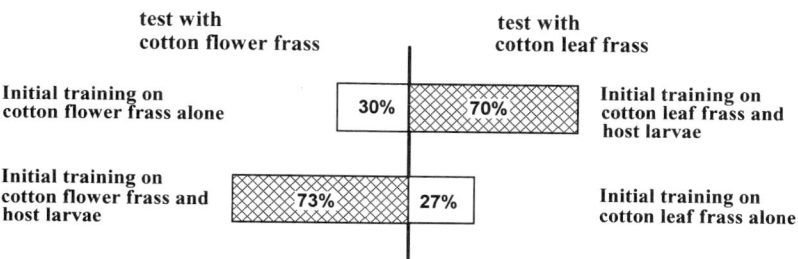

Figure 8.3 Ability of the braconid, *Microplitis croceipes*, to learn to associate a particular odour with hosts, *Helicoverpa zea*. (Redrawn from Wäckers and Lewis, 1994.)

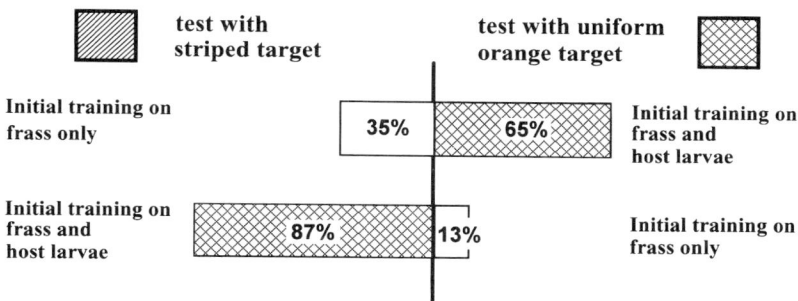

Figure 8.4 Ability of the braconid *Microplitis croceipes* to learn to associate a particular coloured target with hosts, *Helicoverpa zea*. Wasps were trained reciprocally on non-reinforcing targets (host frass only) and rewarding ones (host frass and hosts), and then tested for their preference for their visual target site. (Redrawn from Wäckers and Lewis, 1994.)

The physiological state of the wasp can be important in determining its searching strategy (Wäckers, 1994). Lewis and Takasu (1990) and Takasu and Lewis (1993) showed that parasitic wasps could learn to associate different odour cues with either hosts or food, and in flight-tunnel tests. In these experiments, the authors trained *M. croceipes* to associate particular odours – specifically larval frass and vanilla – with either hosts or sugar solution, respectively. Wasps preferentially responded to the odour cue associated with food if they were hungry or the odour cue associated with hosts if they were well fed (Figure 8.5).

8.18 ADULT DIET

Adult parasitoids seldom feed on solid matter but liquid food sources of both animal and plant origin are often of great importance in determining

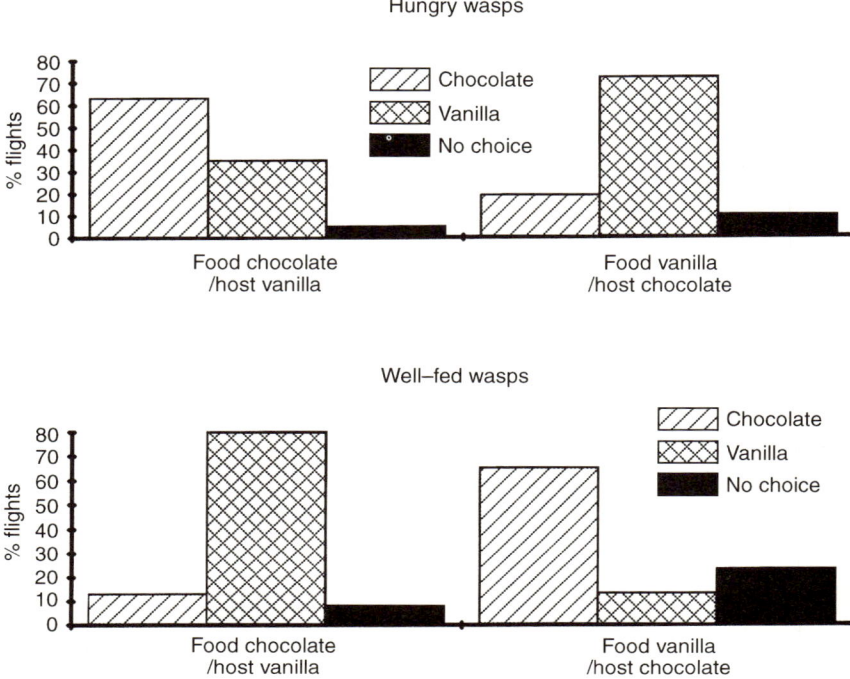

Figure 8.5 Effect of parasitoid hunger status on attraction to odours that it has been trained to associate with either food or hosts. (Redrawn after Lewis and Takasu, 1990.)

the longevity and fecundity of a wasp (Leius, 1961). Adult feeding can be divided for simplicity into host-feeding, which is probably largely associated with gaining proteins, and sugar-feeding for energy; these will be dealt with separately below. In addition, most wasps need to drink regularly, and water availability may be limiting in some places. The effects of adult diet on the longevity, histology and decrepitude of the pteromalid, *Nasonia vitripennis*, were investigated in a series of papers by Davies (1974, 1975, 1977), which, having been published in the non-entomological literature, have since received little attention from entomologists.

8.19 HOST-FEEDING

Many parasitoids, and particularly idiobionts which are often synovigenic, need to feed as adults in order to gain nutrients (Bartlett, 1964). The most frequently used source of such nutrients is host haemolymph, which in most cases of host-feeding is obtained through a hole made by the wasp's ovipositor (e.g. Aubert, 1959). However, in a number of ichneumonids, the puncture is made by the wasp's mandibles (Jervis and Kidd, 1986); for example, members of the metopiine genus, *Chorinaeus* (Aeschlimann, 1974). In some cases where the host is weakly concealed, the wasp may construct a feeding tube to gain access to the host's haemolymph (section 8.20). Brothers (1972) reported that the mutillid, *Pseudomethoca frigida*, also host-feeds, but in this aculeate parasitoid of larval bees the female will consume one or more host pupae or prepupae within a host nest rather than simply drinking some haemolymph. At least some mutillids may live more than a year and so it is not surprising that they host-feed in addition to consuming plant exudates.

Host-feeding can be divided into two types: concurrent and non-concurrent. In concurrent host-feeding the parasitoid feeds from the same host as she oviposits in, and in non-concurrent host-feeding the parasitoid feeds from other individuals (Jervis and Kidd, 1986). Because concurrent hosts will be used as hosts, it is obvious that the host must remain alive and sufficiently 'healthy' for the parasitoid larva to complete its development. In other words, concurrent host-feeding must be non-destructive, at least in the majority of instances. In contrast, non-concurrent host-feeding does not need to leave a viable host, and it is frequently, though not always, destructive. Some wasps only host-feed either concurrently or non-concurrently; others may show both types of feeding, but non-concurrent host-feeding appears to be the most frequently observed strategy among the parasitic Hymenoptera. Non-concurrent host-feeding is not only significant from the point of view of the survival and productivity of a single female but it can also lead to the deaths of far larger number of the host species than the wasp actually parasitizes.

A survey of host-feeding was carried out by Jervis and Kidd (1986) who found references to host-feeding in 26 species of ichneumonoid, 92 chalcidoids, one platygastroid, 19 chrysidoids and three parasitic vespoids, though these represent only a tiny proportion of the wasps that carry out host-feeding. Although Jervis and Kidd were unable to find details of egg type (hydropic versus anhydropic) or egg maturation strategy (pro-ovigenic versus synovigenic) for about a third of these cases, all those that were scorable had anhydropic eggs or were synovigenic. Not surprisingly, therefore, one of the functions of host-feeding seems to be to obtain sufficient proteins and other nutrients in order to mature more eggs. However, in some cases at least, it also seems to be important in supplying metabolic needs. Thus, Collier (1995) showed that in the aphelinid, *Aphytis melinus*, wasps that were provided with honey and water but were deprived of hosts did not simply stop maturing eggs, but resorbed those they already had, and then died younger than wasps that had had access to hosts. Bracken and Nair (1969) have provided evidence that in one ichneumonid, *Exeristes comstockii*, the wasp may need to host-feed in order to obtain juvenile hormone from its host which the wasps needs to stimulate egg maturation (Chapter 4).

It has been postulated that wasps should be more likely to host-feed when their supply of mature eggs is running low. Rosenheim and Rosen (1992) conducted an egg-load manipulation experiment in the aphelinid, *Aphytis lingnanensis*, which made use of the fact that this species is more likely to oviposit in larger hosts. Surprisingly, no effect of egg load was detected in *A. lingnanensis*, though in other wasps the initial prediction has been found to be borne out.

In the great majority of cases, host-feeding is, as its name suggests, limited to actual host species, but several exceptions are known. In one *Itoplectis* (Ichneumonidae: Cryptinae), feeding involves a non-host species, though this lives in the same place as *Itoplectis*'s natural host. Other examples are known among the Dryinidae (Chua and Dyck, 1982; Jervis *et al.*, 1992). It has also been shown that, for some species, hosts used for host-feeding do not actually need to be alive at the time. Thus, Thompson (1975) used coddled (killed by heating in 52°C water bath for 5 minutes) larvae of pink bollworm, *Pectinophora gossypiella*, to feed the ichneumonid, *Exeristes roborator*. When coddled larvae were placed under tissue paper, the ichneumonids would tear the tissue paper with their mouthparts and feed on the larvae.

Some host-feeders kill considerably more hosts by host-feeding than they do by oviposition (DeBach, 1943), and so it might seem likely that it would have a considerable effect on parasitoid–host population dynamics. These possible effects were examined by Kidd and Jervis (1989) using modified Nicholson–Bailey equations. They estimated that as many as a third of hymenopteran species may regularly host-feed, but their modelling suggested that in most situations this would have little impact on host population dynamics. Briggs *et al.* (1995) employed dynamic modelling to examine the likely effects of host-feeding on parasitoid–host populations and came to similar conclusions, i.e. the incorporation of host-feeding had no effect on the stability of the parasitoid–host interaction. Thus from a biological control perspective, despite the fact that in many cases hosts that serve as food for adult wasps are killed, both of the above modelling exercises suggest that host-feeding itself will not make a potential biocontrol agent any more effective – contrary to the common opinion of many practising biocontrol workers. However, in a further and empirical exploration of this, Jervis *et al.* (1996) compared the success of destructive host-feeding versus non-destructive host-feeding parasitoids in past biocontrol programmes for the control of homopterous pests. While recognizing the limitations of such analyses because of inconsistent and inaccurate reporting, they found that destructive host-feeders are in fact superior to non-destructive ones, both in terms of the probability of them becoming established and the level of control achieved. In order to explain this discrepancy between practice and theory, Jervis *et al.* suggested that when alternative non-host food sources are relatively abundant, non-destructive feeders may prefer non-host food sources. Clearly much more work is needed about what adult parasitoids actually are feeding on in the field.

Jervis *et al.* (1992), who considered feeding in adult parasitoids in general, proposed some guidelines for identifying the sources of foods consumed. These ranged from labelling potential sources – for example, with rare, non-toxic elements or radioactive isotopes – and subsequently detecting these in parasitoids, to serological and other analyses of gut contents. Such techniques have been applied little among the parasitic Hymenoptera, though Cate *et al.* (1974) demonstrated host-feeding by one aphelinid by labelling the host aphid with C[14]-labelled artificial diet.

8.20 FEEDING TUBES

In some parasitic wasps, host-feeding is a more complicated matter than simply imbibing haemolymph that oozes from the hole made by the wasp's ovipositor. Such direct feeding is only possible if the wasp is in direct contact with its host and would be impossible for taxa whose hosts are concealed. In these cases many wasps construct feeding tubes, which are formed by secretions released from the ovipositor. These secretions are believed to originate in the Dufour's gland in braconids, but Flanders (1934) and King and Ratcliffe (1969) suggested that in pteromalids they are produced by the large colleterial glands (Chapter 6). Feeding tubes have been described for quite a wide range of parasitoids, including in the Aphelinidae (Rosen and DeBach, 1990), a range of Pteromalidae (King and Ratcliffe 1969; Pupedis, 1978) and the braconid, *Habrobracon* (Génieys, 1925). Construction of the feeding tube in the pteromalids, *Habrocytus cerealellae*, *Nasonia vitripennis* and *Sisyridivora cavigena*, was described in detail by Fulton (1933), Edwards (1954) and Pupedis (1978), respectively, and in the eupelmid, *Eupelmus urozonus*, by Delanoue and Arambourg (1965). In all cases, the authors noted that a clear fluid emerged from the wasp's ovipositor along its entire length and congealed around it. Pupedis also noted that one of his study wasps ate the feeding tube she had constructed after completing host-feeding. This may seem logical, given the investment that the tube represents, but recycling does not appear to be a common practice.

8.21 OTHER TYPES OF FOOD

It has been noted that members of many groups of hymenopteran parasitoids frequent flowers in order to obtain nectar and perhaps also pollen

(Maingay *et al.*, 1991; Jervis *et al.*, 1992). Others have been observed to collect extrafloral nectar (Nishida, 1958; Keeler, 1978; Bugg *et al.*, 1989) or trichome exudates (Olson and Nechols, 1995) or homopteran honeydew (Starý, 1970; Wharton, 1993). Jervis *et al.* (1992) list a number of studies that have shown parasitization rates to be higher in situations where the parasitoids have access to nectar or other food sources.

Although no parasitic wasps are known specifically to consume plant pollen directly, Leius (1963) has shown that pollen included within sugar water can extend the life span or increase the fecundity of the ichneumonid, *Scambus buolianae*. Only females were affected, and different pollens had different effects. Pollen grains have also been detected in the guts of adult tiphiids as well as in a number of various non-parasitic, carnivorous aculeates (Hunt *et al.*, 1991). As pointed out by Jervis *et al.* (1992), however, the mere presence of pollen in an adult hymenopteran's gut does not demonstrate conclusively that pollen-feeding has occurred since pollen grains are ubiquitous and often become trapped in numbers on honeydew, damp surfaces, etc. Hard plant material is apparently consumed by the eulophid, *Pediobius*, adults of which have been seen to consume plant epidermal strips (Askew, 1971).

Many aphidiine braconids, a largely pro-ovigenic group whose members do not host-feed, nevertheless rely heavily on the availability of honeydew for fluid, energy and perhaps other nutrients (Starý, 1970). One species of the aphelinid genus, *Coccophagus*, has even been noted to stroke its coccid host with its antennae, like ants and aphids, to induce it to produce honeydew which it then consumes (Cendana, 1937).

Perhaps the most amazing food source also involves an aphidiine. *Paralipsis enervis*, a parasitoid of ant-attended root aphids, receives food from the ant, *Lasius niger*, by trophallaxis (Völkl *et al.*, 1996). The parasitoid apparently mimics the behavioural signals of the ant in order to stimulate the trophylactic response, and the food so received markedly increases parasitoid longevity.

Jervis *et al.* (1993) discuss the 'conventional wisdom' that many parasitoids commute between sites where hosts are found and sites where food is found, but the available evidence for this is at present largely circumstantial. They suggest that marking individual parasitoids would provide a means of testing this wisdom but as yet no such studies appear to have been carried out with parasitic wasps. Several appropriate marking techniques are available (Chapter 10) and so it is to be hoped that some real data on this possibility will not be too long in coming.

8.22 OVIPOSITION SITE

Precise oviposition siting or egg placement is known to occur in many parasitic wasps, and is generally believed to represent adaptations for overcoming host immune defences. A number of examples of specific egg

placement are provided in Chapter 7, but this is for the purpose of illustrating the range of variation rather than attempting to be comprehensive.
In other cases, wasps may make use of relatively easily penetrated parts of
a host's body in order to get their egg inside. For example, members of the
euphorine braconid genus, *Microctonus*, which attack adult halictid beetles, oviposit into the host's haemocoele through the mouth or the anus
(Wylie, 1985).

8.23 PHORESY AS ADULTS

Members of several groups of small parasitic wasps exhibit phoresy as
adults (Clausen, 1976). This behaviour is perhaps best known in parasites of
the families Scelionidae (Channa Basavanna, 1953) and Trichogrammatidae
(Pinto, 1994), but also occurs in some eulophids, eupelmids, podagrionine
Torymidae, and at least one encyrtid and pteromalid (Naumann and Reid,
1990). All of these are exclusively egg parasitoids. In the case of scelionids,
hosts include the eggs of praying mantids, grasshoppers or bugs; podagri-
onines attack mantid eggs; and those trichogrammatids that exhibit
phoresy attack a wide variety of hosts, including asilid flies (Pinto, 1994), tet-
tigonids and butterflies. In fact, phoresy among egg-parasitoids is not
restricted to parasitic wasps: it is also shown, for example, by mites of the
genus *Acarophenax* (Steinkraus and Cross, 1993).

Channa Basavanna (1953) noted that females of the scelionid, *Lepidoscelio
viatrix*, were only ever found attached to females of its grasshopper host,
Orthacris carli, showing that they were responding to some sex-specific cue.
Release from the female seemed to be triggered by extension of the
grasshopper's abdomen during oviposition. Following oviposition the parasitoid showed no interest in remounting the grasshopper, though she
would hold on if she was carefully placed back on it.

There are two obvious advantages of phoresy for egg parasitoids.
Firstly, the female host will almost certainly carry the parasitoid to a site
where oviposition will take place, and secondly, the eggs reached in this
way will always be freshly laid, which could be important if they subsequently develop a hard coating or if the parasitoid can only develop well
in eggs at an early stage of development. Phoresy in scelionids and poda-
grionines is probably an adaptation to enable them to oviposit into freshly laid eggs. This may be partly due to their telescopic but rather weak
ovipositors, which do not seem to be well adapted to penetrating fully
hardened eggs (Chapter 6), but it may also reflect a physiological need.

8.24 AGGREGATIONS AND HIBERNATION OF PARASITIC WASPS

Several species of parasitic wasps have been found to form aggregations
of uncertain function, apparently as a result of attraction to pheromones
from conspecifics. Others are known to form male mating swarms

(Chapter 2). Starý and Völkl (1988) described aggregation behaviour in a number of species of thelytokous (strictly deuterotokous) aphidiine braconids (*Lysiphlebus* spp.) in cages. The aggregations dispersed when fresh host were introduced and appeared to be largely independent of external factors such as light or food. The authors postulated that they resulted from a specific aggregation pheromone, though it is difficult to see what advantage such a pheromone might confer. Apparently unbeknown to Starý and Völkl, Génieys (1926) and Carver (1984) had described the same behaviour in cultures of *L. fabarum*, and had made the further observation that females would oviposit into other females – what Carver referred to as auto-ovipositional behaviour. This is probably the best interpretation of these observations in that, in culture, densities of wasps reach abnormally high levels so that the resulting extreme concentrations of pheromones lead to atypical behaviours. In the scelionid, *Gryon pennsylvanicum*, aggregations are observed in captivity and are associated with a particular resting posture assumed in the afternoon and referred to as the tuck (Vogt and Nechols, 1991), but whether aggregations of this species occur in the wild is unknown, as is their possible function. Hibernation does occur, however, in another scelionid, *Trissolcus biproruli* (David, 1988).

Several temperate species of ichneumonid hibernate beneath bark and often seem to form aggregations, though it is not known whether this is a matter of coincidence or a response to some pheromones; others hibernate in grass tussocks (Rasnitsyn, 1964). The chalcidid, *Brachymeria lasus*, forms aggregations in the laboratory (Simser and Coppel, 1980a) and overwintering aggregations of *B. intermedia* have been found to occur in various dry places, from a dog-house to an attic window, and comprising from single individuals to a group of two dozen, suggesting that the wasps are merely accumulating in any suitable position (Schaefer, 1993), though again, the possibility that pheromones may also be involved cannot be entirely ruled out.

8.25 ADULT DIAPAUSE

There is a slowly accumulating body of evidence that some parasitic wasps enter diapause as adults. The best-studied examples are the encyrtid, *Ooencyrtus nezarae* (Numata, 1993; Teraoka and Numata, 1995), and the chalcidid, *Brachymeria intermedia* (Barbosa and Frongillo, 1979). It has also been reported in another encyrtid, *Blastothrix confusa* (Sugonjaev, 1963). In *O. nezarae*, the adults can be induced to enter diapause under low-temparature/short-day conditions, both in the laboratory and in field cages. Prior to entering diapause, *O. nezarae* females enter a transient low reproductive state (Numata, 1993) and it has been proposed that whether they go on into true diapause depends on the availability of host eggs. In the field these will not be present late in the year and so the wasps will diapause until the spring.

8.26 DAILY ACTIVITY PATTERNS

Parasitic wasps are usually only active for part of the day. In simple terms, most are either diurnal, crepuscular or nocturnal, but many divide their activity more finely; for example, many emerge at a particular time of day and have preferred times for courtship (van den Assem, 1976), mating (Gordh and Hendrickson, 1976) and host searching. Daily activity patterns have been studied in detail in many insects, though only in relatively few parasitic Hymenoptera. Vogt and Nechols (1991) studied diel activity in the scelionid, *Gryon pennsylvanicum*. As with many other species, most activities (walking, flying, grooming, oviposition, feeding, etc.) were greatest in the mornings. In contrast, in the afternoons, the wasps spent a lot of time resting, but the posture they assumed (the tuck) was quite distinct from the one in which they spent the night. Afternoon resting in this and other species might be associated with avoiding activity during the hottest part of the day.

8.27 HOST GUARDING, MANIPULATION AND MATERNAL CARE

Maternal care is not common among parasitic wasps, and the majority of instances appear to be among the aculeate family, Bethylidae. Other species may drag their host to a secluded, safer site after paralysing it, and some may even conceal it in a manner more typical of higher aculeates (Finnamore and Gauld, 1995). Several species meticulously clean the area where they are going to lay their eggs (e.g. Peter and David, 1991: *Goniozus sensorius*), and one species, *Laelius pedatus*, will shave its hairy host anthrenid larva prior to ovipositing on it so that its larva can feed more easily (Mertens, 1980). The chrysidid, *Chrysis shanghaiensis*, which attacks cocooned lepidopteran larvae, not only chews an access hole through the host cocoon for oviposition, but also seals up the hole afterwards with a mixture of saliva and host silk (Yamada, 1987). Female bethylids lay relatively few eggs on a large paralysed host and often remain with these throughout their development, defending them and their host from other females (Griffiths and Godfray, 1988). Some species belong to the rare category of semi-gregarious parasitoids, laying either a single egg or a small clutch per host.

Once in possession of a host, female bethylids will engage in quite vicious combat with any other female that shows an interest in it. Prior possession is an important factor in the outcome of such contests but the relative size of the combatants is also important (Petersen and Hardy, 1996). If a searching female bethylid encounters a host with unguarded eggs from another conspecific female, she will consume the previous female's eggs and then lay her own clutch. In these brood-guarding bethylids, the total reproduction of a female is in a single brood, and this may comprise very few eggs. However, if a female bethylid is removed from her first host after laying her clutch, she is capable of laying a second, equally large one on a new host. Thus, remaining with the first host to guard one brood means that she is potentially missing opportunities to lay additional broods. Such behaviour could only evolve if the bethylid's total reproductive output was maximized by remaining with her single brood rather than by continuing to forage for hosts and laying multiple broods. As evidence that this is indeed the case, Hardy and Blackburn (1991) demonstrated a clear selective advantage of parental care in this bethylid by showing that survivorship of the original brood is reduced by intrusion of either another conspecific female or other ectoparasitoids such as the braconid, *Habrobracon hebetor*. Further, brood guarding also significantly reduced mortality of the *Goniozus nephantidis* larvae due to multiparasitism by the braconid, *Habrobracon hebetor*.

Recently, Alan Hook (personal communication) has observed brood guarding in the mutillid, *Dasymutilla scaevola*, a parasitoid of the solitary bee, *Cerceris fumipennis*. The bee nests in the ground and if an unguarded nest (that is, unguarded by the bee) is parasitized by a mutillid, the latter will remain with it for a prolonged period. Hook also observed that two mutillids would fight over ownership of a bee nest hole, and that the winner was almost invariably the larger, indicating an absence of ownership asymmetry as is most often found in brood-guarding bethylids. Very little is known of the host associations of mutillids; in fact, host records are available for only about 2% of described species, so it is difficult to tell whether Hook's observations represent one extreme and perhaps unique situation, or whether host guarding may be displayed by many more species.

Among the non-aculeates, the best-known example of maternal care involves the Indian pamboline braconid, *Cedria paradoxa* Wilkinson, which parasitizes the larvae of various pyralids. As with the bethylids that display parental care, *C. paradoxa* lays relatively few eggs, and further, the female braconid does not feed on the host on which it has laid eggs (Beeson and Chaterjee, 1935). M.R. Shaw (1995) has described a similar behaviour in another braconid, *Histeromerus mystacinus* Wesmael, which attacks larvae and pupae of various wood-boring beetles. However, females of *H. mystacinus* that remained with their single brood were not observed to show any aggression towards other females introduced into the vicinity of their hosts, and so whether these wasps are truly displaying parental care, and if so in what form, remains to be determined. Brood guarding also occurs in the pteromalid, *Erixestus winnemana*, an egg parasitoid of the chrysomelid beetle genus, *Calligrapha*, protecting them from super- and hyperparasitoids (Schroder *et al.*, 1996).

Adult females of the chalcidoid, *Signiphora coquilletti*, a hyperparasitoid of the aphelinid, *Encarsia formosa*, on various whiteflies, are unusual in that they spin a web over their host after ovipositing (Woolley and Vet, 1981). The silken strands are produced from the tip of the metasoma; the exact location and nature of the silk glands and spinnerettes has not been determined in this species, but similar strands in some eupelmids are produced from the ovipositor tip (Delanoue and Arambourg, 1965). When the spinning behaviour is completed the host whitefly containing the *Encarsia* pupa is loosely covered with a tangle of threads. The function of this web is uncertain, but Woolley and Vet suggest a number of possibilities, including protection of the *Signiphora* from other competing *Encarsia* females, as an intraspecific marker against superparasitism, or to help prevent the host from falling off its leaf.

Apart from brood-guarding bethylids, parasitic wasps usually show little inter-female aggression, but it does occur in some semi-gregarious egg parasitoids in the Chalcididae, Trichogrammatidae and Scelionidae which fight to protect host patches from intruders (Ohno, 1983; Miura, 1992). Waage (1982) showed that the tendency for a female scelionid to show aggression towards other females depended on the size of the cluster of eggs that she typically parasitizes – those that attack smaller egg clusters being more aggressive. Females of the chalcidid, *Chalcis canadensis*, have also been observed actively defending clusters of host eggs from conspecific intruders using kicks from their enlarged hindlegs (Cowan, 1979). In the scelionid, *Telenomus reynoldsi*, an egg parasitoid of bug eggs, a female that has located a potentially suitable host will usually chase away any other female that intrudes (Cave *et al.*, 1987). Occasional failure to do so allows the second female to superparasitize, and since only one *Telenomus* can develop in a host egg, the superparasitism must cause a reduction in the initial female's fitness.

8.28 SEX PHEROMONES

Many (perhaps the majority of) female parasitic Hymenoptera produce sex pheromones (Vinson, 1972; review by Eller *et al.*, 1984; Table 8.1). As in many other insects, the sex pheromone systems in the parasitic wasps often appear to comprise two or more compounds. One serves for long-range attraction of males by females and can sometimes act over a considerable distance (e.g. Reed *et al.*, 1994); others mediate subsequent stages of courtship, acceptance, mounting and copulation and can be produced by males and females. In the case of some thynnine tiphiids, and even some ichneumonids, female parasitic wasp pheromones have been mimicked by various Australian orchids as part of their pollination system (Peakall, 1990). Female sex pheromone extracts usually initiate positive anemotaxis in males (Bousch and Baerwald, 1967; Cole, 1970; Swedenborg *et al.*, 1994), and exposure of males to female sex pheromones is also often sufficient to elicit courtship responses and sometimes mating attempts, even in the absence of the female herself (Obara and Kitano, 1974; Kitano, 1975; Decker *et al.*, 1993; Gordh and Hendrickson, 1976; Field and Keller, 1993b). Positive electroantennogram responses in males exposed to female sex pheromones have been demonstrated by a number of workers (e.g. Lecomte and Pouzat, 1985; Hidoh *et al.*, 1992).

In general, female sex pheromones are produced only by virgin females and are turned off at the time of courtship or mating. This correlates with the observation that in the majority of parasitic wasps, females only normally mate once (section 8.34). Compared with Lepidoptera, for example, there are relatively few detailed studies of the sources or chemistry of the sex pheromones in parasitic wasps. In a number of species, experiments have indicated that female metasomas or homogenates thereof are particularly attractive to males (Grosch, 1948, 1950; Gordh and Hendrickson, 1976; Vinson, 1978), but the source of the sex pheromone in some species appears to be elsewhere (Kainoh and Oishi, 1993; see below), and in the case of the braconid, *Habrobracon*, the anterior of the female metasoma was found to be more attractive to males than the posterior part (Grosch, 1948). Sometimes extracts of several body parts have appeared to be equally attractive, as in the chalcidids, *Brachymeria lasus* and *B. intermedia* (Simser and Coppel, 1980b). Whilst the possibility of contamination cannot be ruled out in such cases, it is also possible that the sex pheromone glands

Table 8.1 Sex pheromones in parasitic Hymenoptera

Wasp	Family	Sex pheromone	Reference
Alloxysta victrix	Charipidae	6-methyl-5-heptene-2-one	Micha *et al.*, 1993
Ascogaster reticulatus	Braconidae (Cheloninae)	(Z)-9-hexadecenal	Kainoh *et al.*, 1991
Macrocentrus grandii	Braconidae (Macrocentrinae)	(Z, Z)-9, 13-heptacosdiene; (Z)-4-tridecenal; (3R*,5S*,6R*)-3, 5-dimethyl-6-(methylethyl)-3,4,5,6-tetrahydropyran-2-one-	Swedenborg and Jones, 1992a,b; Swedenborg *et al.*, 1993
Itoplectis conquisitor	Ichneumonidae (Pimplinae)	neral; *cis*-37-dimethyl-2,6 octadien-1-al; geraniol *trans* -3,7-dimethyl-2,6-octadien-1-ol	Robacker and Hendry, 1977
Syndipnus rubiginosus	Ichneumonidae (Ctenopelmatinae)	ethyl (Z)-9-hexadecanoate (=ethyl palmitoleate)	Eller *et al.*, 1984
Ereborus terebrans	Ichneumonidae (Campopleginae)	unidentified multiple components	Shu and Jones, 1993
Exetastes cinctipes [*]	Ichneumonidae (Banchinae)	8-dodecenyl acetate; 11-tetradecenyl acetate	Hrdy and Sedivy, 1979

[*] Males are attracted by the two compounds, but it is not certain that they are actually the sex pheromones.

are widely distributed over the wasp's cuticle, which typically is furnished with numerous unitary glands.

Several reports have shown that homogenates of Dufour's glands cause distinct responses in conspecific males, notably antennation. To date, only the work of Syvertsen *et al.* (1995) on the braconid, *Cardiochiles nigriceps*, has conclusively demonstrated that Dufour's gland secretions can elicit a full courtship and mounting response but it seems likely that this is a far more widespread phenomenon, considering the general importance of this gland in pheromone production. This sort of observation led Vinson (1978) to suggest that the Dufour's gland may produce secretions that are involved in the close-range response of males rather than in long-distance mate location or mating initiation. Weseloh (1976a) also suggested that the sex pheromones of the microgastrine braconid, *Cotesia melanoscela*, were produced by the Dufour's gland, but he subsequently revised his opinion (Weseloh, 1980) in agreement with Tagawa (1977, 1983), who had demonstrated that the female sex pheromone gland in a closely related species, *Cotesia* (as *Apanteles*) *glomerata*, is located at the base of the second valvifer of the ovipositor system, near to but distinct from its mechano-sensory setal field. A third species of *Cotesia* was investigated by Field and Keller (1994), who again identified a gland on the second valvifer as a possible pheromone source, but also pointed out the existence of a second female-specific gland slightly more anteriorly located, and were unable to distinguish between them. In the chelonine braconid, *Ascogaster reticulatus*, the sex pheromone is apparently produced by glands in the tibia of the hindlegs (Kainoh and Oishi, 1993) but no ultrastructural studies have been conducted. This correlates quite well with the observations that males follow female trails.

Detailed studies of sex pheromone chemistry have been carried out mostly on larger species of the Ichneumonoidea, though one cynipoid has also been investigated. One of the problems is that the compounds involved may be effective at very low concentrations and therefore present in individual wasps in exceedingly small amounts. This is likely to be particularly true of compounds that act only over small distances. In some taxa the sex pheromone has been shown to be a single component, but in others it is known to be a mixture of several (Swedenborg and Jones, 1992a; Shu and Jones, 1993). Thus, for example, in the braconid *Macrocentrus grandii*, Swedenborg and Jones (1992a) and Swedenborg *et al.* (1993) have shown that the female sex pheromone comprises a mixture of three components. One of these, (Z)-4-tridecenal, is apparently produced by atmospheric oxidation of diene precursors, and because this seems to be an uncontrolled reaction Swedenborg and Jones (1992b) postulated that it could represent an early stage in the evolution of a pheromone. Another interesting feature of this *Macrocentrus* species is that the pheromonal activity of (Z)-4-tridecenal is synergized by a particular lactone which is synthesized by the mandibular glands of both male and female wasps (Jones, 1996). In *Ascogaster reticulatus* the sex pheromone has been identified as (Z)-9-hexa-decenel (Kainoh *et al.*, 1991), which is also a common component in the sex pheromone mixtures of some Lepidoptera, and sensitivity of antennal receptors to this in males has been shown by its dose-dependent electro-antennogram response (Hidoh *et al.*, 1992). Not all female sex pheromones appear to be very volatile. That of the aphelinid, *Aphelinus asychis*, appears to be a contact pheromone which the female lays as a trail (Fauvergue *et al.*, 1995) – males that chance upon such trails respond by intensive local searching.

Numerous observations show that pheromones produced by one sex of a parasitic wasp may be attractive not only to members of the opposite sex (presumed sex pheromones) but also to those of the same sex. Further, newly emerged males are often also attractive to other males. Robacker *et al.* (1976) showed that in the ichneumonid, *Itoplectis conquisitor*, extracts from recently emerged virgin females and males were both attractive to older males, but that the attractiveness of males soon wore off. They were not able to say whether the 'pheromones' involved were derived from the host or were synthesized *de novo*. Micha *et al.* (1993) isolated and identified a chemical that appeared to act as a sex pheromone in the charipid, *Alloxysta victrix*, but the compound was produced by both mated females and males as well as virgin females and it appeared to have other functions too, including acting as a female spacing pheromone (see also Höller *et al.*, 1994a). Pheromone traps containing virgin females of the braconid, *Cardiochiles nigriceps*, attracted males – mostly when traps were placed low down in the corn fields – and other females, especially when placed higher up. It is difficult to see what advantage a female could gain by attracting other females, and it is possible that it is those that attracted that gain advantage, either by increasing the chance of getting mated by accumulating at a place of high pheromonal concentration, or because the presence of one female may indicate a possible source of hosts. Lewis *et al.* (1971) did not dissect or rear from the assembled females to determine whether or not they had already been mated.

It is now becoming clear that males of many parasitic wasps have glands in their antennae (Isidoro and Bin, 1995; Isidoro *et al.*, 1996). The antennal sex glands appear to function specifically during courtship and mating, during which male–female antennal contact is important. Bin *et al.* (1986) have shown that the male antennal gland of the scelionid, *Trissolcus basalis*, which is located on the fifth antennal segment, produces a proteinaceous secretion which is spread over the female's antenna during courtship. Antennectomy removing both glands, or ablation of the glands, leads to females rejecting males, though mating will occur if only one gland is removed. The exact distribution and form of antennal sex glands varies greatly from group to group (Dahms, 1984a; Bin and Vinson, 1986).

8.29 COURTSHIP AND MATING

Courtship has been extensively investigated in many parasitic wasps over a long time, and chalcidoids have been particularly thoroughly investigated, perhaps because their small size makes them easy subjects to work with. Some of the most detailed observations and most of the experimental studies are those of Barrass (1960) on the Pteromalidae and van den Assem and his many co-workers on various small, mostly chalcidoid, wasps (van den Assem, 1974, 1976, 1986; van den Assem *et al.*, 1982; van den Assem and Povel, 1973; van den Assem and Werren, 1994). Courtship usually involves various bouts of wing vibration, antennal movements and/or stroking, and often particular postures. Females often become receptive as a result of a male's courtship, but many species will only become receptive once in their lives (section 8.34). It is probable that females usually signal their receptivity to the male once courtship has provided a sufficient stimulus, but little is known of the signals involved. In one experiment, van den Assem and Jachmann (1982) used some glue to cover the mouthparts (maxillary palps) of a male *Nasonia vitripennis* (Pteromalidae) and found that this prevented him from detecting the female's receptivity signal. Van den Assem and Jachmann also considered the evolution of courtship in the Chalcidoidea, and paid particular attention to the effect of male size and the consequences for courtship position. When several males are competing for a single female, small males might be at a disadvantage because they would take longer to assume a copulation position once the female had signalled her acceptance and so might loose out to a sneaky rival.

In most parasitic wasps, mating itself is a very brief event and it is seldom observed in the wild. Typically, mating lasts only 10–20 seconds (e.g. Wilkes, 1965), and seldom more than a minute or two, though Chrystal and Skinner (1931) noted that mating in the ichneumonid, *Xylonomus brachylabris*, lasts 15–40 minutes. It would be interesting to know how this was related to sperm transfer. Van den Assem and co-workers have carried out numerous studies of courtship in parasitic Hymenoptera, especially in the Chalcidoidea. This work has often been facilitated by the clever use of model female wasps whose movements could be controlled, and by this means it has proved possible to determine exactly what movements elicit what responses. In most taxa, courtship and mating involve a number of stereotypic behaviours in various combinations and sequences, and of various durations. Although many taxa show a broadly similar complement of stereotypic behaviours, small but significant differences can be found between the behaviours of even very closely related species, such as the three sibling species of the pteromalid, *Nasonia*, investigated by van den Assem and Werren (1994). Mackauer (1969) compared the mating behaviour of three closely related species of *Aphidius* that all attack the pea aphid, *Acyrthosiphon pisum*. He found that males were fairly indiscriminate about their mates but females almost invariably refused to mate with allospecific males.

Many parasitic wasps produce sounds during courtship. Van den Assem and Putters (1980) investigated those produced by males of 10 species of chalcidoid. In some species, the courtship song vibrations could be elicited by exposing males to homogenates of their conspecific females. Sonographs showed that the songs of the different species were very distinct, even between members of the same genus. Nevertheless, females of some species were noted to respond at least to some extent to songs of allospecific males. Silencing males by placing a small blob of glue (gum arabic) on the thorax had some disruptive effect on their courtship, but did not seem to block it completely. The sounds produced during courtship and male–male interactions by three species of braconid have been investigated by Sivinski and Webb (1989) – the sounds are produced by wing fanning. In the opiine braconid, *Diachasmimorpha longicaudata*, males produce songs on approaching both male and female conspecifics ('approach songs') and also immediately after mounting a female ('precopulatory songs'). Playing recordings of these songs to wasps demonstrated that they had behavioural significance: upon hearing songs, females became more active. The function of precopulatory songs seems likely to be as part of a final confirmatory signal that ensures that the female will permit mating, but evidence for this is inconclusive in some parasitic wasps (Sivinski and Webb, 1989). Taking the above into account, the role of male courtship songs appears to be to induce receptivity in the females (van den Assem and Putter, 1980).

Field and Keller (1993a) describe an instance of female mimicry by males of the braconid, *Cotesia rubecula*, which represents an addition to the normal mating strategy. Males before mating generally perform either normal courtship behaviour or attempt sneaky matings with females attracted to another courting male, but males after mating frequently (75%) exhibited female-like behaviour in response to the presence of other courting males. The reason for this post-copulatory mimicry appears to lie in the fact that whilst generally females effectively mate only once, there is a brief window of time after their first mating during which they remain receptive (sections 8.32 and 8.34). Therefore, by performing female mimicry a first-mating male probably distracts competitors, so that they do not gain access to his female while she is still receptive.

8.30 MATE CHOICE

Female parasitic wasps vary in the degree of selectivity they show towards courting conspecifics. Van den Assem and Jachmann (1982) say of the pteromalid, *Nasonia vitripennis*: 'A virgin female never rejects a courting conspecific male.' However, females of the braconid, *Habrobracon hebetor*, discriminate in favour of their own sons compared with less closely related males, taking less time before accepting their own offspring (Petters *et al.,* 1985). This has been interpreted as an adaptation that will enable the wasp better to invade new habitats by increasing the likelihood that a single colonizing female will obtain a mating and so be able to produce daughters.

Many tiphiids and a few other parasitic aculeates, such as some bethylids, exhibit phoretic copulation, the females being apterous. Alcock and Gwynne (1987) described the behaviour of two Australian thynnines, *Megalothynnus klugii* and a *Macrothynnus* species. Both have females that sit on twigs, attract males with sex pheromones but resist being pulled off the twigs by the males. Thus larger and stronger males are more likely to succeed in getting a mate. In choice experiments, males of *Megalothynnus klugii* showed a preference for larger females. However, scarcity of partners in the wild seems to be a significant factor in the apparent absence of assortative mating in the field. In the case of these thynnines, males feed females on regurgitated food (nectar) once the male has transported the female to a new mating site away from her initial calling perch.

8.31 MALE MATING SWARMS

A small number of parasitic wasps belonging to the Braconidae, Ichneumonidae, Chalcidoidea and possibly Dryinidae, have been observed to form swarms (Benson, 1944; Southwood, 1957; Starý, 1970; van Achterberg, 1977a; Jervis, 1979; Rotheray, 1981a; Nadel, 1987; Whitfield, 1987). Swarms nearly always occur near some sort of landmark: for example, in the case of the chalcidoids observed by Nadel, swarms are around boulders on a small ridge, whilst in the braconid, *Chelonus hadrogaster*, they are associated with ant hills (Shaw, 1991). In all cases these swarms are strongly male biased or entirely composed of males. Therefore it has been suggested (Southwood, 1957; van Achterberg, 1977a) that they serve to attract females to males in order for mating to take place – in other words, they are lekking sites (Shaw, 1991). The size of the swarms varies considerably: Nadel's swarms comprised several thousands of individuals whilst swarms of the ichneumonid, *Hybrizon*, comprised only about 20 individuals. Southwood's observations of the braconid, *Blacus ruficornis*, showed that the swarming wasps actually perform two distinct types of 'dance': an up-and-down one and a horizontal one. As yet there is no evidence concerning the possible different functions of these two behaviours.

Male mating swarms in parasitic wasps appear to be relatively rare except for those involving *Blacus* species, though as so few field studies of mating have been made in the parasitic Hymenoptera it is difficult to know just how rare swarming is. Thus the discovery of swarms of three species of chalcidoid at a single site is unusual and it led Nadel (1987) to suggest that if the swarms result from male aggregation pheromones, then these may be similar between the species involved. Aggregation pheromones are well known in many parasitic Hymenoptera (Davies and Madden, 1985). These include female aggregation pheromones, male ones and non-sex specific pheromones. Unfortunately, virtually nothing is known about the chemicals involved.

8.32 POST-COPULATION COURTSHIP

In addition to normal precopulatory courtship, many parasitic wasps and other insects have been observed to spend a considerable amount of time engaged in so-called post-copulatory courtship (Barrass, 1960; Eberhardt, 1994; Allen *et al.*, 1994). In these cases, the male dismounts after copulating and inseminating the female and proceeds to display, often in apparently the same way as he did preceding copulation; sometimes he may remount but without attempting copulation. The reasons for this behaviour are not fully understood but the most likely explanation would be that by staying in close proximity with the female the male is effectively mate guarding (Gordh and DeBach, 1978). Allen *et al.* (1994) suggested that the male's behaviour in the case of the aphelinid, *Aphytis melinus*, may serve to keep the female quiescent and they showed that the success of second males in obtaining copulations was reduced significantly in females that had been guarded or courted post-copulation by the first male. When *A. melinus* females mated a second time, the proportion of offspring sired by the second male was independent of whether or not he was allowed to perform post-copulatory courtship. That the situation may sometimes be more complicated is evidenced by the findings of Kajita (1986) in which males of two chalcidoid wasps, *Encarsia* sp. and *Prospaltella* sp., were removed from their mates after copulation but before post-copulatory courtship. In both species, females that had been courted post-copulation laid significantly higher proportions of fertilized eggs than did those from which the males had been removed beforehand. Thus, the attention of the male may also act as a signal to the female that he is of sufficient quality to justify using his sperm. It is possible that post-copulation courtship might therefore represent an instance of run-away sexual selection equivalent to the 'beautiful son' hypothesis.

Males vary greatly in the occurrence of post-copulatory refractory period. Thus Gordh and Hendrickson (1976) reported that males of the ichneumonid showed no signs of interest in females for at least one hour after coitus. The reasons for such refractory periods is not known for certain but it seems likely that it reflects either sperm depletion or the need to construct a spermatophore (see section 4.4).

8.33 ROLE OF COURTSHIP IN SPECIES RECOGNITION

Courtship routines are often highly stereotypic and species specific, and provide an important barrier to hybrid matings. Van den Assem and

Werren (1994) showed that differences in courtship and mating exist between three closely related *Nasonia* species, though they noted that these would probably not constitute a species-isolating mechanism. Courtship behaviour may also provide a useful taxonomic tool in that it may provide a ready and reliable set of species identification characters for some morphologically almost indistinguishable cryptic species – for example, in the difficult pteromalid genus, *Muscidifurax* (van den Assem and Povel, 1973; van den Assem, 1974). Further, courtship features can be used to infer taxonomic affinities, i.e. phylogenetic relationships. For example, in this way, in den Bosch and van den Assem (1986) were able to draw conclusions about the relationships of the enigmatic eulophid, *Aceratoneuromyia granularis*.

In some species, courtship and mating are restricted to particular times of day. In the ichneumonid, *Bathyplectes anurus*, for example, mating took place between 7.45 and 11.00 a.m. and could not be elicited later even when virgin pairs were placed together at other times (Gordh and Hendrickson, 1976). It seems likely that courtship and mating times may constitute part of species isolation mechanisms.

8.34 POLYANDRY

Within the Aculeata, multiple mating is common though not ubiquitous (Page, 1986). The mating strategies of some aculeates – for example, those involving phoretic copulation, as in various tiphiids – makes polyandry particularly unlikely in some cases. Polyandry is also unlikely in species that routinely display sib-mating, such as in many gregarious parasitoids. Within the non-aculeate Apocrita, females of the majority of species investigated appear to mate only once under normal circumstances (Génieys, 1925; van den Assem, 1974; Matthews, 1982; King, 1987).

Although relatively rare, polyandry clearly does occur in a number of parasitic wasps if conditions are right and opportunity arises (Table 8.2). In some species, females become unreceptive immediately after their first copulation, whilst in a few others it has been observed that they may remain receptive for a short while afterwards, allowing a brief window during which polyandry is possible. For example, Hobbs and Krunic (1971) noticed that in the chalcidoid, *Pteromalus venustus*, a female would accept a second mate as long as her vulva remained open, though this only lasted about 5 seconds. Papp (1981) noted that in the braconid, *Doryctes palliatus*, a male may mate repeatedly with the same female. Van den Assem and Feuth-de Bruin (1977) have shown that females of another pteromalid, *Nasonia vitripennis*, will mate twice if a sufficiently long interval has elapsed since the first copulation.

Balfour-Browne (1922) reported that when females of the eulophid, *Melittobia acasta*, ran out of sperm from their first mating, they would apparently mate a second time, as indicated by a resumption of laying

Table 8.2 Some examples of polyandry in parasitic Hymenoptera: the table provides information on polyandry in a wide range of groups but is by no means inclusive (modified after Ridley, 1993)

Family group	Taxa investigated	References
Evaniidae	*Brachygaster minutus*	Brown, 1973
Ibaliidae	*Ibalia drewseni*	Spradbery, 1970b
Eulophidae	*Dahlbominus fuscipennis*	Wilkes, 1966
	Edovum puttleri	Ruberson *et al.*, 1989a
Pteromalidae	*Lariophagus distinguendus*	van den Assem *et al.*, 1989
	Nasonia vitripennis	Barras, 1960; Holmes, 1974; van den Assem and Feuth-de Bruijn, 1977
Torymidae	*Monodontomerus obscurus*	Hobbs and Krunic, 1971
Braconidae	*Aphaereta pallipes*	Salkeld, 1959
	Cotesia flavipes	Gifford and Mann, 1967
	Diachasmimorpha longicaudata	Martínez-Martínez *et al.*, 1993
Ichneumonidae	*Apistephialtes* sp.	Mathur, 1967
	Bathyplectes curculionis	Dowdell and Horn, 1975
	Diadromus pulchellus	El Agoze *et al.*, 1995
	Pseudorhyssa sternata	Spradbery, 1969
Mutilidae	*Pseudomethoca frigida*	Brothers, 1972

female eggs. Hobbs and Krunic (1971) reported that they were unable to repeat Balfour-Browne's observations, but Dahms (1984a) was able to do so with a related species. Strong evidence that sperm-depleted females may mate again to restore their fertilization potential, at least in some species, was provided by Leatemia *et al.* (1995), who worked on *Trichogramma minutum* – but while field longevity would allow some remating after sperm depletion in this species, females apparently remain unattractive to males after mating. There seem to be no other reports within the parasitic Hymenoptera of females apparently monitoring their sperm store.

Ridley (1993) has surveyed mating frequency in 93 species of parasitic Hymenoptera and found that it correlates with whether or not the species is a solitary or a gregarious parasitoid: the former tend to be monandrous whilst the latter are more likely to be polyandrous (though as Ridley himself noted, the results should not be considered as definite proof of a relationship). It was proposed that sib-competition may provide an explanation since the more similar are the genotypes of the developing offspring, the more closely they are likely to be competing with one another. Thus increased genetic diversity within a brood may help to reduce overall competition and hence increase overall fitness.

Female mating behaviour sometimes differs markedly between closely related species (Dowdell and Horn, 1975). For example, females of the ichneumonid, *Bathyplectes curculionis*, apparently indulge freely in multiple matings, remaining receptive to males until they are two days old. In contrast, those of *B. anurus* are monandrous (Gordh and Hendrickson, 1976).

8.35 MALE AGGRESSION

In some taxa, particularly ones that routinely display local mate competition, males will fight with one another for females. Damage to one or other party does not always occur, but in some cases combatants will inflict actual physical damage on each other (e.g. Brown, 1973: Evaniidae; Hamilton, 1979: Agaonidae).

9
Non-physiological host defence strategies

9.1 PHYSICAL BARRIERS

There has been far less work on non-physiological host defence mechanisms than on physical or behavioural ones, and much of what is known has been summarized by Gross (1993). For the most part, the evidence is rather anecdotal. The simplest form of defence may be the thickness and hardness of the potential host's cuticle. For example, Salt (1958) noted that the time spent by a female *Trichogramma* in penetrating a host lepidopteran egg is strongly influenced by the durability of the chorion. Certainly the hardness of many lepidopteran pupae makes penetration by parasitoid ovipositors difficult, and the ovipositor will often slip unless the angle of attack is nearly perpendicular to the cuticle (Cole, 1959b).

Some hosts increase the thickness of their coverings, and hence protection, in various ways. Guershon and Gerling (1994) investigated the protection against parasitoids that the nymphs and adults of the Israeli whitefly, *Aleyrodes singularis*, gain by retaining their exuviae on their backs and by producing waxy fluff. The wax secretion and retained exuviae appear to provide protection against two possible *Encarsia* species (Aphelinidae), but one of these, *E. inaron*, appears to be able to cope with both obstacles whilst *E. transvena* cannot. The thickness of the barrier between the host and its parasitoid can be important. For example, the level of parasitism of the bagworm, *Thyridopteryx ephemeraeformis* (Lep.: Psychididae), by the ichneumonid, *Itoplectis conquisitor*, declines rapidly with host size (Figure 9.1). This is a result of the increased thickness of the bag surrounding the host pupa, which for about 40% of the bagworm population was greater than the length of the *Itoplectis* ovipositor (Cronin and Gill, 1989).

The pupae of many Lepidoptera are equipped with modified abdominal joints which have been termed pupal gin-traps (Cole, 1959b; Bate, 1973). These serve to protect them from oviposition by endoparasitoids

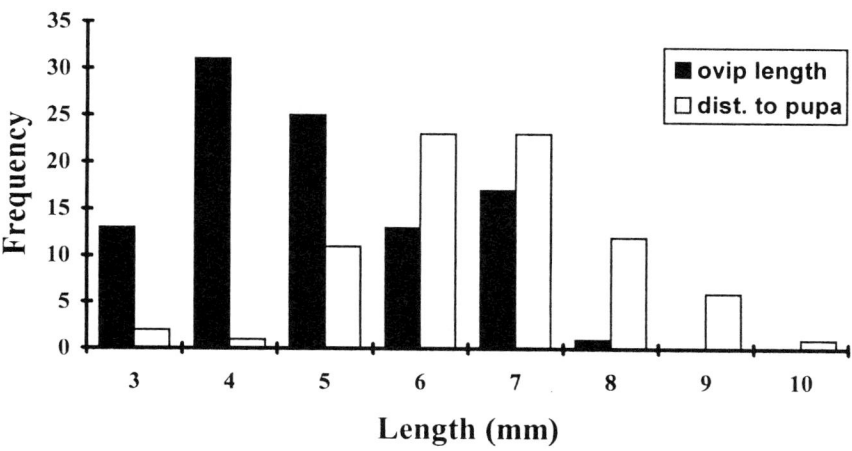

Figure 9.1 Frequency histograms for ovipositor length in the ichneumonid, *Itoplectis conquisitor*, and the thickness to its bagworm host, *Thyridopteryx ephemeraeformis*, pupa through the bag. (Redrawn from Cronin and Gill, 1989.)

through the softer intersegmental membranes. The ovipositor of members of the pimpline ichneumonid genus, *Apechthis*, is crow-bar shaped at the end, an adaptation to prising apart the abdominal segments of its host lepidopteran pupae.

Perhaps one of the commonest supposed anti-parasitoid defensive modifications in Lepidoptera larvae is hairiness. Weseloh (1976b) and Fuester *et al.* (1993) both noted that the increasingly long hairs of later instar larvae of the gypsy moth (*Lymantria dispar*) makes it progressively more difficult for their braconid parasitoids, *Cotesia melanoscelus* and *Meteorus pulchricornis*, to attack them. Gross (1993) suggested that a counter-example was provided by the observation of Sheehan (1991) that, broadly, parasitoids of hairy caterpillars are more generalist than those of hairless hosts. However, it may be that a parasitoid, once it has evolved a mechanism for overcoming the long hairs of a host, can more readily attack other hairy hosts. Perhaps we should ask whether there is any evidence that hairy larvae are less well protected physiologically since their hairs offer them protection against many parasitoids.

Many lepidopterous larvae live concealed or semi-concealed in leaf mines, leaf rolls, etc. It has long been assumed that these excavations and constructions have a protective role, but direct evidence is hard to find. Brandl and Vidal (1987) examined the effect on relative ovipositor length of 21 eulophid parasitoid species of two types of leaf mine: blotch mines, which are thin, and tentiform mines, which are thicker because of contraction of the epidermis on one side of the leaf that causes the other side to bulge. Wasps attacking hosts that form tentiform mines were found to have significantly greater ratios of ovipositor to wing length (Figure 9.2). This suggests that the evolution of tentiform mine formation is probably an adaptation to avoid parasitism.

Following a survey of the parasitoids and predators of Australian spiders, Austin (1985) found that the spider's silken egg cocoon provides an effective protection against generalist predators, but that the coevolution between the spiders and their parasitoids has led to specialist parasitoids that have evolved means of overcoming the spider defences.

9.2 BEHAVIOURAL DEFENCE

Some potential hosts have evolved particular strategies for avoiding parasitism. These include thrashing, kicking, shaking, dropping on silken threads or simply falling off the food plant, and the release of noxious or sticky liquids. Many exposed Lepidoptera and sawfly caterpillars have characteristic head-flicking and thrashing behaviours that are almost certainly involved in deterring parasitoids, and concealed hosts such as leafminers similarly often wriggle violently when they detect a parasitoid (Gross, 1993; Bacher *et al.*, 1996), but research on the effectiveness of these behaviours seems mostly to concern tachinid parasitoids rather than parasitic wasps (e.g. Myers and Smith, 1978). In the case of the eulophid, *Sympiesis sericeicornis*, Bacher *et al.* (1996) have shown that the evasive behaviour of its graciliid host may well be triggered by the strong 5.6 kHz

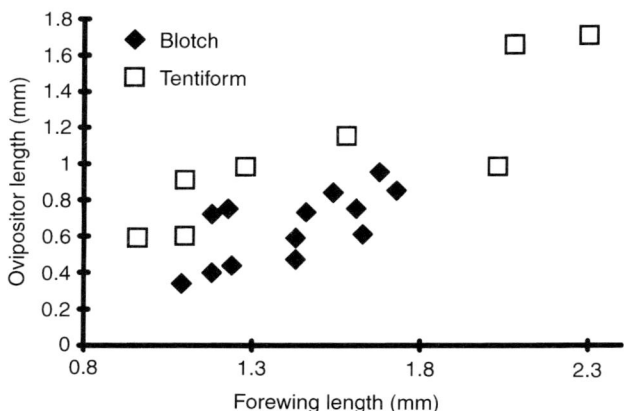

Figure 9.2 Plot of ovipositor length versus forewing length for 21 eulophid parasitoids classified according to host leaf mine type. (Data from Brandl and Vidal, 1987.)

signal (> 10 dB) produced when it probes the host's leaf mine. In some bugs, parental care can also play a significant part in reducing parasitism. In several species of pentatomid the female bug stands guard over her clutch of eggs, and this has been shown to provide considerable protection for most eggs in the batch, though those located at the periphery are subject to higher levels of parasitization (Eberhard, 1975).

Violent rearing and thrashing by caterpillars of the Australian noctuid, *Uraba lugens*, just prior to an encounter with its braconid parasitoids has been shown to be an effective defence, reducing the probability of parasitization by as much as 50% (Allen, 1990). *Uraba*, some nolids, and a few other caterpillars have the unusual feature of retaining shed head capsules so that three or four of them may form a horn-like structure in front of the current head, and it has been suggested that these serve to extend the effective zone of protection afforded by their thrashing behaviour. Many aphids similarly respond to potential danger by violent shaking and bucking, and their defensive behaviour has been shown to protect at least some species from attack by aphidiine parasitoids (Calvert, 1973). However, in many cases, aphids under attack by aphidiines get little opportunity to defend themselves.

Although aphids may generally be poor at defending themselves from aphidiines first time around, Gardner *et al.* (1984) have shown that the proportion of the aphid, *Metopolophium dirhodum*, that exhibit defensive reactions increases with the number of encounters with its parasitoid, *Aphidius rhopalosiphi*, and that the success rate of the parasitoid decreased concomitantly (Figures 9.3a and 9.3b, respectively). Although this progressive increase in defensive reaction does not prevent the aphid from being attacked the first time, it does greatly reduce the incidence of second attacks and also of superparasitism. The potential benefit to the aphid appears to result from the fact that only about 45% of the aphidiine's attacks (ovipositor stabs) are successful in that they result in an egg being laid in the host. So if the aphid is not parasitized at the aphidiine's first attempt, it may manage to avoid parasitization altogether.

More successful defence has been reported in the conifer aphid, *Schizolachnus pineti*, by Völkl and Stadler (1996). Feeding groups in this aphid species surround a pine needle, with those individuals at the colony's periphery facing in towards the colony's centre such that their kicking legs are directed outwards. The aphidiine braconid, *Pauesia unilachni*, was markedly less successful in attacking aphids from the rear and Völkl and Stadler therefore proposed that such defensive arrangements as adopted by *S. pineti* are particularly effective when the aphids surround a linear structure such as a pine needle.

Caterpillars may increase their chance of survival by either dispersing or aggregating, and these strategies may be displayed by different instars of the same species (Allen, 1990). The feeding strategies of caterpillars may

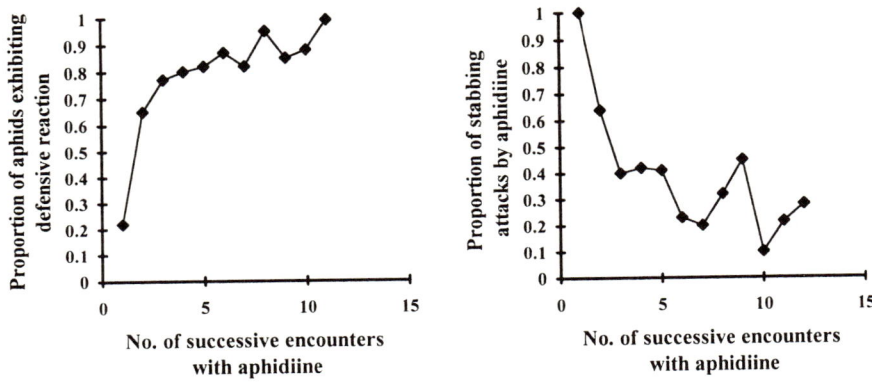

Figure 9.3 Defensive behaviour of aphid in relation to attack by its aphidiine parasitoid and effects on parasitoid success: (a) relationship between encounter number and aphid defensive response; (b) relationship between encounter number and attempted attacks. (Redrawn from Gardner *et al.*, 1984.)

have an effect on the ease with which they are located by parasitoids. Many species feed only in short bouts but may spread the damage they do over several leaves, thus causing lots of small separate plant injuries. These could confuse a visually searching parasitoid and, further, since many plants produce specific volatile compounds in response to insect-feeding (see section 8.10), multiple release sites for such components might also cause parasitoid confusion.

Even some parasitic wasp larvae have behavioural defences against hyperparasitism. The best-studied example involves the megaspilid, *Dendrocerus carpenteri*, a primary hyperparasitoid of aphids via aphidiine braconids. *Dendrocerus* is in turn parasitized by the pteromalid, *Asaphes lucens*, but defends itself by vigorous thrashing in combination with various cuticular protuberances. This strategy is apparently 95% successful in protecting it against hyperparasitism (Carew and Sullivan, 1993).

9.3 HOST ESCAPE RESPONSES

Many arboreal Lepidoptera larvae, especially early instar, respond to potential danger such as attack by birds or parasitoids by dropping down on a silk thread (Askew, 1971). Even pupae of some Lepidoptera, such as the tortricid, *Tortrix viridana*, can shake themselves out of their cocoons in rolled or folded leaves when attacked by parasitic wasps (Cole, 1959b). That dropping on a thread might be a successful means of escaping from many parasitoids was inferred by Gross and Price (1988) who studied two gelechiid leafminers that feed on horsenettle (*Solanum carolinense*). One of the miners is confined to its leaf mine whereas the other ties and folds leaves and drops readily from these shelters if disturbed. The former has a parasitoid assemblage comprising nine species whereas the caterpillar with an escape response is only attacked by two parasitic wasp species.

Although caterpillars that drop on silk threads when disturbed no doubt gain some protection from parasitization, they are by no means necessarily immune from it. One microgastrine braconid, *Microplitis mediator*, is able to attack its host (*Mamestra configurata*) larvae in mid air (Arthur and Mason, 1986). Yeargan and Braman (1986) showed that another microgastrine braconid, *Diolcogaster facetosa*, has evolved to overcome this defensive behaviour in its host, the green clover worm, *Plathypena scabra*, by scrambling down the silk line after it. Yeargan and Braman (1989) subsequently described a different tactic employed in addition to thread descent by the hyperparasitic ichneumonid, *Mesochorus discitergus*, when caterpillars of the green clover worm descend on silk threads. In some cases, particularly when the host is in an earlier instar, this ichneumonid positions itself at the top of the caterpillar's drop-line and reels it back up using its fore tarsi. It is difficult to imagine that such tactics are limited to the parasitoids of just one species of Lepidoptera and most probably the lack of any other reported instances reflects a general dearth of observations of parasitic wasps in their natural environment. Baur and Yeargan (1994b) showed that *Helicoverpa zea* and *Spodoptera frugiperda* caterpillars seldom exhibit escape responses to *M. discitergus*, but that when they did make a vigorous response, it was an effective deterrent. Larvae of *Plathypena scabra* displayed both dropping and vigorous defence movements. Both were effective antiparasitism deterrents, but dropping was more so (68% captured compared with 39%).

9.4 CHEMICAL DEFENCE

There is considerable evidence that host toxins, whether they are produced *de novo* or sequestered from food plants, can have a strong influence on the success of parasitic wasps. Further, since many plants are more heavily loaded with secondary compounds in warmer climates, Gauld *et al.* (1992) and Gauld and Gaston (1994) proposed that this could limit the number of parasitoids per host species and, in so doing, effect a reduction in species richness towards the tropics – the 'nasty host' hypothesis, which is discussed in more detail in Chapter 10.

9.5 DANGEROUS HOSTS

Some hosts pose special problems for parasitoids. Among the most notable are spiders and adult social bees, wasps and ants, all of which are attacked by parasitic Hymenoptera. The parasitic wasps that attack such dangerous hosts do suffer a real risk of injury or death, and even some Lepidoptera larvae may sometimes injure their parasitoids by biting off their appendages (Allen, 1990). Makino and Sayama (1994) showed that when the paper wasp, *Polistes snelleni*, encounters the elasmid, *Elasmus japonicus*, it will sometimes bite and kill the parasitoid – it will also often kill and eat the parasitoid's larvae when it encounters them. Similarly, though perhaps less severely, sapygids (a group of parasitic aculeates that are parasitoids or kleptoparasitoids of vespids and apids) often suffer damage such as loss of antennae at the hands (or rather mandibles) of their hosts (Torchio, 1972). To avoid such risks, the parasitic wasps that attack such dangerous hosts have evolved specialized behaviours that maximize their chances of survival (e.g. Shaw, 1994). For example, the neoneurine braconids which oviposit in adult ants 'stoop' on their targets like a hawk and oviposit in them very quickly before the ant can defend itself (Shaw, 1993).

Some ichneumonids are effectively predators rather than parasitoids in that their larvae will consume multiple 'hosts'. The best examples of such predatory ichneumonids are various pimplines and cryptines that attack spider egg masses. In many cases the ichneumonid larva will completely consume the eggs in the cocoon and it is not surprising therefore that some spiders remain with their eggs to guard them against attack. Morse (1988) showed that the ichneumonid, *Trychosis cyperia*, selectively attacks egg cocoons of smaller individuals of the spider, *Misumena vatia*, thus presumably reducing the risk of being injured by the spider, which no doubt would be able to kill the wasp if it could manage to bite it.

9.6 ASSOCIATION WITH ANTS

Several insects have evolved mutualistic relationships with ants. These include various ant inquilines, many species of aphid and coccid that produce honeydew, and some lycaenid butterflies whose larvae are shepherded by their host ants to and from the nest for feeding, or even remain entirely within the nest. None of these ant-associated species appears to be completely free of hymenopterous parasitoids, but the presence of ants may greatly change the parasitoid composition and lead to reduced overall levels of parasitism. Needless to say, parasitoids that specialize on these ant-attended hosts have evolved their own particular strategies and defences (e.g. Thomas and Elmes, 1993).

Aphids and scale insects that are attended by honeydew-seeking ants may gain from the ants deliberately or inadvertently protecting the colony from attack by parasitoids and predators. For example, populations of the soft brown scale, *Coccus hesperidum*, are controlled by the aphelinids, *Metaphycus flavus* and *Microterys flavus*, when the scale insects are sparse and dispersed; but when they form dense aggregations that are attended by ants, Rosen (1967) found that these parasitoids are both displaced by *Coccophagus* species. Apparently, although the *Metaphycus flavus* and *Microterys* are more efficient at parasitizing the scales than the *Coccophagus*, they are more sensitive to disturbance by ants. Rosen notes that ants are particularly likely to have an adverse effect on parasitoids that either host-feed, or take a long time for oviposition, or are generally more easily disturbed. Some parasitoids of ant-attended aphids have evolved effective defences including noxious chemicals, chemical camouflage, and a general litheness and ability to escape ant-attack (see section 6.45). The susceptibility of some aphid parasites to ant-attack has been suggested as being a limiting factor in some biocontrol programmes (Stechmann *et al.*, 1996).

9.7 ECOLOGICAL FACTORS

Hosts can obviously evade parasitoids if they live in habitats where parasitoids are necessarily scarce or absent. In an interesting study, Ohsaki and Sato (1990) examined the co-occurrence of *Pieris* butterflies and their braconid parasitoids of the genus *Cotesia* (as *Apanteles*). One of the hosts, *P. rapae*, appears to be a specialized colonist species, and its larvae are often in habitats where there are no parasitoids because the latter have not had a chance to colonize. However, *P. melete* and *P. napi* necessarily live in permanent habitats and so experience high levels of parasitoid attack. Correlating well with this is the encapsulation ability of the three hosts. *P. rapae*, the less frequently attacked butterfly, is a poor encapsulator, whereas the other two more regularly attacked pierids are good encapsulators. This implies that these may be a cost associated with the ability to encapsulate the parasitoid (see section 7.19).

10

Ecology and diversity of parasitic wasps

10.1 INTRODUCTION

This chapter reviews some aspects of parasitic wasp community ecology and diversity as well as a number of parasitic wasp features that can be broadly considered ecological, such as fecundity, longevity, and pathogens and predators. It does not attempt to be in any way comprehensive on these matters since they are all the subjects of many more specialized works; rather, it simply draws attention to some features that either have implications for, or stem from, more fundamental features of parasitic wasp biology.

10.2 PARASITOID GUILDS

The ways in which a parasitoid and its host life cycles interact define what have been termed parasitoid guilds – each guild being defined by the developmental stage at which the host is attacked, the stage that is killed and whether the parasitoid is endophagous or ectophagous. Following a survey of the parasitic Hymenoptera, Mills (1994a) concluded that for endopterygote host insects, there is a total of about 12 different life history guilds and these can be represented diagrammatically (Mills 1994a,b) (Figure 10.1), but for any single host group, there are seldom more than six or seven guilds. The number of different guilds that attack a given host is strongly influenced by host phylogeny and by host feeding niche, which is also determined in part by phylogeny. Mills's system is largely based on ecological considerations: thus, all egg parasitoids are classed together because the fact that they attack eggs determines their size, where they search for hosts, how abundant their hosts are, etc. From a physiological point of view, things may be very different. For example, some egg parasitoids attack the early egg, others attack an almost completely developed host larva within the egg shell. The physiological interactions between the parasitoid and host differs greatly according to such factors.

Many hosts that construct cocoons are attacked at the prepupal stage rather than as the fully formed pupa, this being especially true for ectoparasitoids probably because the pupal cuticle is often very hard. As pointed out by Mills, the term egg–larval parasitoid is almost always a misnomer in that these parasitoids do not normally emerge from the larva but rather from the prepupa. This is understandable since the prepupa will normally be formed at a selected site where it would otherwise have had a good chance of successfully completing its development.

Following a large literature survey, Hawkins and Mills (1996) showed that only up to 73% of the potential maximum number of parasitoid guilds that host insects could support were ever occupied. This implies that most parasitoid communities are not saturated.

Parasitoid guilds may be useful in identifying potential agents for neo-biological control – that is, the introduction of parasitoids of closely related pests occurring in other geographical regions (Mills 1994b). For example, *Pissodes strobi*, a species of weevil that attacks spruce in North America, is an important pest, especially in the eastern states. Comparison of its guild of parasitoids with those of two ecologically similar *Pissodes*

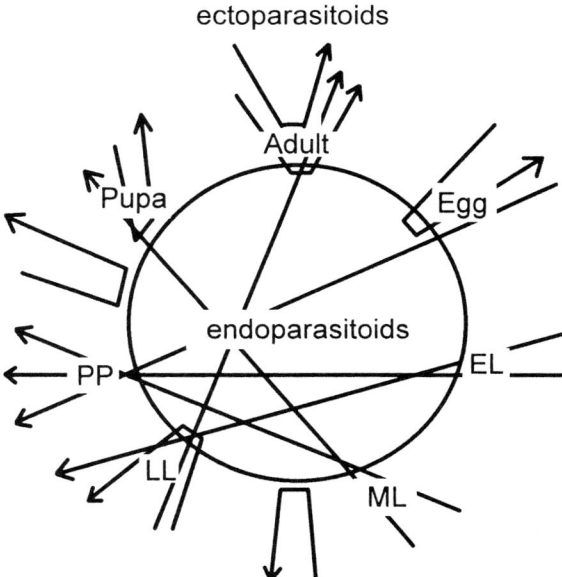

Figure 10.1 Different parasitoid–host relationships. Lines passing into the circle indicate endoparasitic phases of the life cycle. EL, early larva; ML, mid larva; LL, late larva; PP, prepupa. (Based on Mills, 1994a,b.)

species from Europe showed that in the eastern United States, *P. strobi* lacks the egg–larval parasitoid guild that is represented in Europe by members of the braconid genus, *Eubazus*. The latter may therefore be particularly suitable candidates for neo-biological control since they would not be competing with any members of the same guild in North America.

10.3 HOST STAGE ATTACKED

In most cases, a parasitoid only attacks a single host stage – egg, larva, prepupa, pupa or adult – but there are a number of exceptions. Among idiobiont ectoparasitoids of concealed hosts, it is quite common that a parasitoid will attack larvae, prepupae and sometimes even pharate adults (Shaw, 1994). Among endoparasitoids, a number that normally parasitize adult hemimetabolous insects will also attack nymphs of the same species. Aphidiine and some euphorine braconids provide a number of well-known examples (Shaw and Huddleston, 1991). The cosmopolitan euphorine braconid, *Dinocampus coccinellae*, which attacks a large number of aphidophagous coccinellid beetles, normally parasitizes adults but will also sometimes successfully attack immature stages (Shaw, 1988; Orr *et al.*, 1992). The main linking feature of all these situations is that the various host stages attacked occur in the same microhabitat, and often have very similar biologies, and probably kairomones also.

10.4 LIFE EXPECTANCY

At present, there are virtually no reliable data on either the potential or realized longevities of parasitic Hymenoptera in the wild, though in large part due to the efforts of biocontrol workers there have been an enormous number of studies on the longevity of adult parasitic wasps in captivity. These reports often present comparative data for different feeding regimes, temperatures, humidities and photoperiod (Jervis and Copland, 1996). Wasps are typically found to fare better when provided with moisture and even better when dilute honey is provided (e.g. Hawkins and Smith, 1986). As in many other insects, life expectancy in females is nearly always greater than in males. For example, Aubert (1959) found that, in the laboratory, the maximum life span of females of seven species of pimpline ichneumonids was consistently greater than that of their respective males. Olmi (1994) noted that male dryinids tended to die very soon after mating. At the other extreme, adult female mutillids may live more than a year in captivity (Schmidt, 1978). In captivity, and under good conditions, many species will live one or two months; and as many species hibernate as either teneral adults or as mated females, a good number must naturally survive several months as adults in the wild. When kept cold, at 2°C, Aubert found that females of *Pimpla turionellae* could live for up to two years. The effect of adult diet on longevity has been investigated in detail in *Habrobracon* and *Nasonia* (Clark, 1963, and Davies, 1975, respectively). In *Habrobracon's* case, adults live longer when fed honey-water instead of being allowed to host-feed, though ovary atrophy is accelerated under the former circumstance. In the torymid, *Torymus beneficus*, two strains are known which differ in their seasonality and longevity, the early season strain having been found to survive longer in captivity under fairly natural conditions (Piao and Moriya, 1992).

Decisions as to whether or not to lay in a given host will depend on both host quality and life expectancy (Chapter 8). Life expectancy itself depends on many factors – age, health, food availability, etc. One possible major source of mortality in insects in general is thunderstorms, and it might be expected that if insects could predict the imminent arrival of such a life-threatening storm, then they would modify their egg-laying behaviour accordingly. In the case of parasitic wasps, one adaptive change might be to accept lower quality hosts than would ordinarily be accepted if a thunderstorm was likely. In an experiment to test this, Roitberg *et al.* (1993) investigated the effect of a rapid drop in barometric pressure on oviposition behaviour of the eucoilid wasp, *Leptopilina heterotoma*, a parasitoid of *Drosophila* larvae. In accordance with their expectations, wasps exposed to reduced barometric pressure accepted more hosts for oviposition, leading to a very significantly higher level of superparasitism than occurs under conditions of steady pressure (Figure 10.2). Roitberg *et al.* (1992) had previously shown that *Leptopilina* also responds to changes in day length that in the wild would be cues that the wasp's future life expectancy was limited.

Parasitic wasps usually do not out-live their reproductive lives – with the major exception of those that show parental care, which essentially means certain bethylids. Jervis *et al.* (1994) have shown, however, that under conditions of high (perhaps abnormally high) host availability females of the braconid, *Habrobracon* (as *Bracon*) *hebetor*, may live for a considerable while after they have completed laying all of their eggs. During this post-reproductive phase, the wasps continue to paralyse 'hosts' and to feed upon them.

10.5 FECUNDITY

Fecundity ranges enormously in parasitic wasps and depends upon several factors, including the species, the size of individual, and the diet of that individual (Jervis and Copland, 1996). At the lower end are some bethylids such as *Laelius pedatus*, which lays only one to five eggs per host. At the opposite extreme come wasps such as trigonalyids and the ichneumonid, *Euceros*, which rely on some of their myriad eggs being consumed by intermediate hosts that must then be eaten or parasitized by the primary host.

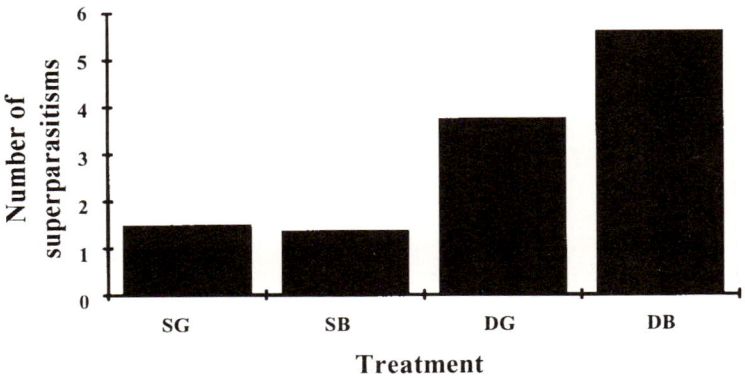

Figure 10.2 The effect of changes in barometric pressure on the number of super-parasitization events by the eucoilid wasp, *Leptopilina heterotoma*. S, steady baro-metric pressure; D, rapidly dropping barometric pressure; G, 'good world', i.e. all hosts unparasitized; B, 'bad world', i.e. all hosts previously parasitized. (Redrawn from data from Roitberg *et al.*, 1993.)

Clausen (1931b) recorded more than 10 000 eggs from a single female of the trigonalyid, *Poecilogonalus thwartsi*, and other trigonalyids often have in excess of 1000 mature eggs at any one time and usually around 500 ovari-oles per ovary (Iwata, 1960b). Following a survey of 474 literature sources, Blackburn (1991) investigated the relationship between adult fecundity, life span and adult size, and various ecological variables across a range of hymenopterous parasitoids. Unlike in vertebrates, Blackburn found no sign of correlation between size and either fecundity or longevity, and conse-quently this drew into question some assumptions about r/K continua – in parasitic Hymenoptera, at least, adult size does not seem to be a significant component of this aspect of life history strategy.

Heritable intraspecific variation in fecundity, as expressed by host infes-tation rate, has been shown to occur in some species of *Trichogramma* (Mimouni, 1992). The inheritance was shown to be via the female line only, suggesting that a cytoplasmic factor might be responsible. Girin and Boulétreau (1994) showed that treatment of *T. bourarachae* with antibiotics reduced the fecundity of the high fecundity strain to a level similar to that of the low fecundity one, implicating possible involvement of endosymbi-otic microorganisms. The presence of such organisms in the eggs of the high fecundity strain and not in the low fecundity or antibiotic-treated strains provided further evidence. These microorganisms appear to further their own survival through enhancing egg production in the female wasp, apparently without other ill effects. Egg production and vitellogenesis in *Trichogramma* are largely completed before wasp eclosion, and thus the causative microorganism must induce their effects before then, presumably in the wasp's pupal phase. Other maternally inherited endosymbiotic microorganisms, such as *Wolbachia* species, have been shown to cause thelytoky in some species and to mediate strain-specific sexual incompatibility in others, including other *Trichogramma* species (Chapter 2). The generic relationships of these *Trichogramma* fecundity-enhancing microorganisms have not yet been determined, but it seems likely that they could also be *Wolbachia* strains.

10.6 EFFECTS OF HOST PLANTS ON PARASITISM SUCCESS

It has been recognized for a long time that host plants can have a strong influence on levels of parasitism by parasitoid Hymenoptera. The effects can occur at the host plant location stage, in searching for hosts once the host plant has been located, and on the development of the parasitoid after parasitization (Steinberg *et al.*, 1993). The effects that plants have include interspecies differences, effects due to physical or chemical differences between plant strains, and even the nutritional status of a particular plant.

Ruberson *et al.* (1989b) showed that the parasitization success of the eulophid, *Edovum puttleri*, an important egg parasitoid of *Leptinotarsa decemlineata*, the Colorado potato beetle, was enhanced by reduced presence of glandular trichomes on the potato plant and in particular, in potato strains that had both enclosed and exposed exudate, droplet-type trichomes – the exposed droplets entrapped many of these small wasps. Physiological effects can also be important. For example, the aphid parasitoid, *Lysiphlebus testaceipes*, attacks the cotton aphid, *Aphis gossypii*, which will grow on both cotton and cucumber (amongst other things). When the aphid is growing on cotton, *L. testaceipes* has a longer developmental time, lower fecundity and lower emergence rate and does not grow so big (Steinberg *et al.*, 1993).

Fertilizers obviously have marked effects on the health and appearance of plants, and it is also well known that fertilizers can have marked effects, both positive and negative, on insect herbivores. Surprisingly few studies have examined whether they also affect the parasitoids of those herbivores, but the potential importance of their effect is now being appreciated more (e.g. Kaneshiro and Johnson, 1996). In one recent piece of work, Bentz *et al.* (1996) found that parasitization of the whitefly, *Bemisia argentifolii*, infecting poinsettias (*Euphorbia pulcherrimus*), by the aphelinid, *Encarsia formosa*, was significantly higher on fertilized plants, and that even the type of fertilizer had an effect, with high nitrogen treatment leading to the highest levels of parasitism. Bentz *et al.* noted that this tritrophic effect was due to the parasitoids preferring to search on the healthiest, most vigorous plants. Such effects are probably widespread.

Some plants show early abscission of leaves that are attacked by insects, as is the case, for example, with American holly (*Ilex opaca*) attacked by the dipterous leafminer, *Phytomyza ilicicola*. Early excision might be expected to adversely affect survival of the phytophage. However, in this case and contrary to expectations, Kahn and Cornell (1989) found that the miner's survival was not significantly affected since abscission usually did not occur until after the fly larva had pupated and because the parasitoids of the leafminer do not search for hosts on the ground so the flies in abscized leaves suffered a lower level of parasitization.

10.7 PARASITOID SPECIES RICHNESS AND DIVERSITY

In terms of numbers of species, the Hymenoptera as a whole, and parasitic wasps in particular, constitute a large proportion of the world's insect fauna. The British list alone includes some 6500 hymenopteran species (so outnumbering the beetles), the majority of which are parasitoids. In the tropics, the numbers are higher, but by how much remains a area of considerable uncertainty. Several studies have compared the parasitic wasp faunas of tropical areas with well-known temperate ones, such as that of Britain. One of the first such studies was that of Owen and Owen (1974) who compared the species diversity of Ichneumonidae collected in malaise traps set up in Britain, Sweden, Uganda and Sierra Leone, and obtained the surprising result that the tropical samples were no more diverse that the temperate ones. In an extension of the study concentrating on the British ichneumonid fauna, Owen *et al.* (1981) reported on the catch of ichneumonids in a malaise trap that was operated in a suburban garden at Leicester, England. The sample displayed remarkably high species richness: over the two-year period, 1972–1973, this trap accumulated 6445 ichneumonid specimens representing a total of 455 species – far higher than would have been anticipated at the time.

A decrease in diversity towards the equator has become known as anomalous diversity, and Janzen (1981) appeared to find additional support for this in ichneumonids by abstracting distribution data from the catalogues of Henry and Marjory Townes. Peak species richness was found in the middle latitudes of the United States rather than in the extreme south. This idea that ichneumonids in particular, and perhaps parasitic Hymenoptera overall, might display anomalous diversity (or at least, a less tropico-centric pattern of diversity) was born out by the general observations of experienced collectors (e.g. Gauld, 1986a), though this view has been challenged (Morrison *et al.*, 1979). Hespenheide (1979) and others, for example, suggested that the sampling in the tropics has simply been too incomplete and conducted over too short a time scale to permit any such conclusions.

Noyes (1989) presented the results of a more extensive and detailed survey of Hymenoptera in Sulawesi based on material collected using five different sampling techniques and at a range of sites. Noyes concluded that virtually all families of Hymenoptera were more speciose in Sulawesi than in the UK, with the possible exceptions of the gall-forming Cynipidae and the sawflies. In a community-based survey, Hawkins and Compton (1992) found a slight trend towards lower species richness of parasitoids living in figs in lower latitudes. Idiobiont egg parasitoids (Mymaridae, Scelionidae, Trichogrammatidae) represent a guild that appeared to be particularly speciose in the tropics, though because of the generally small size of the species involved, they have been far less intensively investigated from a taxonomic point of view than have the larger ichneumonoids. Idiobionts as a whole – not just egg parasitoids – appear to be more species-rich at lower latitudes than koinobionts (Noyes, 1989; Gauld, 1987), and this can be observed within America north of Mexico (Quicke and Kruft, 1995). As being an idiobiont is strongly correlated with attacking concealed hosts, this trend is in agreement with the prediction of Rathcke and Price (1976) who proposed that, due to increased predation pressures in the tropics, groups that are less susceptible to predation should be favoured. It is clear that different families of Hymenoptera have different overall species richness/latitude relationships (Askew, 1990). A similar exercise to Janzen's (1981) was carried out for the Braconidae by Quicke and Kruft (1995) – on the basis of general collecting experience, the Braconidae were believed to be more likely to show normal diversity trends. However, an almost identical pattern to the ichneumonid distribution was found, with a distinct decline in species and genus richness in the southernmost states. In the Braconidae, each subfamily can be categorized as being idiobiont or koinobiont, and so it was an easy exercise to examine trends in the ratio of taxa displaying these two strategies (Figure 10.3). For both genera and species, koinobionts were found to be relatively better represented in the most northern latitudinal zone, and the ratio of taxa displaying these life history strategies declined monotonically with decreasing latitude.

Whilst the insect fauna of Great Britain is probably the best known in the world, with the advent of an increasing interest in biodiversity there has been a growing number of studies aimed at reaching the same level of understanding in other parts of the world. Ahead of most of these is Costa Rica which, despite its small size and developing world status, has been making enormous progress towards inventorying its flora and fauna. Gaston and Gauld (1993) attempted to estimate the number of species of the ichneumonid wasp subfamily Pimplinae that occur in Costa Rica, following analysis of samples from a network of malaise traps set throughout the country and operated between 1986 and 1990. In their study it should be noted that the term Pimplinae refers to the taxon in its more traditional sense, i.e. before the exclusion of the small subfamilies Poemeniinae and Rhyssinae. The final sample totalled more than 100 malaise trap years, and probably represents the most extensive set of malaise trap samples analysed anywhere. Plots of the cumulative number

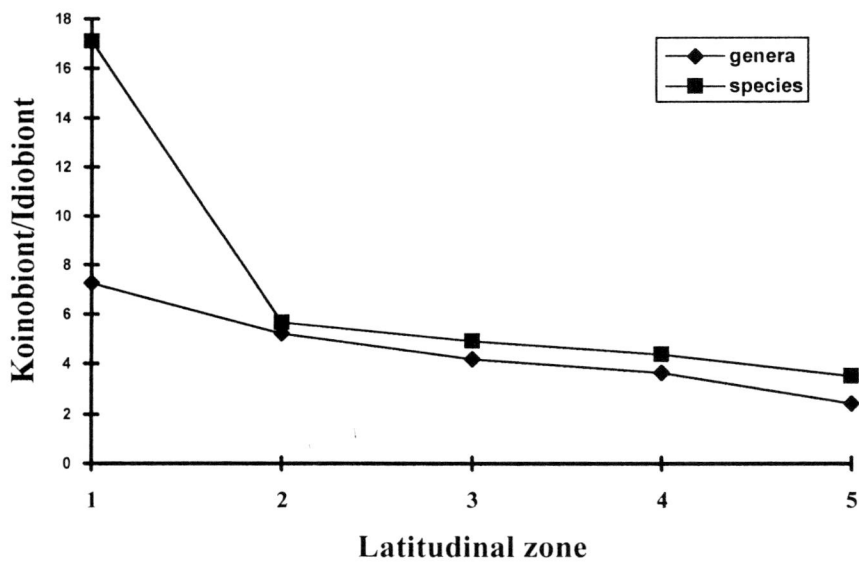

Figure 10.3 Latitudinal trends in the ratios of koinobiont to idiobiont taxa of braconid wasps in America north of Mexico. Latitudinal zone 1 corresponds to North Canada and Alaska; zone 5, the southernmost states of the United States. (Data from Quicke and Kruft, 1995.)

of species versus trap number (Figure 10.4) suggest that the total number of pimpline taxa sampled may be approaching an asymptote. The samples of Pimplinae were not perhaps as enormous as one might imagine: for each malaise trap month, only six individuals were collected on average. Whilst this is not unusual for many groups of insects in the tropics, it does pose considerable problems with interpretation. Nevertheless, extensive hand collecting by parataxonomists failed to yield any species not collected by the malaise traps and therefore Gaston and Gauld felt reasonably confident that the sampling programme had resulted in collection of a large part of the total pimpline species richness of the country. What was perhaps of some surprise was that approximately 40% of Costa Rica's pimpline fauna could be collected at a single site, implying that most species have wide ranges within the country and even throughout the neotropics; this was especially true of the lowland pimpline fauna.

The species richness of pimplines in Costa Rica, when standardized for the country's area, was found to be far greater than that found in any other part of the world with the exception of another small tropical country, Taiwan. Gauld and Gaston (1995) have extended this work to the whole of the hymenopteran fauna and the greater species richness of Costa Rica is very clear (Table 10.1). However, not all families of parasitic wasps show the same trend. In Table 10.2 the relative proportions of the Hymenoptera fauna of Britain and Costa Rica, represented by three superfamilies, are compared. Whereas the Proctotrupoidea and Chalcidoidea constitute a larger proportion of the Costa Rican than the British fauna, the Ichneumonoidea show the opposite relationship.

Parasitic wasp species richness in the tropics is apparently greatest at intermediate altitudes rather than at the warmer lower ones. One possible, and much cited, explanation of this was suggested by Townes (1971): that is, that parasitic wasps in hot climates probably need to drink every day, and for this they rely heavily on the early morning dew, which forms too irregularly at lower altitudes for most species. Host groups no doubt also have a strong effect.

10.8 HOST RANGES AND PARASITOID LOADS

Host range and parasitoid range are both of considerable interest to ecologists, but interpretation of these is fraught with difficulties. Shaw (1994) has emphasized several of these, including misidentifications of parasitoid and host, inadequate sampling and contaminated rearings. Nevertheless, the fact that results of several studies, based on either raw published data or carefully sieved subsets, have been in general agreement may give us some confidence in this approach.

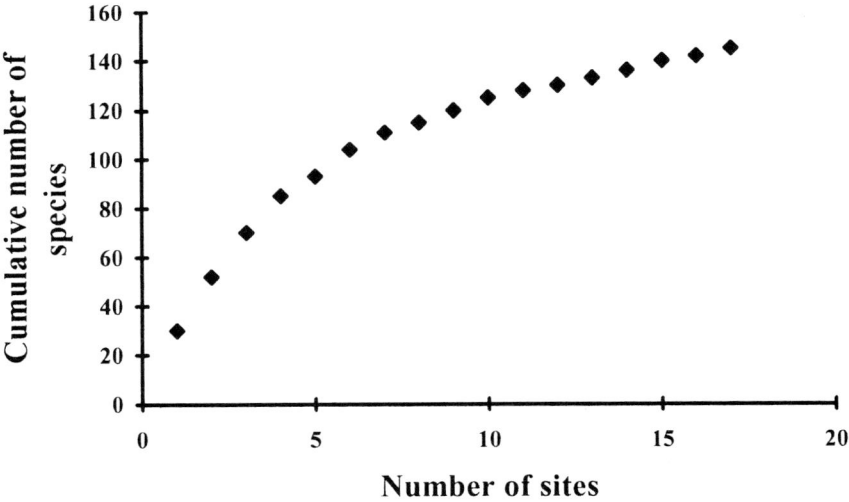

Figure 10.4 Apparent approach to an asymptote of the cumulative number of species of pimpline ichneumonids collected at traps in Costa Rica. (Redrawn from Gaston and Gauld, 1993.)

Table 10.1 Comparisons of species richness and species per unit area for five well-studied regions (modified after Gauld and Gaston, 1995)

Region	Area (10^3 km²)	No. of Hymenoptera species (estimated)	Species per 10^3 km²†
British Isles	314	6 500	1 545
Switzerland	413	9 000	1 997
USA and Canada	19 286	36 000	3 055
Canada alone	9 922	16 665	1 670
Costa Rica	51	20 000	7 484

† Number of species/(area$^{0.25}$)

Table 10.2 Species richness of three large superfamilies of parasitic Hymenoptera in Costa Rica and Britain as percentages of their total hymenoptera faunas (data extracted from Gauld and Gaston, 1995)

Superfamily	Percentage Hymenoptera fauna		Costa Rica/Britain
	Costa Rica	Britain	
Proctotrupoidea	21.2	8.8	2.41
Chalcidoidea	27.8	22.0	1.26
Ichneumonoidea	29.3	47.0	0.62

10.9 COMPETITION

Dean and Ricklefs (1979) showed that for parasitoids of lepidopterous larvae in southern Ontario the abundance of a parasite species on a given host was either uncorrelated with, or even positively correlated with, both the number and abundance of co-occurring parasitoid species attacking the same host. This finding suggests strongly that, at least in the system they studied, the parasitoids were not competing for hosts. If there was competition, one would expect both correlations to be strongly negative because, under conditions of competition, host abundance would be expected to limit parasitoid populations and, all other things being equal, the population of one parasitoid species would be inversely related to the number of competing species. This interpretation was doubted by Force (1980), and a certain amount of debate ensued (Dean and Rickleffs, 1980). Subsequently, Force (1985) presented data on the results of dissection of galls made by the cecidomyiid fly, *Rhopalomyia californica*, over a two-year period. This host is typically subject to high levels of parasitism in the field (with a mean of approximately 70%). Force's results for this system suggested that competition was important, and one species of parasitoid was found to be almost completely inhibited by the presence of others.

Kraaijeveld and van Alphen (1993) have considered the question of whether invasion into a new geographical region by a host may be facilitated or permitted because it is immune to parasitization by the local parasitoids that it encounters in its new range. In their case, the invaders were two palaearctic species of *Drosophila*, *D. subobscura* and *D. ambigua*, which have successfully colonized North America. The dominant larval parasitoid of *Drosophila* species in North America is the alysiine braconid, *Asobara tabida*. Kraaijeveld and van Alphen showed, however, that these two invading *Drosophila* species were just as susceptible to attack by *A. tabida* as were the equivalent native North American species, *D. pseudoobscura* and *D. athabasca*. So in this case, parasitic wasps appear to have had no effect on the colonization potential of an exotic host.

10.10 FACTORS AFFECTING PARASITOID DIVERSITY

Parasitoid assemblage size – that is, the number of species of parasitoid per host species – and the factors that influence it have attracted a great deal of attention of late, particularly by Hawkins and co-workers, and much of this has been summarized in his 1994 book. In a study of the published parasitoid ranges of 191 species of phytophagous cecidomyid gall midges, Hawkins and Gagné (1989) showed that the most significant factor determining the number of parasitoid species per host was the apparency of the galls, with host pupation site and host voltinism having lesser but still detectable effects.

By comparing local and regional community species richness for fig wasp (Agaonidae) parasitoids, Hawkins and Compton (1992) showed a slight drop in community saturation with latitude, although no such drop was found for the galling species associated with figs. Average parasitoid assemblage size decreases towards the tropics for exophytic hosts but does not change for endophytic ones (Hawkins, 1990, 1994). More detailed examination of this trend led Hawkins *et al.* (1992) to conclude that the basic resource fragmentation hypothesis was not adequate to explain parasitoid species richness trends, nor were the host or parasitoid predation hypotheses. A modification of the resource fragmentation could, however, be made to account for the observed relationships.

The rate of accumulation of parasitoids by herbivorous insects that have been introduced into new geographical regions as a result of human activities has been considered in detail by Cornell and Hawkins (1993, 1994). Although their data include both dipterous and hymenopterous parasitoids, the vast majority concern the latter. Whilst one might imagine that it would take a considerable time to build up a full complement of parasitoids, Cornell and Hawkins showed that many herbivore species had acquired a large number of parasitoids quite soon after their initial establishment. Overall, the species richness of parasitoids on introduced hosts was lower than on the host species in their native country (4.04 vs.

7.74 parasites/host; geometric means), but they found no significant relationship between the time elapsed since introduction and the absolute difference between the number of parasitoids in the introduced community and that in the native country. In other analyses a weak trend in parasitoid species richness per host was observed over a 150-year time scale, but the scatter of data points was very large. They concluded that 150 years (the longest period for which reliable data are available) is not long enough for the parasitoid community on the invading herbivore to reach equilibrium. The levels of parasitism found on invading hosts were also lower than on hosts as natives. The parasitoid complexes attacking introduced hosts included a significantly higher proportion of idiobiont taxa – as might be expected, since idiobionts are typically generalists whilst koinobionts tend to be specialists.

The Cornell and Hawkins data came from literature surveys, much of which involved agriculturally important hosts, and there have been very few investigations specifically aimed at examining the rate of parasitoid species accumulation on introduced or naturally invading hosts in non-agricultural systems. One such study, that of Godfray *et al.* (1995), provides support for the conclusions of Cornell and Hawkins. Godfray *et al.* examined the accumulation of parasitoids on two species of leaf-mining lepidopterans, *Phyllonorycter leucographella* and *P. platani*, which invaded the UK in the 1980s and have spread continuously since their first occurrence. They found that by 1994 both invading species had accumulated a total of 16 parasitoid species and that, in this respect, they were indistinguishable from native species of *Phyllonorycter*. Further, and against expectations, the parasitoid species that the two invading hosts had accumulated were not found to have broader host ranges than the parasitoids attacking native *Phyllonorycter* species.

10.11 EFFECTS OF HOST NICHE ON PARASITOID DIVERSITY

The average number of parasitoid species per host insect species is about five, but the total range observed is from zero to 50 or more. Why some species of host support a larger parasitoid community than others has been investigated in detail by Hawkins (1993, 1994). The size of the parasitoid assemblage attacking a range of lepidopterous larvae was shown to be strongly influenced by the host niche. Arranging host feeding niches in an order that approximates the degree of refuge afforded by the niche, the largest number of parasitoids per host species is found to be highest for moderately concealed host taxa rather than for effectively unprotected taxa such as exposed feeders or for heavily concealed taxa such as larvae which bore in wood or feed on plant roots (Figure 10.5). The size and complexity of the plants on which phytophagous hosts are found is also significant, with larger and more complex plants leading to larger assemblages of parasitoids per host (Hawkins *et al.*, 1990). As would be expected, when exo- and endophytic hosts were considered separately, the increase in parasitoid assemblage size with plant size for endophytic hosts was found to be due to an increased number of idiobiont parasitoids, whereas for exophytic hosts the increase was a result of more koinobionts. The basic domed shape of the assemblage size host exposure (feeding niche) plot (Figure 10.5) can be explained theoretically by considering the effect of refuges from parasitization, as shown by Hochberg and Hawkins (1993), with the exception of the observed non-zero value for exposed hosts. This discrepancy could be due to a deficiency of the model, or alternatively, exposed hosts may be associated with some other, non-physical sort of refuge from parasitism. The effect of refuges against parasitism on parasitoid assemblage size and on overall levels of parasitism has implications for classical biocontrol (Hawkins *et al.*, 1993).

10.12 'NASTY HOST' HYPOTHESIS – EFFECTS OF SECONDARY PLANT SUBSTANCES

Gauld *et al.* (1992) proposed that the failure of the diversity of some groups of parasitic Hymenoptera to increase greatly towards the tropics might be explained by tritrophic interactions involving the protective chemistry of their hosts' host plants. This view was further supported by Gauld and Gaston (1994, 1995). Evidence for the so-called nasty host hypothesis came initially from a synthesis of considerations of the host ranges of certain groups of koinobiont ichneumonids in Costa Rica versus the temperate United States, and from a general background of information about the effects of plant secondary compounds on parasitoid development. A further

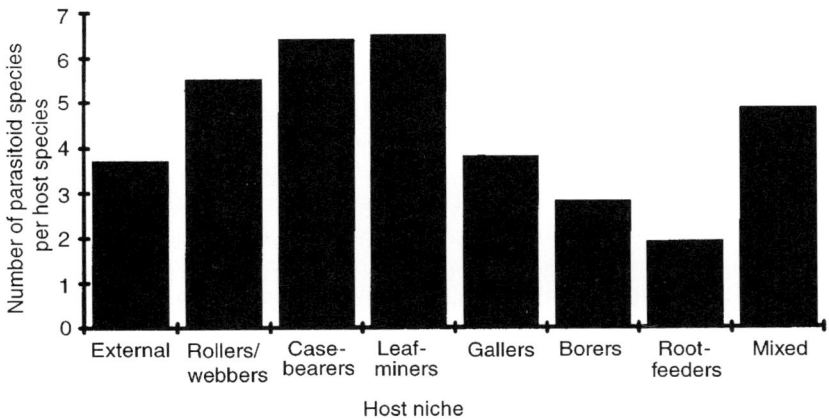

Figure 10.5 Relationship between size of parasitoid guild and host feeding niche. (Redrawn from Hawkins, 1994.)

circumstantial line of evidence came from observations of latitudinal trends in the potency and diversity of secondary plant substances.

Many tropical ichneumonids appear to be strictly monophagous, in marked contrast to their close relatives in the temperate region (Gauld and Janzen, 1994). Thus, for example, Gauld *et al.* (1992) noted that several temperate species of the ophionine ichneumonid genus, *Enicospilus*, each attacks a wide range of species of saturniid moths. After an extensive programme of rearing parasitic wasps from exophytic macrolepidoptera larvae in Santa Rosa, Costa Rica, the 17 endemic *Enicospilus* species there were found to be almost exclusively monophagous, or in a few cases attack a couple of closely related hosts (Gauld, 1988b). In the case of the campoplegine ichneumonid genus, *Cryptophion*, another genus of koinobiont endoparasitoids of lepidopterous larvae, most Costa Rican species are monophagous, and one such species, *C. inaequalipes*, only attacks its sphingid host when it occurs on one of its two host plants, even when the plants and hosts are in close proximity to one another. Many species of exposed potential macrolepidopteran hosts in the tropics appear to have no ichneumonid parasitoids. However, those ichneumonid groups that attack relatively chemically unprotected hosts such as wood-boring insects or vespid wasp larvae, and similarly, encyrtids that attack Homoptera, are all markedly more speciose in the tropics.

Of some 23 000 individual potential host caterpillars collected and reared in the tropical dry forest of Santa Rosa National Park in Costa Rica, some 22% were parasitized – 9% by tachinid flies, 3% by braconid wasps, 3% by ichneumonid wasps and 7% by other hymenopterous parasitoids.

Evidence that plant secondary compounds sequestered by herbivorous insects may play a role in limiting parasitoid species richness in communities comes from a limited number of studies and from a number of anecdotal observations. In particular the work of Barbosa *et al.* (1986, 1991) is often cited. These workers have shown that the level of nicotine in various lepidopteran caterpillars feeding on tobacco plants has a large effect on the developmental success of a range of parasitoid taxa. Whilst it is clear that secondary plant chemicals can provide a barrier that prevents a particular parasitoid from successfully developing on a given host, it is less clear that plant secondary chemicals may play a role in limiting the range of parasitoid species accumulated over evolutionary time on protected host species. Indeed a resource fragmentation argument could be put that the need for parasitoids to become specialized on given 'protected' hosts would lead to higher diversity in the tropics. Wharton (1993) noted that most of the evidence in favour of the 'nasty host' hypothesis derives from studies on the Ichneumonidae, especially species that are larval or pupal endoparasitoids of exophytic Lepidoptera larvae. Several other groups with similar biologies, such as agathidine braconids, are highly speciose in the tropics. All this together might suggest that some groups of endoparasitoid wasps may be more susceptible to sequestered secondary plant compounds than others.

10.13 FOOD WEBS AND TROPHIC INTERACTIONS

The investigation of parasitoid food webs is still in its infancy, and very few systems have been investigated in detail to date. Essentially food webs can be described in either qualitative or quantitative terms. The former webs, referred to as connectance food webs, simply report what attacks what. Quantitative food webs are far more informative since many species of parasitoid may occasionally attack a particular host successfully, but these events may be so rare that they are of little significance to the overall set of interactions in a community. Quantitative food webs allow these rare events to be seen as such.

Memmot and Godfray (1993, 1994) provided a quantitative description of a leafminer community in a region of dry forest regrowth in the Santa Rosa National Park, Costa Rica. Host ranges for ecto- and endophagous parasitoids were calculated and are presented in Figure 10.6. It can be seen that ectoparasitoids include fewer monophagous taxa and, at the opposite extreme, some species attacked up to a dozen host species compared with a maximum of four for endoparasitoids. When more quantitative parasitoid food webs are available, it will enable more aspects of parasitoid community. Surprisingly, one of the most complicated multitrophic parasitoid complexes occurs in a rather small host. Haviland (1920) described a tertiary parasitoid system in which a megaspilid hyper-hyperparasitoid attacks a chalcidoid hyperparasitoid of an aphidiine braconid primary parasitoid of an aphid.

10.14 PREDATORS

As with other insects, adult parasitic wasps are attacked by a range of predators, including birds, lizards and spiders as well as other insects, but there have been few detailed studies of the level of such predation. Several observers have pointed out that ovipositing females that drill through bark or similar thick substrates may be particularly vulnerable. Madden (1982) examined bird predation of the siricid horntail, *Sirex noctilio*, and its parasitoid complex in New Zealand, and found that rhyssine ichneumonid parasitoids often suffered fatal attacks by birds; apparently, in some situations, it is not uncommon to find broken rhyssine ovipositors left sticking out of the bark of *Sirex*-infected trees.

Völkl *et al.* (1994) have shown that female *Alloxysta brevis*, a cynipoid hyperparasitoid of aphidiine braconids in aphids, produces deterrent chemicals (6-methyl-5-hepten-2-one, actinidin and various unidentified iridoids) from their mandibular glands, which protect them from attack by

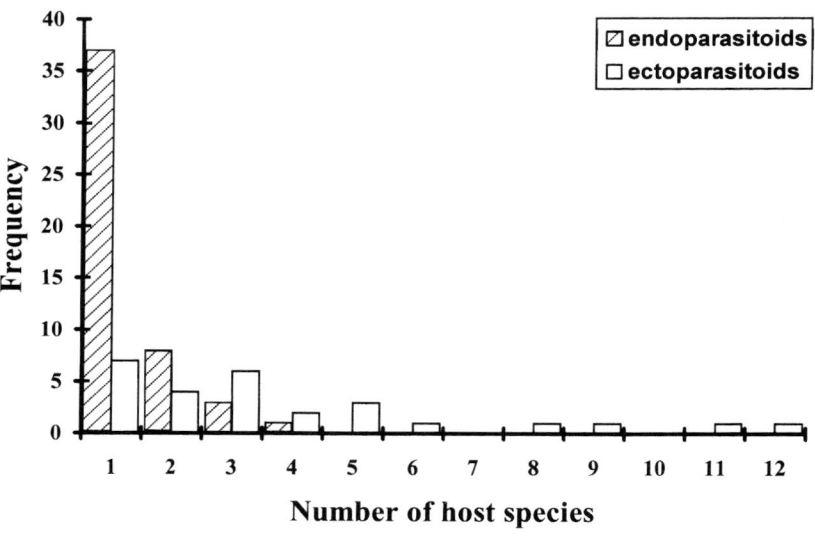

Figure 10.6 Host ranges of ectoparasitic and endophagous leafminer parasitoids in a Costa Rican dry forest. (Redrawn from data from Memmott and Godfray, 1994.)

the ants that attend the aphids amongst which they are searching (see section 6.44). In some contrast to Völkl *et al.*'s study, Vinson and Scarborough (1991) showed that the red imported fire ant, *Solenopsis invicta*, had a negative effect on parasitization of aphids by the aphidiine braconid, *Lysiphlebus testaceipes*. Although in lab conditions the adult wasps made about the same number of attacks in the presence and absence of *Solenopsis*, the ants acted as predators selectively removing parasitized aphids. Aphid honeydew production decreases as a result of parasitism, and this may be the cue for the ant that all is not well. However, the ants can distinguish aphids that are producing less honeydew as a result of parasitism from those that are starving, and the latter they carefully remove and place elsewhere on the plant. Vinson and Scarborough also showed that ants foraged more on plants that were infested with parasitized aphids than those that had equal levels of parasitoid-free aphids. This asymmetry diminished and disappeared as the ants progressively removed the aphid mummies, presumably to eat them in their nest.

Bronstein (1988) found that ants (Pseudomyrmicinae) pose a considerable threat to non-pollinating fig wasps, especially *Idarnes*, while they are ovipositing into figs (*Ficus pertusa*) in Costa Rica – drilling and ovipositions in *Idarnes* often took up to 30 minutes. The pollinating fig wasps which entered the fig through the ostiole did so quickly and so were able to avoid most predation at that stage. Later in development, fig wasp larvae were found to be subject to predation by larvae of a curculionid beetle and a moth, whilst mature wasps inside the figs suffered considerable levels of predation as a result of staphylinid beetles, which were able to gain entry into the fig through the ostiole that had by this time become enlarged by escaping wasps. Another fig wasp/predator system was described by Compton and Disney (1991). They discovered several new species of the phorid fly genus, *Megaselia*, that develop in African figs. The phorid larvae gain access to the figs through the ostiole just as the female pollinating fig wasps do, and after completing their feeding the *Megaselia* larvae again leave the fig to pupate elsewhere. The adult female flies have piercing mouthparts which they use to suck out the contents of adult fig wasps that enter new figs prior to oviposition. When the *Megaselia* were present-ed with both sexes of the fig wasp, they showed a marked preference for preying upon the females, perhaps because in nature they would not nor-mally encounter the apterous males.

Gauld and Gaston (1994) considered that in tropical situations, where individual hosts are perhaps far more widely dispersed than in temperate regions, specialist parasitoids may be particularly long-lived and so may be exposed to far higher lifetime levels of predation than temperate coun-terparts. They may therefore be expected to evolve more defensive fea-tures against predators. Certainly some of the more bizarre defences are found in tropical species, including spines, protective collars, poison claws and urticating setae (See section 6.46). Whether there are proportionately more obviously defended species in the tropics requires careful testing.

10.15 PATHOGENS

As with other phylogenetically recent insect orders, the Hymenoptera as a whole have an efficient system of induciable antibacterial peptides (e.g. cecropins and defensins) that protect them from most potential pathogens (Hoffman *et al.*, 1992). There have been comparatively few studies on pathogens affecting parasitic wasps (Geden *et al.*, 1995); most have involved parasite–host systems maintained in culture and therefore at artificially high population densities. Consequently, very little is known about the occurrence and epidemiology of pathogens in wild host–para-sitoid systems. It is clear, however, that parasitic wasps are prone to infec-tion from host pathogens and in some cases from their own specific ones. The main groups of pathogens that have been investigated are viruses, microsporidian and neogregarine protists, bacteria and (in the case of fig wasps) nematodes. These will be treated separately below.

Virus and other pathogenic infections have potentially important implications for biocontrol programmes in which exotic parasitoids are being introduced into a new area. Hamm *et al.* (1992) recommend that all quarantined parasitic wasp cultures should be routinely surveyed for viruses and other potential pathogens, and if possible efforts should be made to avoid introducing these along with the beneficial agent.

10.16 VIRUSES

Parasitic wasps harbour a number of viruses, but the majority of those studied to date are either neutral or benign (section 7.11). In other cases, difficulties in obtaining virus-free cultures can make it difficult to deter-mine the actual pathogenicity caused by a given virus, and so conclusions have often had to be based on electron microscopic surveys for what looks like tissue damage. When pathogenic effects have been detected, they often bear a close resemblance in their symptomatology to those caused by microsporidian infections. Again, as with microsporidians, some parasitic wasps have been found to become infected by viruses that are normally associated with the wasp's host, such as various polyhedrosis viruses (e.g. Murray *et al.*, 1995).

Probably the best studied instance of virus infecting a parasitic wasp involves the well-studied *Microplitis croceipes*. Some cultures of this bra-conid have been found to be infected with a non-occluded baculovirus, called McNOBV (Hamm *et al.*, 1988). The virus occurs in a wide range of tissues in the adult wasp including fat body, midgut epithelium, Malpighian tubules, tracheal matrix cells, and others. Most tissues show little obvious ultrastructural damage, but fat body and digestive tract show clear histopathology resulting in eventual cell rupture (Chittihunsa and Sikorowski, 1995a). In addition, the virus causes an increase in wasp development time (Chittihunsa and Sikorowski, 1995b). *M. croceipes* also harbours a picorna-like virus (Hamm *et al.*, 1992) that is probably patho-genic, but obtaining clear evidence of this has been difficult.

10.17 MICROSPORIDIANS

The known occurrences of microsporidians, a common group of insect pathogens, infecting parasitic Hymenoptera are summarized in Table 10.3. Most occurrences of micosporidians in parasitic wasps are due to transmis-sion of a host pathogen from the host to the wasp, and to date only two cases are known of wasp-specific species. The degree of pathogenicity expe-rienced by parasitic Hymenoptera when infected with host microsporidia is quite variable and usually depends on the degree of infestation. In *Macrocentrus ancylivorus* cultures, infected individuals can be recognized by conspicuous white patches on the metasomal underside, and in extreme cases the infection results in severely deformed abdomens (Allen and Brunson, 1945). The microsporidian, *Nosema bordati*, was found to be highly pathogenic to the braconid, *Cotesia flavipes*, a parasitoid of *Chilo partellus* (Bordat *et al.*, 1994), but only weakly infected adults were able to transmit the *Nosema* to new hosts. Zchori-Fein *et al.* (1992a) reported that various pteromalid parasitoids of filth-breeding Diptera are often infected by microsporidia, and in the case of *Muscidifurax raptor* the infection causes a serious loss of fitness, with infected females taking longer to develop, living less long and producing only 12–50% of the numbers of offspring as uninfected ones (Geden *et al.*, 1995). The microsporidian was shown to be transmitted maternally, within the egg, with 100% efficiency. These infections accounted for much of the difficulty experienced in culturing these wasps from field-collected samples, and there also appears to be the possibility that the microsporidian may affect sex ratio, as some are known to do in other groups of organisms (Antolin, 1992). Chapman and Hooker (1992) described the ultrastructure and distribution in wasp tissues of two *Nosema* spp. in the eulophid, *Pediobius foveolatus*, a parasitoid of the Mexican bean beetle, *Epilachna varivestis*. In this system, transovarian transmission of the microsporidian by the wasp does occur, but with less efficiency, and although production of infected colonies is reduced, the effects are apparently less than with microsporidian-infected *Muscidifurax* and *Cotesia*. Nevertheless, infection of parasitoid colonies could have serious implications for their potential in biocontrol programmes.

10.18 NEOGREGARINES

Larvae of *Bracon mellitor*, the braconid parasitoid of the boll weevil, *Anthonomus grandis*, can become infected with the neogregarine protist, *Mattesia grandis*, by feeding on infected boll weevil larvae, and adult wasps could also become infected by eating *M. grandis* spores (McLaughlin and Adams, 1966). Infected adult wasps showed a marked reduction in both host stinging behaviour and parasitization rate, so the presence of hosts infected by *M. grandis* is likely to reduce the efficiency of *B. mellitor* in field situations. Infected wasps are not able to transmit the gregarine to healthy boll weevil larvae and they do not act as carriers when they have previously stung an infected host.

10.19 NEMATODES

Many pollinating fig wasps (Agaonidae) are parasitized by species-specific nematodes. These nematodes, which mate inside the fig, attach themselves on the body of newly emerged female fig wasps and subsequently enter them, laying eggs within them, and emerge later ready to infect the fig wasp's offspring or those of other females within the same fig. Herre (1993) examined the virulence of 11 species of *Parasitodiplogaster* that attack pollinator fig wasps (*Pegoscapus* and *Tetrapus* spp.) in South

Table 10.3 Occurence of microsporidian pathogens that infect parasitic wasps

Wasp	Family	Host	Order	Reference
Cotesia flavipes	Braconidae	*Chilo partellus*	Lep.	Bordat *et al.*, 1994
Cotesia glomerata	Braconidae	*Pieris rapae*	Lep.	Tanada, 1955
*Cardiochiles nigriceps**	Braconidae	–	–	Brooks and Cranford, 1972
*Campoletis sonorensis**	Ichneumonidae	–	–	Brooks and Cranford, 1972
Macrocentrus ancylivorus	Braconidae	*Gnorimoschema operculella*	Lep.	Allen and Brunson, 1945
Dahlbominus fuscipennis	Eulophidae	*Pristiphora erichsonii* and other sawflies	Hym.	Smirnoff, 1971
Pediobius foveolatus	Eulophidae	*Epilachna varivestis*	Col.	Own and Brooks, 1986
Muscidifurax raptor, *M. zaraptor and Urolepis rufipes*	Pteromalidae	Various muscids, e.g. *Musca* and *Stomoxys* spp.	Dipt.	Antolin, 1992; Zchori-Fein *et al.*, 1992a†

*　In these two cases, the *Nosema* species were found to be specific to the parasitoids and did not infect the host Lepidoptera larva.

†　In their survey of cultures of other parasitic Hymenoptera attacking muscoid flies, they found no evidence of microsporidia in *Spalangia* (2 spp.), *Nasonia vitripennis*, *Trichomalopsis dubius*, *Muscidifurax raptorellus* or *M. uniraptor*.

America, and showed that nematode virulence was positively correlated with the proportion of multifoundress fig wasp broods and hence increased opportunity for transmission.

10.20 FUNGI

Entomopathogenic fungi are known to attack a few parasitic wasps, but this is likely to be only a fraction of the actual number of associations. Wasps may become infected from their hosts. For example, a strain of the fungus, *Zoophthora radicans*, isolated from the diamondback moth, *Plutella xylostella*, has been shown to be pathogenic to its ichneumonid parasitoid, *Diadegma semiclausum*, though another parasitioid, the braconid, *Cotesia plutellae*, was immune (Furlong and Pell, 1996).

10.21 EPIZOOTIOLOGY

The potential of parasitic wasps to act as vectors for diseases of other insects has been investigated by a number of workers. A considerable potential advantage of such a strategy is that as parasitoids are generally very good at locating their hosts, the disease-causing organism will be very effectively targeted. Raimo *et al.* (1977) demonstrated that the braconid,

Apanteles melanoscelus, carrying gypsy moth nuclear polyhedrosis virus acted as a vector – more hosts died as a result of attack by virus-treated wasps than by untreated ones. Similarly, Levin *et al.* (1983) showed that *Cotesia glomerata* could transmit *Pieris rapae* granulosis virus to hosts. The now widely employed bio-insecticide, *Bacillus thuringiensis* (*Bt*), can be transmitted by several parasitic wasps and the latter are immune to its action (Kurstak, 1966; Mück *et al.*, 1981). However, if the *Bt* infection causes premature death of the host so that the parasitoid's larva does not survive, *Bt* could have a very deleterious effect on the populations of natural parasitoids.

It has been shown that several hymenopterous parasitoids are effective at transmitting microsporidian infections between hosts (Paillot, 1933; Laigo and Tamashiro, 1967; Andreadis, 1980). However, McLaughlin and Adams (1966) showed that the boll weevil parasitoid, *Bracon mellitor*, was unable to transmit neogregarine protist pathogens between hosts. The latter failure may reflect the brief nature of the stinging behaviour of the *Bracon* which gives little opportunity for the pathogen's spores to infect the host beetle larva. In the case of the diamondback moth, *Plutella xylostella*, transmission of the fungal pathogen, *Zoophthora radicans*, was enhanced in the presence of the parasitoid, *Diadegma*, but this did not result from spore transport. Instead, it followed from increased host activity in the presence of parasitoids (Furlong and Pell, 1996).

10.22 COLORATION AND MIMICRY

Many parasitic wasps are brightly coloured and conspicuously patterned, leading one to suspect that either they themselves are unpalatable (warning coloration) or that they are Batesian mimics of other, less palatable taxa (Gauld, 1995). Given the diversity of the parasitic Hymenoptera, it would be surprising if both situations were not the case for various taxa. Certainly, some parasitic wasps are able to deliver a sting to human handlers. This is quite well known for some larger ichneumonids (Ichneumoninae, Ophioninae, Tryphoninae), braconids (Braconinae, Doryctinae) and *Pelecinus polytrurator*, a large parasitoid wasp from the New World, not to mention the parasitic aculeates such as the infamous Mutillidae and Sclerogibbidae (Walton, 1948; Schmidt and Blum, 1977; Schmidt, 1978; Quicke, 1986; Quicke *et al.*, 1992c). Other taxa would not appear well adapted to stinging even a small avian predator, and of course male parasitic wasps lack ovipositors through which venom could be injected.

A few parasitic Hymenoptera, including some small species, produce distinctive and sometimes unpleasant (at least to humans) odours, and these might justify the evolution of aposematism (Townes, 1939; Whitfield, 1988; see sections 5.13 and 6.44). Townes further suggested that this might form the basis for some mimetic associations and concluded that North American ichneumonids, at least, were models in Batesian complexes. The colour patterns shown by many larger parasitic wasps, especially braconids, are often shared by other insects, especially beetles and bugs, but also by some Lepidoptera, Neuroptera and Diptera. That some parasitic wasps specifically act as models for other taxa is clearly shown by mimicry of their distinctive ovipositors. At least one Neotropical moth closely resembles various ichneumonid wasps in that it has evolved a pseudo-ovipositor protruding from the tip of its abdomen as well as a similar slender habitus and typical colour pattern. In another instance, a Neotropical reduviid bug has been photographed holding out one of its hindlegs directly behind its abdomen, causing it to have a particularly braconid-like appearance (Preston-Mafham and Preston-Mafham, 1993). A potentially related mimetic appendage is found in the exceedingly rare Meso-American braconid, *Pheloura dolichoura*, females of which have an extremely long three-piece pseudo-sting that emerges from the region of the anus (Achterberg, 1989) – the true ovipositor of this species is short but the pseudo-ovipositor was for many years assumed to be the real thing, leading to it being illustrated in textbooks as an example of an extremely long-ovipositored parasitoid.

Whilst some parasitic wasps might act as models in mimicry systems, others may be mimics themselves – for example, of aculeates (Mason, 1964). It is also interesting to note that the dominant colour patterns displayed vary from one part of the world to another. The dominant colour patterns in larger Afrotropical braconine braconids are strongly correlated with habitat type – bright red wasps being mostly found in semi-arid areas with red, lateritized soils, whereas predominantly black and dark red species are strongly associated with tropical wet forests (Quicke, 1986). This distribution suggests that coloration in these may be partly cryptic, especially from a distance, although there can be no doubt that, once seen, the patterns are aposematic. A few examples are known in which parasitic wasps and their hosts apparently share the same colour pattern, and several reasons have been suggested to explain this phenomenon, including their necessary co-occurrence (Quicke *et al.*, 1992c).

Some parasitic wasps, particularly those that pupate in exposed situations, construct cryptic or mimetic cocoons. Notable among these are some campoplegine ichneumonids whose black, grey and white cocoons bear a considerable resemblance to bird droppings; the cocoons of some polysphinctine ichneumonids that are constructed in the webs of their host spider (by this time deceased), and are disguised so as to resemble a prey insect wrapped in silk (Gauld, 1991); and one species reported to have cocoons that mimic flowers (Koptur, 1989). One of the most complex examples of mimetic cocoons involves some *Hyposoter* species (Ichneumonidae: Campopleginae) which pupate inside the remains of their host Lepidoptera caterpillar, but before that they spin a clearly visible, thin but firm false cocoon between the forelegs of the dead host larva (Finlayson, 1966). As birds seem to find parasitized larvae unpalatable, the *Hyposoter's* false cocoon may serve to advertise the parasitized state of the caterpillar. That this is the case is supported by the fact that larvae of some species belonging several lepidopteran families, including Gracilariidae, Zygaenidae, Arctiidae and Bombycidae, construct cocoons that mimic ones from which a parasitoid may have emerged or alternatively construct false parasitoid cocoons (Hinton, 1955).

A number of different parasitic wasps exhibit thanatosis (death feigning) when disturbed and in some cases they remain motionless in a characteristic posture for several minutes (Cameron, 1941; Frost, 1959; van Achterberg, 1986). Interestingly, some populations and cultures of the braconid, *Habrobracon hebetor*, are polymorphic for this behaviour (Mettus and Petters, 1981; Grosch, 1988). Breeding experiments have shown that the gene controlling thanatotic behaviour is inherited more or less as a simple, single dominant allele, though it was never possible to obtain a strain in which 100% of individuals exhibited the behaviour.

10.23 FACTORS AFFECTING THE SUCCESS OF BIOCONTROL INTRODUCTIONS

The success of parasitic wasps used in biological control appears to depend to an appreciable extent on phylogeny. Most releases have concerned members of two superfamilies, the Ichneumonoidea and the Chalcidoidea. Of these, the chalcidoids have clearly been more successful than ichneumonoids on average, and within the latter very little difference exists between the Ichneumonidae and Braconidae (Table 10.4; Stouthamer *et al.*, 1992). Townes (1971) suggested several reasons why ichneumonids might not always make good biocontrol agents. For example, he suggested that many need to drink frequently, and that these species would not fair well in situations where there was not a ready day-to-day water source. The sex determination system may also be important (see section 2.38).

Hopper and Roush (1993) analysed a number of factors pertaining to releases of ichneumonoids, chalcidoids and tachinids that may have affected their successful establishment, and in particular they considered whether failures to establish might result from an Allee effect. The analysis, which was restricted to attempts to control lepidopteran pests, showed that establishment was strongly influenced by the total number of parasitoids released in the case of chalcidoids, and the number per release for ichneumonoids. This finding supported the theoretical view that introductions involving few parasitoids are liable to fail because as the insects disperse their density may become too low for individuals to find mates. It follows that the number of females that need to be introduced was

Table 10.4 Comparative success of ichneumonoid and chalcidoid introductions in biological control programmes (based on Stouthamer *et al.*, 1992)

Superfamily/family	Established	Failed to establish	Success (%)
Ichneumonoidea	82	231	26
Ichneumonidae	61	147	29
Braconidae	21	84	20
Chalcidoidea	40	29	58

found to decrease with increasing reproductive rate and with increasing mate-finding distance. Another interesting finding from their model was that arrhenotokous species (i.e. parasitic wasps) in which unfertilized females still produce progeny, though all males, increase the minimum number of females needed in an introduction compared with normal sexual species in which females produce no offspring at all. Thus thelytokous species might well have an advantage (Stouthamer, 1993; see section 2.37).

Biocontrol programmes have used both novel host–parasitoid associations and long-standing ones that have been subject to long periods of coevolution. Hokkanen and Pimentel (1984) surveyed the relative successes of these in 286 cases, and found that novel associations had a significantly higher success rate; they therefore recommended that, whenever possible, new associations should be employed in control programmes.

It has been a common practice to make multiple introductions of hymenopteran parasitoids if the first introduction does not succeed, and in particular to try to release a different strain of the parasitoid in the hope that it will be better adapted to the new environment. This may seem logical, but empirical evidence suggests that if an initially released strain has failed, then subsequent ones have little better chance of succeeding. This conclusion comes from a survey carried out by Clarke and Walter (1995). Whilst the exact figures might be misleading because of the inadvertent releases of cryptic species rather than 'strains' of the initially released species, the data suggest that if the first introduction fails, it may be better to concentrate on finding an alternative species of parasitoid for release.

10.24 SUCCESSION AND COLONIZATION

There is a great need for studies of colonization by parasitic wasps of new habitats – including regenerating fields, areas after burns or deforestation and new islands – that take into account the different biologies and life history strategies of the taxa concerned. Maetô and Thornton (1993) presented the results of a preliminary analysis of the braconid fauna of Anak

Krakatau, an Indonesian volcanic island that began to emerge from the submerged caldera of Krakatau in 1930, and which has been slowly colonized by plants, birds and insects ever since. Contrary to expectations, the early-phase colonists were dominated by taxa that are koinobiont endoparasitoids of Lepidoptera – though not a single koinobiont endoparasitoid of Diptera was collected, despite their being a very significant part of the parasitoid faunas of adjacent islands. The absence of parasitoids of Diptera may reflect the dominance of grassland on Anak Krakatau. However, as koinobionts are generally specialists (Chapter 3), it would have been imagined by most workers that they would have been less successful as colonists than the more generalist idiobiont species. The relative abundance of koinobionts might similarly reflect the stage in vegetational succession, but why is not immediately apparent.

10.25 DISPERSAL AND LABELLING OF PARASITOIDS

Most of what is known about the dispersal abilities of parasitic wasps comes from studies of field releases of biocontrol agents, and even that is not a great amount (Keller *et al.*, 1985; Smith, 1988). Almost nothing is known about natural movements. It is clear that parasitic wasps, especially small species, are able to disperse over large distances and to colonize islands and other remote habitats (Askew, 1968), and it has been known for a long time that small parasitic hymenopterans often occur in the aerial plankton, but whether these individuals are there more by choice than accident is unclear. The small mymarid, *Anagrus delicatus*, regularly disperses a kilometre or more to offshore Floridan islets (Antolin and Strong, 1987), and the majority of these dispersing individuals are females, which suggests that the behaviour is part of a deliberate strategy, albeit a high risk one. When it comes to larger species that cannot become part of the normal aerial plankton for dispersal, little is known.

An ability to recognize individual parasitoids as having originated from a particular site or release would facilitate or even enable many different sorts of studies on both their pure and applied ecology. In an impressive early study, radiolabelled phosphorus was used by Stern *et al.* (1965) to label more than 2 million individuals of *Trichogramma* wasps, and to follow their dispersal from their release site. Most were found to spread quickly up to 200 ft (60 m), and after two and a half days some were recovered 3500 ft (*c.* 1 km) from the release site. Several other labelling techniques have been reported but to date relatively little use has been made of them. The most frequently reported labelling procedure involves the non-toxic element, rubidium. Jackson *et al.* (1988) used rubidium introduced into *Lygus* bug hosts through an artificial diet, and showed that it was passed through the bug's eggs into its egg parasitoid, the mymarid, *Anaphes ovijentatus*. However, the label was only detectable for four days, and at high concentrations it caused a significant increase in parasitoid mortality. Jackson and Debolt (1990) used rubidium to label the euphorine braconid, *Leiophron uniformis*, an idiobiont endoparasitoid of nymphal stages of *Lygus* bugs. Hopper and Woolson (1991) compared the retention of rubidium (Rb), caesium (Cs), dysprosium (Dy) and strontium (Sr) in the braconid, *Microplitis croceipes*, and showed that labelling with either Cs or Dy remained detectable above background for 20 days post-emergence, whereas that with Rb and Sr declined rapidly. Most of these labels have effects on adult longevity and developmental time, but these are small and probably will have little effect on their use in field studies. Further, techniques for detecting these labels are improving all the time so doses will not need to be so high in future studies. Other substances that have been used for labelling include ^{65}Zn (Soenjaro, 1979). As an alternative to labelling with heavy metals, Strand *et al.* (1990b) used acridine orange, a fluorescent dye, to mark parasitoids. The increased sensitivities of modern detection equipment will no doubt greatly increase the utility of some of these labels, and hopefully lead to studies of natural systems as well.

11

Parasitic wasp phylogeny and taxonomy

11.1 INTRODUCTION

The Hymenoptera have traditionally been classified into three suborders: the primarily phytophagous 'Symphyta' or sawflies; the stinging wasps, bees and ants or Aculeata; and the primarily parasitic remainder called the 'Parasitica' or 'Terebrantia'. In accordance with modern cladistic ideology, two of these, the 'Symphyta' and the 'Parasitica', are now disapproved of as formal classificatory units since neither is a monophyletic (or holophyletic) group that includes all of the descendants of their closest common ancestor (Quicke, 1993) in the same way that the 'Reptilia' does not include all the descendants of the first reptile. The 'Symphyta' is not monophyletic because the Apocrita (wasp-waisted Hymenoptera) evolved from within that group, and the same argument applies to the 'Parasitica' because the Aculeata are now widely accepted as having evolved from a parasitic ancestor. For these reasons it is preferable to use less formal terminology wherever possible, such as parasitic wasp, but this can cause confusion because many aculeates are also parasitoids. On the other hand, the Apocrita, which comprises the 'Parasitica' plus the Aculeata, as well as the Aculeata itself, are both monophyletic and can therefore be used as formal taxonomic categories. It will be apparent that strict adherence to cladistic classificatory principles which forbid formal recognition of paraphyletic groups such as 'Symphyta' and 'Parasitica', whilst best representing the phylogeny of the group as we understand it at present, also leads to some awkwardness when it comes to labelling grade taxa such as the sawflies and parasitic wasps.

In addition to the three suborders mentioned above, various other groups have been singled out for special treatment by various workers. For example, the Orussidae, which share many morphological features with sawflies but are parasitoids, have sometimes been allotted their own suborder, the Idiogastra. This and other such treatments have been discussed in some detail by Malyshev (1968), and it is not necessary to go into any greater detail of the pros and cons of such treatments here.

The parasitic Hymenoptera comprises approximately 59 extant families divided among some 13 superfamilies. A further 14 or so families are known only from fossils (Rasnitsyn, 1988) (Tables 11.1, 11.2). The higher-level relationships between the various families and superfamilies are far from understood and even the compositions of some superfamilies have been subject to considerable change in recent times. These changes have been due to past failures to base classification on synapomorphies and this problem has been particularly apparent in the case of the Proctotrupoidea. At family level matters are, if anything, even more uncertain. Several new extant families have been described in the recent past, and it is likely that at least a few others remain to be discovered. The most recent new addition is the Renyxidae from the Russian far east (Kozlov, 1994; Lelej, 1994), and a very strange family of small wasps belonging to the Proctotrupoidea s.l. is shortly to be described from Australia and New Zealand (Earley, Naumann and Masner, in preparation).

11.2 PHYLOGENETIC HYPOTHESES

To date, evidence for phylogenetic relationships between the higher taxa of Hymenoptera has come principally from morphological investigations, albeit using a wide range of character systems. There is a general consensus based on biology and the fossil record that the Apocrita (the narrow-waisted wasps) have evolved from sawflies. This means that the traditional suborder Symphyta is a paraphyletic group and, as such, is rejected as a taxon by many pure cladists, though to my mind the concept is still a useful handle as long as it is recognized that it represents a grade rather than a monophyletic taxon.

The sawflies are clearly paraphyletic with respect to the Apocrita, but there has been considerable debate as to what group of sawflies is closest to the common ancestor of the Apocrita. Both the stem sawflies, Cephoidea (e.g. Oeser, 1961; Malyshev, 1968; Königsmann, 1977; Zessin, 1985) and Orussoidea, have been strong contenders but Siricidae and Xiphydriidae have also had their supporters. With few exceptions, most workers have never seriously questioned the monophyly of the Apocrita, but it should be borne in mind that many of the derived morphological features that unite them could be expected to arise convergently in parasitic taxa. However, a growing number of morphological studies have been providing stronger and stronger support for both a monophyletic Apocrita and for an Orussidae + Apocrita clade (e.g. Gibson, 1985; Whitfield *et al.*, 1989; Basibuyuk and Quicke, 1995), and most workers now accept this as being the most likely hypothesis. In the past the Orussidae have been variously associated with the sawflies and with the

Table 11.1 Current classification of the Hymenoptera, based largely on Rasnitsyn (personal communication), giving original authors and dates for each family

'SYMPHYTA'
  Xyeloidea
    Xyelidae Newman 1834
  Megalodontoidea
    †Xyelydidae Rasnitsyn 1968
    Megalodontidae Konow 1897
    Pamphiliidae Cameron 1890
    †Praesiricidae Rasnitsyn 1968
  Tenthredinoidea
    Argidae Konow 1897
    Blasticotomidae Thomson 1871
    Cimbicidae Leach 1817
    Diprionidae Rohwer 1911
    †Electrotomidae Rasnitsyn 1977
    Pterygophoridae Cameron 1878
      (=Pergidae Ashmead 1898)
    Tenthredinidae Latreille 1802
    †Xyelotomidae Rasnitsyn 1968
  Cephoidea
    Cephidae Newman 1834
    †Sepulcidae Rasnitsyn 1968
  Siricoidea
    Anaxyelidae Martynov 1925
    †Gigasiricidae Rasnitsyn 1968
    Siricidae Billbergh 1820
    Xiphydriidae Leach 1815
  Orussoidae
    Orussidae Newman 1834
    †Paroryssidae Martynov 1925

APOCRITA
  'PARASITICA'
    Megalyroidea
      Megalyridae Schletterer 1889
    Stephanoidea
      Stephanidae Leach 1815
    Ephialtitoidea
      †Ephialtitidae Handlirsch 1906
      †Karatavitidae Rasnitsyn 1963
      †Maimetshidae Rasnitsyn 1975
    Trigonalyoidea
      Trigonalyidae Cresson 1867
    Evanioidea s.l.
      Aulacidae
      †Cretevaniidae Rasnitsyn 1975
      Evaniidae Latreille 1802

Gasteruptiidae Ashmead 1890
†Praeaulacidae Rasnitsyn 1972
Ceraphronoidea
  Ceraphronidae Haliday 1833
  Megaspilidae Ashmead 1893
  †Stigmaphronidae Kozlov 1975
Platygastroidea
  Platygastridae Westwood 1840
  Scelionidae Haliday 1839
Proctotrupoidea
  Austroniidae Kozlov 1975
  Diapriidae Haliday 1833
  Heloridae Foerster 1856
  †Jurapriidae
  †Mesoserphidae Kozlov 1970
  Monomachidae Szépligeti 1889
  Pelecinidae Haliday 1840
  Peradeniidae Naumann and Masner 1985
  Proctotrupidae Latreille 1802
  Renyxidae Kozlov 1994
  Roproniidae Viereck 1916
  Vanhornidae
Mymarommatoidea
  Mymarommatidae Debauch 1948
Chalcidoidea
  Agaonidae Billbergh 1820
  Aphelinidae Thomson 1876
  Chalcididae Spinola 1811
  Elasmidae Foerster 1856
  Encyrtidae Walker 1837
  Eucharitidae Leach 1815
  Eulophidae Westwood 1840
  Eupelmidae Walker 1846
  Eurytomidae Walker 1833
  Leucospidae Walker 1834
  Mymaridae Haliday 1833
  Ormyridae Foerster 1856
  Perilampidae Foerster 1856
  Pteromalidae Dalman 1820
  Rotoitidae Boucek & Noyes 1987
  Signiphoridae Howard 1894
  Tanaostigmatidae Ashmead 1904
  Tetracampidae Thomson 1878
  Torymidae Walker 1833
  Trichogrammatidae Foerster 1856
Cynipoidea
  †Archaeocynipidae[1] Rasnitsyn and Kovalev 1988
  Austrocynipidae
  Charipidae[2]

Cynipidae Latreille 1802
Eucoilidae[2]
Figitidae Hartig 1840
Ibaliidae Thomson 1862
Liopteridae
Ichneumonoidea[3]
  Ichneumonidae Latreille 1802
  Braconidae Nees 1814
  †Praeichneumonidae Rasnitsyn 1983
  †Eoichneumonidae Jell and Duncan 1986
Aculeata
  Bethylonymoidea
    †Bethylonymidae Rasnitsyn 1975
  Chrysidoidea
    Bethylidae Haliday 1833
    Chrysididae Latreille 1802
    Dryinidae Haliday 1833
    Embolemidae Foerster 1856
    Plumariidae Bischoff 1920
    Sclerogibbidae Ashmead 1902
    Scolebythidae Evans 1963
  Vespoidea
    †Armaniidae Dlussky 1993
    Bradynobaenidae Saussure 1892
    Formicidae Latreille 1802
    Masaridae
    Mutillidae Latreille 1802
    Pompilidae Latreille 1804
    Rhopalosomatidae Ashmead 1896
    Sapygidae Latreille 1810
    Scoliidae Latreille 1802
    Sierolomorphidae Krombein 1951
    †Sphecomyrmidae Wilson and Brown 1967
    Tiphiidae Leach 1915
    Vespidae Laicharting 1781
  Apoidea
    Apidae Latreille 1802
    Sphecidae Latreille 1802

---

† Extinct family.
* Does not include a number of new families proposed or elevated by Kovalev (1994) largely on the basis of extinct taxa, specifically: Brachycleistogastridae†, Emarginidae†, Gerocynipidae††, Palaeocynipidae†, Pycnostigmatidae†, Rasnitsyniidae†, and Thrasorusidae†.
[1] Ronquist's (1995) re-examination of *Archaeocynips* led him to suggest that they are not actually cynipoids and so should be treated as *incertae cedis*.
[2] In Ronquist (1995) the Eucoilidae and Charipidae are included within the Figitidae.
[3] The Aphidiinae and Paxylommatinae (Braconidae and Ichneumonidae respectively) are sometimes afforded separate family status though this is clearly unjustified in the light of recent detailed studies (Sharkey & Wahl, 1992).

Table 11.2 Extinct families of presumed parasitic Hymenoptera* and their known geological time scales (courtesy of Alexander Rasnitsyn)

Family	Known range of occurence
Paroryssidae	Upper Jurassic–Lower Cretaceous
Ephialtitidae	Lower Jurassic–Lower Cretaceous
Karatavitidae	Lower Jurassic–Upper Jurassic
Maimetshidae	Middle Cretaceous
Stigmatophronidae	Lower Cretaceous–Middle Cretaceous
Praeaulacidae	Upper Jurassic–Lower Cretaceous
Cretevaniidae	Lower Cretaceous–Middle Cretaceous
Mesoserphidae	Upper Jurassic–Lower Cretaceous
Jurapriidae	Upper Jurassic
Serphitidae	Lower Cretaceous–Middle Cretaceous
Archaeocynipidae	Lower Cretaceous
Praeichneumonidae	Lower Cretaceous
Eoichneumonidae	Lower Cretaceous
Bethylonymidae	Upper Jurassic

* Trupochalcididae and Pelecinopteridae are considered as a subfamily and genus in the Austroniidae and Pelecinidae respectively (Rasnitsyn, personal communication)

'Parasitica'. Much of the uncertainty about the relationship of orussids stems from the fact that they are inadequately known and also that they are highly apomorphic – so much so, in fact, that Rohwer and Cushman (1917) created a whole new suborder, the Idiogastra, for them. Despite a long history of debate, it is now clear that members of the Orussidae are parasitoids. Nuttall (1980) showed that *Guiglia schauinslandi*, a New Zealand species, oviposits on siricid woodwasp larvae. The orussid larva starts off as an ectoparasitoid but later on may feed internally on the decaying host remains. It is also worth noting that the host appears to start to decay, whereas in the case of most other ectoparasitic Hymenoptera the host remains in good condition even for a long period, due no doubt to secretions from the parasitoid. It is tempting to speculate that the biology of *Guiglia* is relatively primitive and that this represents an early stage in the evolution of the parasitic Hymenoptera. However, it should be noted that *Sirex*, the host, is not native to New Zealand; this is therefore not a natural host–parasitoid association and no observations have yet been made of *Guiglia* attacking any native New Zealand hosts (wood-boring beetle larvae).

Although many workers have developed hypotheses about various aspects of hymenopteran phylogeny over the years, most of these have based their discussions on considerations of one or relatively few character systems. Königsmann (1977, 1978) was effectively the first worker to attempt to resolve hymenopteran phylogeny based on a consideration of all information available at the time. His hand-constructed cladograms (summarized in Figure 11.1a) are severely lacking in resolution for the Apocrita, reflecting his uncertainties about their relationships, and of course he did not have available to him at the time any of the modern computerized parsimony analysis packages that we rely upon today. In his schema, it should be noted that the Apocrita are considered monophyletic and are placed as the sister group of the stem sawflies

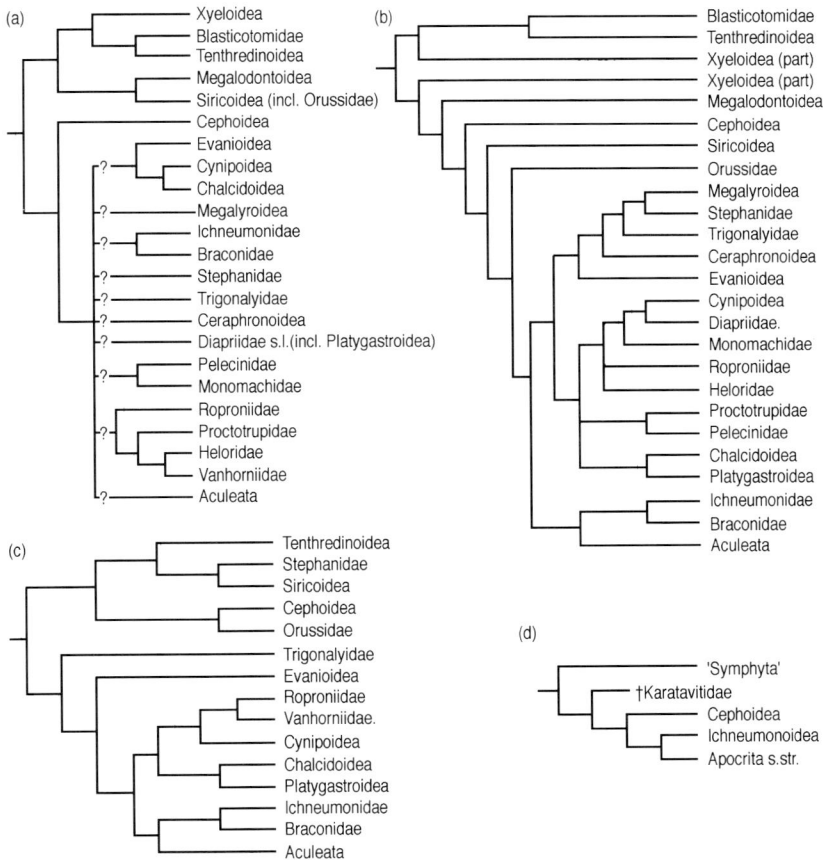

Figure 11.1 Phylogenetic relationships of apocritan Hymenoptera according to: (a) Königsmann (1977); (b) Rasnitsyn (1988); (c) Dowton and Austin (1994); (d) Zessin (1985).

(Cephoidea), and that the parasitic Orussidae are included within the Siricoidea, necessitating two independent origins of the parasitoid way of life. His cladogram therefore accords well with Malyshev's (1968) rather controversial views on the origins of parasitoidism (Chapter 1). Zessin (1985), basing his arguments to a considerable extent on fossil Hymenoptera, proposed a partly resolved phylogeny suggesting that the Cephoidea form the sister group of the remaining Ichneumonoidea + Apocrita (Figure 11.1d), with the latter two clades regarded as sister groups. However, Zessin's use of the potentially available evidence was extremely limited, and a great deal of evidence now points to a sister group relationship between ichneumonoids and aculeates. In particular the presence of valvilli – small articulated flaps within the ovipositor or sting (Chapter 6; Figures 6.6f and 6.9d) suggest that the Aculeata are probably the sister group to the Ichneumonoidea (Königsmann, 1978; Mason, unpubl.; Rasnitsyn, 1980, 1988; Zessin, 1985; Quicke *et al.*, 1992a).

The best known and most widely used phylogeny of the Hymenoptera is that of the Russian worker Alexandr P. Rasnitsyn, who combined a lifetime's research on fossil insects, and Hymenoptera in particular, with a great deal of information on the morphology of extant forms (Rasnitsyn, 1980, 1988). His phylogeny (summarized in Figure 11.1b) was based on cladistic principles but was constructed by hand and so may not be the most parsimonious interpretation of the data it represents. Rasnitsyn's hypothesis differs from that of most workers in that he considered the sawflies to form two clades, with the typical tenthredinoid sawflies originating from the macroxyeline xyelids, and the other sawflies together with the Apocrita originating from the xyeline xyelids. As regards his views on the origin and evolution of the parasitic wasps, Rasnitsyn has no doubt that the Orussidae occupy a key position and form a sister group to the Apocrita.

The important feature as regards the parasitic wasps is that the Apocrita is seen to comprise three major clades. The most basally derived clade comprises the Ichneumonoidea, now widely believed to form the sister group of the Aculeata. The remaining parasitic wasps comprise a clade more or less composed of the traditional 'microhymenopteran' families (Chalcidoidea s.l. Proctotrupoidea s.s., Diapriidae, Platygastroidea and Cynipoidea), and a second clade comprising a number of mostly small superfamilies, many of which appear early in the fossil record – such as the Megalyroidea and Evanioidea. Various workers (e.g. Heraty *et al.*, 1994) have presented slightly different versions of Rasnitsyn's tree, modified in the light of new evidence, but apart from rejecting the diphyletic origin of the higher sawflies, and incorporating additional taxa, the basic plan remains essentially the same, but with the Ichneumonoidea + Apocrita derived less basally.

11.3 RELATIONSHIPS AT FAMILY LEVEL

There have been a number of phylogenetic proposals for family level relationships within various major groups: Cynipoidea (Ronquist, 1994a); Ichneumonoidea (Sharkey and Wahl, 1992); Aculeata (Brothers, 1975; Brothers and Carpenter, 1993). Even the monophyly of some superfamilies is questionable. For example, the Evanioidea, traditionally comprising the Aulacidae, Evaniidae and Gasteruptiidae, share a highly inserted metasoma and an apparently apomorphic mid-coxal articulation (Johnson, 1988), but the biology of each family is very different and other features such as the thoracic musculature and internal skeletal structure (Gibson, 1985; Whitfield *et al.*, 1989), antenna cleaner (Basibuyuk and Quicke, 1994, 1995; see below) and ovipositor steering mechanisms (Quicke and Fitton, 1995) fail to provide any evidence for monophyly.

Shaw (1990) presented a phylogenetic analysis of genera comprising the ancient family Megalyridae which today are largely restricted to the tropics and are generally very rare and poorly known insects. Closely related genera were found not to occupy the same continents, suggesting that the group had already radiated before the break-up of Pangea.

The Trigonalyidae is a particularly enigmatic group within the parasitic Hymenoptera, sharing a number of morphological features with both sawflies and Apocrita. They are parasitoids, have a narrow waist and a propodeum, and have distinct trochantelli, but they also have the most complete venation of any Apocrita, save for a few sphecids, and they possess plantar lobes and very primitive mouthparts. The latter feature in particular has led them to be placed in their own suborder, the Archiglossata (Weinstein and Austin, 1991), though this has not become widely accepted. Trigonalyids also possess a highly derived antenna cleaner (Figure 6.13d) with the basitarsal comb composed of plates rather than setiform structures and with the lamella of the fore tibial spur having microsculpture (Basibuyuk and Quicke, 1994, 1995). The same two modifications occur in the Evaniidae, another family whose relationships have often been questioned (e.g. Gibson, 1985; Whitfield *et al.*, 1989). However, single character systems can be very misleading, and far more work on the relationships of these two taxa will be required before their placement can be finally settled.

The monophyly of the Cynipoidea has never seriously been doubted, but as Ronquist (1995) points out, there are few obvious synapomorphies for the group. His phylogenetic analyses have helped to elucidate the biological transitions that have occurred within the superfamily (Figure 11.2). The three basal clades (Austrocynipidae, Ibaliidae and Liopteridae) are all parasitoids of wood-boring insect larvae, whereas among the 'microcynipoidea', in addition to gall-formers and gall inquilines, there are pri-

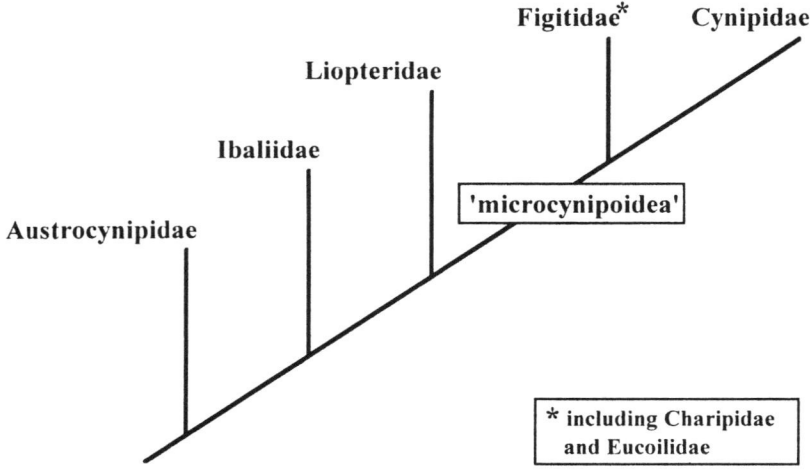

Figure 11.2 Phylogenetic hypothesis of family level relationships within the Cynipoidea derived from analysis of a large morphological data set by Ronquist (1995).

mary parasitoids of Diptera and Neuroptera, and hyperparasitoids of aphidiine braconids and aphelinids in aphids.

The Proctotrupoidea s.l. has long been considered as a heterogeneous assemblage, and in the last few years four families have been removed to other superfamilies, namely the Platygastridae and Scelionidae to the Platygastroidea and the Ceraphronidae and Megaspilidae to the Ceraphronoidea. The relationship of the Diapriidae to the remaining 'proctos' (as they are often called) is also debatable. If the Diapriidae are excluded, then the rest of the proctos may form a monophyletic group – as indicated by the remarkable mode of pupation that some families are known to display. The Proctotrupidae, Heloridae and Pelecinidae, whilst endoparasitic as larvae, emerge partly from their hosts to pupate but their posterior ends are left embedded in the host's shrivelled body (Eastham, 1929; Clancy, 1946; Lim *et al.*, 1980; Table 5.2). Unfortunately, the biology is known either not at all or not in sufficient detail for the other families.

The monophyly of the Chalcidoidea (including the Mymaridae) was strongly supported by the work of Gibson (1986a), who showed that the whole group possessed a particular type of placoid sensilla on the antennae that apparently does not occur elsewhere in the Hymenoptera. Chalcidoidea, excluding Mymaridae (which appear to be the sister group of the remaining chalcidoids), are also strongly supported as being monophyletic. In addition to the characters given by Gibson may be mentioned: (i) the possession of a laminated bridge (a transverse striate cuticular strand at the base of the upper ovipositor valve) (Fulton, 1933; King and Copland, 1969); and (ii) the asymmetrical upper ovipositor valve (Quicke *et al.*, 1994). As regards relationships among the many other families of Chalcidoidea, these have so far received only rather informal treatment with a few suggested groupings, though even these are not in complete agreement because of the sparsity of reliable characters (Trjapitzin, 1978; LaSalle, 1987; Boucek, 1988; Woolley, 1988; Noyes, 1990; Gibson, 1990). In addition, several of the chalcidoid families may prove to be paraphyletic or even polyphyletic as relationships become better understood, mono-phyly of the Pteromalidae, Aphelinidae and Encyrtidae being particularly dubious. As van den Assem (1974) noted, the huge chalcidoid family Pteromalidae has been a 'dumping ground' for many genera that do not show any obvious affinities to other groups, and it is impossible to define any synapomorphy for them as a whole. Such groups obviously need a great deal more work involving surveys of new character systems and the preparation of molecular data sets in order to be able to derive something approaching a natural classification. Few subfamilies have been subjected to explicit cladistic analysis but the number of such studies is growing rapidly.

The family level relationships among the parasitic Aculeata have been intensively investigated, and much is owed to the seminal work of Brothers (1975) that brought together a large amount of data for phyloge-netic analysis. Since then, Carpenter (1986) has provided a detailed analy-sis of the families of the Chrysidoidea, and Brothers and Carpenter (1993) have carried out a particularly thorough analysis of the Chrysidoidea and Vespoidea, including many aberrant taxa. All these works support essen-tially the same picture, with the Chrysidoidea being a well-supported monophyletic taxon, and with the Vespoidea also monophyletic and the sister group of the Apoidea (including Sphecoidea).

11.4 MORPHOLOGICAL CHARACTERS

Over the years there have been numerous surveys of organ systems across the Hymenoptera which have revealed a multitude of potentially phylo-genetically informative characters. As might be expected, only a few of these studies have attempted to survey the entire order rigorously, and therefore additional original research will be needed before most can be plugged into a comprehensive phylogenetic analysis. Among the more comprehensive surveys that have been conducted are mid-coxal articula-tion (Johnson, 1988), thoracic musculature (Gibson, 1986b), ovipositor structure and function (Oeser, 1961; Quicke *et al.*, 1992a, 1994; Quicke and Fitton, 1995), sperm ultrastructure (Quicke *et al.*, 1992b; Newman and Quicke, in preparation), antenna cleaner (Basibuyuk and Quicke, 1994, 1995), spiracles (Tonapi, 1958a), and tracheal system (Tonapi, 1958b-c) and karyology (Gokhman and Quicke, 1995; Quicke and Gokhman, 1996). Evidence from grooming behaviour provides some additional support for several relationships (Basibuyuk and Quicke, in preparation). However, many more character systems await detailed investigation, many of which have been mentioned earlier in this book. Some of these require little more than basic microscopy but others demand very sophisticated procedures. Whilst the following list of some potentially informative novel character systems for Hymenoptera phylogeny is not comprehensive, it will give a flavour of the sorts of possible sources of additional data:

- sex determination system
- steroid chemistry
- vitellogenin chemistry
- venom chemistry
- silk chemistry
- retro element insertions
- larval internal anatomy
- adult rectal pad ultrastructure
- heavy metal utilization in cuticle hardening
- antennal sensilla ultrastructure and ontogeny.

11.5 THE FOSSIL RECORD AND EXTINCT FAMILIES OF PARASITIC HYMENOPTERA

A number of hymenopteran families are known only from fossils (Table 11.2). The origins of the Apocrita, and therefore presumably of the parasitic way of life among the Hymenoptera, dates back to the Jurassic era, some 135 million years ago. Two families in particular possess features indicating that they are close to the origins of the Apocrita, namely the Karatavitidae from the early to mid-Jurassic and the Ephialtitidae from the late Jurassic. Other important early fossils include *Cleistogaster*, placed in the extant Megalyridae, and *Paraulacus* in the extinct Mesoserphidae, both also from the Jurassic. The Karatavitidae share many features in common with the sawflies, but they have reduced cenchri, a distinct constriction between the first and second abdominal segments with a fusion of the two halves of the first tergite to form a single plate, and a needle-like ovipositor. Phylogenetic arguments about extinct families such as the Karatavitidae are complicated by the fact that there is little or no evidence that the taxa placed in them actually form a monophyletic group. Without doubt, future work will have to take into account each of the representative genera separately until higher-level monophyletic groups can be discerned unambiguously. Also, inclusion within particular superfamilies is largely an *ad hoc* decision, and no great significance should be read into it.

The fossil record of the Hymenoptera is moderately complete. Nevertheless, the composition of fossil faunas is highly biased in favour of larger, harder bodied taxa. The great bulk of the earlier material studied to date is made up of impression fossils in fine shales and schists (see cover illustration), much of it originating in Mongolia and the countries of the former Soviet Union, though there are also important deposits in South America (Darling and Sharkey, 1990) and Australia (Jell and Duncan, 1986), though the latter are particularly in need of further study.

Darling and Sharkey (1990) have tabulated published data on the Cretaceous fossil Hymenoptera. What is quite surprising is that the family-level diversity and composition of the order has changed little since the beginning of the Cretaceous. All major groups were present then, including aculeates, and although the oldest known fossils of Chalcidoidea, Mymarommatoidea and Proctotrupoidea (Austroniidae) are from the Upper Cretaceous, about 70 million years ago, this is probably attributable largely to their generally smaller size and consequent improbability of being preserved in the earlier sedimentary deposits.

Many of the taxa described from Baltic and Dominican amber, 15–40 million years old, are readily referable to extant genera. Of course, a number of apparently extinct genera have been described from these and other fossil deposits of similar age, but the available evidence suggests that most families of parasitic wasp had undergone their major radiations by the beginning of the Palaeocene. Some taxa appear to have changed little over very long periods of time. Poinar (1992) specifically mentions the records of *Serphites paradoxus* in both Canadian and Siberian amber of ages that apparently differ by some 8 million years. However, he did not draw special attention to Doutt (1973) who reported two extant species of Mymaridae, *Alaptus psocidivorus* and *A. globosicornis* from Mexican amber, deposits that are from the Oligocene–Miocene border some 24 millions years ago. Whilst it is perhaps doubtful that these fossil taxa represent the same biological species as their apparent extant counterparts, it does indicate a remarkable degree of morphological stasis. Such constancy of form is also found in other groups of insects – for instance, in some Strepsiptera from Dominican amber deposits 15–40 million years old (Kathirithamby and Grimaldi, 1993).

Nothing much can be said about the biologies of these extinct groups, though judging from ovipositor lengths many were probably parasitic on concealed insects. On the basis of ovipositor length and head shape, Rasnitsyn (1980) postulated that if the Karatavitidae were indeed parasitic, they were probably restricted to parasitizing relatively weakly concealed hosts such as subcortical larvae or twig borers. Rarely, however, we can get a more detailed glimpse of extinct parasitic wasp biology as, for example, in the case of the amber fossil spider illustrated by Poinar (1986) which has a koinobiont ectoparasitoid larva on it, probably belonging to the pimpline ichneumonid tribe Polysphinctini, which have similar biologies today.

Fossils have thus far played only a minor role in parasitic Hymenoptera phylogeny below the family level. Inclusion of data from fossils in phylogenetic analyses often causes problems because of missing data and also because extinct taxa may display character state combinations that are not observed in extant taxa. Nevertheless, as they do provide information about lineages they should not be dismissed and methods are being developed that will facilitate their incorporation in phylogenetic studies (Wilkinson, 1995; Wilkinson and Benton, 1996). Shaw (1990) investigated the inclusion of data from fossil genera of the family Megalyridae in his parsimony analyses. The effect was minimal, with the tree obtained agreeing perfectly with one of the two equally most parsimonious ones retrieved from analysis of extant taxa alone. Thus in this case, fossil evidence reduced the number of preferred trees, but unfortunately this is not often likely to be the case.

The preservation state of some fossils is remarkably good. Examination by environment chamber scanning electron microscopy can often reveal details that are indistinguishable (or not convincingly so) using optical equipment; and since this technique does not require coating of the specimen it is relatively benign and can be employed with unique material. For example, Figure 11.3b shows details of antennal sensilla on *Manlaya* and Figure 11.3c shows the antennal cleaner of *Cleistogaster* (Megalyridae) from the early–mid Jurassic, complete with setae. Application of such techniques have the potential to reveal far more detail from hymenopteran fossils than might hitherto have been thought possible, and makes their incorporation into cladistic analyses less problematic.

11.6 MOLECULAR PHYLOGENETIC STUDIES

In common with many other groups of organisms, there is now an increasing number of molecular phylogenetic studies on the Hymenoptera at all levels from species to suborder (Cameron *et al.*, 1991). Fortunately, it has proved possible to extract good quality DNA from alcohol-preserved wasps as might be caught in malaise traps, and also, though generally with less success, from dried specimens. It appears that some of the past apparent lack of reproducibility in working with DNA from preserved material may have been due to details of exactly how the insects were killed and preserved (Dillon *et al.*, 1996) – for example, the popular killing agent, ethyl acetate, appears to block the polymerase chain reaction, at least for Hymenoptera specimens. Increasingly, molecular phylogenetics is being used in order to provide independent data sets from which phylogenetic hypotheses can be derived (Dowton and Austin, 1994). However, analysis of molecular sequence data is far from straightforward, and it is now generally accepted that phylogenetic analyses based on relatively short sequences, and applying the parsimony criteria advocated by modern cladistics, seems unlikely to yield cladograms whose relationships are any more certain than those derived from analyses of non-molecular characters. This is not to say that molecular data sets will not ultimately provide the best estimates of evolutionary history once sufficient data have been obtained (Purvis and Quicke, in press) and

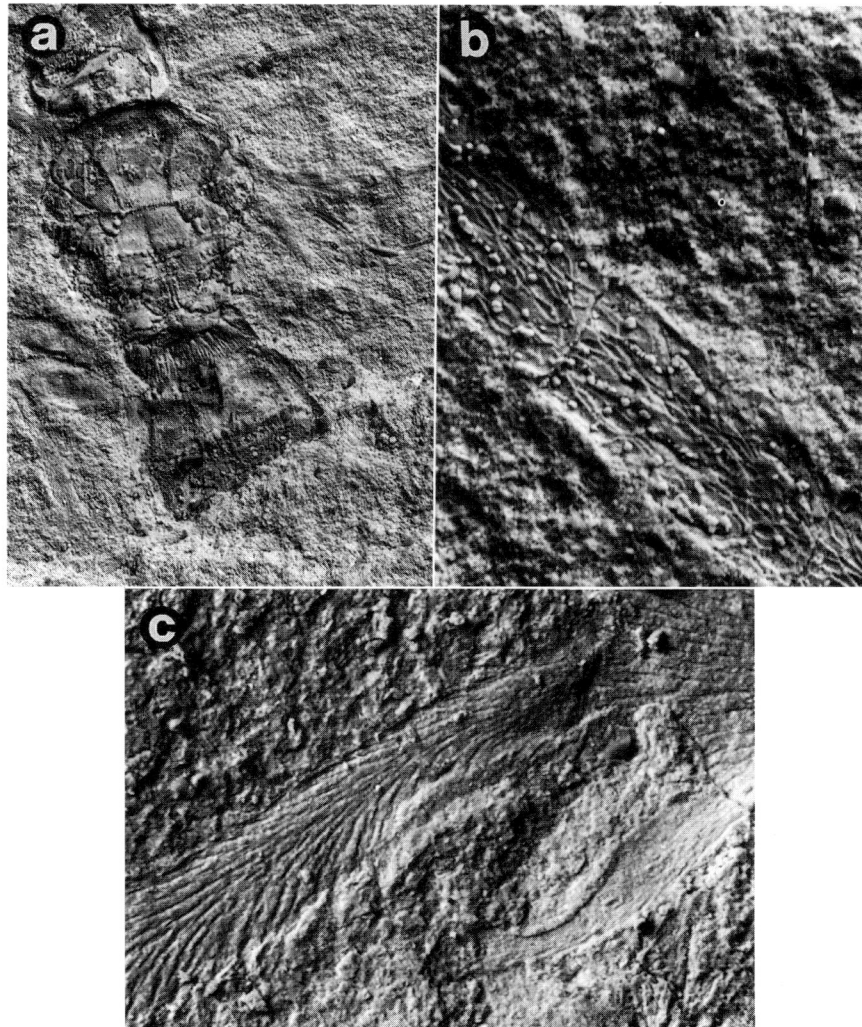

Figure 11.3 Environment chamber scanning electron micrographs of Hymenoptera impression fossils: (a) mesosoma and anterior metasoma of a megaspilid; (b) single flagellomere of *Manlaya* (Evanioidea) showing internal cuticle of antennal sensilla; (c) antenna cleaner on foreleg of *Cleistogaster* (Megalyroidea).

means of their interpretation more fully understood (Purvis and Quicke, 1997). After an early study by Derr *et al.* (1992b) using the large mitochondrial ribosomal gene which was found subsequently to be flawed by the presence of contaminants (Derr *et al.*, 1992a) – highlighting the need for caution in using universal primers – several other more detailed surveys have now been carried out. Carmean *et al.* (1992) published an analysis of 10 families based on variation in the 18S ribosomal DNA gene. The results, however, did not seem to agree particularly well with expectations based on morphology, and it is possible that the rate of evolutionary change in this gene is unsuitable for resolving relationships within the Hymenoptera at superfamily or family level because of too high a level of homoplasy. Other studies using the sequence of the 16S ribosomal DNA gene by Dowton and Austin (1994, and in preparation) have seemingly been more successful. The cladogram presented by Dowton and Austin (Figure 11.1c) agrees quite well with expectations based on non-molecular character systems. Their work has, for example, provided evidence that the Chalcidoidea are closely related to the Platygastroidea; however, the placement of the Stephanidae seems intuitively less likely.

Dowton and Austin (1995) have shown that the evolution of the Apocrita in the early Jurassic was correlated with two genomic features: a marked increase in the AT richness of mitochondrial genes and an increased rate of molecular divergence. These authors postulate that these features are causally related to the evolution of a parasitic way of life and they suggest a number of possible reasons, including the relative frequencies of founder events, greater selection pressure and genetic arms races. However, without knowing more about the exact sister group relationship of the Apocrita and their molecular organization, this cannot be stated with certainty.

Preliminary data are now appearing for a number of other potentially informative genes such as the D1 and D2 divergent domains of the 28S rDNA gene (Schmitz and Moritz, 1994; Belshaw and Quicke, in press), and it is hoped that within the next few years a combined approach using molecular, morphological and fossil evidence will lead to a well-supported cladogram for the order. A marked difference in AT content is also apparent between the nucleur 28S rDNA genes of the closely related Braconidae and Ichneumonidae, the latter being far richer in G and C (Belshaw *et al.*, in press). Why these families should differ so is not at all obvious.

11.7 SPECIATION

Whilst everyone recognizes that the parasitic Hymenoptera are a very speciose group, and will generally attribute this to the variety of hosts and their many ways of utilizing them, rather less thought has gone into the actual mechanisms of speciation. Godfray (1993) summarizes the three main ways that parasitic wasps can speciate: allopatrically, sympatrically and by co-cladogenesis with their speciating hosts. Of these, the latter appears to be of minor importance in parasitic Hymenoptera, unlike in some other host–parasite systems. Parasitic Hymenoptera, even koinobiont species, seem fairly frequently to accrue new hosts to their host ranges, and because they are mobile organisms they have plenty of opportunity to do so.

Parasitic wasps are probably a good candidate group for sympatric speciation because of a number of features of their biology such as their physiological and phenological adaptations to their hosts (Godfray, 1993). Askew (1968) emphasized the likely significance of sib-mating in speciation and further pointed out that this isolating force will be increased by the monoandry displayed in many groups (Chapter 8). Learning may also be important, since wasps emerging from one host or one host plant are likely to concentrate their searching in the same habitat and look for the same host. This does not prevent many from attacking other hosts, but it will tend to favour selection for genes that confer an advantage on one particular host. Haplodiploidy is also likely to play an important role in the speciation rate of parasitic wasps, since advantageous alleles will become fixed in the population at a greater rate because they are exposed, unbuffered by other alleles in the males. This is the converse of the argument that deleterious mutations will be more rapidly eliminated in haplodiploid organisms, which has been used to explain apparently low levels of heterozygosity found in some taxa (Chapter 2).

Gibbons (1979) proposed that ovipositor length provided a means of dividing a single host resource in the case of the ichneumonid genus, *Megarhyssa*, which parasitizes larvae of the horntail, *Tremex columba*. As *Tremex* larvae of similar size occupy a large range of different depths within rotting wood (between approximately 20 and 140 mm) and since these larvae maintain their given depth for much of their development, *Megarhyssa* that attack hosts at different depths will not be competing. Gibbons consequently proposed a new sympatric speciation mechanism – competitive speciation – that could have led to the existence of three separate *Megarhyssa* species that differ largely in the lengths of their ovipositors. The proposal is similar to that of Pimm (1979).

11.8 EFFECT OF HOST AND ENVIRONMENT ON WASP MORPHOLOGY

Knowing whether or not one is dealing with one or several species is a problem all taxonomists have to face sooner or later. Coloration in many insects can be strongly influenced by temperature (Askew, 1961; Liu and Carver, 1982), the early pupal stage being particularly sensitive. In bivoltine species this often means that members of the overwintering generation will be darker than those of the summer one. In the case of many parasitic wasps, the problem is particularly difficult since their coloration and often also their morphology can be affected by the host they developed in (Huber and Rajakulendran, 1988; Johnson *et al.*, 1987; Pintureau and Daumal, 1995). Thus a taxonomist may be presented with wasps of one morphotype reared from one host species and wasps of a different though similar morphotype reared from a different host. This has been well documented, for example, in some polyphagous aphidiine braconids (Starý, 1970), and in many other groups such as trigonalyids (Weinstein and Austin, 1991). An extreme example involves the egg parasitoid, *Trichogramma semblidis*, which produces winged males in one host and apterous ones in another (Salt, 1937). Under such circumstances it is simply not possible to say with total confidence how many species are involved; sometimes it will be just one, sometimes it will be two or more. In order to resolve these difficulties it is generally necessary either to conduct rearing experiments from which it will be possible to examine the effects of host species or other rearing conditions on the parasitoid's morphology, or to examine the degree of genetic interchange between wild populations attacking the different hosts – for example, by allozyme electrophoresis or RAPD analysis.

11.9 EVIDENCE FOR GENETIC ISOLATION

Alloenzyme electrophoresis has been employed in a number of cases to obtain information about the genetic isolation within parasitic wasps. In these cases, the data obtained enable one to determine whether or not two populations belong to a single genetic system with free flow of genes between them. Alloenzyme studies have been particularly useful in the case of the taxonomically difficult groups, and have helped to determine whether other variation is due to the existence of putative sibling species. Alloenzyme electrophoresis has been particularly useful in the trichogrammatid genus, *Trichogramma*, in which small size and a paucity of morphological variation conspire against the taxonomist, and in the aphidiine braconids, in which host species can induce morphological changes in the adult wasps and therefore make determination of species boundaries difficult (Pungerl, 1986; Unruh *et al.*, 1989). In *Trichogramma*, for example, two very similar species – *T. deion* and *T pretiosum* – were investigated using this technique by Pinto *et al.* (1993). Importantly, they examined material of each putative taxon from across a large part of their geographical ranges. Out of 13 enzyme loci examined, only one, glucose-6-phosphate dehydrogenase (*G6pd*), was found to be fixed for a different allele in the two species across their entire ranges. However, at one site the two species differed in a fixed way at two other loci. The latter observation provided additional evidence that there is no gene flow occurring between them on a local basis. To have relied on data from just one site would have given a misleading impression of the degree of difference between them. Another important consideration when using alloenzyme electrophoresis to investigate gene flow between potential populations is that the sampling of individuals must reflect the population as a whole. Employing individuals that might disproportionately represent the offspring of a small subset of the population – for example, by collecting from single host plants – could readily invalidate the results.

Hung and Schaefer (1990) tried to apply alloenzyme electrophoresis to the identification of geographical strains of the eulophid egg parasitoid,

Pediobius foveolatus, from the Oriental and Indo-Australian regions. Although they found allelic differences at two enzyme loci between stocks from different regions, their results for the population known to have been introduced into Guam did not correspond to biocontrol records. The discrepancy might be due to the fact that they used samples from laboratory cultures, and it is quite possible that they were not representative of the source populations because of founder effects or that initial polymorphisms at loci had been lost due to genetic drift or selection.

11.10 CROSSING EXPERIMENTS

One of the most widely accepted definitions of a species is the biological species concept as championed by Ernst Mayr, and it has been applied with *some* success in the parasitic Hymenoptera (Ryan and Yoshimoto, 1975; Khasimuddin and DeBach, 1976; Walker, 1979). Some of the difficulties that can be encountered in applying such tests to parasitic wasp species are highlighted, for example, by the findings of Pinto *et al.* (1991) who performed 132 crosses between various populations of three closely related species of *Trichogramma*. Unlike Ryan and Yoshimoto's results, those of Pinto *et al.* were far from straightforward. Many pairs of populations of apparently the same species were reproductively incompatible, or only partly compatible, or only compatible in one direction. Pinto *et al.* used their findings to caution against the common practice of relying on testing compatibility between laboratory cultures of taxonomically difficult groups as the sole test of species separation. In a similar study, Haardt and Holler (1992) found that in a series of crosses between different strains of *Aphelinus abdominalis*, reproductive barriers were greatest between strains that differed most in various life history parameters, but whether this indicated the presence of sibling species was unclear. At the very least, therefore, separate species status should only be accepted after testing for compatibility among a range of geographically separate populations of each putative species. Even within taxa that are well known to be single species, incompatibility may be found between various strains – for example, in *Nasonia vitripennis* (Connor and Saul, 1986). Such incompatibility may result from the presence of *Wolbachia* or similar endosymbionts (Chapter 2). In the case of *Wolbachia*-induced incompatibility, as in *Nasonia vitripennis*, elimination of the bacterial symbiont using antibiotic treatment has been shown to restore compatibility (Breeuwer and Werren, 1990).

It is not uncommon to find that the degree of reproductive isolation between closely related species is dependent on the sex combination of the pairing. For example, Mackauer (1969) investigated the success of mating attempts between three closely related species of aphidiine braconid parasitic on the pea aphid, *Acyrthosiphon pisum*, in North America. Some pairings produced no female offspring, indicating that either mating had failed or the male's sperm had failed to fertilize the female's eggs.

However, in the case of *Aphidius ervi* × *A. pulcher*, female hybrids were obtained when the male was *pulcher*, but only males were produced when *ervi* was the male. Of course, if this sort of result has been obtained, it still has to be interpreted, which may not always be so simple. As in a number of other cases, the males appear to be rather indiscriminate in their selection of partners, and it is the females that appear to be most responsible for enforcing species isolation.

Not all crossing experiments are so difficult to interpret. For example, Hoy and Marsh (1979) were able to demonstrate that two 'species' of the microgastrine braconid genus, *Cotesia* – *C. solitarius* and *C. melanosceles* – that had been imported into North America to control satin moth (*Stilpnotia salicis*) and gypsy moth (*Lymantria dispar*), respectively, in fact belong to the same species and so are synonyms. As the two strains had different hosts, host distributions and phenologies, they might effectively be reproductively isolated in North America, despite being completely interfertile.

11.11 SIBLING AND CRYPTIC SPECIES

In an increasingly large number of cases it is becoming apparent that what were once thought to be single species are actually pairs or groups of closely related and almost indistinguishable biological species, at least using conventional external morphological features. In some groups, especially of the very small parasitic wasps such as the important egg parasitoids, *Trichogramma* and *Telenomus*, external morphology provides few characters that are useful taxonomically when it comes to separating closely related species. Characters that may at first have appeared to be diagnostic have subsequently been revealed to be so plastic as to be almost useless (Pinto *et al.*, 1989). As these authors pointed out, the tendency of taxonomists is to describe new species; but they argue that in such difficult and complex situations reference to particular populations, strains, cultures, etc. might be best done using more neutral terms such as 'form' or 'race' rather than formally describing new species level taxa until their status as such has been satisfactorily determined. However, some sibling species are valid species, and van den Assem (1974) noted that in some cases they can be more easily distinguished on the basis of behaviour than from their morphology, e.g. the taxonomically difficult pteromalid genus, *Muscidifurax*.

11.12 TAXONOMIC IMPLICATIONS OF THELYTOKY

Many thelytokous 'species' appear to have normal arrhenotokous sister species, suggesting that the thelytokous populations may have been derived from the latter. The taxonomy of some of the parasitic wasps concerned is difficult and identification sometimes depends on having males available for study. Some thelytokous species or strains do produce occasional males, but these are normally so rare as to make experimentation with them impractical. Rössler and DeBach (1973) investigated the potential of males from arrhenotokous 'species' of the aphelinid, *Aphytis*, to inseminate females of the thelytokous sibling species. Taking advantage of eye-colour mutants they found that, once inseminated, the females could make use of the arrhenotokous male's sperm and produce fertile hybrid offspring. The ability to produce males in some of the thelytokous strains by antibiotic or heat treatment has meant that their species identification could be confirmed. A similar result was found by Pijls *et al.* (1996) for the encyrtid, *Apoanagyrus diversicornis*. The results of curing are not, unfortunately, always straightforward to interpret. In the case of the *Aphytis* species, for example, the arrhenotokous males were not very efficient at recognizing and mating with the thelytokous females. In contrast, in *Apoanagyrus*, cured males displayed displayed normal behaviour, but thelytokous females were less readily inseminated than arrhenotokous ones, independent of male type. The age of the thelytokons strain is probably very important here.

Stouthamer and Luck (1991) investigated a reputed case of conversion of arrhenotokous strains of the aphelinid, *Encarsia perniciosi*, to thelytoky through rearing at constant temperature. They were unable to repeat the previous findings and concluded that such 'transitions' probably reflected replacement of arrhenotokous strains with thelytokous ones in originally mixed cultures as a consequence of the culture technique. Stouthamer and Luck also demonstrated that fixed allelic differences at an alloenzyme locus characterized the thelytokous and arrhenotokous strains, indicating that both had probably been separate for a long time and could not readily be interconverted.

11.13 MOLECULAR TOOLS AT THE SPECIES LEVEL

As molecular (DNA and alloenzyme) studies become more common among parasitic wasps, it is to be expected that more and more cases of cryptic or sibling species will be uncovered. A particularly relevant example – because it involves one of the most widely investigated species of all parasitic wasps, the pteromalid, *Nasonia vitipennis* – was described by Darling and Werren (1990). *N. vitipennis* has always been assumed to be the sole representative of its genus and it has often been brought into culture from its hosts, various cyclorrhaphous Diptera. Darling and Werren, using a combination of careful morphometric comparison, biological information, alloenzyme data and results from experimental crosses, have revealed the existence of two, not uncommon, cryptic species of *Nasonia* in North America. Indeed, they note on the basis of morphological data alone that it would have been unlikely that either of the new species would have been definitely recognized, as the morphological differences between the species are so slight. The importance of maintaining voucher specimens in any ecological or physiological studies cannot be over-emphasized (Yoshimoto, 1978).

Several new molecular tools are now being employed to facilitate recognition of cryptic species. They include DNA sequencing, restriction fragment length polymorphism (RFLP) analysis, random amplified polymorphic DNA (RAPD) analysis, and single-strand conformation polymorphism (SSCP) analysis. Of these, RAPD has been particularly popular: it makes use of relatively short, randomly constructed primers and provides a quick method for comparing individuals. Using RAPDs, it has often proved possible to find RAPD bands that are diagnostic for particular species which might otherwise be difficult to separate, at least by non-specialists, and even between different strains which may be truly indistinguishable morphologically but still have important biological differences (Landry *et al.*, 1993). Edwards and Hoy (1995) used RAPD markers to follow the fate of laboratory selected, insecticide resistant strains of the aphidiine braconid, *Trioxys pallidus*, following release. Sappel *et al.* (1995) have used RFLP analysis following selective polymerase chain reaction (PCR) of LTS and 28S rDNA to distinguish three *Trichogramma* species.

Hiss *et al.* (1994) have employed SSCP analysis to identify differences between two species of aphidiine braconid, and more recently it has been used to provide supporting evidence that the biologically different but morphologically almost indistinguishable forms of the pteromalid genus, *Muscidifurax*, are indeed separate species. SSCP analysis makes use of the tendency of single-stranded DNA to fold on itself, giving a stable and consistent conformation. The electrophoretic mobility of these folded molecules in non-denaturing gels is highly sensitive to conformation, which is in turn highly sensitive to sequence, such that it is frequently possible to detect differences at single positions. The technique has the advantage over sequencing in that it permits rapid comparison of many samples with less cost and effort. Once differences have been detected, they can be further explored by sequence analysis.

11.14 SPECIES DISCRIMINATION AND IDENTIFICATION

In some taxa, recognition of closely related species may depend on features of a particular sex – for example, male or female genitalia, or male antennae. Identification of thelytokous taxa, for which only females are known, can therefore be problematical. Such is particularly the case with trichogrammatids, where male genitalia provide the only known reliable means of distinguishing some species. In some cases, in which thelytoky is the result of infection with the endosymbiotic microorganism, *Wolbachia* (Chapter 2), treatment with antibiotics or high temperatures now provides a means of obtaining previously unobtainable males. Stouthamer *et al.* (1990) used this technique to induce male production in thelytokous strains of four species of *Trichogramma*. By crossing the cured males produced with females of the similar arrhenotokous strains, they found that in each case the thelytokous strains were indeed strains rather than distinct, entirely thelytokous species.

For many purposes, alloenzyme electrophoresis offers the quickest and cheapest way of investigating gene flow between populations and can provide diagnostic characters (fixed allelic differences) for species identification. However, for very closely related species, there may be rather few polymorphic enzyme loci or interspecific differences. This appears to be especially the situation with the Hymenoptera, which often tend to show especially low levels of heterozygosity (Chapter 2). Powell and Walton (1989) provided a detailed survey of allozyme techniques for the identification of adult parasitoids and in some cases for the identification of parasitoid larvae feeding on or within their hosts. On occasion, morphotaxonomy may be very difficult or the necessary expertise unavailable, and in these situations allozymes can provide the best identification tool.

Some DNA procedures are relatively simple and more sensitive tools for this species-level work. Vanlerberghe-Masutti (1994a) tested the utilities of both restriction fragment length polymorphism (RFLP) and random amplified polymorphic DNA (RAPD) analyses for distinguishing between seven morphologically similar species of *Trichogramma*. With RFLP, using a total of 14 restriction enzymes, considerable interspecific variation was discovered, with the proportion of shared restriction sites in pairwise comparisons between species ranging from 30% to 83%. Similarly useful variation was found with RAPD, and having tested seven random primers Vanlerberghe-Masutti found apparently species-specific banding patterns for all the *Trichogramma*s investigated. However, RAPD analysis also revealed high levels of intraspecific variation within *T. evanescens*, with 20 different banding patterns being detected among 40 isofemale lines with just a single primer (Vanlerberghe-Masutti, 1994b). Further work in this area is needed, but potentially both RFLP and RAPD could provide useful tools for rapid identification of taxonomically difficult wasps such as *Trichogramma*.

Cuticular hydrocarbons have been used extensively in many groups of insects for distinguishing between species and even between members of different strains, and among social aculeates colonies (Quicke, 1993), but they have been little studied among the parasitic Hymenoptera (see section 6.45). This is a pity since it has been demonstrated that they are frequently useful for revealing the existence of sibling species. One such example involving parasitic wasps was provided by Espelie *et al.* (1990) who used them to show that the pteromalid, *Rhopalicus pulchripennis*, actually comprises a sibling species pair.

Identification of parasitoid larvae is typically more difficult than identifying the adults, even for experts. Thus molecular techniques offer a number of potential solutions that can be employed when monitoring pests and their parasitoids is involved and there is neither the time nor facilities to rear the parasitoids through. Allozymes provide one possible solution (see above) but other techniques are also worth investigating. For example, monoclonal antibodies have been shown by Stuart and Burkholder (1991) to be capable of detecting and distinguishing between immature stages of *Habrobracon hebetor* and *Laelius pedatus* attacking stored product pests. Significantly, the monoclonal antibodies were not influenced by the developmental stage; eggs through to adults could be detected equally well. The sensitivity of this procedure lends itself particularly to the detection of the early parasitoid instars, and the procedure can be automated relatively easily.

11.15 SEX ASSOCIATIONS

In some groups, the very different morphologies of males and females can make associating males and females of the same species very difficult. Notorious among these are the Mutillidae and Tiphiidae, though sex association problems are by no means limited to these aculeate groups. Phoretic copulation within the thynnine tiphiids has helped greatly in the association of sexes in this subfamily, but other means have to be found to work out true associations for others in which mating is brief and may occur in some hiding place or after dark. In one interesting case, it was possible to associate males and females of the mutillid, *Dasymutilla gloriosa*, through the chance discovery of a gynandromorph individual, but such specimens are rare and have seldom proved of use. Assembling males to attract females, rather as lepidopterists do to attract males, is one possibility. Further, the lack of rearing data for many of these groups – hosts are only known for 2% of mutillid species – makes this of little value. Perhaps more work using molecular techniques will be of value in some cases, but in the end it seems probable that the only definite answer is more fieldwork and collecting of pairs *in copulo*.

11.16 IMPORTANCE OF CORRECT IDENTIFICATION IN BIOCONTROL

A very important aspect of any biocontrol programme is the associated taxonomy (Delucchi *et al.*, 1976; Rosen, 1986). There have been numerous occasions when either searches for potential control agents or releases have failed because of inaccurate identification. Inadequate taxonomy can manifest itself in various ways (Noyes, 1994). Specimens may simply be misidentified by a poor taxonomist when a more competent one could have readily provided a correct name. However, the group may be too poorly studied to permit identification at the present, even by a good taxonomist. This may be because no revision has yet been carried out and no worthwhile keys are available, or because the group is too poorly collected to make for a meaningful revision. Both of these situations can lead to the need to describe anew the potentially important biocontrol agent in isolation from any revisionary work. In addition to these problems, cryptic and sibling species crop up with alarming regularity and cannot be ignored, as they often differ significantly in their host associations and other biological characters (Rosen, 1986). Working out the true nature of the taxonomy of such cryptic species may require the integration of classical taxonomy with biological, and perhaps also with molecular studies – such work necessarily cannot be done in a hurry.

Whilst incorrect identifications of both hosts and parasitoids have at various times been responsible for failures or at least delays in control programmes, the relatively poor state of the taxonomy of many groups of parasitic Hymenoptera as well as other insects has been the major contributing factor – for example, in the case of the cassava mealy bug, *Phaenococcus manihoti*, when it was first starting to become a serious pest in Africa. The mealy bug was initially identified as a species from Central America, but searches there failed to yield any suitable parasitoids (Noyes, 1994). Subsequent re-examination revealed the true identity of the pest, which was known to originate from South America. When searches for parasitoids were instigated, the encyrtid, *Apoanagyrus lopezi* (previously often referred to as *Epidinocarsus lopezi*), was soon discovered, and in a relatively short time a very successful release programme led to the near perfect control of a major pest that was threatening the lives of many millions of Africans who rely on cassava as a dietary staple. Another well-documented example is that of the California red scale, *Aonidella aurantii*, which was originally misidentified as a species of *Chrysomphalus* and accordingly believed to originate in South America. Only when its true identity as a member of the East Asian genus *Aonidella* was recognized, following proper taxonomic revision of the group, was an expedition to find biocontrol agents successful mounted.

11.17 SPECIMEN PREPARATION AND VOUCHERS

For taxonomic studies, especially those involving small, delicate taxa such as many chalcidoids, the mode of preservation is very important: badly prepared material may be unidentifiable, and many of the problems faced by taxonomists today stem from poor practice in the past. Whilst it is understandable how workers 50–100 years ago may have been unaware of the problems they were giving rise to and how to overcome them, it is still far too often the case that much of the material sent to taxonomists for identification or even material that forms the basis for new taxonomic descriptions is often in a shockingly poor state. Given the cost implications of many biocontrol and IPM programmes, it is false economy to fail to prepare material to an adequate standard. Further, even when it is believed that a parasitoid has been correctly identified there is still a need to preserve an adequate number of voucher specimens for future reference. All too often what was once believed to be a single species of parasitoid turns out to be a complex of two or more cryptic species. Lack of adequately preserved voucher material means that it may be impossible to trace exactly what was released in a control programme, and just because one member of a sibling species pair may have failed to exercise control it does not follow that the other will be equally unsuccessful.

Many works provide descriptions of how particular taxa should be mounted and preserved (Noyes, 1982; Gauld and Bolton, 1988). Some procedures are simple; others are rather more complicated, especially those for dealing adequately with the small and fragile groups, which may necessitate making microscopic slide preparations. Dry mounting of some of these is possible but unless they are properly treated they will collapse on drying and become nigh on impossible to identify. Keeping small, weakly sclerotized taxa in alcohol does not offer a long-term solution either, as they will eventually disintegrate, especially if they are exposed to strong light and high temperatures, though useful alcohol storage time can be extended by keeping them in a fridge or deep-freeze.

Appendix A

Parasitic Hymenoptera frequently studied in culture

Over the past 60 or so years, biologists have cultured and studied a very large number of parasitic wasps, for biocontrol, for pure ecological, behavioural or physiological research or simply to investigate their zoocoenoses. In part this reflects the enormous range of biologies and adaptations collectively displayed by the parasitic Hymenoptera and which have made different species ideal for investigating different types of biological problems. To list and describe all of these would be a daunting task. Nevertheless, a relatively few species have received a disproportionate amount of attention and are very frequently referred to in publications. Below is an annotated list of a number of these in the hope that it will prove useful for those who are starting to get to grips with the wealth of parasitoid literature.

Anagrus optabilis (Perkins)
Classification: Superfamily Chalcidoidea; Family Mymaridae.
Biology: An egg parasitoid of *Perkinsiella saccharicida*, a pest of sugarcane.

Anisopteromalus calandrae (Howard)
Classification: Superfamily Chalcidoidea; Family Pteromalidae.
Biology: A parasitoid of the bruchid beetles *Callosobruchus maculatus* Fabricius and *C. chinensis* (L.) but can also be reared on the rice weevil, *Sitophilus oryzae* (L.).

Apanteles spp.
Many species have now been transferred to other genera, especially *Cotesia* (q.v.).

Aphaereta pallipes (Say)
Classification: Superfamily Ichneumonoidea; Family Braconidae; Subfamily Alysiinae.
Biology: A koinobiont larvo-pupal endoparasitoid of the onion maggot, *Halemya antiqua* (Meig.).

Aphidius ervi Haliday
Classification: Superfamily Ichneumonoidea; Family Braconidae; Subfamily Aphidiinae.
Biology: A common Eurasian koinobiont endoparasitoid of various aphids but often cultured on *Acyrthosiphon pisum* (Harris). Other frequently studied species include *A. colemani* Viereck, *A. matricariae* Haliday, *A. nigripes* Ashmead, *A. smithi* Sharma & Subba Rao.

Aphytis lingnanensis Compere
Classification: Superfamily Chalcidoidea; Family Aphelinidae; Subfamily Aphelinae.
Biology: Facultatively gregarious ectoparasitoid of armoured diaspidid scales, e.g. California red scale, *Aonidiella aurantii* (Maskell). Other frequently studied species include members of the *A. mytilaspidis* species complex which includes both arrhenotokous and thelytokous sibling species. *A. melinus* DeBach has also been used in many studies.

Ascogaster reticulatus Watanabe
Classification: Superfamily Ichneumonoidea; Family Braconidae; Subfamily Cheloninae.
Biology: An ovo-larval endoparasitoid of the tortricid moth, *Adoxophyes* sp. *Ascogaster quadrimaculata* Wesmael, which attacks *Cydia pomonella* L., has also been intensively studied.

Asobara tabida Nees
Classification: Superfamily Ichneumonoidea; Family Braconidae; Subfamily Alysiinae.
Biology: *Asobara* species are common koinobiont endoparasitoids of *Drosophila* species, especially *D. melanogaster* Meigen and *D. subobscura* Collis. Other species that are occasionally cultured include *A. persimilis* Papp and *A. rufescens* (Förster).

Brachymeria intermedia (Nees)
Classification: Superfamily Chalcidoidea; Family Chalcididae; Subfamily Brachymeriinae.
Biology: A polyphagous idiobiont pupal endoparasitoid of lepidopterous pupae including the gypsy moth, *Lymantria dispar* (L.), and spruce budworm, *Choristoneura fumiferana* (Clemens), but can also be cultured on the greater wax moth, *Galleria mellonella* (L.). Other cultured species include *B. lasus*.

Bracon hebetor: see *Habrobracon hebetor*

Bracon mellitor Say
Classification: Superfamily Ichneumonoidea; Family Braconidae; Subfamily Braconinae.
Biology: A solitary idiobiont ectoparasitoid of boll weevil larvae, *Anthonomus grandis* Boheman, but also attacks a range of other curculionid beetles and Lepidoptera larvae in similar concealed situations.

Campoletis sonoriensis (Cameron)
Classification: Superfamily Ichneumonoidea; Family Ichneumonidae; Subfamily Campopleginae.
Biology: This species has been widely employed in studies of host parasitoid physiological interactions and in particular the roles of polydnaviruses and virus-like particles in overcoming the host's immune defences. Sometimes erroneously referred to as *C. perdistinctus* Viereck in earlier biocontrol literature.

Cardiochiles nigriceps Viereck
Classification: Superfamily Ichneumonoidea; Family Braconidae; Subfamily Cardiochilinae.
Biology: A koinobiont endoparasitoid of *Heliothis virescens* (Fabricius) larvae. This braconid has been extensively studied from behavioural, ecological and physiological perspectives.

Catolaccus grandis (Burks)
Classification: Superfamily Chalcidoidea; Family Pteromalidae; Subfamily Pteromalinae.
Biology: Normally an idiobiont ectoparasitoid of boll weevil, *Anthonomus grandis* Boheman, prepupae and pupae; originally from Mexico.

Cephalonomia gallicola (Ashmead)
Classification: Superfamily Chrysidoidea; Family Bethylidae; Subfamily Epyrinae.
Biology: A gregarious ectoparasitoid of dermestid beetle larvae. It has been a popular study subject because it displays parental care (broodguarding). *C. waterstoni* (Gahan), a parasitoid of the rusty grain beetle, *Cryptolestes ferrugineus* (Stephens), has also been used in many studies.

Chelonus sp. nr *curvimaculatus*
Classification: Superfamily Ichneumonoidea; Family Braconidae; Subfamily Cheloninae.
Biology: All species are ovo-larval endoparasitoid of various moths, including noctuids. Several *Chelonus* species have been studied intensively in terms of pure biology and their physiological interactions with their hosts. One undescribed (or unidentified) species, *C.* Sp. nr *curvimaculatus*, which has been widely investigated, is a parasitoid of *Trichoplusia ni* (Hübner). Other studied species include *C. insularis* Cresson (= *C. texanus* Cresson), which attacks *Heliothis virescens* Fabricius, and *C. inanitus* (L.) on *Spodoptera* spp.

Coccygomimus turionellae: see *Pimpla turionellae*

Colpoclypeus florus (Walker)
Classification: Superfamily Chalcidoidea; Family Eulophidae; Subfamily Eulophinae.

Biology: A larval ectoparasitoid of the torticid moth, *Adoxophyes orana* (Fischer von Röslerstamm). The wasp oviposits near but not on the host, making it suitable for egg manipulation experiments.

Copidosoma floridanum (Ashmead)
Classification: Superfamily Chalcidoidea; Family Encyrtidae; Subfamily Copidosomatinae.
Biology: A polyembryonic, ovo-larval endoparasitoid of various plusiine moths usually reared on *Trichoplusia ni* (Hübner). This species has been intensively stuidied because of its polyembryonic developments and its embryology, and also as a model for sex allocation studies. *C. truncatellus* (Dalman) has also been extensively investigated.

Cotesia spp.
Classification: Superfamily Ichneumonoidea; Family Braconidae; Subfamily Microgastrinae.
Biology: Many species are treated under the generic name *Apanteles*. All are koinobiont endoparasitoids of Lepidoptera larvae, feeding internally at first but emerging in the last instar to complete feeding and to pupate externally; most are also gregarious. Commonly studied species include: *C. glomerata* (L.), a gregarious endoparasitoid of pierid butterfly larvae, especially *Pieris brassicae*; *C. marginiventris* (Cresson); *C. rubecula* (Marshall), solitary on the small cabbage white butterfly, *Pieris rapae* (L.); *C. congregata* (Say), on the tobacco hornworm, *Manduca sexta* (L.); *C. karyai* (Watanabe), on the common armyworm, *Pseudaletia* (= *Leucania*) *separata* (Walker); *C. melanoscela* (Ratzeburg), solitary on the gypsy moth, *Lymantria dispar* (L.), *Leucoma salicis* (L.) and others.

Dahlbominus fuscipennis (Zetterstedt)
Classification: Superfamily Chalcidoidea; Family Eulophidae; Subfamily Eulophinae.
Biology: An ectoparasitoid of diprionid sawfly prepupae and pupae, especially *Pristiphora erichsonii* (Hartig), and will also attack various *Neodiprion* spp.

Dendrocerus carpenteri Curtis
Classification: Superfamily Ceraphronoidea; Family Megaspilidae.
Biology: A hyperparasitoid of aphidiine braconids.

Diachasmimorpha longicaudata (Ashmead)
Classification: Superfamily Ichneumonoidea; Family Braconidae; Subfamily Opiinae.
Biology: Solitary larvo-pupal endoparasitoid of the Caribbean fruit fly, *Anastrepha suspensa* (Loew) (Tephritidae).

Diadegma chrysostictos Gmelin
Synonym: *Horogenes chrysostictos*

Classification: Superfamily Ichneumonoidea; Family Ichneumonidae; Subfamily Campopleginae.
Biology: *Diadegma* species are koinibiont endoparasitoids of Lepidoptera larvae. *D. chrysostictos* attacks flour moth larvae.

Diadromus pulchellus Wesmael
Classification: Superfamily Ichneumonoidea; Family Ichneumonidae; Subfamily Ichneumoninae.
Biology: A monophagous, idiobiont, pupal endoparasitoid of the lepidopteran, *Acrolepiopsis assectella* Zeller.

Diglyphus begini (Ashmead)
Classification: Superfamily Chalcidoidea; Family Eulophidae; Subfamily Eulophinae.
Biology: A parasitoid of *Liriomyza* leafminers (Diptera), used in glasshouse control programmes.

Edovum puttleri Grissell
Classification: Superfamily Chalcidoidea; Family Eulophidae; Subfamily Entedoninae.
Biology: An idiobiont egg parasitoid of Colorado potato beetle, *Leptinotarsa decemlineata* (Say).

Encarsia formosa Gahan
Classification: Superfamily Chalcidoidea; Family Aphelinidae; Subfamily Coccophaginae.
Biology: *Encarsia* species are heteronomous endoparasitoids of scale insects, cultured on the greenhouse whitefly, *Trialeuroides vaporariorum* (Westwood), and used widely for its control. *E. formosa* and *E. deserti* Gerling & Rivnay are important for the control of *Bemisia* whiteflies. Other species commonly cultured include *E. berlesei* (Howard), *E. pergandiella* Howard and *E. tricolor* Förster.

Eretmocerus mundus Mercet
Classification: Superfamily Chalcidoidea; Family Aphelinidae; Subfamily Aphelinae.
Biology: An important parasitoid of nymphs of the whitefly, *Bemisia tabaci* (Gennadius); eggs are laid externally under the host but the larva bites and then becomes engulfed by host tissue.

Exeristes roborator (F.)
Classification: Superfamily Ichneumonoidea; Family Ichneumonidae; Subfamily Pimplinae.
Biology: An ectoparasitoid of larval pink bollworm, *Pectinophora gossypiella* (Saunders). *E. comstockii* Cresson has also been much studied.

Goniozus legneri Gordh
Classification: Superfamily Chrysidoidea; Family Bethylidae; Subfamily Bethylinae.

Biology: An ectoparasitoid of various gelechiids, including the navel orange worm, *Amyelois transitella*, and the pink bollworm, *Pectinophora gossypiella* (Saunders). *G. nephantidis* Muesebeck, a gregarious parasitoid of *Opisina arenosella* Walker (= *Nephantis serinopa* Meyrick), has also been intensively studied.

Habrobracon hebetor (Say)
Classification: Superfamily Ichneumonoidea; Family Braconidae; Subfamily Braconinae.
Synonyms: *Bracon hebetor, Microbracon hebetor.*
Biology: Perhaps the most intensively investigated parasitic wasp of all time, this is a gregarious idiobiont ectoparasitoid of any of a range of lepidopterous larvae infecting stored products (e.g. *Plodia interpunctella* (Hübner), *Anagasta kuehniella* (Zeller)). It has played major roles in the study of hymenopteran sex determination and sex allocation, and has also been advocated as a suitable bioassay species for mutagens. Other closely related and frequently studied species include *H. juglandis* and *H. serinopiae*. The taxonomy of the genus is in a mess and it is quite possible that several species are synonyms of one another (as has often been proposed) and also that there may be previously unrecognized species.

Heterospilus prosopidis Viereck
Classification: Superfamily Ichneumonoidea; Family Braconidae; Subfamily Doryctinae.
Biology: An idiobiont larval ectoparasitoid of bruchid beetles such as *Callosobruchus maculatus* F. and *C. chinensis* (L.).

Itoplectis conquisitor (Say)
Classification: Superfamily Ichneumonoidea; Family Ichneumonidae; Subfamily Pimplinae.
Biology: A highly polyphagous, idiobiont endoparasitoid of lepidopterous pupae. *I. naranyae* (Ashmead) and *I. maculator* (F.) also frequently studied.

Laelius pedatus (Say)
Classification: Superfamily Chrysidoidea; Family Bethylidae; Subfamily Epyrinae.
Biology: A semi-gregarious idiobiont ectoparasitoid (with a brood size of 1–5) of various dermestid beetle larvae, especially *Anthrenus verbasci* and *A. albipes*. Other studied species include *L. utilis* Cockerell, also a dermestid parasitoid.

Leptomastix dactylopii Howard
Classification: Superfamily Chalcidoidea; Family Encyrtidae; Subfamily Anagyrini.
Biology: An economically important endoparasitoid of the citrus mealybug, *Plannococcus citri* (Risso). Although it will oviposit in other mealybugs, its successful development appears to be limited to its natural host. Its biology has been described in detail by Zinna (1960).

Leptopilina heterotoma (Thomson)
Classification: Superfamily Cynipoidea; Family Eucoilidae.
Synonym: *Pseudeucoila bochei* Weld.
Biology: A koinobiont endoparasitoid of *Drosophila* larvae. Other species of the genus that are fairly commonly cultured include *L. boulardi* (Barbotin, Carton & Kelner-Pillault) and *L. clavipes* (Hartig).

Macrocentrus grandii Goidanich
Classification: Superfamily Ichneumonoidea; Family Braconidae; Subfamily Macrocentrinae.
Biology: A koinobiont endoparasitoid of larvae of the European corn borer moth, *Ostrinia nubilalis* (Hübner).

Melittobia chalybii Ashmead
Classification: Superfamily Chalcidoidea; Family Eulophidae; Subfamily Tetrastichinae.
Biology: *Melittobia* species are gregarious larval/pupal ectoparasitoids of various solitary bee and wasp larvae. *M. australica* Girault and *M. acasta* (Walker) (= *Cirrospilus acasta*) have also been studied in some detail.

Microplitis croceipes (Cresson)
Classification: Superfamily Ichneumonoidea; Family Braconidae; Subfamily Microgastrinae.
Biology: As with other microgastrines, *Microplitis* species are koinobiont larval endoparasitoids of Lepidoptera larvae (see also *Cotesia*). This much studied solitary species attacks lepidopterous larvae belonging to the genera *Heliothis* and *Helicoverpa*. Other studied species include *M. demolitor* Wilkinson on *Prodenia litura* (F.) and *Pseudoplusia includens*, and *M. rufiventris* Kokujev on *Spodoptera littoralis* (Boids).

Mormoniella vitripennis: see *Nasonia vitripennis*

Muscidifurax raptor Girault & Sanders
Classification: Superfamily Chalcidoidea; Family Pteromalidae; Subfamily Pteromalinae.
Biology: *Muscidifurax* species are idiobiont parasitoids of muscid fly pupae such as *Musca domestica* L. and *Stomoxys calcitrans* (L.). Other species fairly commonly cultured include *Muscidifurax raptorellus* Kogan & Legner, *M. zaraptor* Kogan & Legner and *M. uniraptor* Kogan & Legner. Species identification in the genus is problematic, but valid biological differences exist.

Nasonia vitripennis (Walker)
Classification: Superfamily Chalcidoidea; Family Pteromalidae; Subfamily Pteromalinae.
Biology: A gregarious idiobiont endoparasitoid of muscid fly puparia, especially *Musca domestica* L. and *Lucilia sericata* (Meigen), but actually is ectoparasitic on the fly's pupa. This species is another contender for the most studied parasitoid prize, and is certainly the most studied chalcidoid.

Nemeritis canescens: see *Venturia canescens*

Ooencyrtus nezarae Ishii
Classification: Superfamily Chalcidoidea; Family Encyrtidae.
Biology: Parasitic on the eggs of the bean bug, *Riptortus clavatus*. Other widely studied members of the genus include *O. papilionis* Ashmead and *O. telenomicida* (Vassiliev).

Opius fletcheri: see *Psyttalia fletcheri*

Pachycrepoides vindemiae (Rondani)
Classification: Superfamily Chalcidoidea; Family Pteromalidae; Subfamily Sphegigastrinae.
Biology: A primary puparial endoparasitoid of *Drosophila* and *Piophila*, and facultatively a hyperparasitoid, e.g. through *Asobara tabida* Nees. Other cultured species include *P. dubius* Ashmead.

Pimpla turionellae (L.)
Classification: Superfamily Ichneumonoidea; Family Ichneumonidae; Subfamily Pimplinae.
Biology: An idiobiont endoparasitoid of lepidopterous pupae, including gypsy moth. In laboratory culture it is often maintained on wax moth (*Galleria mellonella* (L.)) or mealworm (*Tenebrio mollitor*) pupae.

Pseudeucoila bochei: see *Leptopilina heterotoma*

Psyttalia fletcheri (Silvestri)
Classification: Superfamily Ichneumonoidea; Family Braconidae; Subfamily Opiinae.
Biology: This and other species of *Psyttalia* are larval–pupal endoparasitoids of Diptera.

Pteromalus puparum (L.)
Classification: Superfamily Chalcidoidea; Family Pteromalidae; Subfamily Pteromalinae.
Biology: A gregarious pupal endoparasitoid of numerous butterfly species including *Pieris*, *Nymphalis*, *Aglais*, *Papilio* and *Vanessa* species. *Pteromalus venustus* Walker is a commonly studied parasitoid of Megachilid bees.

Spalangia cameroni Perkins
Classification: Superfamily Chalcidoidea; Family Pteromalidae; Subfamily Spalangiinae.
Biology: An ectoparasitoid of synanthropic Diptera pupae within the puparium, and is particularly active in humid subtropical and tropical regions.

Telenomus heliothidis Ashmead
Classification: Superfamily Scelionoidea; Family Scelionidae; Subfamily Telenominae.

Biology: A solitary egg parasitoid of the noctuid, *Helicoverpa virescens* (F.), as well as some other *Heliothis* species; much used in studies of host location.

Trichogramma evanescens Westwood
Classification: Superfamily Chalcidoidea; Family Trichogrammatidae.
Biology: All *Trichogramma* species are egg parasitoids; some are solitary; others, like *T. embryophagum* Hartig, are gregarious. *T. evanescens* is a polyphagous species, commonly cultured in cabbage moth (*Mamestra brassicae* L.) and in *Anagasta kuehniella* (Zeller) eggs. Other commonly cultured species include *T. confusum* Viggiani, *T. dendrolimi* Matsumura, *T. maidis* Pintureau & Voegele, *T. minutum* Riley, *T. papilionis* Nagarkatti and *T. pretiosum* (Riley).

Trissolcus basalis (Wollaston)
Classification: Superfamily Scelionoidea; Family Scelionidae; Subfamily Telenominae.
Biology: An egg parasitoid of the green stink bug, *Nezara viridula* (L.). Other commonly cultured species include *T. grandis* (Thomson), all of which attack Pentatomidae.

Venturia canescens (Gravenhorst)
Classification: Superfamily Ichneumonoidea; Family Ichneumonidae; Subfamily Campopleginae.
Synonyms: *Nemeritis canescens*, *Devorgilla canescens*.
Biology: A solitary thelytokous koinobiont endoparasitoid of larvae of various phycitid moths, commonly cultured on the flour moth *Anagasta* (= *Ephestia*) *kuehniella* (Zeller) but can also be reared on *E. cautella*, the navel orange worm, *Paramyelois transitella* and *Plodia interpunctella* (Hübner). It has been extensively used in physiological investigations as well as in oviposition decision-making experiments.

Appendix B

Utilization of host groups by different families of parasitic Hymenoptera

Appendix B Utilization of host groups by different families of parasitic Hymenoptera[1-3]

	Plants	Solpug	Aran	Acar	Odon	Orth	Blatt	Mant	Phasm	Isop	Emb	Derm	Hem	Psoch	Thys	Netur	Mecop	Tric	Lep	Strep	Col	Hym	Siph
Orussidae	—	—	—	—	—	—	—	—	—	—	—	—	—	—	—	—	—	—	—	—	◆	◆	—
Trigonalyidae	—	—	—	—	—	—	—	—	—	—	—	—	—	—	—	—	—	—	—	—	—	◆	—
Megalyridae	—	—	—	—	—	—	—	—	—	—	—	—	—	—	—	—	—	—	—	—	◆	◆	—
Stephanidae	—	—	—	—	—	—	—	—	—	—	—	—	—	—	—	—	—	—	—	—	◆	—	—
Gasteruptiidae	*	—	—	—	—	—	—	—	—	—	—	—	—	—	—	—	—	—	—	—	◆	◆	—
Aulacidae	—	—	—	—	—	—	—	—	—	—	—	—	—	—	—	—	—	—	—	—	—	◆	—
Evaniidae	—	—	—	—	—	—	◆	—	—	—	—	—	—	—	—	—	—	—	—	—	—	—	—
Ceraphronidae	—	—	—	—	—	—	—	—	—	—	—	—	◆	—	—	◆	—	—	•	—	—	◆	—
Megaspilidae	—	—	—	—	—	—	—	—	—	—	—	—	◆	—	—	•	•	—	—	—	—	•	—
Scelionidae	—	—	•	—	—	◆	—	•	—	—	•	—	◆	—	•	•	—	—	◆	—	•	•	—
Platygastridae	—	—	—	—	—	—	—	—	—	—	—	—	◆	—	—	•	—	—	—	—	•	◆	—
Figitidae	—	—	—	—	—	—	—	—	—	—	—	—	—	—	—	◆	—	—	—	—	—	—	—
Eucoilidae	—	—	—	—	—	—	—	—	—	—	—	—	—	—	—	—	—	—	—	—	—	—	—
Charipidae	—	—	—	—	—	—	—	—	—	—	—	—	◆	—	—	—	—	—	—	—	—	—	—
Ibaliidae	—	—	—	—	—	—	—	—	—	—	—	—	—	—	—	—	—	—	—	—	—	—	—
Cynipidae	◆	—	—	—	—	—	—	—	—	—	—	—	—	—	—	—	—	—	—	—	—	◆	—
Diapriidae	—	—	—	—	—	—	—	—	—	—	—	—	—	—	—	—	—	—	—	—	•	•	—
Heloridae	—	—	—	—	—	—	—	—	—	—	—	—	—	—	—	◆	—	—	—	—	—	•	—
Pelecinidae	—	—	—	—	—	—	—	—	—	—	—	—	—	—	—	—	—	—	—	—	◆	—	—
Proctotrupidae†	—	—	—	—	—	—	—	—	—	—	—	—	—	—	—	—	—	—	•	—	◆	—	—
Roproniidae	—	—	—	—	—	—	—	—	—	—	—	—	—	—	—	—	—	—	—	—	◆	—	—
Vanhorniidae	—	—	—	—	—	—	—	—	—	—	—	—	—	—	—	—	—	—	—	—	◆	—	—
Agaonidae	◆	—	—	—	—	—	—	—	—	—	—	—	◆	—	—	—	—	—	—	—	◆	—	—
Aphelinidae	—	—	—	—	—	•	—	—	—	—	—	—	—	—	—	•	—	—	•	—	—	◆	—
Chalcididae	—	—	—	—	—	•	—	—	—	—	—	—	—	—	—	•	—	—	◆	•	•	•	—
Elasmidae	—	—	—	—	—	—	—	—	—	—	—	—	—	—	—	—	—	—	◆	—	—	—	—
Encyrtidae	—	—	•	•	—	•	•	—	—	—	—	—	◆	—	—	•	—	—	◆	—	•	◆	—
Eucharitidae	—	—	—	—	•	—	—	—	—	—	—	—	—	—	•	—	—	—	◆	—	—	◆	—
Eulophidae‡	•	—	◆	•	•	•	•	—	—	—	—	—	◆	—	—	•	—	—	◆	—	◆	◆	—
Eupelmidae	—	—	•	—	—	•	•	•	—	—	—	—	•	—	—	•	—	—	◆	—	◆	◆	—

	Plantis	Solpug	Aran	Acar	Odon	Orth	Blatt	Mant	Plasm	Isop	Emb	Derm	Hem	Psoch	Thys	Neur	Mecop	Tric	Lep	Strep	Col	Hym	Siph
Eurytomidae	◆	–	–	–	–	•	•	–	–	–	–	–	◆	–	–	–	–	–	◆	–	◆	◆	–
Leucospidae	–	–	•	–	–	–	–	–	–	–	–	–	–	–	–	–	–	–	–	–	–	◆	–
Mymaridae	–	–	–	–	•	◆	–	–	–	–	–	–	◆	◆	•	–	–	–	–	–	◆	–	–
Ormyridae	–	–	–	–	–	•	–	–	–	–	–	–	–	–	–	•	–	–	◆	–	◆	◆	–
Perilampidae	–	–	•	–	–	•	–	–	–	–	–	–	•	–	–	•	–	–	◆	–	◆	◆	◆
Pteromalidae	•	–	•	–	–	•	•	–	–	–	–	•	◆	–	–	•	–	–	◆	–	◆	◆	–
Signiphoridae	–	–	–	–	–	–	•	–	–	–	–	–	◆	–	–	–	–	–	–	–	•	•	–
Tanaostigmatidae	◆	–	–	–	–	–	–	–	–	–	–	–	–	–	–	–	–	–	–	–	–	•	–
Tertacampidae	•	–	–	–	–	–	–	–	–	–	–	–	–	–	–	–	–	–	•	–	•	•	–
Torymidae	–	–	–	–	–	•	–	•	–	–	–	–	•	–	•	•	–	–	◆	–	◆	◆	–
Trichogrammatidae	–	–	•	–	•	•	–	–	–	–	–	–	◆	–	•	•	–	–	◆	–	◆	◆	–
Braconidae	•	–	•	–	–	–	–	–	–	?	–	•	◆	–	–	•	•	•	◆	–	◆	•	–
Ichneumonidae††	–	–	◆	–	–	–	–	–	–	–	–	–	•	–	–	•	–	•	•	–	◆	◆	–
Bethyliidae	–	–	–	–	–	–	–	–	–	–	–	–	–	–	–	–	–	–	◆	–	◆	–	–
Chrysididae	–	–	–	–	–	–	–	–	•	–	–	–	–	–	–	–	–	–	•	–	•	•	–
Dryinidae	–	–	–	–	–	–	–	–	–	–	–	–	◆	–	–	–	–	–	–	–	–	–	–
Embolemidae	–	–	–	–	–	–	–	–	–	–	–	–	◆	–	–	–	–	–	–	–	–	–	–
Sclerogibbidae	–	–	–	–	–	–	–	–	–	–	◆	–	–	–	–	–	–	–	–	–	–	–	–
Scolebythidae	–	–	–	–	–	–	–	–	–	–	–	–	–	–	–	–	–	–	–	–	◆	–	–
Bradynobaenidae	–	◆	–	–	–	–	–	–	–	–	–	–	–	–	–	–	–	–	–	–	•	–	–
Mutillidae	–	–	–	–	–	–	•	–	–	–	–	–	–	–	–	–	–	–	–	–	•	◆	–
Rhopalosommatidae	–	–	–	–	–	◆	–	–	–	–	–	–	–	–	–	–	–	–	•	–	–	◆	–
Sapygidae	–	–	–	–	–	–	–	–	–	–	–	–	–	–	–	–	–	–	–	–	–	◆	–
Scoliidae	–	–	–	–	–	–	–	–	–	–	–	–	–	–	–	–	–	–	–	–	◆	–	–
Tiphiidae	–	–	–	–	–	•	–	–	–	–	–	–	–	–	–	–	–	–	–	–	◆	–	–

[1] Updated from an unpublished collation by J. S. Noyes.

[2] Principal hosts are indicated by large bullets, minor ones by small bullets.

[3] Only primary hosts are indicated for hyperparasitoids.

* Gasteruptiids are initially egg predators and then continue development by consuming their host bee's pollen store.

† One species has also been recorded as parasitoid of centipedes (Chilopoda: Lithobiidae).

‡ One species has also been recorded as eating nematodes.

†† Some cryptines are predators on pseudoscorpion egg masses.

Glossary

accessory nuclei Small nuclei that bud off from the main nucleus of the developing oöcyte in at least some Diptera, Hymenoptera and Mallophaga, and probably in Mecoptera.

acid gland The main venom gland of apocritan Hymenoptera; cf. alkaline gland (q.v.).

adelphoparasitism A parasitoid of a closely related taxon, including as an extreme case, obligate autoparasitism, cf. alloparasitoid (q.v.); sometimes used only in the latter more restricted sense.

aedeagus The phallus or penis of an insect.

agastoparasitism Term coined to described parasitism when it occurs between two closely related species (Ronquist, 1994b).

alecithal Eggs without or with only very meagre yolk content.

alkaline gland Usually smaller than the acid gland (q.v.) this gland is homologous with Dufour's gland (q.v.) of aculeates.

Allee effect Strictly suboptimal effects of overcrowding or undercrowding, but most often used to refer to difficulties associated with living at low densities such that, for example, individuals may have a low probability of finding mates and therefore will fail to reproduce.

allomone Any chemical substance produced by, acquired by, or produced as a result of an action of, an individual of one species (organism A), that evokes a response in a member of another species (organism B); as defined by Nordlund (1981), allomones specifically favour organism A but not organism B; cf. kairomone (q.v.), synomone (q.v.).

alloparasitoid A parasitoid of members of an unrelated taxon; cf. adelphoparasitoid (q.v.).

allotopic Occupying different macrohabitats.

amnion An extraembryonic membrane located between the embryo and the serosal membranes of some species.

amphimixis The fusion of egg and sperm pronuclei to form the zygote nucleus.

amphitoky Another name for deuterotoky (q.v.)

anal vesicle A swollen, protruding vesicle at the posterior end of early instar larvae of various endoparasitoids formed partly or wholly by an evagination of the rectum.

anautogenous Insects whose females must feed before they can produce mature eggs (Jervis and Kidd, 1986).

aneuploid Cells with non-integer multiples of the haploid chromosome number.

anhydropic Eggs that are rich in yolk (Flanders, 1942a); equivalent to lecithal (q.v.).

apneustic Without functional spiracles or tracheal system; cf. holopneustic (q.v.), metapneustic (q.v.).

Apocrita A probably monophyletic division of the Hymenoptera comprising the traditional (paraphyletic) Parasitica and the Aculeata, i.e. all Hymenoptera except for the Symphyta.

apolysis The intermoult period during which the epidermis secretes moulting fluid and fresh cuticle.

apomixis Thelytoky in which meiosis is absent; cf. automixis (q.v.).

apomorphy A derived character state, cf. plesiomorphy (q.v.). *See also* synapomorphy; symplesiomorphy.

apparent parasitism *See* pseudoparasitoid.

aptery The complete absence of wings.

arrhenotoky A form of parthenogenesis dominant amongst the Hymenoptera in which males develop from unfertilized eggs and females from fertilized ones.

aulax The groove that runs along the length of each of the lower ovipositor valves (= gonapophyses of abdominal segment 8) and which when coupled with the rhachis (q.v.) comprises the olistheter (q.v.) (Smith, 1969; Quicke *et al.*, 1994).

autapomorphy A derived (advanced) character state exhibited by the members of a taxon that indicates that group's monophyly and provides no evidence regarding its relationships (Quicke, 1993); sometimes restricted to such character states occurring within a single species.

autogenous Species whose females do not need to feed before they lay eggs.

automixis Thelytoky in which meiosis occurs, cf. apomixis (q.v.).

autoparasitoid A species that parasitizes a conspecific, as in many heteronomous aphelinids (see Chapter 2).

axoneme The microtubular organelle comprising the motile element of a sperm tail (flagellum, q.v.).

B chromosome A very small, supernumary chromosome.

biotype A genetically distinct race adapted to a particular biological or physical environment.

bivoltine Having two generations per year.

brachyptery The possession of short or very reduced wings; brachypterous species are usually unable to fly, certainly in all of the more extreme cases.

calcar The modified fore tibial spur used in antennal grooming.

calyx gland A gland located at the distal end of each lateral oviduct (Salt, 1968).

carnivoroid Another name for parasitoid (q.v.).

caudate Of parasitoid larvae which have a tail-like posterior projection; usually most well developed in 1st instar larvae (Clausen, 1932).

cecidogenic Causing the formation of galls.

cenchri Roughened pads on the metanotum in most sawflies which interlock with roughened paches on the anal region of the forewings when the wings are folded.

cerci Typically small, unsegmented pair of appendages at the posterior margin of the 10th+11th abdominal tergite, originally at least with a sensory role.

Chalastogastra Another name for Symphyta.

chimaeric Of mosaic (q.v.) individuals: individuals that have tissues with two or more different genotypes. (*See also* gynandromorph).

chorion The outermost non-cellular layer of an insect egg secreted by the follicle cells.

Clistogastra Another name for Apocrita.

coccidivorous Preying on (parasitizing) Coccoidea (i.e. mealy-bugs and scale insects).

colleterial gland Generally a pouch-like gland at the anterior end of the common oviduct; also referred to as accessory gland and vaginal gland.

complementary sex determination (CSD) Sex determination system in which homozygous or hemizygous individuals at a specific gene locus (or loci) become males whilst heterozygotes develop into females (Whiting, 1943; Cook and Crozier, 1995).

cornus A tail-like projection from the posterior larval tergites of some sawflies.

corpora allata Endocrine gland that secretes juvenile hormone.

crop An expanded posterior part of the foregut.

CSD *See* complementary sex determination.

cystoblast The daughter cell from the division of a stem cell (germ line cell) that will go on to form an egg and associated nurse cells.

cystocyte A cell arising from division of a cystoblast (q.v.).

deuterotoky A form of parthenogenesis in which both males and females develop from unfertilized eggs (Luck *et al.*, 1992).

diapause A reversible, hormonally mediated state of reduced development and metabolic activity.

diphagous Of parasitoids in which both sexes feed on the same host species but in different ways; for example, in some Aphelinidae in which males are ectoparasitoids whilst females are endoparasitoids (Walter, 1983a, 1988a; Williams and Polaszek, 1996).

direct hyperparasitoid A hyperparasitoid (secondary parasitoid) which only attacks (oviposits into or near) its parasitoid host, cf. indirect hyperparasitoid (q.v.) (Flanders, 1943; Sullivan, 1987).

discrimination The ability of a parasitoid to distinguish between suitable and unsuitable host individuals (such as ones already parasitized) and to regulate oviposition accordingly.

Dufour's gland An exocrine gland which discharges into the female reproductive tract close to the base of the ovipositor; sometimes called the alkaline gland.

Dyar's law Larval head width increases from instar to instar in a regular geometric progression (Dyar, 1890).

eclosion (1) Hatching of an adult insect from the pupal case; (2) hatching of a larval insect from the egg.

ectoparasitoid A parasitoid whose larvae feed externally on a host.

egg–larval/pupal parasitism A koinobiont life history strategy in which the parasitoid lays its egg within the egg of the host, but its development is not completed until the host has reached a (usually late) larval/pupal stage (Clausen, 1954).

egg parasitoid A species that completes its development entirely within the egg stage of its host.

egress Escape from something such as a newly emerged parasitoid from its pupation site.

elaterium The flexible, mid-dorsal part of the dorsal ovipositor valve in some species (Smith, 1969; Quicke *et al.*, 1994).

Emery's rule The observation that social parasite species are usually similar to their host species and are therefore presumed to be closely related to them (Ronquist, 1994b).

encapsulation A process in insects in which foreign particles such as pathogens or parasitoids become surrounded by haemocytes which subsequently flatten, toughen and melanize.

endoparasitoid A parasitoid whose larva develops for most of its life inside a host organism.

epistasis Interaction between genes at two different loci in which the gene at one locus masks the effects of the other.

euparasitism Actual parasitism of a host; cf. pseudoparasitism (q.v.).

factitious host A species not attacked by a parasitoid under natural conditions but which will serve as a host for it in laboratory conditions.

feeding tube A secreted tubular structure produced by an adult female parasitoid to enable it to imbibe host body fluids when the host is not directly accessible.

flagellomere A segment of a flagellum.

flagellum (1) That part of the antenna distal to the scapus and pedicel; (2) the mobile post-nuclear part of a sperm.

follicle cells Cells, derived from prefollicular tissue in the germarium of the ovary, which surround an oöcyte and its associated nurse cells.

gaster Usually meaning that part of the metasoma (q.v.) behind the petiole (q.v.).

germ cell A cell that only divides to give rise either to gametes (via intermediates) or other germ cells; *see also* cystoblast; cystocyte.

germarial zone *See* germarium.

germarium Anterior part of an ovariole where oögonia (cystoblasts) differentiate into oöcytes and nurse cells (q.v.), or the equivalent part in the testis containing the spermatogonia.

germ line A lineage of cells which are determined at an early stage of embryogenesis to become gametes.

giant cells Another name for teratocytes (q.v.)

gonobase Internal sclerite of male genitalia; also referred to as basal ring, gonocardo, cardo.

gonocoxa The metasomal sclerite from which the ovipositor sheath arises.

gonoplac Alternative name for ovipositor sheaths (q.v.) in some groups (e.g. Galloway *et al.*, 1992)

gonapophyses A pair of appendages on both the 8th and 9th abdominal segments of a female hymenopteran (and many insects) that together form the ovipositor.

gonapophysis 8 Lower ovipositor valve; also called valvula 1 (q.v.).

gonapophysis 9 Upper ovipositor valve; also called valvula 2 (q.v.).

gregarious parasitoid A parasitoid that typically lays multiple eggs in a single host that develop into multiple adults, or in the case of polyembryonic species, lays one egg that gives rise to multiple individuals from a single host; cf. quasi-gregarious (q.v.).

gynandromorph An individual that comprises a mosaic (q.v.) of male and female tissue.

gynogenesis Also called pseudogamy, the activation of embryogenesis in an egg by a sperm whose genetic material then degenerates and plays no part in the genetic make-up of the developing individual.

haemocyte A general term for any of several classes of free-floating, circulating cells in the haemolymph; *see also* lamellocyte.

Hallez's law This 'law' expresses the near universal relationship between the orientaion of an egg within an insect's ovariole and the orientaion of the developing embryo such that the part of the egg that gives rise to the embryonic head is orientated towards the anterior of the ovariole in the insect (Hallez, 1886).

haplodiploid Organisms in which the sex of an individual is determined by whether it possesses a haploid or a diploid genetic complement.

heterogony The alternation of sexual reproducing generations with one or more asexual ones, as in some Cynipidae and the ichneumonid, *Sphecophaga vesparum*.

heteronomy Different ontogeny, as for example in some Aphelinidae; a classificatory system is provided by Walter (1988a).

Heterophaga Alternative name for the paraphyletic 'Parasitica' that takes note of the fact that many taxa in that group are phytophagous rather than parasitic.

heterosis Hybrid vigour – increased fitness displayed by some hybrids or by heterozygotes.

heterotrophic parasitoid A subset of heteronomous (q.v.) ontogenies in which the different sexes utilize very different hosts (Walter, 1988a; Williams and Polaszek, 1996).

holopneustic Of larvae, having a full complement of functional spiracles (two thoracic and eight abdominal pairs); cf. apneustic (q.v.), metapneustic (q.v.).

homoplasy Character state reversals and convergent character state evolution that cause problems with phylogeny reconstruction.

host-feeding Use of a host insect as a food source by an adult parasitoid (Jervis and Kidd, 1986; Kidd and Jervis, 1989).

hydropic Of eggs that have little or no yolk and that swell in, and absorb most of their nutrients from, the host.

hymenopteriform Of larvae that show clear segmentation and no conspicuous derived features (Figure 5.2a).

hypermetamorphosis The occurrence of more than one distinct larval stage (with different biology) as, for example, when the first larval instar is a planidium (q.v.); problems with this definition are discussed by Darling and Miller (1991).

hyperparasitoid A parasitoid of another parasitoid.

idiobiont A life history strategy in which the host does not develop any further following the parasitization event (Askew and Shaw, 1986); cf. koinobiont (q.v.).

Idiogastra A name employed occasionally for a hymenopteran suborder comprising only the Orussoidea.

idiophyte Superceded term for idiobiont (q.v.).

iliac glands A pair of tubular glands in the larvae of parasitic Hymenoptera which open into the hindgut along with the generally less well-developed Malpighian tubules – quite possibly they are a specialized pair of Malpighian tubules, and they are referred to as such in many papers.

ilium The region of hindgut (proctodaeum) between the Malpighian tubule insertions and the rectum.

indirect hyperparasitoid A hyperparasitoid (q.v.) which attacks its secondary host irrespective of whether at that time it is parasitized by the primary parasitoid host, cf. direct hyperparasitoid (q.v.) (Flanders, 1943; Sullivan, 1987).

inquiline A species that lives within the domicile of another, such as a gall formed by another species, strictly without causing harm to the original inhabitant.

isofemale line A group of related individuals all of which are descended from a single (usually fertilized) female, usually obtained originally from the wild.

kairomone A chemical substance produced by, or produced as a result of an action of an individual of, one species (organism A), that evokes a response in a member of another species (organism B) which is detrimental to organism A but is beneficial to organism B (Nordlund, 1981; Dicke and Sabelis 1988); cf. allomone (q.v.), kairomone (q.v.).

kleptoparasite A parasitoid that consumes the host or food reserve of another, which it usually kills in the process.

koinobiont A parasitoid that allows its host to continue developing more or less normally for a period following the parasitization event; cf. idiobiont (q.v.).

koinophytes Superceded term for koinobiont (q.v.).

labial gland Another name for salivary gland (q.v.) or silk gland.

lamellocyte A recognizable category of haemocytes involved in capsule fomation.

lamellolycin An enzyme that renders lamellocytes (q.v.) ineffective for encapsulation.

lamnium The primitively segmented distal part of a hymenopterous ovipositor (Smith, 1969, 1970); cf. radix (q.v.).

lecithal Of eggs that contain yolk (usually substantial quantities); cf. alecithal (q.v.).

life table A table detailing all the mortality factors of a population or cohort divided into age or life stage categories.

local mate competition Situation in which males mate, or compete for mates, among a small group of related individuals, usually siblings.

meconium The waste material excreted by an insect upon eclosion as an adult (a few parasitic wasps) or, in the case of the majority of parasitic wasps, just prior to pupation.

meroistic ovary An ovary in which oocytes are accompanied by nurse cells.

mesenteron Midgut.

mesosoma The thorax and 1st abdominal segment (propodeum, q.v.) of an Apocritan hymenopteran.

metapneustic Having only the last pair of abdominal spiracles present or functional; cf. apneustic (q.v.), holopneustic (q.v.).

metasoma The abdomen excluding its 1st segment (propodeum, q.v.) of an Apocritan hymenopteran.

microhymenoptera Informal term for a group of parasitic superfamilies dominated by species of small size, usually comprising the Chalcidoidea, Platygastroidea and Cynipoidea, approximating a monophyletic group, but sometimes with other taxa added such as the Stephanoidea.

micropylar appendage or **micropyle** The specialized part of an insect egg shell (chorion, q.v.) through which sperm enter.

monootene Of ovarioles that can store only a single mature egg (Flanders, 1950).

monophagic Consuming only one type of host. Usually used to mean attacking a single host species, but sometimes referring to taxa that attack only members of a particular genus. Several distinctions were proposed by Frilli (1981).

monophyletic Of clades that have a single common ancestor and include all of its descendents

mosaic An individual comprising cells of more than one genotype.

multiple parasitism or **multiparasitism** Parasitization of a single host individual by members of two or more parasitoid species.

multivoltine Having more than one generation per year; plurivoltine (q.v); cf. bivoltine (q.v.), univoltine (q.v.).

mummy The usually expanded, dried-out cuticular remains of some hosts left after an endoparasitoid has completed feeding and has pupated; the term is usually encountered in association with aphids that have been attacked by aphidiine braconids or by aphelinids, and lepidopterous larvae that have been parasitized by rogadine braconids, though a few other examples are known.

novel In biocontrol, term used to indicate a new host–parasite interaction.

nurse cell Member of the group of cystocytes (q.v.) connected directly or via other nurse cells to the developing oocyte which it helps provide with components.

obligatory multiparasitism The absolute requirement that a host has been parasitized by one species before a second parasitoid species can successfully use it as a host (Askew, 1971; see also Guzo and Stoltz, 1985)

olistheter The longitudinal tongue-and-groove mechanism which interlocks the upper and lower ovipositor (or sting) valves and comprises the aulax (q.v.) and the rhachis (q.v.)

ontogeny The sequence of developmental stages of an individual from egg to adult.

oösome A special structure that forms at the posterior pole of many developing insect eggs; in many groups they are important for the subsequent development of the germ plasm.

oösorption The resorption of (usually) mature ovarian eggs by a female.

ovariole A fully functional subunit of an ovary encased in its own membrane; most parasitic wasps have two or more ovarioles but only one occurs in some aphidiine braconids.

ovicide Deliberate killing of eggs of other individuals.

ovipositor sheaths A pair of structures arising from the distal end of gonocoxite 9 that normally protect the ovipositor proper.

ovo-larval/pupal parasitism *See* egg–larval/pupal parasitism.

panmixis Random mating/fertilization within a population.

parameres Paired appendages of the male external genitalia on either side of the aedeagus (q.v.) which are involved in clasping the female during copulation.

paraphyletic Referring to a taxon, all of whose members have a single, most recent, common ancestor, but which does not include all of the descendents of that ancestor.

parasitic Hymenoptera Used variously, sometimes including the parasitic aculeates, sometimes excluding the phytophagous groups that have evolved from parasitic ancestors.

parasitoid An insect or other organism that develops as a parasite of a single host which it (alone or with others) consistently ultimately kills.

peritreme A small distinct sclerite surrounding a spiracle, sometimes more or less complexly modified.

peritrophic membrane A secreted chitinous tube (or concentric set of tubes) that occupies the midgut and separates the food-containing lumen from direct contact with the midgut cells.

permissive Allowing development of a parasitoid.

petiolar segment and **petiole** The second abdominal segment in apocritan hymenoptera, named because of its narrow base which gives rise to the wasp-waist.

pharynx In adult wasps, the cephalic part of the foregut.

pheromone An infochemical released by one individual of a species and detected by another to the benefit of both.

placoidea An alternative name for placodiform sensillum.

placodiform sensillum An elongate shallow-domed antennal sensillum. Similar structures are found in many parasitic wasp taxa, but ultrastructural studies suggest that they are not all homologous.

placoid sensillum An alternative name for placodiform sensillum.

planidial larva *See* planidium.

planidiform larva A more general term than planidium (q.v.) recommended to be used for a range of mobile, host-seeking 1st instar larvae (Darling and Miller, 1991).

planidium A term used to describe certain free-living, active, flattened and usually legless 1st larval instars which are involved in host location (etymology from Greek 'a small wanderer'); cf. planidiform larva (q.v.) (Darling and Miller, 1991).

plantar lobes Apico-ventral membranous pad-like protuberances from the sub-apical tarsal articles of many sawflies and members of the parasitic family, Trigonalyidae.

plurivoltine Having multiple generations per year.

polar body Structure formed by the fusion of the three polar nuclei (q.v.).

polar nuclei The three nuclei that result from meiosis in oögenesis that do not take part in the zygotic genome; usually they degenerate but sometimes they are involved in the production of serosal membranes.

polydnavirus A member of the virus family Polydnaviridae, which are characterized by possession of multiple circular DNA molecules per virus particle.

polyembryony Development of from two to many larvae from a single parasitoid egg as a result of fragmentation of the polygerm (q.v.).

polygerm A mass of cells derived clonally from the egg of a polyembryonic parasitoid, each cell of which will ultimately develop into a separate larva.

polyphyletic An artificial grouping of taxa that do not share a single common closest ancestor; cf. monophyletic (q.v.), paraphyletic (q.v.).

polytene Of ovarioles that can store two or more mature eggs (Flanders, 1950).

polytrophic An ovariole in which more than one nurse cell is involved in maturing an oöcyte (King and Ratcliffe, 1969).

prepupa The immobile last instar larva prior to pupation.

primary host The host directly attacked (consumed) by a parasitoid.

primary parasitoid A parasitoid that survives in and gains all its nourishment from its primary host; cf. hyperparasitoid (q.v.), secondary parasitoid (q.v.).

proctodaeum Hindgut.

pronucleus Either the oöcyte nucleus that will fuse with the sperm nucleus, or the sperm nucleus in the fertilized egg before fusion.

pro-ovigenic Of species that have matured all their eggs at the time of eclosion, or very soon thereafter (Flanders, 1950); cf. synovigenic (q.v.).

propodeum The specially modified 1st abdominal tergum in apocritan Hymenoptera.

protelean parasite Insect in which only the larval stage is parasitic.

prothoracic gland Endocrine gland in insects that secretes ecdysone.

proventriculus The short, narrow, posterior region of the foregut which forms a valve separating fore and midgut.

pseudogerm Groups of cells derived from the paranucleus of the egg, or by fragmentation of the trophamnium (q.v.) in some polyembryonic species (Tremblay, 1966; Salt, 1968).

pseudohyperparasitoid A species that parasitizes another parasitoid after the latter has already completed development in (or on) and killed its primary host.

pseudoparasitoid Term used for instances in which the parasitoid does not develop in a potential host individual for any of a variety of reasons, but either kills it or castrates it (also referred to as apparent parasitism); in experimental work, referring to situations in which a female parasitoid is allowed to envenomate a host but is prevented from laying an egg.

pseudopodia Fleshy protruberances from the body segments of some parasitoid larvae.

pseudovirginity Females that have paired but have no sperm in their spermathecae and therefore can only lay male eggs. Can result from interrupted copulation or depletion of sperm store in species that mate only once.

pterostigma A usually conspicuous, swollen vein/cell complex at the antero-distal margin of the forewing.

pygostyles Another name for cerci (q.v.).

quasi-gregarious A term applied to taxa that lay a single egg in each host but whose hosts typically form dense patches, e.g. parasitoids of eggs that are laid in clusters; cf. gregarious (q.v.).

radix The basal, unsegmented part of the hymenopteran ovipositor (Smith, 1969, 1970); cf. lamnium (q.v.).

rectal glands, rectal pads Discrete, specialized areas on the rectum of many adult insects that are involved in fluid and ion balance.

retroelement A large group of RNA viruses and other mobile genetic elements that includes retroviruses and retrotransposons (q.v.), and which can insert their genome into that of the their host.

retrotransposon A retroelement (q.v.) that lacks envelope genes.

reunion biocontrol The bringing back into contact of a host and its parasitoid (or other pest/enemy pair) after a period of separation; for example, after a pest has previously been introduced into a new country without its natural enemies.

revision A taxonomic study of a group of organisms in which all species are treated, described and keyed as necessary.

rhachis The T-section, tongue-like longitudinal ridge on either side of the upper ovipositor valve (gonapophyses of 9th abdominal segment) that interlocks with the aulax (q.v.) forming the olistheter mechanism (q.v.).

saprophagous Of organisms that feed on dead or decaying matter.

secondary host The host of the primary host in hyperparasitism situations; cf. primary host.

secondary parasitoid A parasitoid of an insects primary parasitoid (q.v.); a hyperparasitoid of that insect.

semigregarious Facultatively gregarious; that is, capable of choosing whether to be a solitary or gregarious parasite, depending on the situation.

seminal glomerulus A differentiated, coiled part of the vas deferens in the male internal reproductive system where sperm are stored prior to transfer to the female.

seminal receptacle A differentiated, swollen part of the vas deferens in the male internal reproductive system where sperm are stored prior to transfer to the female.

semiochemicals Any chemicals that carry information about the emitter to a receiver.

serosa The layer of extraembryonic tissue that surrounds the egg (Dahlman, 1990).

sex ratio By convention, the ratio of male to female offspring.

socii Another name for cerci (q.v.).

solitary parasitoid A species that routinely or obligatorily develops only a single individual within a single host.

spermatheca A female internal reproductive organ that arises as an outpocketing of the genital duct and which is used to store sperm.

sperone A pre-apical, medial spur-like process arising from the ventral floor of the dorsal ovipositor valve in some chalcidoids (Zinna, 1960).

stem cell Germ-line cell that divides to produce cytoblasts which subsequently divide to produce the nurse cell (q.v.) cluster and oöcyte (= primary oögonia).

stomodaeum Foregut.

stylet Another name for the lower ovipositor valve, especially in the case of the aculeate Hymenoptera (see Table 6.1).

superparasitism Attack of a host by more than one primary parasitoid resulting in the laying of more than one, or more than the normal complement of, egg(s) in it – the term as originally proposed by Fiske (1910) did not distinguish between self-superaparsitism or superparasitism by two individuals (Chacko, 1964).

synapomorphy A shared derived character state indicating common descent.

synkary Fusion of the male and female pronuclei in a fertilized egg.

synomone Any chemical substance produced by, acquired by, or produced as a result of an action of, an individual of one species (organism A), that evokes a response in a member of another species (organism B) which will be beneficial to both organism A and organism B (Nordlund, 1981); cf. allomone (q.v.), kairomone (q.v.).

synovigenic Of an organism which continuously matures its eggs throughout its adult life (Flanders, 1950; cf. pro-ovigenic, q.v.).

teratocytes Cells derived from the extraembryonic membranes of some endoparasitic species which separate and continue to grow within the host.

teratoid larva Another name for precocious larva.

Terebrantia The parasitic Hymenoptera excluding the parasitic aculeates.

terminal filament An often thin, connective tissue thread that connects the anterior ends of the ovaries (and testes) to the anterior metasomal body wall.

tertiary host The host of the host of a parasitoid.

tertiary parasitoid A hyper-hyperparasitoid; a parasitoid of a hyperparasitoid (q.v.); a secondary parasitoid (q.v.) of a primary parasitoid (q.v.).

thelytoky A form of parthenogenesis in which unfertilized eggs always give rise to female offspring (Luck *et al.* 1992).

thylacium A sac formed from moulted larval skins of dryinid larvae which adhere to the host's integument and form a protective case for that part of the parasitoid larva that is external to the host (Olmi, 1994).

trophamnion A layer of cells, surrounding an egg and (usually) subsequently the embryo and even the larval instars, which are derived from the serosal membrane (q.v.) and may give rise to teratocytes (q.v.).

trophic cells Another name for both teratocytes (q.v.) and nurse cells (q.v.).

trophocyte Also called nurse cell (q.v.).

univoltine Having only one generation per year.

vagina A term frequently employed as a synonym of the common oviduct, but restricted by some to refer just to the posterior part of the latter which receives the aedeagus during copulation.

valve The dorsal or paired ventral components of the ovipositor; sometimes used for the valvillus (q.v.) homologues in the Aculeate sting.

valvifers Female genital sclerites from which the ovipositor and ovipositor sheaths arise.

valvillus Articulated flaps arising from the lower ovipositor or sting valves and protruding into the egg or poison canal in many ichneumonoids and aculeates (also called Hemmplätchen, cogs, valves).

valvulae 1 Lower ovipositor valves (see Table 6.1).

valvulae 2 Upper ovipositor valves (see Table 6.1).

valvulae 3 Ovipositor sheaths (see Table 6.1).

vesiculate larva A typically early instar larva that has the rectum or whole proctodaeum evaginated to give the rear end a vesiculate appearance.

vitellarium The main part of the ovariole in which the oöcyte develops and acquires yolk.

vitelline membrane A membrane, surrounding the oöcyte, secreted by follicle cells before secretion of the chorion.

vitellogenesis Formation of the yolk of an egg.

References

Abbott, B.D. and Grosch, D.S. (1984) Developmental anomalies in *Habrobracon hebetor* exposed to volatilized agents. *Ann. Entomol. Soc. Am.* **77**, 597–603.

Abbott, B.D. and Grosch, D.S. (1987) Antennal bud development in *Bracon hebetor* (Hymenoptera: Braconidae) examined by light and electron microscopy during the third and fourth instars. *Ann. Entomol. Soc. Am.* **80**, 353–360.

Abbott, C.E. (1934) How *Megarhyssa* deposits her eggs. *J. N.Y. Entomol. Soc.* **42**, 127–133.

Abe, Y. and Koyama, K. (1991) Embryonic development and selective oviposition of a dryinid wasp, *Hoplogonatopus atratus* Esaki et Hashimoto (Hymenoptera: Dryinidae). *Jpn. J. Appl. Ent. Zool.* **35**, 57–63. (In Japanese.)

Achtelig, M. and Krause, G. (1971) Experimente am ungefurchten Ei von *Pimpla turionellae* L. (Hymenoptera) zur Funktionsanalyse des Oosombereichs. *Wilhelm Roux arch. EntwMech. Org.* **167**, 164–182.

Achterberg, C. van (1977a) The function of swarming in *Blacus* species (Hymenoptera, Braconidae, Helconinae). *Ent. Ber. Amst.* **37**, 151–152.

Achterberg, C. van (1977b) Sensory bristle-fields of the petiolar segment in some Hymenoptera. *Entomol. Bericht.* **37**, 101–102.

Achterberg, C. van (1986) The oviposition behaviour of parasitic Hymenoptera with very long ovipositors (Ichneumonoidea: Braconidae). *Ent. Ber. Amst.* **46**, 113–115.

Achterberg, C. van (1989) *Pheloura* gen. nov., a neotropical genus with an extremely long pseudo-ovipositor (Hymenoptera: Braconidae). *Ent. Ber. Amst.* **49**, 105–108.

Adams, J., Rothman, E.D., Kerr, W.E. and Paulino, Z.L. (1977) Estimation of the number of sex alleles and queen matings from diploid male frequencies in a population of *Apis mellifera*. *Genetics* **86**, 581–596.

Aeschlimann, J.-P. (1974) Biologie et comportement de *Chorinaeus funebris* Gravenhorst (Hymenoptera: Ichneumonidae). *Ann. Zool.-Ecol. Animale* **6**, 529–538.

Aeschlimann, J.-P. (1990) Simultaneous occurrence of thelytoky and bisexuality in Hymenopteran species, and its implications for the biological control of pests. *Entomophaga* **35**, 3–5.

Agelopoulos, N.G. and Keller, M.A. (1994) Plant natural enemy association in tritrophic system, *Cotesia rubecula* – *Pieris rapae* – Brassicaceae (Cruciferae). 3. Collection and identification of plant and frass volatiles. *J. Chem. Ecol.* **20**, 1955–1967.

Alauzet, C. (1987) Bioecologie de *Eubazus semirugosus, Coeloides abdominalis* et *C. sordidator* (Hym.: Braconidae) parasites de *Pissodes notatus* (Col.: Curculionidae) dans le sud de la France. *Entomophaga*, **32**, 39–47.

Alcock, J. and Gwynne, D.T. (1987) Courtship feeding and mate choice in thynnine wasps (Hymenoptera: Tiphiidae). *Aust. J. Zool.* **35**, 451–458.

Aldrich, J.R. (1995) Chemical communication in the true bugs and parasitoid exploitation. In *Chemical Ecology of Insects 2* (eds R.T. Cardé and W.J. Bell), Chapman & Hall, New York, pp. 318–363.

Aldrich, J.R., Kochansky, J.P. and Abrams, C.B. (1984) Attractant for a beneficial insect and its parasitoids: Pheromones of the predatory spined soldier bug, *Podisus maculiventris* (Hem.: Pentatomidae). *Environ. Entomol.* **13**, 1031–1036.

Alford, D.V. (1968) The biology and immature stages of *Syntretus splendidus* (Marshall) (Hymenoptera: Braconidae, Euphorinae), a parasite of adult bumblebees. *Trans R. Entomol. Soc. Lond.* **120**, 375–393.

Allen, G.R. (1990) Influence of host behavior and host size on the success of oviposition of *Cotesia urabae* and *Dolichogenidea eucalypti* (Hymenoptera: Braconidae). *J. Insect Behav.* **3**, 733–749.

Allen, G.R., Kazmer, D.J. and Luck, R.F. (1994) Postcopulatory male behaviour, sperm precedence and multiple mating in a solitary parasitoid wasp. *Anim. Behav.* **48**, 635–644.

Allen, H.W. and Brunson, M.H. (1945) A microsporidian in *Macrocentrus ancylivorus*. *J. Econ. Ent.* **38**, 393.

Alphen, J.J.M. van and Janssen, A.R.M. (1982) Host selection by *Asobara tabida* (Braconidae; Alysiinae), a larval parasitoid of fruit inhabiting *Drosophila* species. II. Host species selection. *Neth. J. Zool.* **32**, 215–231.

Alphen, J.J.M. van and Visser, M.E. (1990) Superparasitism as an adaptive strategy in insect parasitoids. *Annu. Rev. Entomol.* **35**, 59–79.

Alvi, S.M. and Momoi, S. (1994) Environmental regulation and geographical adaptation of diapause in *Cotesia plutellae* (Hymenoptera: Braconidae), a parasitoid of the diamondback moth larvae. *Appl. Entomol. Zool.* **29**, 89–95.

Amos, W.B. and Salt, G. (1974) An atlas of the development of eggs of an ichneumon wasp. *J. Entomol.* (B) **43**, 11–18.

Amy, R.L. (1961) The embryology of *Habrobracon juglandis* (Ashmead). *J. Morphol.* **109**, 199–217.

Andreadis, T.G. (1980) *Nosema pyrausta* infection in *Macrocentrus grandi*, a braconid parasite of the European corn borer, *Ostrinia nubilalis*. *J. Invert. Pathol.* **35**, 229–233.

Antolin, M.F. (1992) Sex ratio variation in a parasitic wasp. II. Diallele cross. *Evolution* **46**, 1511–1524.

Antolin, M.F. and Strand, M.R. (1992) Mating system of *Bracon hebetor* (Hymenoptera: Braconidae). *Ecol. Ent.* **17**, 1–7.

Antolin, M.F. and Strong, D.R. (1987) Long-distance dispersal by a parasitoid (*Anagrus delicatus*, Mymaridae) and its host. *Oecologia* **73**, 288–292.

Antolin, M.F., Ode, P.J. and Strand, M.R. (1995) Variable sex-ratios and ovicide in an outbreeding parasitic wasp. *Anim. Behav.* **49**, 589–600.

Arakaki, N., Wakamura, S. and Yasuda, T. (1996) Phoretic egg parasitoid, *Telenomus euproctidis* (Hymenoptera, Scelionidae), uses sex pheromone of tussock moth *Euproctidis taiwana* (Lepidoptera, Lymantriidae) as a kairomone. *J. Chem. Ecol.* **22**, 1079–1085.

Arakawa, R. (1987) Attack on the parasitized host by a primary solitary parasitoid, *Encarsia formosa* (Hymenoptera: Aphelinidae): The second female pierces with her ovipositor the egg laid by the first one. *Appl. Ent. Zool.* **22**, 644–645.

Arthur, A.P. (1961) The cleptoparasitic habits and the immature stages of *Eurytoma pini* Bugbee (Hymenoptera: Chalcidae), a parasite of the European pine shoot moth, *Rhyacionia buoliana* (Schiff.) (Lepidoptera: Olethreutidae). *Canad. Entomol.* **93**, 655–660.

Arthur, A.P. and Ewen, Al.B. (1975) Cuticular encystment: A unique and effective defense reaction by cabbage looper larvae against parasitism by *Banchus flavescens* (Hymenoptera: Ichneumonidae). *Ann. Entomol. Soc. Am.* **68**, 1091–1094.

Arthur, A.P. and Mason, P.G. (1986) Life history and immature stages of the parasitoid *Microplitis mediator* (Hymenoptera: Braconidae), reared on the Bertha armyworm *Mamestra configurata* (Lepidoptera: Noctuidae). *Canad. Entomol.* **118**, 487–491.

Arthur, A.P., Stainer, J.E.R. and Turnbull, A.L. (1964) The interaction between *Orgilus obscurator* (Nees) (Hymenoptera: Braconidae) and *Temelucha interruptor* (Grav.) (Hymenoptera: Ichneumonidae), parasites of the pine shoot moth, *Rhyacionia buoliana* (Schiff.) (Lepidoptera: Olethreutidae). *Canad. Entomol.* **96**, 1030–1034.

Arthur, A.P., Hegdekar, B.M. and Batsch, W.W. (1972) A chemically defined synthetic medium that induces oviposition in the parasite *Itoplectis conquisitor* (Hymenoptera: Ichneumonidae). *Canad. Entomol.* **104**, 1251–1258.

Asgari, S. and Schmidt, O. (1994) Passive protection of eggs from the parasitoid, *Cotesia rubecula*, in the host, *Pieris rapae*. *J. Insect Physiol.* **40**, 789–795.

Askew, R.R. (1961) The biology of the British species of the genus *Olynx* Forster (Hymenoptera: Eulophidae), with a note on seasonal colour forms in the Chalcidoidea. *Proc. R. Entomol. Soc., Lond (A)* **36**, 103–112.

Askew, R.R. (1968) Considerations on speciation in Chalcidoidea (Hymenoptera). *Evolution* **22**, 642–645.

Askew, R.R. (1971) *Parasitic Insects*, Heinemann Educational Books Ltd, London. 316 pp.

Askew, R.R. (1990) Species diversity of hymenopteran taxa in Sulawesi. In *Insects and the Rain Forests of South East Asia (Wallacea)* (eds W.J. Knight and J.D. Holloway), Royal Entomological Society, London. pp. 255–260.

Askew, R.R. and Shaw, M.R. (1986) Parasitoid communities: Their size, structure and development. In *Insect Parasitoids* (eds J. Waage and D. Greathead), Academic Press, London. pp. 225–264.

Assem, J. van den (1974) Male courtship patterns and female receptivity signal of Pteromalinae (Hym., Pteromalidae), with a consideration of some evolutionary trends and a comment on the taxomic position of *Pachycrepoideus vindemiae*. *Neth. J. Zool.* **24**, 253–278.

Assem, J. van den [(1975)/1976] Temporal patterning of courtship behaviour in some parasitic Hymenoptera, with special reference to *Melittobia acasta*. *J. Entomol.*, Series A **50**, 137–146.

Assem, J. van den (1986) Mating behaviour in parasitic wasps. In *Insect Parasitoids* (eds J. Waage and D. Greathead), Academic Press, London. pp. 137–167.

Assem, J. van den and Feuth-de Bruin, E. (1977) Second matings and their effect on the sex ratio of the offspring in *Nasonia vitripennis* (Hymenoptera: Pteromalidae). *Entomol. Exp. Appl.* **21**, 23–28.

Assem, J. van den and Jachmann, F. (1982) The coevolution of receptivity signalling and body-size dimorphism in the Chalcidoidea. *Behaviour*, **80**, 96–105.

Assem, J. van den and Povel, G.D. (1973) Courtship behaviour of some *Muscidifurax* species (Hym., Pteromalidae): A possible example of a recently evolved ethological isolating mechanism. *Neth. J. Zool.* **23**, 465–487.

Assem, J. van den and Putters, F.A. (1980) Patterns of sound produced by courting chalcidoid males and its biological significance. *Entomol. Exp. Appl.* **27**, 293–302.

Assem, J. van den and Werren, J.H. (1994) A comparison of the courtship and mating behaviour of three species of *Nasonia* (Hymenoptera, Pteromalidae). *J. Insect Behav.* **7**, 53–66.

Assem, J. van den, Gijswijt, M.J. and Nubel, B.K. (1982) Characteristics of courtship and mating behaviour used as classificatory criteria in Eulophidae–Tetrastichinae, with special reference to the genus *Tetrastichus*. *Tijdschr. Ent.* **125**, 205–220.

Assem, J. van den, Iersal, J.A. van and Los-den Hartogh, R.L. (1989) Is being large more important for female than male parasitic wasps? *Behaviour* **108**, 160–195.

Aubert, J.-F. (1959) Biologie de quelques Ichneumonidae Pimplinae et examen critique de la theorie de Dzierzon. *Entomophaga* **4**, 1–188.

Auger, J., Lecomte, C., Paris, J. and Thibout, E. (1989) Identification of leek-moth and diamondback-moth frass volatiles that stimulate parasitoid, *Diadromus pulchellus*. *J. Chem. Ecol.* **15**, 1391–1398.

Austin, A.D. (1983) Morphology and mechanics of the ovipositor system of *Ceratobaeus* Ashmead (Hymenoptera: Scelionidae) and related genera. *Int. J. Insect Morphol. Embryol.* **12**, 139–155.

Austin, A.D. (1984) The fecundity, development and host relationships of *Ceratobaeus* spp. (Hymenoptera: Scelionidae), parasites of spider eggs. *Ecol. Ent.* **9**, 125–138.

Austin, A.D. (1985) The function of spider egg sacs in relation to parasitoids and predators, with special reference to the Australian fauna. *J. Nat. Hist.* **19**, 359–376.

Austin, A.D. (1989) Revision of the genus *Buluka* de Saeger (Hymenoptera: Braconidae: Microgastrinae). *Syst. Ent.* **14**, 149–163.

Austin, A.D. and Browning, T.O. (1981) A mechanism for movement of eggs along insect ovipositors. *Int. J. Insect Morphol. Embryol.* **10**, 93–108.

Baaren, J. van and Nénon, J.P. (1994) Factors involved in host discrimination by *Epidinocarsis lopezi* and *Leptomastix dactylopii* (Hym., Encyrtidae). *J. Appl. Ent.* **118**, 76–83.

Baaren, J. van, Boivin, G. and Nénon, J.P. (1994) Intraspecific and interspecific host discrimination in two closely-related egg parasitoids. *Oecologia* **100**, 325–330.

Baaren, J. van, Boivin, G. and Nénon, J.P. (1995) Intraspecific hyperparasitism in a primary hymenopteran parasitoid. *Behav. Ecol. Sociobiol.* **36**, 237–242.

Baccetti, B. (1958) Ghiandole labialie fabbricazione del bozzolo negli Imenotteri ricerche Ichneumonidi e Braconidi. *Redia* **43**, 215–293.

Bacher, S., Casas, J. and Dorn, S. (in press) Parasitoid vibrations as potential releasing stimulus of evasive behaviour in a leafminer. *Physiol. Ent.*

Baehrecke, E.H. and Strand, M.E. (1990) Embryonic morphology and growth of the polyembryonic parasitoid *Copidosoma floridanum* (Ashmead) (Hymenoptera: Encyrtidae). *Int. J. Insect Morphol. Embryol.* **19**, 165–175.

Baehrecke, E.H., Grbic, M. and Strand, M.R. (1992) Serosa ontogeny in two embryonic morphs of *Copidosoma floridanum*: The influence of host hormones. *J. Exp. Zool.* **262**, 30–39.

Bahadur, J. and Reddy, K.K. (1966) Structure and function of the rectal pads in Hymenoptera. *Zool. Anz.* **178**, 262–268.

Baker, J.E. (1995) Stability of malathion resistance in two hymenopterous parasitoids. *J. Econ. Ent.* **88**, 232–236.

Bakker, K., Alphen, J.J.M. van, Batenburg, F.H.D. van *et al.* (1985) The function of host discrimination and superparasitization in parasitoids. *Oecologia* **67**, 572–576.

Baldwin, W.F. (1969) Congenital body malformation and eye-color mutations in progeny from irradiated female wasps (*Dahlbominus fuscipennis*: Hym., Chalcidoidea). *Rad. Res.* **38**, 569–579.

Baldwin, W.F. (1972) Linear dose–effect response for mutations in mature oocytes of the wasp (*Dahlbominus*). *Rad. Res.* **49**, 190–196.

Baldwin, W.F., Shaver, E. and Wilkes, A. (1964) Mutants of the parasitic wasp *Dahlbominus fuscipennis* (Zett.) (Hymenoptera: Eulophidae). *Canad. J. Genet. Cytol.* **6**, 453–466.

Balfour-Browne, E. (1922) On the life-history of *Melittobia acasta* Walker; a chalcid parasite of bees and wasps. *Parasitology* **14**, 349–370.

Barbier, R. (1975) Différenciation de structures ciliaires et mise en place des canaux au cours de l'organogenèse des glandes collétériques de *Galleria mellonella* L. (Lépidoptère, Pyralide). *J. Microscopie. Biol. Cell.* **24**, 315–326.

Barbosa, P. and Frongillo, E.A. Jr (1979) Photoperiod and temperature influences on egg number in *Brachymeria intermedia* (Hymenoptera: Chalcididae), a pupal parasitoid of *Lymantria dispar* (Lepidoptera: Lymantriidae). *J. New York Entomol. Soc.* **87**, 175–180.

Barbosa, P., Saunders, J.A., Kemper, J. *et al.* (1986) Plant allelochemicals and insect parasitoids. Effects of nicotine on *Cotesia congregata* (Say) (Hymenoptera: Braconidae) and *Hyposoter annulipes* (Cresson) (Hymenoptera: Ichneumonidae). *J. Chem. Ecol.* **12**, 1319–1328.

Barbosa, P., Gross, P. and Kemper, J. (1991) Influence of plant allelochemicals on the tobacco hornworm and its parasitoid, *Cotesia congregata*. *Ecology* **72**, 1567–1575.

Barlin, M.R. and Vinson, S.B. (1981a) Multiporous plate sensilla in the antenna of the Chalcidoidea (Hymenoptera). *Int. J. Insect Morphol. Embryol.* **10**, 29–42.

Barlin, M.R. and Vinson, S.B. (1981b) The multiporous plate sensillum and its potential use in braconid systematics (Hymenoptera: Braconidae). *Canad. Entomol.* **113**, 931–938.

Barrass, R. (1960) The courtship behaviour of *Mormoniella vitripennis* Walk. (Hymenoptera, Pteromalidae). *Behaviour* **15**, 185–209.

Barrass, R. (1961) A quantitative study of the behaviour of the male *Mormoniella vitripennis* towards two constant stimulus situations. *Behaviour* **18**, 288–312.

Barrera, J.F., Gomez, J. and Alauzet, C. (1994) Evidence for a marking pheromone in host discrimination by *Cephalonomia stephoderis* (Hym.: Bethylidae). *Entomophaga* **39**, 363–366.

Bartlett, B.R. (1964) Patterns in the host feeding habit of adult parasitic Hymenoptera. *Ann. Entomol. Soc. Am.* **57**, 344–350.

Basibuyuk, H.H. and Quicke, D.L.J. (1994) Evolution of antennal cleaner structure in the Hymenoptera (Insecta). *Norw. J. Agric. Sci. Suppl.* **16**, 199–206.

Basibuyuk, H.H. and Quicke, D.L.J. (1995) Morphology of the antenna cleaner in the Hymenoptera with particular reference to non-aculeate families (Insecta). *Zool. Scripta* **24**, 157–177.

Basibuyuk, H.H. and Quicke, D.L.J. (in press) Hamuli in the Hymenoptera (Insecta) and their phylogenetic implications. *J. Nat. Hist.*

Bate, C.M. (1973) The mechanism of the pupal gin trap. I. Segmental gradients and the connections of the triggering sensilla. *J. Exp. Biol.* **59**, 95–108.

Battaglia, D., Pennacchio, F., Marincola, G. and Tranfaglia, A. (1993) Cornicle secretion of *Acyrthosiphon pisum* (Homoptera, Aphididae) as a contact kairomone for the parasitoid *Aphidius ervi* (Hymenoptera, Braconidae). *Eur. J. Entomol.* **90**, 423–428.

Baumann, C. (1924) Uber den Bau des Abdomens und die Funktion des Legeapparates von *Thalesa leucographa* Grav. *Zool. Anz.* **58**, 149–162.

Baur, M.E. and Yeargan, K.V. (1994a) Developmental stages and kairomones from the primary parasitoid *Cotesia marginiventris* (Hymenoptera: Braconidae) affect the response of the hyperparasitoid *Mesochorus discitergus* (Hymenoptera: Ichneumonidae) to parasitized caterpillars. *Ann. Entomol. Soc. Am.* **87**, 954–961.

Baur, M.E. and Yeargan, K.V. (1994b) Behavioural interactions between the hyperparasitoid *Mesochorus discitergus* (Hymenoptera: Ichneumonidae) and four species of noctuid caterpillars: Evasive tactics and capture efficiency. *J. Entomol. Sci.* **29**, 420–427.

Beard, R. L. (1952) The toxicity of *Habrobracon* venom: A study of a natural insecticide. *Connecticut Exp. St. Bull.* **562**, 3–27.

Beard, R.L. (1964) Pathogenic stinging of housefly pupae by *Nasonia vitripennis* (Walker). *J. Insect. Pathol.* **6**, 1–7.

Beard, R.L. (1971) Production and use of venom by *Bracon brevicornis* (Wesm.). In *Toxins of Animal and Plant Origin*, Vol. 1 (eds A. de Vries and E. Kockva), Gordon and Breach, New York. pp. 181–190.

Beard, R. L. (1972) Effectiveness of paralysing venom and its relation to host discrimination by braconid wasps. *Ann. Entomol. Soc. Am.* **65**, 90–93.

Beard, R.L. (1978) Venoms of the Braconidae. In *Handbuch der experimentellen Pharmakologie, Volume 48: Arthropod Venoms* (ed. S. Bettini) Springer-Verlag, Berlin. pp. 773–800.

Beaver, R.A. (1966) The biology and immature stages of *Entedon leucogramma* (Ratzeburg) (Hymenoptera: Eulophidae), a parasite of bark beetles. *Proc. R. Entomol. Soc. Lond.* (A) **41**, 37–41.

Beckage, N.E. (1985) Endocrine interactions between endoparasitic insects and their hosts. *Annu. Rev. Entomol.* **30**, 371–413.

Beckage, N.E. (1991) Host–parasite hormonal relationships: A common theme? *Exp. Parasitol.* **72**, 332–338.

Beckage, N.E. and DeBuron, I. (1994) Extraembryonic membranes of the endoparasitic wasp *Cotesia congregata*: Presence of a separate amnion and serosa. *J. Parasitol.* **80**, 389–396.

Beckage, N.E. and Riddiford, L.M. (1982) Effects of parasitism by *Apanteles congregatus* on the endocrine physiology of the tobacco hornworm *Manduca sexta*. *Gen. Comp. Endocrinol.* **47**, 308–322.

Beckage, N.E. and Templeton, T.J. (1986) Physiological effects of parasitism by *Apanteles congregatus* in terminal-stage tobacco hornworm larvae. *J. Insect Physiol.* **32**, 299–314.

Beckage, N.E., Metcalf, J.S., Nesbit, D.J. *et al.* (1990) Host hemolymph monophenoloxidase activity in parasitized *Manduca sexta* larvae and evidence for inhibition by wasp polydnavirus. *Insect Biochem.* **20**, 285–294.

Beckage, N.E., Tan, F.F., Schleifer, K.W. *et al.* (1994) Characterization and biological effects of *Cotesia congregata* polydnavirus on host larvae of the tobacco hornworm, *Manduca sexta*. *Arch. Insect Biochem. Physiol.* **26**, 165–195.

Beeson, C.F.C. and Chatterjee, S.N. (1935) On the biology of the Braconidae (Hymenopt.). *Indian For. Records (N.S.)* **1**, 105–138.

Bekkaoui, A. and Thibout, E. (1992) Rôle de substances cuticulaires non volatiles d'*Acrolepiopsis assectella* (Lep.: Hyponomeutoidea) dans la reconnaissance de l'hôte par les parasitoïdes *Diadromus pulchellus* et *D. collaris* (Hym.: Ichneumonidae). *Entomophaga* **37**, 627–639.

Bekkaoui, A. and Thibout, E. (1993) Role of the cocoon of *Acrolepiopsis assectella* (Lep., Hyponomeutoidea) in host-recognition by the parasitoid *Diadromus pulchellus* (Hym., Ichneumonidae). *Entomophaga* **38**, 101–113.

Bell, W.J. and Bohm, M.K. (1975) Oosorption in insects. *Biol. Rev.* **50**, 373–396.

Belshaw, R. and Quicke, D.L.J. (in press) A molecular phylogeny of the Adidiinae (Hymenoptera: Braconidae), *Mol. Phylogenet. Evol.*

Bender, J.C. (1943) Anatomy and histology of the female reproductive organs of *Habrobracon juglandis* (Ashmead) (Hym., Braconidae). *Ann. Entomol. Soc. Am.* **36**, 537–545.

Benn, M., DeGrave, J., Gnanasunderam, C. and Hutchins, R. (1979) Host plant pyrrolizidine alkaloids in *Nyctemera annulata* Boisduval: Their persistence through the life-cycle and transfer to a parasite. *Experientia* **35**, 731–732.

Bennett, A.W. and Sullivan, D.J. (1978) Defensive behaviour against tertiary parasitism by the larva of *Dendrocerus carpenteri*, an aphid hyperparasitoid. *J. New York Entomol. Soc.* **86**, 153–160.

Bennett, F.D. (1993) Do introduced parasitoids displace native ones? *Florida Ent.* **76**, 54–63.

Benson, R.B. (1944) Swarming flight of *Blacus tripudians* Haliday. *Entomol. Mon. Mag.* **80**, 21.

Bentz, J.-A., Reeves, J. III, Barbosa, P. and Francis, B. (1996) The effect of nitrogen fertilizer applied to *Euphorbia pulcherrima* on the parasitization of *Bemisia argentifolii* by the parasitoid *Encarsia formosa*. *Entomol. Exp. Appl.* **78**, 105–110.

Berg, E. van der, Prinsloo, G.L. and Neser, S. (1990) An unusual host association: *Aprostocetus* sp. (Eulophidae), a hymenopterous predator of the nematode *Subanguina mobilis* (Chit and Fischer, (1975) Brzeski, (1981) (Anguinidae)). *Phytophylactica*, **22**, 125–127.

Berkelhamer, R.C. (1983) Intraspecific genetic variation and haplodiploidy, eusociality, and polygyny in the Hymenoptera. *Evolution*, **37**, 540–545.

Berland, L. (1951) Superfamille des Ichneumonidae. In *Traité de Zoologie* (ed. P. Grassé) **10**, 773–975.

Beukeboom, L.W. (1994) Phenotypic fitness effects of the selfish B-chromosome, paternal sex-ratio (PSR) in the parasitic wasp *Nasonia vitripennis*. *Evol. Ecol.* **8**, 1–24.

Beukeboom, L.W. and Werren, J.H. (1992) Population genetics of a parasitic chromosome – Experimental analysis of PSR in subdivided populations. *Evolution* **46**, 1257–1268.

Bigot, Y., Hamelin, M.H. and Periquet, G. (1991) Molecular analysis of the genomic organization of the Hymenoptera *Diadromus pulchellus* and *Eupelmus vuilleti*. *J. Evol. Biol.* **4**, 541–556.

Bigot, Y., Hamelin, M.H., Capy, P. and Periquet, G. (1994) Mariner-like elements in hymenopteran species – Insertion site and distribution. *Proc. Natn. Acad. Sci. USA* **91**, 3408–3412.

Bigot, Y., Drezen, J.M., Sizaret, P.Y. *et al.* (1995) The genome segments of DPRV, a commensal reovirus of the wasp *Diadromus pulchellus* (Hymenoptera). *Virology*, **210**, 109–119.

Bilinski, S. (1991a) Are accessory nuclei involved in the establishment of developmental gradients in hymenopteran oocytes? *Wilhelm Roux Arch. Dev. Biol.* **199**, 423–426.

Bilinski, S. (1991b) Morphological markers of anteroposterior and dorsoventral polarity in developing oocytes of the hymenopteran, *Cosmoconus meridionator* (Ichneumonidae). *Wilhelm Roux Arch. Dev. Biol.* **200**, 330–335.

Bilinski, S., Klag, J. and Kubrakiewicz, J. (1993) Morphogenesis of accessory nuclei during final stages of oögenesis in *Cosmoconus meridionator* (Hymenoptera, Ichneumonidae). *Wilhelm Roux Arch. Devel. Biol.* **203**, 100–103.

Bin, F. (1981) Definition of female antennal clava based on its plate sensilla in Hymenoptera Scelionidae Telenominae. *Redia* **64**, 245–261.

Bin, F. and Vinson, S.B. (1986) Morphology of the antennal sex gland in male *Trissolcus basalis* (Woll.) (Hymenoptera: Scelionidae), an egg parasitoid of the green stink bug, *Nezara viridula* (Hemiptera: Pentatomidae). *Int. J. Insect Morphol. Embryol.* **15**, 129–138.

Bin, F., Strand, M.R. and Vinson, S.B. (1986) Antenna structures and mating behavior in *Trissolcus basalis* (Woll.) (Hym.: Scelionidae), egg parasitoid of the green stink bug. In *Trichogramma and Other Egg Parasites. 2nd International Symposium,*

Guangzhou (China), Nov. 10–15, 1986 (ed. INRA), Les Colloques de l'INRA No. 43, Paris, pp. 144–151.

Bin, F., Colazza, S., Isidoro, N. *et al.* (1989) Antennal chemosensilla and glands, and their possible meaning in the reproductive behavior of *Trissolcus basalis* (Woll.) (Hym.: Scelionidae). *Entomologica* **24**, 33–97.

Bischof, C. (1996) Effects of heavy metal stress on free amino acids in the haemolymph and proteins in haemolymph and total body tissue of *Lymantria dispar* larvae parasitized by *Glyptapanteles liparidis*. *Entomol. Exp. Appl.* **79**, 61–68.

Blackburn, T.M. (1991) A comparative examination of lifespan and fecundity in the parasitoid Hymenoptera. *J. Anim. Ecol.* **60**, 151–164.

Blackwell, J. and Weih, M.A. (1980) Structure of chitin protein complexes: Ovipositor of the ichneumon fly *Megarhyssa*. *J. Mol. Biol.* **137**, 49–60.

Blass, S. and Ruthmann, A. (1989) Fine structure of the accessory glands of the female genital tract of the ichneumonid *Pimpla turionellae* (Insecta, Hymenoptera). *Zoomorphology* **108**, 367–377.

Bledowski, R. and Krainska, M.K. (1926) Die Entwicklung von *Banchus femoralis* Thoms. (Hymenoptera, Ichneumonidae). *Varsaviae* **16**, 1–5 + Tables I–VIII.

Blumberg, D. and Luck, R.F. (1990) Differences in the rates of superparasitism between two strains of *Comperiella bifasciata* (Howard) (Hymenoptera: Encyrtidae) parasitizing California red scale (Homoptera: Diaspididae): An adaptation to circumvent encapsulation? *Ann. Entomol. Soc. Am.* **83**, 591–597.

Boato, A. and Battisti, A. (1996) High genetic variability despite haplodiploidy in primitive sawflies of the genus *Cephalcia* (Hymenoptera: Pamphiliidae). *Experientia* **52**, 516–521.

Bocchino, F.J. and Sullivan, D.J. (1981) Effects of venoms from two aphid hyperparasitoids, *Asaphes lucens* and *Dendrocerus carpenteri* (Hymenoptera: Pteromalidae and Megaspilidae), on larvae of *Aphidius smithii* (Hymenoptera: Aphidiidae). *Canad. Entomol.* **113**, 887–889.

Boenisch, A. and Jürgens, M. (1994) Enhanced reproduction of an aphid in the presence of its hyperparasitoid. *Norw. J. Agric. Sci. Suppl.* **16**, 123–129.

Boivin, G., Picard, C. and Auclair, J.L. (1993) Preimaginal development of *Anaphes* n. sp. (Hymenoptera: Mymaridae), an egg parasitoid of the carrot weevil (Coleoptera: Curculionidae). *Biol. Control* **3**, 176–181.

Bordat, D., Goudegnon, A.E. and Bouix, G. (1994) Relationships between *Apanteles flavipes* (Hym.: Braconidae) and *Nosema bordati* (Microspora, Nosematidae) parasites of *Chilo partellus* (Lep.: Pyralidae). *Entomophaga* **39**, 21–32.

Borden, J.H., Chong, L. and Rose, A. (1978) Morphology of the elongate placoid sensillum on the antenna of *Itoplectis conquistor*. *Ann. Entomol. Soc. Am.* **71**, 223–227.

Borstel, R.C. von (1960) Sulla natura della letalita dominante indotta dalle radiazioni. *Atti Assoc. Genet. Italiana* **5**, 35–50.

Borstel, R.C. von (1968) (plus 20 other authors) Mutational response of *Habrobracon* in the Biosatellite II experiment. *Bioscience* **18**, 598–601.

Borstel, R.C. von and Smith, R.H. (1977) Measuring dominant lethality in *Habrobracon*. In *Handbook of Mutagenicity Test Procedures* (eds B.J. Kilbey, M. Legator, W. Nichols and C. Ranel), Elsevier, Amsterdam. pp. 375–387.

Bosch, H.A.J. in den and Assem, J. van den (1986) The taxonomic position of *Aceratoneuromyia granularis* Domenichini (Hymenoptera: Eulophidae) as judged by characteristics of its courtship behaviour. *Syst. Entomol.* **11**, 19–23.

Boucek, Z. (1988) An overview of the higher classification of the Chalcidoidea (Parasitic Hymenoptera). In *Advances in Parasitic Hymenoptera Research* (ed. V.K. Gupta), E. J. Brill, Leiden. pp. 11–23.

Bouchard, Y. and Cloutier, C. (1984) Honeydew as a source of host-searching kairomones for the aphid parasitoid *Aphidius nigripes* (Hymenoptera: Aphidiidae). *Canad. J. Zool.* **62**, 1513–1520.

Boulétreau, M. (1972) Development et croissance larvae varies en conditions semi-artificielles et artificielles chez un hyménoptère entomophage: *Pteromalus puparum* L. (Chalc.). *Entomophaga* **17**, 265–273.

Boulétreau, M. (1986) The genetic and coevolutionary interactions between parasitoids and their hosts. In *Insect Parasitoids* (eds J. Waage and D. Greathead), Academic Press, London. pp. 169–200.

Boulétreau, M. and Quiot, J.M. (1972) Effet toxique des larves d'un hyménoptère parasite *Pteromalus puparum* L. (Chalc.) sur les cultures cellulaires de lépidoptères. *Comptes Rend. Acad. Sci.* **275**, 233–234.

Bousch, G.M. and Baerwald, R.A. (1967) Courtship behavior and evidence for a sex pheromone in the apple maggot parasite, *Opius alloecus. Ann. Entomol. Soc. Am.* **60**, 865–866.

Boush, R.T. and Hopper, K.R. (1995) Use of single family lines to preserve genetic variation in laboratory colonies. *Ann. Entomol. Soc. Am.* **88**, 713–717.

Boyce, H.R. (1936) Laboratory breeding of *Ascogaster carpsocapsae* Vier. with notes on biology and larval development. *Canad. Entomol.* **68**, 241–246.

Bracken, G.K. (1966) Role of ten dietary vitamins on fecundity of the parasitoid *Exeristes comstockii* (Cress.) (Hymenoptera: ichneumonidae). *Canad. Entomol.* **98**, 918–922.

Bracken, G.K. and Nair, K.K. (1969) Stimulation of yolk deposition in an ichneumonid parasitoid by feeding synthetic juvenile hormone. *Nature* **216**, 483–484.

Braman, S.K. and Yeargan, K.V. (1989) Reproductive strategy of *Trissolcus euschisti* (Hymenoptera: Scelionidae) under conditions of partially used host resources. *Ann. Entomol. Soc. Am.* **82**, 172–176.

Brandl, R. and Vidal, S. (1987) Ovipositor length in parasitoids and tentiform leaf mines: adaptations in eulophids (Hymenoptera: Chalcidoidea). *Biol. J. Linn. Soc.* **32**, 351–355.

Breeuwer, J.A.J. and Werren, J.H. (1990) Microorganisms associated with chromosome destruction and reproductive isolation between two insect species. *Nature* **346**, 558–560.

Breeuwer, J.A.J., Stouthamer, R., Werren, J.H. and Weisburg, W.G. (1992) Phylogeny of cytoplasmic incompatibility microorganisms in the parasitoid wasp genus *Nasonia* (Hymenoptera: Pteromalidae) based on 16S ribosomal DNA sequences. *Insect Molec. Biol.* **1**, 25–36.

Bressac, C. and Rousset, F. (1993) The reproductive incompatibility system in *Drosophila simulans*: Dapi-staining analysis of the *Wolbachia* symbionts in sperm cysts. *J. Invert. Pathol.* **61**, 226–230.

Briggs, C.J., Nisbet, R.M., Murdoch, W.W. *et al.* (1995) Dynamic effects of host-feeding in parasitoids. *J. Anim. Ecol.* **64**, 403–416.

Brodeur, J. and McNeil, J.N. (1989a) Seasonal microhabitat selection by an endoparasitoid through adaptive modification of host behavior. *Science* **244**, 226–228.

Brodeur, J. and McNeil, J.N. (1989b) Biotic and abiotic factors involved in diapause induction of the parasitoid, *Aphidus nigripes* (Hymenoptera: Aphidiidae) *J. Insect Physiol.* **35**, 969–974

Brodeur, J. and McNeil, J.N. (1990) Overwintering microhabitat selection by an endoparasitoid (Hym., Aphidiidae): Induced phototactic and thigmokinetic responses in dying hosts. *J. Insect Behav.* **3**, 751–763.

Brodeur, J. and McNeil, J.N. (1992) Host behaviour modified by the endoparasitoid *Aphidius nigripes*: A strategy to reduce hyperparasitism. *Ecol. Ent.* **17**, 97–104.

Brodeur, J. and McNeil, J.N. (1994) Life-history of the aphid hyperparasitoid *Asaphes vulgaris* Walker (Pteromalidae) – Possible consequences on the efficacy of the primary parasitoid *Aphidius nigripes* Ashmead (Aphidiidae). *Canad. Ent.* **126**, 1493–1497.

Bronner, R. (1985) Anatomy of the ovipositor and oviposition behavior of the gall wasp *Diplolepis rosae* (Hymenoptera, Cynipidae). *Canad. Entomol.* **117**, 849–858.

Bronskill, J.F. (1959) Embryology of *Pimpla turionellae* (L.) (Hymenoptera: Ichneumonidae). *Canad. J. Zool.* **37**, 655–688.

Bronskill, J.F. (1964) Embryogenesis of *Mesoleius tenthredinis* Morl. (Hymenoptera: Ichneumonidae). *Canad. J. Zool.* **42**, 439–453.

Bronskill, J.F. and House, H.L. (1957) Notes on rearing a pupal endoparasite, *Pimpla turionellae* (L.) (Hymenoptera: Ichneumonidae), on unnatural food. *Canad. Entomol.* **89**, 483.

Bronstein, J.L. (1988) Predators of fig wasps. *Biotropica* **20**, 215–219.

Broodryk, S.W. and Doutt, R.L. (1966) II. The biology of *Coccophagoides utilis* Doutt (Hymenoptera: Aphelinidae). *Hilgardia* **37**, 233–254.

Brooks, W.M. and Cranford, J.D. (1972) Microsporidioses of the hymenopterous parasites, *Campoletis sonorensis* and *Cardiochiles nigriceps*, larval parasites of *Heliothis* species. *J. Invert. Pathol.* **20**, 77–94.

Brothers, D.J. (1972) Biology and immature stages of *Pseudomethoca f. frigida*, with notes on other species (Hymenoptera: Mutillidae). *Univ. Kans. Sci. Bull.* **50**, 1–38.

Brothers, D.J. (1975) Phylogeny and classification of the aculeate Hymenoptera, with special reference to Mutillidae. *Kans. Univ. Sci. Bull.* **50**, 483–648.

Brothers, D.J. and Carpenter, J.M. (1993) Phylogeny of the Aculeata: Chrysidoidea and Vespoidea. *J. Hym. Res.* **2**, 227–302.

Brown, J.J. and Kainoh, Y. (1992) Host castration by *Ascogaster* spp. (Hymenoptera: Braconidae). *Ann. Entomol. Soc. Am.* **85**, 67–71.

Brown, J.J. and Reed-Larsen, D. (1991) Ecdysteroids and insect host/parasitoid interactions. *Biol. Control* **1**, 136–143.

Brown, J.J., Kiuchi, M., Kainoh, Y. and Takeda, S. (1993) *In vitro* release of ecdysteroids by an endoparasitoid, *Ascogaster reticulatus* Watanabe. *J. Insect Physiol.* **39**, 229–234.

Brown, V.K. (1973) The biology and development of *Brachygaster minutus* Olivier (Hymenoptera: Evaniidae), a parasite of the oothecae of *Ectobius* spp. (Dictyoptera: Blattidae). *J. Nat. Hist.* **7**, 665–674.

Browning, H.W. and Oatman, E.R. (1984) Intra- and interspecific relationships among some parasites of *Trichoplusia ni* (Lepidoptera: Noctuidae). *Env. Ent.* **13**, 551–556.

Buckell, E.R. (1928) Notes on the life-history and habits of *Melittobia chalybii* Ashmead. (Chalcidoidea: Elachertidae). *Pan-Pacif. Ent.* **5**, 14–22.

Buckingham, G.R. (1975) The parasites of walnut husk flies (Diptera: Tephritidae: *Rhagoletis*) including comparative studies on the biology of *Biosteres juglandis* Mues. (Hymenoptera: Braconidae) and on the male tergal glands of Braconidae (Hymenoptera). Unpublished PhD thesis, University of California, Berkeley. 282 pp.

Buckingham, G.R. and Sharkey, M.J. (1988) Abdominal exocrine glands in Braconidae (Hymenoptera). In *Advances in Parasitic Hymenoptera Research* (ed. V.K. Gupta), E.J. Brill, Leiden. pp 199–242.

Bugg, R.L., Ellis, R.T. and Carlson, R.W. (1989) Ichneumonidae (Hymenoptera) using extrafloral nectar of faba bean (*Vicia faba* L., Fabiaceae) in Massachusetts. *Biol. Agric. Hort.* **6**, 107–114.

Bull, A.L. (1982) Stages of living embryos in the jewel wasp *Mormoniella* (*Nasonia*) *vitripennis* (Walker) (Hymenoptera: Pteromalidae). *Int. J. Insect Morphol. and Embryol.* **11**, 1–23.

Büning, J. (1994) *The Insect Ovary. Ultrastructure, Previtellogenic Growth and Evolution*, Chapman & Hall, London. 400 pp.

Burke, W.D., Eickbush, D.G., Xiong, Y. *et al.* (1993) Sequence relationship of retro-transposable elements R1 and R2 within and between divergent insect species. *Mol. Biol. Evol.* **10**, 163–185.

Burks, B.D. (1938) A study of chalcidoid wings (Hymenoptera). *Ann. Entomol. Soc. Am.* **31**, 157–161.

Burt, A., Bell, G. and Harvey, P.H. (1991) Sex differences in recombination. *J. Evol. Biol.* **4**, 259–277.

Butterfield, A. and Anderson, M. (1994) Morphology and ultrastructure of antennal sensilla of the parasitoid, *Trybliographa rapae* (Westw.) (Hymenoptera: Cynipidae). *Int. J. Insect Morphol. and Embryol.* **23**, 11–20.

Buyck, E.J.E. (1949) Recherches sur un dryinide *Aphelopus indivisus*, parasite de cicadines. *La Cellule* **52**, 63–155.

Cals-Usciati, J., Cals, P. and Pralavorio, R. (1985) Adaptations fonctionelles des structures liées à la prise alimentaire chez la larve primaire de *Trybliographa daci* Weld. (Hymenoptera Cynipoidea), endoparasitoïde de la mouche des fruits *Ceratitis capitata*. *C. R. Acad. Sci. Paris* **300**, 103–108.

Calvert, D.J. (1973) Experimental host preferences of *Monoctonus paulensis* (Hymenoptera: Braconidae) including a hypothetical scheme of host selection. *Ann. Entomol. Soc. Am.* **66**, 28–33.

Cameron, E. [(1939)/1941] The holly leaf-miner (*Phytomyza ilicis* Curt.) and its parasites. *Bull. Entomol. Res.* **30**, 173–208.

Cameron, S.A., Derr, J.N., Austin, A.D. *et al.* (1992) The application of nucleotide sequence data to phylogeny of the Hymenoptera: A review. *J. Hym. Res.* **1**, 63–79.

Carew, W.P. and Sullivan, D.J. (1993) Interspecific parasitism between two aphid hyperparasitoids, *Dendrocerus carpenteri* (Hymenoptera: Megaspilidae) and *Asaphes lucens* (Hymenoptera: Pteromalidae). *Ann. Entomol. Soc. Am.* **86**, 794–798.

Carmean, D., Kimsey, L.S. and Berbee, M.L. (1992) 18S rDNA sequences and the holometabolous insects. *Mol. Phylog. Evol.* **1**, 270–278.

Carpenter, J.M. (1986) Cladistics of the Chrysidoidea (Hymenoptera). *J. N.Y. Entomol. Soc.* **94**, 303–330.

Carrillo, S. and Caltagirone, L.E. (1970) Observations on the biology of *Solierella peckhami*, *S. blaisdelli* (Sphecidae), and two species of Chrysididae (Hymenoptera). *Ann. Entomol. Soc. Am.* **63**, 672–681.

Carton, Y. and David, J.R. (1983) Reduction of fitness in *Drosophila* adults surviving parasitization by a cynipid wasp. *Experientia*, **39**, 231–233.

Carver, M. (1984) The potential host ranges in Australia of some imported aphid parasites (Hym.: Ichneumonoidea: Aphidiidae). *Entomophaga* **29**, 351–359.

Casas, J. (1989) Foraging behaviour of a leafminer parasitoid in the field. *Ecol. Ent.* **14**, 257–265.

Cassidy, J.D. and King, R.C. (1972) Ovarian development in *Habrobracon juglandis* (Ashmead) (Hymenoptera, Braconidae). I. The origin and differentiation of the oocyte–nurse cell complex. *Biol. Bull.* **143**, 483–505.

Cate, R.H., Sauer, J.R. and Eikenbary, R.D. (1974) Demonstration of host-feeding by the parasitoid *Aphelinus asychis* (Hymenoptera: Eulophidae). *Entomophaga* **19**, 479–482.

Cave, R.D., Gaylor, M.J. and Bradley, J.T. (1987) Host handling and recognition by *Telenomus reynoldsi* (Hymenoptera: Scelionidae), an egg parasitoid of *Geocoris* spp. (Heteroptera: Lygaeidae). *Ann. Entomol. Soc. Am.* **80**, 217–223.

Cendana, S.M. (1937) Studies on the biology of *Coccophagus* (Hymenoptera) a genus parasitic on nondiaspidine Coccidae. *Univ. Cal. Publ. Ent.* **6**, 337–400.

Ceresa-Gastaldo, L. and Chiappini, E. (1994) Observations on the cocoon of *Oligosita krygeri* Girault (Hymenoptera: Trichogrammatidae) oophagous parasitoid of *Cicadella viridis* (L.) (Homoptera: Cicadellidae). *Norw. J. Agric. Sci. Suppl.* **16**, 131–140.

Chacko, M.J. (1964) The term 'superparasitism' in insect parasitology. *Curr. Sci.* **33**, 2–3.

Channa Basavanna, G.P. (1953) Phoresy exhibited by *Lepidoscelio viatrix* Brues (Scelionidae, Hymenoptera). *Ind. J. Ent.* **15**, 264–266, 384–385.

Chapman, G.B. and Hooker, M.E. (1990) Ultrastructural features of the musculo-epidermal and epiderm–cuticular junctions in the abdomen of *Pediobius foveolatus* (Hymenoptera: Eulophidae). *Ann. Entomol. Soc. Am.* **83**, 215–219.

Chapman, G.B. and Hooker, M.E. (1992) A light and electron microscopic investigation of the occurrence of *Nosema* sp. (Microsporidia: Nosematidae) in the abdomen of the parasitic wasp *Pediobius foveolatus* (Hymenoptera: Eulophidae). *Trans. Am. Microsc. Soc.* **111**, 314–326.

Chapman, G.B. and Hooker, M.E. (1994) Occurrence of ferritin crystals in thoracic and abdominal cells of the adult parasitic wasp *Pediobius foveolatus* (Hymenoptera: Eulophidae). *Trans. Am. Microsc. Soc.* **113**, 52–58.

Chapman, R.F. (1982) *The Insects. Structure and Function*, 3rd edn, Hodder and Stoughton, London, 919pp.

Charnov, E.L. and Skinner, S.W. (1984) Evolution of host selection and clutch size in parasitoid wasps. *Flor. Ent.* **67**, 5–21.

Chauvin, G., El Agoze, M., Hamon, C. and Huignard, J. (1987) Ultrastructure des spermatozoïdes diploïdes de *Diadromus pulchellus* Wesmael (Hymenoptera: Ichneumonidae). *Int. J. Insect Morphol. and Embryol.* **17**, 358–366.

Chelliah, J. and Jones, D. (1990) Biochemical and immunological studies of proteins from polydnavirus *Chelonus* sp. near *curvimaculatus*. *J. Gen. Virol.* **71**, 2353–2359.

Cherian, M.C. and Israel, P. (1941) *Rhaconotus roslinensis* (Braconidae). A larval parasite of the sugarcane borer, *Scirpophaga rhodoproctalis*. *Indian J. Entomol.* **3**, 173–176.

Chihrane, J. and Laugé, G. (1994) Effects of high temperature shocks on male germinal cells of *Trichogramma brassicae* (Hymenoptera: Trichogrammatidae). *Entomophaga* **39**, 11–20.

Chiri, A.A. and Legner, E.F. (1986) Response of three *Chelonus* (Hymenoptera: Braconidae) species to kairomones in scales of six Lepidoptera. *Canad. Entomol.* **118**, 329–333.

Chittihunsa, T. and Sikorowski, P.P. (1995a) Occurrence of nonoccluded baculovirus (Baculoviridae) in *Microplitis croceipes* (Hymenoptera, Braconidae). *Biol. Control* **5**, 622–628.

Chittihunsa, T. and Sikorowski, P.P. (1995b) Effects of nonoccluded baculovirus (Baculoviridae) infection on *Microplitis croceipes* (Hymenoptera, Braconidae). *Envir. Ent.* **24**, 1708–1712.

Chow, A. and Mackauer, M. (1996) Sequential allocation of offspring sexes in the hyperparasitoid wasp, *Dendrocerus carpenteri. Anim. Behav.* **51**, 859–870.

Chow, F.J. and Mackauer, M. (1985) Multiple parasitism of the pea aphid: Stage of development of parasite determines survival of *Aphidius smithi* and *Praon prequodorum* (Hymenoptera: Aphidiidae). *Canad. Entomol.* **117**, 133–134.

Chow, F.J. and Mackauer, M. (1986) Host discrimination and larval competition in the aphid parasite, *Ephedrus californicus. Entomol. Exp. Appl.* **41**, 243–254.

Christiansen-Weniger, P. (1994) Morphological observations on the preimaginal stages of *Aphelinus varipes* (Hym., Aphelinidae) and the effects of this parasitoid on the aphid *Rhopalosiphum padi* (Hom., Aphididae). *Entomophaga* **39**, 267–274.

Christiansen-Weniger, P and Hardie J. (in press) Development of the aphid parasitoid, *Aphidus ervi*, in asexual and sexual females of the pea aphid, *Acyrthosiphon pisum*, and the blackberry-cereal aphid, *Sitobion frogariae. Entomophaga Suppl.*

Chrystal, R.N. (1930) Studies of the *Sirex* parasites. *Oxford Forestry Memoirs* **11**, 1–63.

Chrystal, R.N. and Skinner, E.R. (1931) Studies in the biology of *Xylonomus brachylabris* Kr. and *X. irrigator* F., parasites of the larch longhorn beetle, *Tetropium gabrieli* Weise. *Forestry* **5**, 21–33.

Chua, T.H. and Dyck, V.A. (1982) Assessment of *Pseudogonatopus flavifemur* E. and H. (Dryinidae: Hymenoptera) as a biological control agent of the rice brown planthopper. *Proc. Int. Conf. Pl. Prot. in Tropics* **March**, 253–265.

Chumakova, B.M. (1968) Regulative mechanisms of fertilization and sexual determination in progenies of parasitic Hymenoptera. Biological methods of plant protection. *Proc All-Union Sci.-Res. Inst. Plant. Protect. Leningrad.* **31**, 121–163. (In Russian.)

Clancy, D.W. (1946) The insect parasites of the Chrysopidae (Neuroptera). *Univ. Calif. Publ. Ent.* **7**, 403–496.

Claret, J., Porcheron, P. and Dray, F. (1978) La teneur en ecdysones circulantes au cours du dernier stade larvaire de l'Hyménoptère endoparasite *Pimpla instigator*, et l'entrée en diapause. *Comptes Rendus Acad. Sci. Paris* D **286**, 639–641.

Clark, A.M. (1963) The influence of diet upon the adult life span of two species of *Bracon. Ann. Entomol. Soc. Am.* **56**, 616–619.

Clark, A.M. and Egen, R.C. (1975) Behaviour of gynandromorphs of the wasp *Habrobracon juglandis. Develop. Biol.* **45**, 251–259.

Clark, A.M. and Gould, A.B. (1972) Evidence for post-cleavage fertilization among mosaics in *Habrobracon juglandis. Genetics* **72**, 63–68.

Clarke, A.R. and Walter, G.H. (1995) Strains and the classical biological control of insect pests. *Canad. J. Zool.* **73**, 1777–1790.

Clausen, C.P. (1931a) Biological observations in *Agriotypus* (Hymenoptera). *Proc. Entomol. Soc. Wash.* **33**, 29–37.

Clausen, C.P. (1931b) Biological notes on the Trigonalyidae (Hymenoptera). *Proc. Entomol. Soc. Wash.* **33**, 72–81.

Clausen, C.P. (1932) The early stages of some tryphonine Hymenoptera parasitic on sawfly larvae. *Proc. Entomol. Soc. Wash.* **34**, 49–60.

Clausen, C.P. (1940) *Entomophagous Insects*. McGraw-Hill, London, 688pp.

Clausen, C.P. (1950) Respiratory adaptations in the immature stages of parasitic insects. *Arthropoda* **1**, 199–223.

Clausen, C.P. (1954) The egg–larval host relationship among parasitic Hymenoptera. *Boll. Lab. Zool. Gen. Agr. Portici* **33**, 119–133.

Clausen, C.P. (1976) Phoresy among entomophagous insects. *Annu. Rev. Entomol.* **21**, 343–368.

Clausen, C.P., King, J.L. and Teranishi, C. (1927) The parasites of *Popillia japonica* in Japan and Choser (Korea) and their introduction into the US. *US Dep. Agric. Bull. No. 1429*, pp. 1–55.

Cloutier, C. (1986) Amino acid utilization in the aphid *Acyrthosiphon pisum* infected by the parasitoid *Aphidius smithi. J. Insect Physiol.* **32**, 263–267.

Cloutier, C. and Mackauer, M. (1979) The effect of parasitism by *Aphidius smithi* (Hymenoptera: Aphidiidae) on the food budget of the pea aphid, *Acyrthosiphon pisum* (Homoptera: Aphididae). *Canad. J. Zool.* **57**, 1605–161.

Cohen, A.C. and Debolt, J.W. (1984) Fatty acid and amino acid composition of teratocytes from *Lygus hesperus* (Miridae: Hemiptera) parasitized by two species of parasites, *Leiophron uniformis* (Braconidae: Hymenoptera) and *Peristenus stygicus* (Braconidae: Hymenoptera). *Comp. Biochem. Physiol.* **79B**, 335–337.

Colazza, S. and Bin, F. (1992) Introduction of the oophage *Edovum puttleri* Griss. (Hymenoptera: Eulophidae) in Italy for the biological control of Colorado potato beetle. *Redia* **75**, 203–225.

Cole, L.R. (1959a) On a new species of *Syntretus* Foerster (Hym., Braconidae) parasitic on an adult ichneumonid, with a description of the larva and notes on its life history and that of its host, *Phaeogenes invisor* (Thunberg). *Entomol. Mon. Mag.* **95**, 18–21.

Cole, L.R. (1959b) On the defenses of lepidopterous pupae in relation to the oviposition behaviour of certain Ichneumonidae. *J. Lepidopt. Soc.* **13**, 1–10.

Cole, L.R. (1970) Observations on the finding of mates by *Phaeogenes invisor* and *Apanteles medicaginis* (Hymenoptera: Ichneumonidae). *Anim. Behav.* **18**, 184–192.

Cole, L.R. (1981) A visible sign of a fertilization act during oviposition by an ichneumonid wasp, *Itoplectis maculator*. *Anim. Behav.* **29**, 299–300.

Collier, T.R. (1995) Host feeding, egg maturation, resorption, and longevity in the parasitoid *Aphytis melinus* (Hymenoptera, Aphelinidae). *Ann. Entomol. Soc. Am.* **88**, 206–214.

Compton, S.G. and Disney, R.H.L. (1991) New species of *Megaselia* (Diptera: Phoridae) whose larvae live in fig syconia (Urticales: Moraceae), and adults prey on fig wasps (Hymenoptera: Agaonidae). *J. Nat. Hist.* **25**, 203–219.

Compton, S.G. and McLaren, F.A.C. (1989) Respiratory adaptations in some male fig wasps. *Proc. Konink. Ned. Akad. Wetensch.* **92C**, 57–71.

Compton, S.G. and Nefdt, R. (1988) Extra-long ovipositors in chalcid wasps: Some examples and observations. *Antenna*, **12**, 102–105.

Conner, G.W. and Saul II, G.B. (1986) Acquisition of incompatibility by inbred wild-type stocks of *Mormoniella*. *J. Hered.* **77**, 211–213.

Cook, D. and Stoltz, D.B. (1983) Comparative serology of viruses isolated from ichneumonid parasitoids. *Virology* **130**, 215–220.

Cook, J.M. (1991) Sex determination and sex ratios in parasitoid wasps. Unpublished PhD Thesis, University of London.

Cook, J.M. (1993a) Experimental tests of sex determination in *Goniozus nephantidis*. *Heredity* **71**, 130–137.

Cook, J.M. (1993b) Sex determination in the Hymenoptera – a review of models and evidence. *Heredity* **71**, 421–435.

Cook, J.M. (1993c) Inbred lines as reservoirs of sex alleles in parasitoid rearing programs. *Environ. Ent.* **22**, 1213–1216.

Cook, J.M. and Crozier, R.H. (1995) Sex determination and population biology in the Hymenoptera. *TREE* **10**, 281–286.

Coop, L.B. and Croft, B.A. (1990) Diapause and life history attributes of *Phytodietus vulgaris* (Hymenoptera: Ichneumonidae), a parasitoid of *Argyrotaenia citrana* (Lepidoptera: Tortricidae). *Ann. Entomol. Soc. Am.* **83**, 1148–1151.

Cooper, K.W. (1953) Egg gigantism, oviposition and genital anatomy: Their bearing on the biology and phylogenetic position of *Orussus* (Hymenoptera: Siricoidea). *Proc. Roch. Acad. Sci.* **10**, 38–68.

Cooper, K.W. (1954) Biology of eumenine wasps, IV. A trigonalid wasp parasitic on *Rygchium rugosum* (Saussure) (Hymenoptera, Trigonalidae). *Proc. Entomol. Soc. Wash.* **56**, 280–288.

Cooper, K.W. and Dessart, P. (1975) Adult, larva and biology of *Conostigmus quadratogenalis* Dessart and Cooper, sp. n., (Hym. Ceraphronoidea), parasite of *Boreus* (Mecoptera) in California. *Bull. Ann. Soc. R. Belge Ent.*, **111**, 37–53.

Copland, M.J.W. (1976) Female reproductive system in the Aphelinidae (Hymenoptera: Chalcidoidea). *Int. J. Insect Morphol. Embryol.* **5**, 151–166.

Copland, M.J.W. and King, P.E. (1971a) The structure and possible function of the reproductive system in some Eulophidae and Tetracampidae. *Entomologist* **1971**, 4–28.

Copland, M.J.W. and King, P.E. (1971b) The structure of the female reproductive system in the Chalcididae. *Entomol. Mon. Mag.* **107**, 230–239.

Copland, M.J.W. and King, P.E. (1972a) The structure of the female reproductive system in the Pteromalidae (Chalcidoidea: Hymenoptera). *Entomologist* **105**, 77–96.

Copland, M.J.W. and King, P.E. (1972b) The structure of the female reproductive system in the Eurytomidae (Hymenoptera: Chalcidoidea). *J. Zool. Lond.* **166**, 185–212.

Copland, M.J.W. and King, P.E. (1972c) The structure of the female reproductive system in the Torymidae (Hymenoptera: Chalcidoidea). *Trans. R. Entomol. Soc. Lond.* **124**, (191–212.

Copland, M.J.W., King, P.E. and Hill, D.S. (1973) The structure of the female reproductive system in the Agaonidae (Chalcidoidea, Hymenoptera). *J. Ent.* (A) **48**, 25–35.

Corbet, S.A. (1968) The influence of *Ephestia kuehniella* on the development of its parasite *Nemeritis canescens*. *J. Exp. Biol.* **48**, 291–304.

Corbet, S.A. (1971) Mandibular gland secretion of larvae of the flour moth, *Anagasta kuehniella*, contains an epideictic pheromone and elicits oviposition movements in a hymenopteran parasite. *Nature* **232**, 481–484.

Corbet, S.A. (1985) Insect chemosensory response: A chemical legacy hypothesis. *Ecol. Ent.* **10**, 143–153.

Cornell, H.V. (1988) Solitary and gregarious brooding, sex ratios and the incidence of thelytoky in the parasitic Hymenoptera. *Am. Midl. Nat.* **119**, 63–70.

Cornell, H.V. and Hawkins, B.A. (1993) Accumulation of native parasitoid species on introduced herbivores: A comparison of hosts as natives and hosts as invaders. *Amer. Nat.* **141**, 847–865.

Cornell, H.V. and Hawkins, B.A. (1994) Patterns of parasitoid accumulation on introduced herbivores. In *Parasitoid Community Ecology* (eds B.A. Hawkins and W. Sheehan), Oxford University Press, Oxford, pp. 77–89.

Cortesero, A.M. and Monge, J.P. (1994) Influence of preemergence experience on response to host and host-plant odors in the larval parasitoid *Eupelmus vuilleti*. *Entomol. Exp. Appl.* **72**, 281–288.

Coudron, T.A. and Brandt. S.L. (in press). Characteristics of a developmental arrestant in the venom of the ectoparasitoid wasp *Euplectrus comstockii Toxicon*.

Coudron, T.A. and Puttler, B. (1988) Response of natural and factitious hosts to the ectoparasite *Euplectrus plathypenae* (Hymenoptera: Eulophidae). *Ann. Entomol. Soc. Am.* **81**, 931–937.

Coustau, C., Carton, Y., Nappi, A. *et al.* (1996) Differential induction of antibacterial transcripts in *Drosophila* susceptible and resistant to parasitism by *Leptopilina boulardi*. *Ins. Mol. Biol.* **5**, 167–172.

Couturier, A. (1949) Observations sur *Rhyssa approximator* F. cleptoparasite de *Rh. persuasoria* L. (Hym. Ichneumonidae). *Bull. Soc. Entomol. Fr.* **54**, 62–63.

Cowan, D.P. (1979) The function of enlarged hind legs in oviposition and aggression by *Chalcis canadensis* (Hymenoptera: Chalcididae). *Great Lakes Entomologist* **12**, 133–136.

Cox, J.A. (1932) *Ascogaster carpocapsae* Viereck, an important larval parasite of the codling moth and oriental fruit moth. *Tech. Bull. N.Y. St. Agric. Exp. Stn.* **188**, 26 pp.

Cronin, J.T. and Gill, D.E. (1989) The influence of host distribution, sex and size on the level of parasitism by *Itoplectis conquisitor* (Hymenoptera: Ichneumonidae). *Ecol. Ent.* **14**, 163–173.

Cronin, J.T. and Strong, D.R. (1996) Genetics of oviposition success of a thelytokous fairyfly parasitoid, *Anagrus delicatus*. *Heredity* **76**, 43–54.

Cross, J.E. and Simpson, R.G. (1972) Cocoon construction by *Bathyplectes curculionis* and attack by secondary parasites. *Env. Ent.* **1**, 631–633.

Crozier, R.H. (1971) Heterozygosity and sex determination in haplodiploidy. *Amer. Nat.* **105**, 399–412.

Crozier, R.H. (1975) *Animal Cytogenetics*, Vol. 3, Part 7, Gebrüder Borntraeger, Berlin. pp. 1–95.

Cruz, Y.P. (1981) A sterile defender morph in a polyembryonic hymenopterous parasite. *Nature*, **294**, 446–447.

Cruz, Y.P. (1986) Development of the polyembryonic parasite *Copidosoma tanytmenus* (Hymenoptera: Encyrtidae). *Ann. Entomol. Soc. Am.* **79**, 121–127.

Cruz, Y.P., Oelhaf, R.C. Jr, and Jockusch, E.L. (1990) Polymorphic precocious larvae in the polyembryonic parasitoid *Copidosomopsis tanytmema* (Hymenoptera: Encyrtidae). *Ann. Entomol. Soc. Am.* **83**, 549–554.

Cunha, A.B. and Kerr, W.E. (1957) A genetical theory to explain sex determination by arrhenotokous parthenogenesis. *Forma Functio* **1**, 33–36.

Cushman, R.A. (1916) *Thersilochus conotracheli*, a parasite of the plum curculio. *J. Agric. Res.* **6**, 847–855.

Cutler, J.R. (1955) The morphology of the head of the final instar larva of *Nasonia vitripennis* Walker (Hymenoptera: Chalcidoidea). *Proc. R. Ent. Soc. Lond. A* **30**, 73–81.

Dahlman, D.L. (1990) Evaluation of teratocyte functions: An overview. *Arch. Insect Bioch. Physiol.* **13**, 159–166.

Dahlman, D.L. and Vinson, S.B. (1975) Trehalose and glucose levels in hemolymph of *Heliothis virescens* parasitized by *Microplitis croceipes* or *Cardiochiles nigriceps*. *Comp. Biochem. Physiol.* **52B**, 465–468.

Dahms, E.C. (1984a) A review of the biology of species in the genus *Melittobia* (Hymenoptera: Eulophidae) with interpretations and additions using observations on *Melittobia australica*. *Mem. Queensl. Mus.* **21**, 337–360.

Dahms, E.C. (1984b) An interpretation of the structure and function of the antennal sense organs of *Melittobia australica* (Hymenoptera: Eulophidae) with the discovery of a large dermal gland in the male scape. *Mem. Queensl. Mus.* **21**, 361–385.

Dallai, R. (1975) Fine structure of the spermatheca in *Apis mellifera*. *J. Insect Physiol.* **21**, 89–109.

Daly, H.V. (1963) Close-packed and fibrillar muscles of the Hymenoptera. *Ann. Entomol. Soc. Am.* **56**, 295–306.

Daly, H.V. (1964) Skeleto-muscular morphogenesis of the thorax and wings of the honey bee, *Apis mellifera* (Hymenoptera, Apidae). *Univ. Calif. Publ. Ent.* **39**, 1–77.

Danforth, B.N. (1989) The evolution of hymenopteran wings: The importance of size. *J. Zool., Lond.* **218**, 247–276.

Danforth, B.N. and Michener, C.C. (1988) Wing folding mechanisms in the Hymenoptera. *Ann. Entomol. Soc. Am.* **81**, 342–349.

Daniel, D.M. (1932) *Macrocentrus ancylivorus* Rohwer, a polyembryonic braconid parasite of the oriental fruit moth. *Tech. Bull. N.Y. St. Agric. Stn* **187**, 1–101.

Danyk, T.P. and Mackauer, M. (1993) Discrimination between self- and conspecific-parasitized hosts in the aphid parasitoid *Praon pequodorum* Viereck (Hymenoptera: Aphidiidae). *Canad. Entomol.* **125**, 957–964.

Darling, D.C. (1991) *Bohpa maculata*, a new genus and species of Ceinae from South Africa (Hymenoptera: Chalcidoidea: Pteromalidae). *Proc. Entomol. Soc. Wash.* **93**, 622–629.

Darling, D.C. (1992) The life history and larval morphology of *Aperilampus* (Hymenoptera: Chalcidoidea: Philodrominae), with a discussion of the phylogenetic affinities of the Philodrominae. *Syst. Entomol.* **17**, 331–339.

Darling, D.C. and Hanson, P.E. (1986) Two new species of *Spalangiopelta* from Oregon (Hymenoptera: Chalcidoidea), with a discussion of wing length variation. *Pan-Pac. Ent.* **62**, 153–164.

Darling, D.C. and Miller, T.D. (1991) Life-history and larval morphology of *Chrysolampus* (Hymenoptera, Chalcidoidea, Chrysolampinae) in western North America. *Canad. J. Zool.* **69**, 2168–2177.

Darling, D.C. and Sharkey, M.J. (1990) Order Hymenoptera. In: Insects from the Santana Formation. *Bull. Amer. Mus. Nat. Hist.* **195**, 123–153.

Darling, D.C. and Werren, J.H. (1990) Biosystematics of *Nasonia* (Hymenoptera: Pteromalidae): Two new species reared from birds' nests in North America. *Ann. Entomol. Soc. Am.* **83**, 352–370.

David, G.J. (1988) Fecundity, longevity and overwintering of *Trissolcus biproruli* Girault (Hymenoptera: Scelionidae), a parasitoid of *Biproulus bibax* Breddin (Hemiptera: Pentatomidae). *J. Austr. Entomol. Soc.* **27**, 297–301.

Davies, D.H., Burghardt, R.L. and Vinson, S.B. (1986) Oogenesis of *Cardiochiles nigriceps* Viereck (Hymenoptera: Braconidae): Histochemistry and development of the chorion with special reference to the fibrous layer. *Int. J. Insect Morphol. Embryol.* **15**, 363–374.

Davies, D.H., Strand, M.R. and Vinson, S.B. (1987) Changes in differential haemocyte count and *in vitro* behaviour of plasmatocytes from host *Heliothis virescens* caused by *Campoletis sonorensis* polydnavirus. *J. Insect Physiol.* **33**, 143–153.

Davies, D.H. and Vinson, S.B. (1986) Passive evasion by eggs of braconid parasitoid *Cardiochiles nigriceps* of encapsulation *in vitro* by haemocytes of host *Heliothis virescens*. Possible role for fibrous layer in immunity. *J. Insect Physiol.* **32**, 1003–1010.

Davies, I. (1974) The effect of age and diet on the the ultrastructure of hymenopteran flight muscle. *Exp. Gerontol.* **9**, 215–219.

Davies, I. (1975) A study of the effect of diet on the life-span of *Nasonia vitripennis* (Walk.) (Hymenoptera, Pteromalidae). *J. Gerontol.* **30**, 294–298.

Davies, I. (1977) The effect of diet on the ultrastructure of the mid gut cells of *Nasonia vitripennis* (Walk.) (Insecta: Hymenoptera) at various ages. *Cell. Tiss. Res.* **184**, 529–538.

Davies, I. and King, P.E. (1975) The structure of the rectal papilla in a parasitoid hymenopteran *Nasonia vitripennis* (Walker) (Hymenoptera Pteromalidae). *Cell. Tiss. Res.* **161**, 413–419.

Davies, N.W. and Madden, J.L. (1985) Mandibular gland secretions of two parasitic wasps (Hymenoptera: Ichneumonidae). *J. Chem. Ecol.* **11**, 1115–1127.

Day, M.C. (1971) A new species of *Platygaster* Latreille (Hym., Proctotrupoidea, Platygasteridae) reproducing by thelytokous parthenogenesis. *Ent. Gaz.* **22**, 37–42.

Day, M.C. (1979) The affinities of *Loboscelidia* Westwood (Hymenoptera: Chrysididae, Loboscelidiinae). *Syst. Ent.* **4**, 21–30.

Day, W.H. (1970) The survival value of its jumping cocoons to *Bathyplectes anurus*, a parasite of the alfalfa wevil. *J. Econ. Ent.* **63**, 586–589.

Day, W.H. and Hedlung, R.C. (1988) Biological comparisons between arrhenotokous and thelytokous biotypes of *Mesochorus nigriceps*. *Entomophaga* **33**, 201–210.

Dean, J.M. and Ricklefs, R.E. (1979)) Do parasites of Lepidoptera larvae compete for hosts? No! *Amer. Nat.* **113**, 302–306.

Dean, J.M. and Ricklefs, R.E. (1980) Do parasites of Lepidoptera larvae compete for hosts? No evidence. *Amer. Nat.* **116**, 882–884.

DeBach, P. (1943) The importance of host feeding by adult parasitoids in the reduction of host population. *J. Econ. Entomol.* **36**, 647–658.

DeBuron, I. and Beckage, N.E. (1992) Characterization of a polydnavirus (PDV) and virus-like filamentous particle (VFLP) in the braconid wasp *Cotesia congregata* (Hymenoptera: Braconidae). *J. Inv. Pathol.* **59**, 315–327.

Decker, U.M., Powell, W. and Clark, S.J. (1993) Sex pheromones in the cereal aphid parasitoids *Praon volucre* and *Aphidius rhopalosiphi*. *Entomol. Exp. Appl.* **69**, 33–39.

Delanoue, P. and Arambourg, Y. (1965) Contribution à l'étude en laboratoire d'*Eupelmus urozonus* Dalm. (Hym. Chalcidoidea Eupelmidae). *Ann. Soc. Entomol. Fr. (N.S.)* **1**, 817–842.

Delanoue, P. and Arambourg, Y. (1967) Contribution à l'étude en laboratoire de *Pnigalio mediterraneus* (Hym. Chalcidoidea Eulophidae). *Ann. Soc. Entomol. Fr. (N.S.)* **3**, 909–927.

DeLoof, A., Van Voon, J. and Vanderroost, C. (1979) Influence of ecdysterone, precocene and compounds with juvenile hormone activity on induction, termination and maintenance of diapause in the parasitoid wasp, *Nasonia vitripennis*. *Physiol. Ent.* **4**, 319–328.

Delucchi, V., Rosen, D. and Schlinger, E.I. (1976) Relationship of systematics to biological control. In *Theory and Practice of Biological Control* (eds C.B. Huffaker and P.S. Messenger), Academic Press, New York. pp. 81–91.

Derr, J.N., Davis, S.K., Woolley, J.B. and Wharton, R.A. (1992a) Reassessment of the 16s rRNA nucleotide sequence from members of the parasitic Hymenoptera. *Mol. Phylog. Evol.* **1**, 338–341.

Derr, J.N., Davis, S.K., Woolley, J.B. and Wharton, R.A. (1992b) Variation and the phylogenetic utility of the large ribosomal sub-unit of mitochondrial DNA from the insect order Hymenoptera. *Mol. Phylog. Evol.* **2**, 136–147.

Dib-Hajj, S.D., Webb, B.A. and Summers, M.D. (1993) Structure and evolutionary implications of a 'cysteine-rich' *Campoletis sonorensis* polydnavirus gene family. *Proc. Natl. Acad. Sci.* **90**, 3765–3769.

Dicke, M. (1994) Local and systemic production of volatile herbivore-induced terpenoids – their role in plant–carnivore mutualism. *J. Plant. Physiol.* **143**, 465–472.

Dicke, M. and Sabelis, M.W. (1988) Infochemical terminology: Based on cost–benefit analysis rather than origin of compounds? *Funct. Ecol.* **2**, 131–139.

Dicke, M., Vet, L.E.M., Wiskerke, J.S.C. and Stapel, O. (1994) Parasitoid of *Drosophila* larvae solves foraging problem through infochemical detour: Conditions affecting employment of this strategy. *Norw. J. Agric. Sci. Suppl.* **16**, 227–232.

Digilio, M.C., Pennacchio, F. and Tremblay, E. (in press) Host regulation effects of ovary fluid and venom of *Aphidius ervi* Haliday (Hymenoptera: Braconidae). *J. Insect Physiol.*

Dijken, M.J. van (1991) A cytological method to determine primary sex ratio in the solitary parasitoid *Epidinocarsis lopezi*. *Entomol. Exp. Appl.* **60**, 301–304.

Dijken, M.J. van and Waage, J.K. (1987) Self and conspecific superparasitism by the egg parasitoid *Trichogramma evanescens*. *Entomol. Exp. Appl.* **43**, 183–192.

Dijken, M.J. van, Stratum, P. van and Alphen, J.J.M. van (1992) Recognition of individual-specific marked parasitized hosts by the solitary parasitoid *Epidinocarsis lopezi*. *Behav. Ecol. Sociobiol.* **30**, 77–82.

Dijkerman, H.J. (1988) Notes on the parasitisation behaviour and larval development of *Trieces tricarinatus* and *Triclistus yponomeutae* (Hymenoptera, Ichneumonidae), endoparasitoids of the genus *Yponomeuta* (Lepidoptera, Yponomeutidae). *Proc. Kon. Ned. Akad. Wet. C* **91**, 19–30.

Dijkstra, L.J. [(1986)/1987] Optimal selection and exploitation of hosts in the parasitic wasp *Colpoclypeus florus* (Hym., Eulophidae). *Neth. J. Zool.* **36**, 177–301.

Dillon, N., Austin, A.D. and Bartowsky, E. (1996) Comparison of preservation techniques for DNA extraction from hymenopterous insects. *Insect Molec. Biol.* **5**, 21–24.

Dindo, M.L., Sama, C. and Farneti, R. (1994) *In vitro* rearing of *Brachymeria intermedia* (Nees) (Hymenoptera Chalcididae) on veal homogenate-based diets. *Norw. J. Agric. Sci. Suppl.* **16**, 395.

Ding, D., Swedenborg, P.D. and Jones, R.L. (1989) Chemical stimuli in host-seeking behavior of *Macrocentrus grandis* (Hymenoptera: Braconidae). *Ann. Entomol. Soc. Am.* **82**, 232–236.

Dmoch, J. and Rutkowska-Ostrowska, Z. (1978) Host-finding and host-acceptance mechanism in *Trichomalus perfectus* Walker (Hymenoptera, Pteromalidae). *Bull. Acad. Polon. Sci. Biol.* **26**, 317–323.

Donaldson, J.S. and Walter, G.H. (1984) Sex ratios of *Spalangia endius* (Hymenoptera: Pteromalidae), in relation to current theory. *Ecol. Ent.* **9**, 395–402.

Donovan, B.J. (1991) Life cycle of *Sphecophaga vesparum* (Curtis) (Hymenoptera: Ichneumonidae), a parasitoid of some vespid wasps. *N.Z. J. Zool.* **18**, 181–192.

Doutt, R.L. (1947) Polyembryony in *Copidosoma koehleri* Blanchard. *Am. Nat.* **81**, 435–453.

Doutt, R.L. (1951) The phenomenon of autonarcosis in a parasitic wasp. *Canad. Entomol.* **83**, 132–132.

Doutt, R.L. (1952) The teratoid larva of polyembryonic Encyrtidae. *Canad. Entomol.* **84**, 247–250.

Doutt, R.L. (1959) The biology of parasitic Hymenoptera. *Annu. Rev. Entomol.* **4**, 161–182.

Doutt, R.L. (1973) The fossil Mymaridae (Hymenoptera: chalcidoidea). *Pan Pac. Entomol.* **49**, 221–228.

Dover, B.A., Davies, D.H. and Vinson, S.B. (1988) Degeneration of last instar *Heliothis virescens* prothoracic glands by *Campoletis sonorensis* polydnavirus. *J. Invert. Pathol.* **51**, 80–91.

Dowdell, R.V. and Horn, D.J. (1975) Mating behavior of *Bathyplectes curculionis* (Hym.: Ichneumonidae) a parasitoid of the afalfa weevil, *Hypera postica* (Col.: Curculionidae). *Entomophaga* **23**, 271–273.

Dowden, P.B. (1941) Parasites of the birch leaf-mining sawfly (*Phyllotoma nemorata*). *Tech. Bull. USDA.* **757**, 1–55.

Dowry, G., Rojas-Rousse, D. and Periquet, G. (1995) Ability of *Eupelmus orientalis* ectoparasitoid larvae to develop on an unparalysed host in the absense of female stinging behaviour. *J. Insect Physiol.* **41**, 287–296.

Dowton, M. and Austin, A.D. (1994) Molecular phylogeny of the insect order Hymenoptera: Apocritan relationships. *Proc. Natl. Acad. Sci. USA*, **91**, 9911–9915.

Dowton, M. and Austin, A.D. (1995) Increased genetic diversity in mitochondrial genes is correlated with the evolution of parasitism in the Hymenoptera. *J. Mol. Evol.* **41**, 958–965.

Drenth, D. (1974) Susceptibility of different species of insects to an extract of the venom gland of the wasp *Microbracon hebetor* (Say). *Toxicon* **12**, 189–192.

D'Rozario, A.M. (1942) On the development and homologies of the genitalia and their ducts in Hymenoptera. *Trans. R. Entomol. Soc. Lond.* **92**, 363–415.

Du, Y.-J., Poppy G.M. and Powell, W. (1996) Relative importance of semiochemicals from the first and second trophic level in host foraging behaviour of *Aphidius ervi*. *J. Chem. Ecol.* **22**, 1591–1605.

Duchateau, M.J., Hoshiba, H. and Velthuis, H.H.W. (1994) Diploid males in the bumble bee *Bombus terrestris*. *Entomol. Exp. Appl.* **71**, 263–269.

Duodu, Y.A. and Antoh, F.F. (1984) Effects of parasitism by *Apanteles sagax* (Hym.: Braconidae) on growth, food consumption and food utilization in *Sylepta derogata* larvae (Lep.: Pyralidae). *Entomophaga* **29**, 63–71.

Duodu, Y.A. and Davis, D.W. (1974) A comparison of growth, food consumption, and food utilization between unparasitized alfalfa weevil larvae and those parasitized by *Bathyplectes curculionis* (Thomson). *Env. Ent.* **3**, 705–710.

Dyar, H.G. (1890) The number of molts of lepidopterous larvae. *Psyche* **5**, 420–422.

Eastham, L.E.S. (1929) The post–embryonic development of *Phaenoserphus viator* Hal. (Proctotrypoidea), a parasite of the larva of *Pterostichus niger* (Carabidae), with notes on the anatomy of the larva. *Parasitology* **21**, 1–21.

Eberhard, W.G. (1975) The ecology and behaviour of a subsocial pentatomid bug and two scelionid wasps: Strategy and counterstrategy in a host and its parasites. *Smithson. Contrib. Zool.* **205**, 1–39.

Eberhard, W.G. (1994) Evidence for widespread courtship during copulation in 131 species of insects and spiders, and implications for cryptic female choice. *Evolution*, **48**, 711–733.

Edson, K.M. and Vinson, S.B. (1976) The function of the anal vesicle in respiration and excretion in the braconid wasp, *Microplitis croceipes. J. Insect Physiol.* **22**, 1037–1043.

Edson, K.M. and Vinson, S.B. (1977) Nutrient absorption by the anal vesicle of the braconid wasp, *Microplitis croceipes. J. Insect Physiol.* **23**, 5–8.

Edson, K.M. and Vinson, S.B. (1979) A comparative morphology of the venom apparatus of female braconids (Hymenoptera, Braconidae). *Canad. Entomol.* **111**, 1013–1024.

Edson, K.M., Vinson, S.B., Stoltz, D.B. and Summers, M.D. (1981) Virus in a parasitoid wasp: Suppression of the cellular immune response in the parasitoid's host. *Science*, **211**, 582–583.

Edson, K.M., Barlin, M.R. and Vinson, S.B. (1982) Venom apparatus of braconid wasps: Comparative ultrastructure of reservoirs and gland filaments. *Toxicon* **20**, 553–562.

Edwards, O.R. and Hoy, M.A. (1995) Random amplified polymorphic DNA markers to monitor laboratory-selected, pesticide-resistant *Trioxys pallidus* (Hymenoptera, Aphidiidae) after release into three California walnut orchards. *Env. Ent.* **24**, 487–496.

Edwards, R.L. (1954) The host finding and oviposition behaviour of *Mormoniella vitripennis* (Walker) (Hym., Pteromalidae), a parasite of muscoid flies. *Behaviour* **7**, 88–112.

Eggleton, P. (1989) The phylogeny and evolutionary biology of the Pimplinae (Hymenoptera: Ichneumonidae). Unpublished PhD Thesis, University of London.

Eggleton, P. and Belshaw, R. (1992) Insect parasitoids: An evolutionary overview. *Phil. Trans. R. Soc. Lond. B*, **337**, 1–20.

Eggleton, P. and Belshaw, R. (1993) Comparisons of dipteran, hymenopteran and coleopteran parasitoids: Provisional phylogenetic explanations. *Biol. J. Linn. Soc.* **48**, 213–226.

Eickbush, D.G., Eickbush, T.H. and Werren, J.H. (1992) Molecular characterization of repetitive DNA sequences from a B chromosome. *Chromosoma* **101**, 575–583.

El Agoze, M. (1985) Aspects biologiques et génétiques de l'activité reproductrice de l'hymenoptère *Diadromus pulchellus*, en relation avec le processus de deploidisation des males. Thèse doctorat, Université François-Rabelais de Tours.

El Agoze, M. and Périquet, G. (1993) Viability of diploid males in the parasitic wasp, *Diadromus pulchellus* (Hymenoptera: Ichneumonidae). *Entomophaga* **38**, 111–118.

El Agoze, M., Drezen, J.M., Renault, S. and Périquet, G. (1994) Analysis of the reproductive potential of diploid males in the wasp *Diadromus pulchellus* (Hymenoptera: Ichneumonidae). *Bull. Entomol. Res.* **84**, 213–218.

El Agoze, M., Poirié, M. and Périquet, G. (1995) Precedence of the first male sperm in successive matings in the Hymenoptera *Diadromus pulchellus. Entomol. Exp. Appl.* **75**, 251–255.

Eller, F.J., Bartelt, R.J., Jones, R.L. and Kulman, H.M. (1984) Ethyl (Z)-9-hexadecanoate a sex pheromone of *Syndipnus rubiginosus*, a sawfly parasitoid. *J. Chem. Ecol.* **10**, 291–300.

Eller, F.J., Tumlinson, J.H. and Lewis, W.J. (1990) Intraspecific competition in *Microplitis croceipes* (Hymenoptera: Braconidae), a parasitoid of *Heliothis* species (Lepidoptera: Noctuidae). *Ann. Entomol. Soc. Am.* **83**, 504–508.

Ellington, C.P. (1975) Non-steady-state aerodynamics of the flight of *Encarsia formosa*. In *Swimming and Flying in Nature*, Vol. 2 (eds T.Y. Wu, C. Brokaw and C. Brennan), Academic Press, London. pp. 783–796.

Elliott, J.M. (1982) The life cycle and spatial distribution of the aquatic parasitoid *Agriotypus armatus* (Hymenoptera: Agriotypidae) and its caddis host, *Silo pallipes* (Trichoptera: Goeridae). *J. Anim. Ecol.* **51**, 923–941.

El-Sawaf, B.M. and Zohdy, N.Z.M. (1976) Host–parasite relationship. 2 – Cholinesterase activity of the larvae of the rice moth *Corcyra cephalonica* (Lep.: Pyralidae) parasitized by *Bracon hebetor* (Hym.: Braconidae). *Entomophaga* **21**, 99–101.

El-Sufty, R. and Führer, E. (1981) Wechsebbeziehungen zwischen *Pieris brassicae* L. (Lep., Pieridae), *Apanteles glomeratus* L. (Hym., Braconidae) und dem Pilz *Beauveria bassiana* (Bals.) Vuill. *Z. Angew. Entomol.* **92**, 321–329.

Elzen, G.W., Williams, H.J. and Vinson, S.B. (1984) Isolation and identification of cotton synomones mediating searching behavior by parasitoid *Campoletis sonorensis*. *J. Chem. Ecol.* **10**, 1251–1254.

Englemann, F. (1970) *The Physiology of Insect Reproduction*, Pergamon Press, Oxford. 307 pp.

English-Loeb, G.M., Karban, R. and Brody, A.K. (1990) Arctiid larvae survive attack by a tachinid parasitoid and produce viable offspring. *Ecol. Ent.* **15**, 361–362.

Eslin, P. and Prevost, G. (1996) Variation in *Drosophila* concentration of hemocytes associated with different ability to encapsulate *Asobara tabida* larval parasitoid. *J. Insect Physiol.* **42**, 549–555.

Espelie, K.E., Berisford, C.W. and Dahlsten, D.L. (1990) Cuticular hydrocarbons of geographically isolated populations of *Rhopalicus pulchripennis* (Hymenoptera: Pteromalidae) – Evidence for 2 species. *Comp. Biochem. Physiol.* **96B**, 305–308.

Evans, A.C. (1933) Comparative observations on the morphology and biology of some hymenopterous parasites of carrion-infesting Diptera. *Bull. Entomol. Res.* **24**, 385–405.

Evans, H.E. (1969) Phoretic copulation in Hymenoptera. *Entomol. News* **80**, 113–124.

Evans, H.E. and Eberhard, M.J.W. (1970) *The Wasps*, David and Charles, Newton Abbot, 265pp.

Ewen, A.B. and Arthur, A.P. (1976) Cuticular encystment in three noctuid species (Lepidoptera): Induction by acid gland secretion from an ichneumonid parasite (*Banchus flavescens*). *Ann. Entomol. Soc. Am.* **69**, 1087–1090.

Fabres, G. and Reymonet, C. (1991) L'induction maternelle de la diapause larvaire chez *Dinarmus acutus* (Hym.: Pteromalidae). *Entomophaga* **36**, 121–129.

Farish, D.J. (1972) The evolutionary implications of qualitative variation in the grooming behaviour of the Hymenoptera (Insecta). *Anim. Behav.* **20**, 662–676.

Fauvergue, X., Hopper, K.R. and Antolin, M.F. (1995) Mate finding via a trail sex-pheromone by a parasitoid wasp. *Proc. Natn. Acad. Sci. USA* **92**, 900–904.

Feddersen, I., Sander, K. and Schmidt, O. (1986) Virus-like particles with host protein-like antigenic determinants protect an insect parasitoid from encapsulation. *Experientia* **42**,1278–1281.

Federici, B.A. (1991) Viewing polydnaviruses as gene vectors of endoparasitic Hymenoptera. *Redia* **74**, 387–392.

Fergusson, N.D.M. (1988) A comparative study of the structures of phylogenetic significance of female genitalia of the Cynipoidea (Hymenoptera). *Syst. Ent.* **13**, 12–30.

Fergusson, N.D.M. (1990) A phylogenetic study of the Cynipoidea (Hymenoptera). Unpublished PhD Thesis, City of London Polytechnic.

Ferkovich, S.M. and Dillard, C.R. (1987) A study of uptake of radiolabeled host proteins and protein synthesis during development of eggs of the endoparasitoid, *Microplitis croceipes* (Cresson) (Braconidae). *Insect Biochem.* **16**, 337–345.

Field, S.A. and Austin, A.D. (1994) Anatomy and mechanics of the telescopic ovipositor system of *Scelio* Latreille (Hymenoptera: Scelionidae) and related genera. *Int. J. Insect Morphol. and Embryol.* **23**, 135–158.

Field, S.A. and Keller, M.A. (1993a) Alternative mating tactics and female mimicry as postcopulatory mate-guarding behaviour in the parasitic wasp *Cotesia rubecula*. *Anim. Behav.* **46**, 1183–1189.

Field, S.A. and Keller, M.A. (1993b) Courtship and intersexual signalling in the parasitic wasp *Cotesia rubecula* (Hymenoptera, Braconidae). *J. Insect Behav.* **6**, 737–750.

Field, S.A. and Keller, M.A. (1994) Localization of the female sex pheromone gland in *Cotesia rubecula* Marshall (Hymenoptera: Braconidae). *J. Hym. Res.* **3**, 151–156.

Finlayson, T. (1964) The caudal appendage of final-instar larvae of some Porizontinae (Hymenoptera: Ichneumonidae). *Canad. Entomol.* **96**, 1155–1158.

Finlayson, T. (1966) The false cocoon of *Hyposoter parorgyiae* (Vier.) (Hymenoptera: Ichneumonidae). *Canad. Entomol.* **98**, 139.

Finnamore, A.T. and Gauld, I.D. (1995) Bethylidae. In *The Hymenoptera of Costa Rica* (eds P. Hanson and I.D. Gauld), Oxford University Press, Oxford. pp. 470–479.

Fisher, J.P. (1970) The biology and taxonomy of some chalcidoid parasites (Hymenoptera) of stem-living larvae of *Apion* (Coleoptera: Curculionidae). *Trans. R. Entomol. Soc. Lond.* **122**, 293–322.

Fisher, K. (1932) *Agriotypus armatus* (Walk.) (Hymenoptera) and its relations with its hosts. *Proc. Zool. Soc. Lond.* **1932** (2), 451–461.

Fisher, R.A. (1930) *The Genetical Theory of Natural Selection*, 1st edn, Oxford University Press, Oxford.

Fisher, R.C. (1958) An experimental study of multiparasitism in Ichneumonidae. Unpublished PhD Thesis, University of Cambridge.

Fisher, R.C. (1963) Oxygen requirements and the physiological suppression of supernumerary insect parasitoids. *J. Exp. Biol.* **40**, 531–540.

Fisher, R.C. (1971) Aspects of the physiology of endoparasitic Hymenoptera. *Biol. Rev.* **46**, 243–278.

Fiske, W.F. (1910) Superparasitism: An important factor in the natural control of insects. *J. Econ. Ent.* **3**, 88–97.

Fitton, M.G., Shaw, M.R. and Austin, A.D. (1987) The Hymenoptera associated with spiders in Europe. *Zool. J. Linn. Soc.* **90**, 65–93.

Flanders, S.E. (1934) The secretion of the colleterial glands in the parasitic chalcids. *J. Econ. Ent.* **27**, 861–862.

Flanders, S.E. (1938) Cocoon formation in endoparasitic chalcidoids. *Ann. Entomol. Soc. Am.* **31**, 167–180.

Flanders, S.E. (1939) Environmental control of sex in hymenopterous insects. *Ann. Entomol. Soc. Am.* **32**, 11–26.

Flanders, S.E. (1942a) Oösorption and ovulation in relation to oviposition in the parasitic Hymenoptera. *Ann. Entomol. Soc. Am.* **35**, 251–266.

Flanders, S.E. (1942b) The larval meconium of parasitic Hymenoptera as a sign of the species. *J. Econ. Ent.* **35**, 456–457.

Flanders, S.E. (1942c) The sex-ratio in the Hymenoptera: A function of the environment. *Ecology*, **23**, 120–121.

Flanders, S.E. (1943) Indirect hyperparasitism and observations on three species of indirect hyperparasites. *J. Econ. Ent.* **36**, 921–926.

Flanders, S.E. (1945) The role of the spermatophore in the mass propagation of *Macrocentrus ancylivorus*. *J. Econ. Ent.* **38**, 323–327.

Flanders, S.E. (1950) Regulation of ovulation and egg disposal in the parasitic Hymenoptera. *Canad. Entomol.* **82**, 134–140.

Flanders, S.E. (1964) Dual ontogeny of the male *Coccophagus gurneyi* Comp. (Hymenoptera: Aphelinidae): A phenotypic phenomenon. *Nature (London)* **204**, 944–946.

Flanders, S.E. (1966) Unique biological aspects of the genus *Casca* and a description of a new species (Hymenoptera: Aphelinidae). *Ann. Entomol. Soc. Am.* **59**, 79–82.

Flanders, S.E. (1967) Deviate-ontogenies in the aphelinid male (Hymenoptera) associated with the oviposition behavior of the parental female. *Entomophaga* **12**, 415–427.

Fleming, J.A.G.W. and Summers, M.D. (1991) Polydnavirus DNA is integrated in the DNA of its parasitoid wasp host. *Proc. Natn. Acad. Sci.* **88**, 9770–9774.

Fleury, F., Allemand, R., Fouillet, P. and Boulétreau, M. (1994) Geographic variation in locomotor activity rhythms of *Leptopilina heterotoma*: Inheritance and role in species richness of the *Drosophila* parasitoid community. *Norw. J. Agric. Sci. Suppl.* **16**, 191–197.

Force, D.C. (1980) Do parasitoids of Lepidoptera larvae compete for hosts? Probably! *Amer. Nat.* **116**, 873–875.

Force, D.C. (1985) Competition among parasitoids of endophytic hosts. *Amer. Nat.* **126**, 440–444.

Ford, E.B. (1947) A murexide test for the recognition of pterins in intact insects. *Proc. R. Entomol. Soc. Lond. (A)* **22**, 72–76.

Francke-Grosmann, H. (1967) Ectosymbiosis in wood-inhabiting insects. In *Symbiosis* (ed. S.M. Henry), Academic Press, New York. pp. 142–205.

Frank, S.A. (1984) The behaviour and morphology of the fig wasps *Pegoscapus assuetus* and *P. jimeneza*. Descriptions and suggested behavioural characters for phylogenetic studies. *Psyche* **91**, 289–308.

Frazier, J.L. (1985) Nervous system: Sensory system. In *Fundamentals of Insect Physiology* (ed. M.S. Blum), John Wiley and Sons, New York, pp. 287–356.

Freeman, B.E. and Ittyeipe, K. (1982) Morph determination in *Melittobia*, a eulophid wasp. *Ecol. Ent.* **7**, 355–363.

Frilli, F. (1965a) Studi sugli imenotteri icneumonidi. I. *Devorgilla canescens* (Grav.). *Entomologica* **1**, 119–197.

Frilli, F. (1965b) Studi sugli imenotteri icneumonidi. III. *Latibulus argiolus* (Rossi) parassita delle larve di *Polistes* spp. *Entomologica* **2**, 21–53.

Frilli, F. (1981) Gli imenotteri icneumonidi ed il loro comportamento parassitario. *Atti Accad. Naz. Ital. Entomol.* **1981**, 1–20 + 2 plates.

Frost, S.W. (1959) Death feigning. In *Insect Life and Natural History*, Dover, New York. pp. 468–470.

Frühauf, E. (1924) Legeapparat und Eiablage bei Gallwespen (Cynipidae). *Zeitschr. Wissenschaft. Zool.* **121**, 656–723.

Fuester, R.W., Taylor, P.B., Peng, H. and Swan, K. (1993) Laboratory biology of a uniparental strain of *Meteorus pulchricornis* (Hymenoptera: Braconidae), an exotic larval parasite of the gypsy moth (Lepidoptera: Lymantriidae). *Ann. Entomol. Soc. Am.* **86**, 298–304.

Führer, E. and El-Sufty, R. (1979) Produktion fungistatischer Metabolite durch Teratocyten von *Apanteles glomeratus* L. (Hym., Braconidae). *Z. Parasitenkd.* **59**, 21–25.

Führer, E. and Kilincer, N. (1972) Die motorische aktivität der endoparasitischen Larven von *Pimpla turionellae* L. und *Pimpla flavicoxis* Ths. (Hym., Ichneum.) in der Wirtspuppe. *Entomophaga* **17**, 149–163.

Führer, E. and Willers, D. (1986) The anal secretion of the endoparasitic larva *Pimpla turionellae*: Sites of production and effects. *J. Insect Physiol.* **32**, 361–367.

Fukushima, J.I., Kuwahara, Y. and Suzuki, T. (1989) Isolation and identification of a kairomone responsible for the stinging behaviour of *Bracon hebetor* Say (Hymenoptera: Braconidae) from frass of the almond moth *Cadra cautella* Walker. *Agric. and Biol. Chem.* **53**, 3057–3059.

Fukushima, J.I., Kuwahara, Y., Yamada, A. and Suzuki, T. (1990) New non-cyclic homo-diterpene from the sting glands of *Bracon hebetor* Say (Hymenoptera, Braconidae). *Agric. and Biol. Chem.* **54**, 809–810.

Fulton, B.B. (1933) Notes on *Habrocytus cerealellae*, a parasite of Angoumois grain moth. *Ann. Entomol. Soc. Am.* **26**, 536–553.

Fulton, B.B. (1940) The hornworm parasite, *Apanteles congregatus* Say, and the hyperparasite *Hypopteromalus tabacum* (Fitch). *Ann. Entomol. Soc. Am.* **33**, 231–244.

Fung, S.Y. (1988) Butenolides in parasitoids and adults of small ermine moths, *Yponomeuta* spp. (Lepidoptera: Yponomeutidae). *Proc. Kon. Ned. Acad. Wet.* **91C**, 363–367.

Furlong, M.J. and Pell, J.K. (1996) Interactions between the fungal ento-mopathogen *Zoophthora radicans* Brefeld (Entomophthorales) and two hymenopteran parasitoids attacking the diamondback moth, *Plutella xylostella* L. *J. Invert. Pathol.* **68**, 15–21.

Fursov, V.N. and Kostyukov, V.V. (1987) New species of the genus *Tetrastichus* (Hymenoptera, Eulophidae), egg parasites of damselflies and dragonflies and of predaceous diving beetles. *Zool. Zhurnal* **66**, 217–228. (In Russian with English summary.)

Galloway, I.D., Austin, A.D. and Masner, L. (1992) Revision of the genus *Neuroscelio* Dodd, primitive scelionids (Hymenoptera: Scelionidae) from Australia, with a discussion of the ovipositor system of the tribe Gryonini. *Invertebr. Taxon.* **6**, 523–545.

Ganesalingam, V.K. (1972) Anatomy and histology of the sense organs of the ovipositor of the ichneumonid wasp *Devorgilla canescens*. *J. Insect Physiol.* **18**, 1857–1868.

Ganeshaiah, K.N., Kathuria, P., Uma Shaanker, R. and Vasudeva, R. (1995) Evolution of style-length variability in figs and optimization of ovipositor length in their pollinator wasps: a coevolutionary model. *J. Genet.* **74**, 25–39.

Gardiner, L.M. (1966) A photographic record of oviposition by *Rhyssa lineolata* (Kirby) (Hymenoptera: Ichneumonidae). *Canad. Entomol.* **98**, 95–97.

Gardner, S.M., Ward, S.A. and Dixon, A.F.G. (1984) Limitation of superparasitism by *Aphidius rhopalosiphi*: A consequence of aphid defensive behaviour. *Ecol. Ent.* **9**, 149–155.

Gargiulo, G., Malva, C., Pennacchio, F. and Tremblay, E. (1988) Structure of *Aphidius* Nees (Hymenoptera, Braconidae) rDNA: A molecular tool in biosys-tematic research. *Boll. Lab. Ent. Agr. Filippo Silvestri* **45**, 203–219.

Gaston, K.J. and Gauld, I.D. (1993) How many species of pimplines (Hymenoptera: Ichneumonidae) are there in Costa Rica? *J. Trop. Ecol.* **9**, 491–499.

Gatenby, J.B. (1917) The embryonic development of *Trichogramma evanescens* Westw., monoembryonic egg parasite of *Donacia simplex*. *Quart. J. Microscop. Sci.* **62**, 149–187.

Gauld, I.D. (1976) The taxonomy of the genus *Heteropelma* Wesmael (Hymenoptera: Ichneumonidae). *Bull. Br. Mus. (Nat. Hist.) Entomol.* **34**, 155–219.

Gauld, I.D. (1986a) Latitudinal gradients in ichneumonid species-richness in Australia. *Ecol. Ent.* **11**, 155–161.

Gauld, I.D. (1986b) Taxonomy, its limitations and its role in understanding para-sitoid biology. In *Insect Parasitoids* (eds J. Waage and D. Greathead), Academic Press, London. pp. 1–21.

Gauld, I.D. (1987) Some factors affecting the composition of tropical ichneu-monoid faunas. *Biol. J. Linn. Soc.* **30**, 299–312.

Gauld, I.D. (1988a) Evolutionary patterns of host utilization by ichneumonoid par-asitoids (Hymenoptera: Ichneumonidae and Braconidae). *Biol. J. Linn. Soc.* **35**, 351–377.

Gauld, I.D. (1988b) A survey of the Ophioninae (Hymenoptera: Ichneumonidae) of tropical Mesoamerica with special reference to the fauna of Costa Rica. *Bull. Br. Mus. (Nat. Hist.) Entomol.* **57**, 1–309.

Gauld, I.D. (1991) The Ichneumonidae of Costa Rica, 1. *Mem. Amer. Entomol. Inst.* **47**, 1–589.

Gauld, I.D. (1995) Ichneumonidae, in *The Hymenoptera of Costa Rica*, (eds P. Hanson and I.D. Gauld), Oxford University Press, Oxford, pp. 389–431.

Gauld, I.D. and Bolton, B. (eds) (1988) *The Hymenoptera*, Oxford University Press/The Natural History Museum, London, Oxford. 332 pp.

Gauld, I.D. and Fitton, M.G. (1987) Sexual dimorphism in the Ichneumonidae: A response to Hurlbutt. *Biol. J. Linn. Soc.* **31**, 291–300.

Gauld, I.D. and Gaston, K.J. (1994) The taste of enemy-free space: Parasitoids and nasty hosts. In *Parasitoid Community Ecology* (eds B.A. Hawkins and W. Sheehan), Oxford University Press, Oxford. pp. 279–299.

Gauld, I.D. and Gaston, K.J. (1995) The Costa Rican Hymenoptera fauna. In *The Hymenoptera of Costa Rica* (eds P. Hanson and I.D. Gauld), Oxford University Press, Oxford. pp. 13–19.

Gauld, I.D. and Hanson, P.E. (1995) Carnivory in the larval Hymenoptera. In *The Hymenoptera of Costa Rica* (eds P. Hanson and I.D. Gauld), Oxford University Press, Oxford. pp. 40–45.

Gauld, I.D. and Huddleston, T. (1976) The nocturnal Ichneumonoidea of the British Isles, including a key to the genera. *Entomologist's Gaz.* **27**, 35–49.

Gauld, I.D. and Janzen, D.H. (1994) The classification, evolution and biology of the Costa Rican species of *Cryptophion* (Hymenoptera, Ichneumonidae). *Zool. J. Linn. Soc.* **110**, 297–324.

Gauld, I.D., Gaston, K.J. and Janzen, D.H. (1992) Plant allochemicals, tritrophic interactions and the anomalous diversity of tropical parasitoids: The 'nasty' host hypothesis. *Oikos*, **65**, 353–357.

Geden, C.J., Long, S.I., Rutz, D.A. and Becnel, J.J. (1995) Nosema disease of the parasitoid *Muscidifurax raptor* (Hymenoptera, Pteromalidae) – Prevalence, patterns of transmission, management, and impact. *Biol. Control* **5**, 607–614.

Geervliet, J.B.F., Vet, L.E.M. and Dicke, M. (1994) Volatiles from damaged plants as major cues in long-range host-searching by the specialist parasitoid *Cotesia rubecula. Entomol. Exp. Appl.* **73**, 289–297.

Génieys, P. (1925) *Habrobracon brevicornis* Wesm. *Ann. Entomol. Soc. Am.* **18**, 143–202.

Génieys, P. (1926) Aberration de la ponte d'un hyménoptère parasite. *Feuille Nat.* **47**, 121–122.

Gennerich, J. (1922) Morphologische und biologische Untersuchungen der Putzapparate der Hymenopteren. *Arch. Naturgesch.* **A 12**, 1–63.

Gerling, D. and Legner, E.F. (1968) Developmental history and reproduction of *Spalangia cameroni*, parasite of synanthropic flies. *Ann. Entomol. Soc. Am.* **61**, 1436–1443.

Gerling, D. and Orion, T. (1972) The giant cells produced by *Telenomus remus* Nixon (Hym., Scelionidae). *Bull. Entomol. Res.* **61**, 385.

Gerling, D. and Rotary, N. (1974) Structure and function of the seminal vesicles and the spermatheca in *Bracon mellitor* (Hym., Braconidae). *Int. J. Insect Morphol. Embryol.* **3**, 159–162.

Gerling, D., Spivak, D. and Vinson, S.B. (1987) Life history and host discrimination of *Encarsia deserti* (Hymenoptera: Aphelinidae), a parasitoid of *Bemisia tabaci* (Homoptera: Aleyrodidae). *Ann. Entomol. Soc. Am.* **80**, 224–229.

Gerling, D., Orion, T. and Delarea, Y. (1990) *Eretmocerus* penetration and immature development: A novel approach to overcome host immunity. *Archiv. Insect Bioch. Physiol.* **13**, 247–253.

Gerling, D., Tremblay, E. and Orion, T. (1991) Initial stages of the vital capsule formation in the *Eretmocerus–Bemisia tabaci* association. *Redia* **74**, 411–415.

Gherna, R.L., Werren, J.H., Weisburg, W. *et al.* (1991) *Arsenophonus nasoniae* gen. nov., sp. nov., the causative agent of the son-killer trait in the parasitic wasp *Nasonia vitripennis. Int. J. Syst. Bact.* **41**, 563–565.

Gibbons, J.R.H. (1979) A model for sympatric speciation in *Megarhyssa* (Hymenoptera: Ichneumonidae): Competitive speciation. *Am. Nat.* **114**, 719–741.

Gibson, D.O. and Mani, G.S. (1984) An experimental investigation of the effects of selective predation by birds and parasitoid attack on the butterfly *Danaus chrysippus* (L.). *Proc. R. Soc. Lond. B*, **221**, 31–51.

Gibson, G.A.P. (1985) Some pro- and meso-thoracic structures important for phylogenetic analysis of Hymenoptera, with a review of terms used for the structures. *Canad. Entomol.* **117**, 1395–1443.

Gibson, G.A.P. (1986a) Evidence for monophyly and relationships of Chalcidoidea, Mymaridae, and Mymarommatidae (Hymenoptera: Terebrantes). *Canad. Entomol.* **118**, 205–240.

Gibson, G.A.P. (1986b) Mesothoracic skeletomusculature and mechanics of flight and jumping in Eupelminae (Hymenoptera, Chalcidoidea: Eupelmidae). *Canad. Entomol.* **118**, 691–728.

Gibson, G.A.P. (1990) A word on chalcidoid classification. *Chalcid Forum* **13**, 7–9.

Gifford, J.R. and Mann, G.A. (1967) Biology, rearing, and a trial release of *Apanteles flavipes* in the Florida Everglades to control sugarcane borer. *J. Econ. Ent.* **60**, 44–47.

Giri, M.K. and Freytag, P.H. [(1986)/1989] Development of *Dicondylus americanus* (Hymenoptera: Dryinidae). *Frust. Entomol. (N.S.)* **9**, 215–222.

Girin, C. and Boulétreau, M. (1994) Maternal inheritance of infestation efficiency in a parasitoid wasp, *Trichogramma bourarachae*: The role of symbionts. *Norw. J. Agric. Sci. Suppl.* **16**, 177–184.

Godfray, H.C.J. (1987) The evolution of clutch size in parasitic wasps. *Amer. Nat.* **129**, 221–233.

Godfray, H.C.J. (1988) Virginity in haplodiploid populations: A study on fig wasps. *Ecol. Ent.* **13**, 283–291.

Godfray, H.C.J. (1993) *Parasitoids: Behavioural and Evolutionary Ecology*, Princeton University Press, New Jersey.

Godfray, H.C.J. and Hunter, M.J. (1992) Sex ratio of heteronomous hyperparasitoids: Adaptive or non-adaptive? *Ecol. Ent.* **17**, 89–90.

Godfray, H.C.J. and Shaw, M.R. (1987) Seasonal variation in the reproductive strategy of the parasitic wasp *Eulophus larvarum* (Hymenoptera: Chalcidoidea: Eulophidae). *Ecol. Ent.* **12**, 251–256.

Godfray, H.C.J. and Hardy, I.C.W. (1993) Virginity in haplodiploid animals. In *Insect Sex Ratios* (eds S. Wrensch and D. Krainacker), Chapman & Hall, New York, pp. 404–417.

Godfray, H.C.J., Agassiz, D.J.L., Nash, D.R. and Lawton, J.H. (1995) The recruitment of parasitoid species to two invading herbivores. *J. Anim. Ecol.* **64**, 393–402.

Gokhman, V.E. (1989) Karyotypes of ichneumon flies of the *Tycherus osculator* group (Hymenoptera: Ichneumonidae). *Entomol. Obozren.* **68**, 710–714. (In Russian.)

Gokhman, V.E. (1991) New species of Phaeogenini (Hymenoptera: Ichneumonidae) from the European part of the USSR. *Zoologichesky Zh.* **70**, 73–80. (In Russian.)

Gokhman, V.E. (1993) New data on the karyology of Ichneumonina (Hymenoptera: Ichneumonidae). *Zoologichesky Zh.* **72**, 85–91. (In Russian.)

Gokhman, V.E. and Quicke, D.L.J. (1995) The last twenty years of parasitic Hymenoptera karyology: An update and phylogenetic implications. *J. Hym. Res.* **4**, 41–63.

Gordh, G. (1976) *Goniozus gallicola* Fouts, a parasite of moth larvae, with notes on other bethylids (Hymenoptera: Bethylidae; Lepidoptera: Gelechiidae). *Tech. Bull. USDA* **1524**, 1–27.

Gordh, G. and DeBach, P. (1976) Male inseminative potential in *Aphytis lingnanensis* (Hymenoptera: Aphelinidae). *Canad. Entomol.* **108**, 583–589.

Gordh, G. and DeBach, P. (1978) Courtship behaviour in the *Aphytis lingnanensis* group, its potential usefulness in taxonomy, and a review of sexual behaviour in the parasitic Hymenoptera (Chalc.; Aphelinidae). *Hilgardia* **46**, 37–75.

Gordh, G. and Hendrickson, R. (1976) Courtship behaviour in *Bathyplectes anurus* (Thomson) (Hymenoptera: Ichneumonidae). *Entomol. News*, **87**, 271–274.

Goulet, H. and Huber, J.T. (eds) (1993) *Hymenoptera of the World: An Identification Guide to Families*, Agriculture Canada, Research Branch, Ottawa. 668 pp.

Graham, M.W.R. de V. (1969) The Pteromalidae of North-Western Europe (Hymenoptera: Chalcidoidea). *Bull. Br. Mus. (Nat. Hist.), Ent. Suppl.* **16**, 1–908.

Graur, D. (1985) Gene diversity in Hymenoptera. *Evolution*, **39**, 190–199.

Grbíc, M., Ode, P.J. and Strand, M.R. (1992) Sibling rivalry and brood sex ratios in polyembryonic wasps. *Nature* **360**, 254–257.

Greany, P. (1986) *In vitro* culture of hymenopterous larval endoparasitoids. *J. Insect Physiol.* **32**, 409–419.

Greany, P., Hawke, S.D., Carlysle, T.C. and Anthony, D.W. (1977) Sense organs in the ovipositor of *Biosteres (Opius) longicaudatus*, a parasite of the Caribbean fruit fly *Anastrepha suspensa*. *Ann. Entomol. Soc. Am.* **70**, 319–321.

Greathead, D.J. (1986) Parasitoids in classical biological control. In *Insect Parasitoids* (eds J. Waage and D. Greathead), Academic Press, London. pp. 289–318.

Green, R.F., Gordh, G.C. and Hawkins, B.A. (1982) Precise sex ratios in highly inbred parasitic wasps. *Amer. Nat.* **120**, 653–655.

Grenier, S. (1994) Rearing *Trichogramma* and other egg parasitoids on artificial diets. In *Biological Control with Egg Parasitoids* (eds E. Wajnberg and S.A. Hassan), CAB International, Wallingford. pp. 73–92.

Griffiths, D.C. (1960) The behaviour and specificity of *Monoctonus paludum* Marshall (Hym., Braconidae), a parasite of *Nasonovia ribis-nigri* (Mosley) on lettuce. *Bull. Entomol. Res.* **51**, 303–319.

Griffiths, D.C. (1961) The development of *Monoctonus paludum* Marshall (Hym., Braconidae) in *Nasonovia ribis-nigri* (Mosley) on lettuce, and immunity reactions in other lettuce aphids. *Bull. Entomol. Res.* **52**, 147–163.

Griffiths, N. and Godfray, H.C.J. (1988) Local mate competition, sex ratio and clutch size in bethylid wasps. *Behav. Ecol. Sociobiol.* **22**, 211–217.

Grissell, E.F. and Goodpasture, C.E. (1981) A review of Nearctic Podagrionini, with description of sexual behavior of *Podagrion mantis* (Hymenoptera: Torymidae). *Ann. Entomol. Soc. Am.* **74**, 226–241.

Gromyszkalkowska, K. and Grochowska, M. (1992) Respiration rates of some developmental stages of *Polemochartus liparae* (Gir.) (Hymenoptera) and its host *Lipara similis* Schin (Diptera). *Comp. Biochem. Physiol.* **102A**, 473–476.

Grosch, D.S. (1948) Experimental studies on the mating reaction of male *Habrobracon*. *J. Comp. Physiol. Psychol.* **41**, 188–195.

Grosch, D.S. (1950) Olfactometer experiments with male braconids. *Ann. Entomol. Soc. Am.* **43**, 334–342.

Grosch, D.S. (1951) Octonucleate and uninucleate structural units: Cytological contrasts in the larval and adult midguts of the parasitic wasp *Habrobracon*. *J. Elisha Mitchell Sci. Soc.* **67**, 184–185.

Grosch, D.S. (1952) The spinning glands of impaternate (male) *Habrobracon* larvae: Morphology and cytology. *J. Morphol.* **91**, 221–236.

Grosch, D.S. (1988) Genetic research on braconid wasps. *Adv. Genetics* **35**, 109–184.

Grosch, D.S., Kratsas, R.G. and Petters, R.M. (1977) Variation in *Habrobracon juglandis* ovariole number. I. Ovariole number increase induced by extended cold shock of fourth-instar larvae. *J. Embryol. Exp. Morph.* **40**, 245–251.

Gross, P. (1993) Insect behavioral and morphological defenses against parasitoids. *Annu. Rev. Entomol.* **38**, 251–273.

Gross, P. and Price, P.W. (1988) Plant influences of parasitism of two leafminers: A test of enemy-free space. *Ecology* **69**, 1506–1516.

Grossniklaus-Buergin, C. and Lanzrein, B. (1990) Endocrine relationship between the parasitoid *Chelonus* sp. and its host *Trichplusia ni. Arch. Insect Biochem. Physiol.* **14**, 201–216.

Grout, T.G. and Brothers, D.J. (1982) Behaviour of a parasitic pompilid wasp (Hymenoptera). *J. Entomol. Soc. South Afr.* **45**, 217–220.

Grubbs, S.C. and Conner, G.W. (1976) Visible mutation induction in *Mormoniella* by low frequency ultrasonic energy. *J. Hered.* **67**, 191–193.

Guerra, A.A. and Martinez, S. (1994) An *in vitro* rearing system for the propogation of the ectoparasitoid *Catolaccus grandis. Entomol. Exp. Appl.* **72**, 11–16.

Guerra, A.A., Robacker, K.M. and Martinez, S. (1993) Free amino acid and protein titers in *Anthonomus grandis* larvae venomized by *Bracon mellitor. Entomophaga* **38**, 519–525.

Guerra, A.A., Robacker, K.M. and Martinez, S. (1994) *In vitro* rearing of *Bracon mellitor* and *Catolaccus grandis* with artificial diets devoid of insect components. *Entomol. Exp. Appl.* **68**, 303–307.

Guershon, M. and Gerling, D. (1994) Defence of a sessile host against parasitoids: *Aleyrodes singularis* vs. *Encarsia* spp. *Norw. J. Agric. Sci. Suppl.* **16**, 255–260.

Guertin, D.S., Ode, P.J., Strand, M.R. and Antolin, M.F. (1996) Host-searching and mating in an outbreeding parasitoid wasp. *Ecol. Ent.* **21**, 27–33.

Guillot, F.S. and Vinson, S.B. (1972) Sources of substances which elicit a behavioural response from the insect parasitoid, *Campoletis perdistinctus. Nature* **235**, 169–170.

Gurney, A.B. (1953) Notes on the biology and immature stages of a cricket parasite of the genus *Rhopalosoma. Proc. US Natn. Mus.* **103**, 19–34.

Gutierrez, A.P. (1970) Studies on host selection and host specificity of the aphid hyperparasite *Charips victrix* (Hymenoptera: Cynipidae). 6. Description of sensory structures and a synopsis of host selection and host specificity. *Ann. Entomol. Soc. Am.* **63**, 1705–1709.

Guzo, D. and Stoltz, D.B. (1985) Obligatory multiparasitism in the tussock moth, *Orgyia leucostigma. Parasitology* **90**, 1–10.

Haardt, H. and Holler, C. (1992) Differences in life history traits between isofemale lines of the aphid parasitoid *Aphelinus abdominalis* (Hymenoptera, Aphelinidae). *Bull. Entomol. Res.* **82**, 479–484.

Haeselbarth, E. (1979) Zur Parasitierung der Puppen von Forleuhe (*Panolis flammea* (Schiff.)), Kiefernspanner (*Bupalus piniarius* (L.)) und Heidelbeerspanner (*Boarmia bistortana* (Goezel)) in bayerischen Kiefernwaldern. *Z. Angew. Entomol.* **87**, 186–202, 311–322.

Hagen, K.S. (1964) Developmental stages of parasites. In *Biological Control of Insect Pests and Weeds* (ed. P. DeBach), Chapman & Hall, London. pp. 168–246.

Hailemichael, Y., Smith, J.W. and Wiedenmann, R.N. (1994) Host finding behavior, host acceptance, and host suitability of the parasite *Xanthopimpla stemmator. Entomol. Exp. Appl.* **71**, 155–166.

Hallez, P. (1886) Loi de l'orientation de l'embryon chez les insectes. *Compt. Rend.* **103**, 606–608.

Halstead, J.A. (1988) A gynandromorph of *Hockeria rubra* (Ashmead) (Hymenoptera: Chalcididae). *Proc. Entomol. Soc. Wash.* **90**, 258–259.

Hamilton, W.D. (1967) Extraordinary sex ratios. *Science* **156**, 477–488.

Hamilton, W.D. (1979) Wingless and fighting males in fig wasps and other insects. In *Sexual Selection and Reproductive Competition in Insects* (eds M.S. Blum and A.A. Blum), Academic Press, New York. pp. 167–200.

Hamm, J.J. and Styer, E.L. (1985) A new virus associated with the reproductive tract of the hymenopterous parasitoid *Cotesia marginiventris* and its replication in noctuid host larvae. *Proc. 18th Ann. Mtg Soc. Invert. Pathol.* (abstract).

Hamm, J.J., Styer, E.L. and Lewis, W.J. (1988) A baculovirus pathogenic to the parasitoid, *Microplitis croceipes* (Hymenoptera, Braconidae). *J. Invert. Pathol.* **52**, 189–191.

Hamm, J.J., Styer, E.L. and Lewis, W.J. (1992) Three viruses found in the braconid parasitoid *Microplitis croceipes* and their implications in biological control programmes. *Biol. Control* **2**, 329–336.

Hamm, J.J., Styer, E.L. and Steiner, W.W. (1994) Reovirus-like particle in the parasitoid *Microplitis croceipes* (Hymenoptera, Braconidae). *J. Invert. Pathol.* **63**, 304–306.

Handlirsch, A. (1907) *Die Fossilen Insekten und die Phylogenie der rezenten Formen. Ein Handbuch fur Pälaontologen und Zoologen*, Leipzig. 1430 pp.

Hanson, P. and Gauld, I.D. (eds) (1995) *The Hymenoptera of Costa Rica*, Oxford University Press. 893pp.

Hardy, I.C.W. (1994) Sex ratio and mating structure in the parasitoid Hymenoptera. *Oikos*, **69**, 3–20.

Hardy, I.C.W. (1996) Precocious larvae in the polyembryonic parasitoid *Copidosoma sosares* (Hymenoptera: Encyrtidae). *Ent. Ber., Amst.* **56**, 88–92.

Hardy, I.C.W. and Blackburn, T.M. (1991) Brood guarding in a bethylid wasp. *Ecol. Ent.* **16**, 55–62.

Hardy, I.C.W. and Cook, J.M. (1995) Brood sex ratio variance, developmental mortality and virginity in a gregarious parasitoid wasp. *Oecologia* **103**, 162–169.

Hardy, I.C.W., Hick, A.J., Höller, C. *et al.* (1994) The responses of *Praon* spp. parasitoids to aphid sex pheromone components in the field. *Entomol. Expt. Appl.* **71**, 95–99.

Harris, V.E. and Todd, J.W. (1980) Male-mediated aggregation of male, female and 5th-instar southern green stink bugs and concomitant attraction of a tachinid parasite, *Trichopria pennipes*. *Entomol. Exp. Appl.* **27**, 117–126.

Hatakeyama, M., Sawa, M. and Oishi, K. (1994) Production of haploid–haploid chimeras by sperm injection in the sawfly, *Athalia rosae* (Hymenoptera). *Roux Arch. Devel. Biol.* **203**, 450–453.

Haviland, M.D. (1920) On the bionomics and development of *Lygocerus testaceimanus*, Kieffer, and *Lygocerus cameroni*, Kieffer, (Proctotrypoidea – Ceraphronidae) parasites of *Aphidius* (Braconidae). *Quart. J. Microsc. Sci.* **65**, 101–127.

Havron, A. and Rosen, D. (1988) Selection for pesticide resistance in *Aphytis*. In *Proceedings of the Sixth International Citrus Congress* (eds R. Goren and K. Mendel), Balaban Publishers, Tel Aviv. pp. 1187–1193.

Havron, A., Rosen, D. and Rössler, Y. (1987a) A test method for pesticide tolerance in minute parasitic Hymenoptera. *Entomophaga* **32**, 83–95.

Havron, A., Rosen, D., Rössler, Y. and Hillel, J. (1987b) Selection on the male hemizygous genotype in arrhenotokous insects and mites. *Entomophaga* **32**, 261–268.

Hawke, S.D., Farley, R.D. and Greany, P.D. (1973) The fine structure of sense organs in the ovipositor of the parasitic wasp, *Orgilus lepidus* Muesebeck. *Tissue and Cell*, **5**, 171–184.

Hawkins, B.A. (1990) Global patterns of parasitoid assemblage size. *J. Anim. Ecol.* **59**, 57–72.

Hawkins, B.A. (1993) Refuges, host population dynamics and the genesis of parasitoid diversity. In *Hymenoptera and Biodiversity* (eds J. LaSalle and I.D. Gauld), C.A.B. International, Wallingford. pp. 235–256.

Hawkins, B.A. (1994) *Pattern and Process in Host–Parasitoid Interactions*, Cambridge University Press, Cambridge. 190 pp.

Hawkins, B.A. and Mills, N.J. (1996) Variability in parasitoid community structure. *J. Anim. Ecol.* **65**, 501–516.

Hawkins, B.A. and Compton, S.G. (1992) African fig wasp communities: Undersaturation and latitudinal gradients. *J. Anim. Ecol.* **61**, 361–372.

Hawkins, B.A. and Gagné, R.J. (1989) Determinants of assemblage size for the parasitoids of Cecidomyiidae (Diptera). *Oecologia* **81**, 75–88.

Hawkins, B.A. and Lawton, J.H. (1987) Species richness for parasitoids of British phytophagous insects. *Nature* **326**, 788–790.

Hawkins, B.A. and Smith, J.W. Jr (1986) *Rhaconotus roslinensis* (Hymenoptera: Braconidae), a candidate for biological control of stalkboring sugarcane pests (Lepidoptera: Pyralidae): Development, life tables, and intraspecific competition. *Ann. Entomol. Soc. Am.* **79**, 905–911.

Hawkins, B.A., Askew, R.R. and Shaw, M.R. (1990) Influences of host feeding-niche and foodplant type on generalist and specialist parasitoids. *Ecol. Ent.* **15**, 275–280.

Hawkins, B.A., Shaw, M.R. and Askew, R.R. (1992) Relations among assemblage size, host specialization, and climatic variability in North American parasitoid communities. *Am. Nat.* **139**, 58–79.

Hawkins, B.A., Thomas, M.B. and Hochberg, M.E. (1993) Refuge theory and biological control. *Science,* **262**, 1429–1437.

Hawlitzky, N. (1972) Mode de pénétration d'un parasite ovo-larvaire *Phanerotoma flavitestacea* Fisch. (Hym.: Braconidae) dans son hôte embryonnaire, *Anagasta kuehniella* Zell. (Lep.: Pyralidae). *Entomophaga* **17**, 375–389.

Hawlitzky, N. (1979) Devenir de l'oeuf et comportement de la larve de *Phanerotoma flavitestacea* (Hym.: Braconidae) lorsque la femelle pond dans des oeufs d'*Anagasta kuehniella* (Lep.: Pyralidae) ayant, atteint des stades de développement variés. *Entomophaga* **24**, 273–245.

Hawlitzky, N. and Boulay, C. (1986) Effects of the egg–larval parasite *Phanerotoma flavitestacea* Fisch. (Hymenoptera, Braconidae) on the dry weight and chemical composition of its host *Anagasta kuehniella* Zell. (Lepidoptera, Pyralidae). *J. Insect Physiol.* **32**, 269–274.

Hayakawa, Y., Yazaki, K., Yamanaka, A. and Tanaka, T. (1994) Expression of polydnavirus genes from the parasitoid wasp *Cotesia kariyai* in two noctuid hosts. *Insect Molec. Biol.* **3**, 97–103.

Hays, D.B. and Vinson, S.B. (1971) Acceptance of *Heliothis virescens* (F.) (Lepidoptera, Noctuidae) as a host by the parasite *Cardiochiles nigriceps* Viereck (Hymenoptera, Braconidae). *Anim. Behav.* **19**, 344–352.

Heatwole, H., Davis, D.M. and Wenner, A.M. (1962) The behaviour of *Megarhyssa*, a genus of parasitic hymenopterans (Ichneumonidae: Ephialtinae). *Z. Tierpsychol.* **19**, 652–664.

Hegazi, E.M., Schopf, A., Führer, E. and Fouad, S.H. (1988) Developmental synchrony between *Spodoptera littoralis* (Boids.) and its parasite *Microplitis rufiventris* Kok. *J. Insect Physiol.* **34**, 773–778.

Hegazi, E.M., Shaaban, M.A. and El-Singaby, N.R. (1991) Development of *Microplitis rufiventris* (Hymenoptera: Braconidae) in superparasitized *Spodoptera littoralis* (Lepidoptera: Noctuidae). *Ann. Entomol. Soc. Am.* **84**, 571–574.

Heimpel, G.E. (1994) Virginity and the cost of insurance in highly inbred Hymenoptera. *Ecol. Ent.* **19**, 299–302.

Heinz, K.M. (1991) Sex specific reproductive consequences of body size in the solitary ectoparasitoid, *Diglyphus begini*. *Evolution,* **45**, 1511–1515.

Hellqvist, S. (1994) Biology of *Synacra* sp. (Hym., Diapriidae), a parasitoid of *Bradysia paupera* (Dipt., Sciaridae) in Swedish greenhouses. *J. Appl. Ent.* **117**, 491–497.

Henaut, A. (1990) Study of sound produced by *Pimpla instigator* (Hymenoptera, Ichneumonidae) during host selection. *Entomophaga*, **35**, 127–139.

Henaut, A. and Guerdoux, J. (1982) Location of a lure by the drumming insect *Pimpla instigator* (Hymenoptera, Ichneumonidae). *Experientia*, **38**, 346–347.

Hennessey, R.D. (1981) At-rest setal wing coupling and restraining mechanisms in the Encyrtidae and Aphelinidae (Hymenoptera: Chalcidoidea). *Ann. Entomol. Soc. Am.* **74**, 172–176.

Henriksen, K.L. (1922) Notes upon some aquatic Hymenoptera. *Ann. Biol. Lacustre* **11**, (19–37.

Henriquez, N.P. and Spence, J.R. (1993) Host location by the gerrid egg parasitoid *Tiphodytes gerriphagus* (Marchal) (Hymenoptera: Scelionidae). *J. Insect Behav.* **6**, 455–466.

Henter, H.J. and Via, S. (1995) The potential for coevolution in a host–parasitoid system. I. Genetic variation within an aphid population in susceptibility to a parasitic wasp. *Evolution* **49**, 427–438.

Heraty, J.M. and Barber, K.N. (1990) Biology of *Obeza floridanum* (Ashmead) and *Pseudochalcura gibbosa* (Provancher) (Hymenoptera: Eucharitidae). *Proc. Entomol. Soc. Wash.* **92**, 248–258.

Heraty, J.M. and Darling, D.C. (1984) Comparative morphology of the planidial larvae of Eucharitidae and Perilampidae (Hymenoptera: Chalcidoidea). *Syst. Entomol.* **9**, 309–328.

Heraty, J.M., Woolley, J.B. and Darling, D.C. (1994) Phylogenetic implications of the mesofurca and mesopostnotum in Hymenoptera. *J. Hym. Res.* **3**, 241–277.

Hermann, H.R. and Douglas, M.E. (1976) Comparative survey of the sensory structures on the sting and the ovipositor of hymenopterous insects. *J. Georgia Entomol. Soc.* **11**, 223–239.

Hermann, H.R. and Willer, D.E. (1986) Resilin distribution and its function in the venom apparatus of the honey bee, *Apis mellifera* L. (Hymenoptera: Apidae). *Int. J. Insect Morphol. Embryol.* **15**, 107–114.

Herre, E.A. (1989) Coevolution of reproductive characteristics in 12 species of New World figs and their pollinator wasps. *Experientia* **45**, 637–647.

Herre, E.A. (1993) Population structure and the evolution of virulence in nematode parasites of fig wasps. *Science* **259**, 1442–1445.

Hespenheide, H.A. (1979) Are there fewer parasitoids in the tropics? *Amer. Nat.* **113**, 766–769.

Hidoh, O., Kawashima, T., Fukami, J.-I. and Kainoh, Y. (1992) EAG responses of parasitoids, *Ascogaster reticulatus* Watanabe (Hymenoptera: Braconidae), to the female sex pheromone. *Appl. Entomol. Zool.* **27**, 587–589.

Hill, C.C. (1926) *Platygaster hiemalis* Forbes, a parasite of the hessian fly. *J. Agric. Res. Wash. DC* **32**, 261–275.

Hinton, H.E. (1955) Protective devices of endopterygote pupae. *Trans Soc. Br. Ent.* **12**, 49–92.

Hinton, H.E. (1981) *Biology of Insect Eggs*, 3 vols, Pergamon Press, Oxford. 1125pp.

Hinz, R. (1981) Die europäischen Arten der Gattung *Stilbops* Foerster (Hymenoptera, Ichneumonidae). *Nachricht. Bayerisch. Entomol.* **30**, 62–64.

Hinz, R. and Short, J. (1983) Life-history and systematic position of the European *Alomya* species (Hymenoptera: Ichneumonidae). *Ent. Scand.* **14**, 462–466.

Hirose, Y. (1994) Determinants of species richness and composition in egg parasitoid assemblages of Lepidoptera. In *Parasitoid Community Ecology* (eds B.A. Hawkins and W. Sheehan), Oxford University Press, Oxford. pp. 19–29.

Hiss, R.H., Norris, D.E., Dietrich, C.H. *et al.* (1994) Molecular taxonomy using single-strand conformation polymorphism (SSCP) analysis of mitochondrial ribosomal DNA genes. *Insect Molec. Biol.* **3**, 171–182.

Hobbs, G.A. and Krunic, M.D. (1971) Comparative behavior of three chalcidoid (Hymenoptera) parasites of the alfalfa leafcutter bee, *Megachile rotundata*, in the laboratory. *Canad. Entomol.* **103**, 674–685.

Hochberg, M.E. and Hawkins, B.A. (1992) Refuges as a predictor of parasitoid diversity. *Science* **255**, 973–976.

Hochberg, M.E. and Hawkins, B.A. (1993) Predicting parasitoid species richness. *Am. Nat.* **142**, 671–693.

Hoebeke, E.R. and Wheeler, Q.D. (1990) Notes on the biology of *Brachyserphus barberi* Townes (Hymenoptera, Serphidae), a parasitoid of the fungus beetle *Mycetophagus melsheimeri* LeConte (Coleoptera, Mycetophagidae). *J. New York Entomol. Soc.* **98**, 376–378.

Hoffman, J.D. and Ignoffi, C.M. (1974) Growth of *Pteromalus puparum* in a semisynthetic medium. *Ann. Entomol. Soc. Am.* **67**, 524–525.

Hoffman, J.D., Ignoffi, C.M. and Dickerson, W.A. (1975) *In vitro* rearing of the endoparasitic wasp, *Trichogramma pretiosum*. *Ann. Entomol. Soc. Am.* **68**, 335–336.

Hoffmann, J.A., Dimarcq, J.L. and Bulet, P. (1992) Inducible antibacterial peptides of insects. *Med. Sci.* **8**, 432–439. (In French)

Hogge, T.R. and King, P.E. (1975) The ultrastructure of spermatogenesis in *Nasonia vitripennis* (Walker) (Hymenoptera: Pteromalidae). *J. Submicrosp. Cytol.* **7**, 81–96.

Hokkanen, H. and Pimentel, D. (1984) New approaches for selecting biological control agents. *Canad. Entomol.* **116**, 1109–1121.

Höller, C. (1991) Movement away from the feeding site in parasitized aphids: Host suicide or an attempt by the parasitoid to escape hyperparasitism. In *Behaviour and Impact of Aphidophaga* (eds L. Polgar, R.J. Chambers, A.F.G. Dixon and I. Hodek), SPB Academic Publishing bv., The Hague. pp. 45–49.

Höller, C., Williams, H.J. and Vinson, S.B. (1991) Evidence for a two-component external marking pheromone system in an aphid hyperparasitoid. *J. Chem. Ecol.* **17**, 1020–1035.

Höller, C., Bargen, H., Vinson, S.B. and Witt, D. (1994a) Evidence for the external use of juvenile hormone for host marking and regulation in a parasitic wasp, *Dendrocerus carpenteri*. *J. Insect Physiol.* **40**, 317–322.

Höller, C., Micha, S.G., Schulz, S. *et al.* (1994b) Enemy-induced dispersal in a parasitic wasp. *Experientia* **50**, 182–185.

Holmes, H. B. (1974) Patterns of sperm competition in *Nasonia vitripennis*. *Canad. J. Genet. Cytol.* **16**, 789–795.

Hood, W.G. and Tschinkel, W.R. (1990) Desiccation resistance in arboreal and terrestrial ants. *Physiol. Ent.* **15**, 23–35.

Hooker, M.E. and Barrows, E.M. (1989) Clutch sizes and sex ratios in *Pediobius foveolatus* (Hymenoptera: Eulophidae), primary parasites of *Epilachna varivestis* (Coleoptera: Coccinellidae). *Ann. Entomol. Soc. Am.* **82**, 460–465.

Hopkins, C.R. and King, P.E. (1964) Egg resorption in *Nasonia vitripennis* (Walker) (Hymenoptera: Pteromalidae). *Proc. R. Entomol. Soc. Lond* (A) **39**, 101–107.

Hopper, K.R. and Roush, R.T. (1993) Mate finding, dispersal, number released, and the success of biological control introductions. *Ecol. Ent.* **18**, 321–331.

Hopper, K.R. and Woolson, E.A. (1991) Labeling a parasitic wasp, *Microplitis croceipes* (Hymenoptera: Braconidae), with trace elements for mark–recapture studies. *Ann. Entomol. Soc. Am.* **84**, 255–262.

Howard, R.W. (1992) Comparative analysis of cuticular hydrocarbons from the ectoparasitoids *Cephalonomia waterstoni* and *Laelius utilis* (Hymenoptera: Bethylidae) and their respective hosts, *Cryptolestes ferrugineus* (Coleoptera:

Cucujidae) and *Trogoderma variabile* (Coleoptera: Dermestidae). *Ann. Entomol. Soc. Am.* **85**, 317–325.

Howard, R.W. and Flinn, P.W. (1990) Larval trails of *Cryptolestes ferrugineus* (Coleoptera: Cucujidae) as kairomonal host-finding cues for the parasitoid *Cephalonomia waterstoni* (Hymenoptera: Bethylidae). *Ann. Entomol. Soc. Am.* **83**, 239–245.

Howard, R.W. and Liang, Y. (1993) Cuticular hydrocarbons of winged and wingless morphs of the ectoparasitoid *Choetospila elegans* Westwood (Hymenoptera: Pteromalidae) and its host, larval lesser grain borer (*Rhyzopertha dominica*) (Coleoptera: Bostrrichidae). *Comp. Biochem. Physiol.* **106B**, 407–414.

Howell, J. and Fisher, R.C. (1977) Food conversion efficiency of a parasitic wasp, *Nemeritis canescens*. *Ecol. Ent.* **2**, 143–151.

Hoy, M.A. and Marsh, P.M. (1979) Breeding tests support synonymy of *Apanteles melanoscelus* and *Apanteles solitarius* (Hymenoptera: Braconidae). *Proc. Entomol. Soc. Wash.* **81**, 75–81.

Hrdy, I. and Sedivy, J. (1979) Males of *Exetastes cinctipes* (Hymenoptera, Ichneumonidae) attracted to 8-dodecenyl and 11-tetradecenyl acetates. *Acta Entomol. Bohem.* **76**, 59–61.

Hubbard, S.F., Marris, G.C., Reynolds, A.J. and Rowe, G.W. (1987) Adaptive patterns in the avoidance of superparasitism by solitary parasitic wasps. *J. Anim. Ecol.* **56**, 387–404.

Huber, J.T. and Rajakulendran, V.K. (1988) Redescription and host-induced antennal variation in *Anaphesiole* Girault (Hymenoptera, Mymaridae), an egg parasite of Miridae (Hemiptera) in North America. *Canad. Entomol.* **120**, 893–901.

Huddleston, T. (1975) Variation in *Encardia* (Hym., Ichneumonidae) with a key to the species. *J. Entom. (B)* **49**, 15–20.

Huddleston, T. and Gauld, I.D. (1988) Parasitic wasps in British light-traps. *Entomologist* **107**, 134–154.

Huger, A.M., Skinner, S.W. and Werren, J.H. (1985) Bacterial infections associated with the son-killer trait in the parasitoid wasp *Nasonia* (= *Mormoniella*) *vitripennis* (Hymenoptera: Pteromalidae). *J. Invert. Pathol.* **46**, 272–280.

Huggert, L. (1979) *Cryptoserphus* and Belytinae wasps (Hymenoptera: Proctotrupoidea) parasitizing fungus- and soil-inhabiting Diptera. *Notulae Entomol.* **59**, 139–144.

Hung, A.C.F. (1985) Tandem gene duplication and fixed heterozygosity in the parasitic wasp, *Trichogramma marylandense*. *Experientia* **41**, 508–509.

Hung, A.C.F. (1986) Chromosomes of three *Brachymeria* species (Hymenoptera: Chalcidoidea). *Experientia* **42**, 579–580.

Hung, A.C.F. (1990) Scale-like structures on the tibia of the parasitic wasps, *Trichogramma* spp. (Hymenoptera: Trichogrammatidae). *Proc. Entomol. Soc. Wash.* **92**, 548–551.

Hung, A.C.F., Day, W.H. and Hedlung, R.C. (1988) Genetic variability in arrhenotokous and thelytokous forms of *Mesochorus nigripes* Ratzeburg. (Hym.: Ichneumonidae). *Entomophaga*, **33**, 7–15.

Hung, A.C.F. and Schaefer, P.W. (1990) Isozyme analysis in six populations of *Pediobius foveolatus* (Crawford) (Hymenoptera: Eulophidae). *Proc. Entomol. Soc. Wash.* **92**, 160–165.

Hunt, J.H., Brown, P.A., Sago, K.M. and Kerker, J.A. (1991) Vespid wasps eat pollen. *J. Kansas Entomol. Soc.* **64**, 127–130.

Hunter, M.S. (1993) Sex allocation in a field population of an autoparasitoid. *Oecologia*, **93**, 421–428.

Hunter, M.S., Nur, U. and Werren, J.H. (1993) Origin of males by genome loss in an autoparasitoid wasp. *Heredity*, **70**, 162–171.

Hurlbutt, B. (1987) Sexual size dimorphism in parasitoid wasps. *Biol. J. Linn. Soc.* **30**, 63–89.

Hutchison, W.D., Moratorio, M. and Martin, J.M. (1990) Morphology and biology of *Trichogrammatoidea bactrae* (Hymenoptera: Trichogrammatidae), imported from Australia as a parasitoid of pink bollworm (Lepidoptera: Gelechiidae) eggs. *Ann. Entomol. Soc. Am.* **83**, 46–54.

Imms, A.D. (1935) *Textbook of Entomology*, Methuen, London.

Infante, F., Hanson, P. and Wharton, R. (1995) Phytophagy in the genus *Monitoriella* (Hymenoptera: Braconidae) with description of new species. *Ann. Entomol. Soc. Am.* **88**, 406–415.

Ishii, M. (1990) An observation on the oviposition behaviour of the parasite moth, *Epipomponia nawai* (Dyar) (Lepidoptera, Epipyropidae). *Jpn. J. Ent.* **58**, 441–442.

Isidoro, N. (1992) Fine structure of the sensillum coeloconicum in *Trissolcus basalis* (Woll.) (Hymenoptera, Scelionidae) antennae. *Redia* **75**, 169–178.

Isidoro, N. and Bin, F. (1994) Fine structure of the preocellar pit in *Trissolcus basalis* (Woll.) (Hymenoptera: Scelionidae). *Int. J. Insect Morphol. Embr.* **23**, 189–196.

Isidoro, N. and Bin, F. (1995) Male antennal gland of *Amitus spiniferus* (Brethes) (Hymenoptera: Platygastridae), likely involved in courtship behaviour. *Int. J. Insect Morphol. Embr.* **24**, 365–373.

Isidoro, N., Bin, F., Colazza, S. and Vinson, S.B. (1996) Antennal gustatory sensilla and glands in some parasitic Hymenoptera: A critical morpho-functional approach. *J. Hym. Res.* **5**, 206–239.

Ivanova-Kasas, O.M. (1956) Comparative study of embryonal development in aphidiids (*Aphidius* and *Ephedrus*). *Entomol. Obozrenie* **35**, 245–261. (In Russian.)

Ivanova-Kasas, O.M. (1958) Biology and embryonic development of *Eurytoma aciculata* Ratz. (Hymenoptera, Eurytomidae). *Rev. Entomol l'URSS* **37**, 5–23. (In Russian.)

Ivanova-Kasas, O.M. (1960) Embryologische Entwicklung von *Angitia vestigialis* Ratz. (Hymenoptera, Ichneumonidae) – des Endoparasiten von *Pontania capreae* L. (Hymenoptera, Tenthredinidae). *Entomol. Obozrenie* **39**, 284–295. (In Russian.)

Ivanova-Kasas, O.M. (1970) Polyembryony in insects. In *Developmental Systems: Insects* (eds S.J. Counce and C.H. Waddington), Academic Press, London. pp. 243–271.

Iwata, I. (1960a) The comparative anatomy of the ovary in Hymenoptera. Part V. Ichneumonidae. *Acta Hymenopt.* **1**, 115–169.

Iwata, I. (1960b) The comparative anatomy of the ovary in Hymenoptera. Supplement on Aculeata with descriptions of ovarian eggs of certain species. *Acta Hymenopt.* **1**, 205–211.

Jackson, C.G. and Debolt, J.W. (1990) Labeling of *Leiophron uniformis*, a parasitoid of *Lygus* spp., with rubidium. *Southwest. Entomol.* **15**, 239–243.

Jackson, C.G., Cohen, A.C. and Verdugo, C.L. (1988) Labeling *Anaphes ovijentatus* (Hymenoptera: Mymaridae), an egg parasite of *Lygus* spp. (Hemiptera: Miridae), with rubidium. *Ann. Entomol. Soc. Am.* **81**, 919–922.

Jackson, D.J. (1928) The biology of *Dinocampus* (*Perilitus*) *rutilis* Nees, a braconid parasite of *Sitona lineata* L. Part 1. *Proc. Zool. Soc. Lond.* **1928**, 597–630.

Jackson, D.J. (1958) Observations on the biology of *Caraphractus cinctus* Walker (Hymenoptera: Mymaridae), a parasitoid of the eggs of Dytiscidae (Coleoptera). I. Methods of rearing and numbers bred on different host eggs. *Trans R. Entomol. Soc. Lond.* **110**, 533–554.

Jackson, D.J. (1966) Observations on the biology of *Caraphractus cinctus* Walker (Hymenoptera: Mymaridae), a parasitoid of the eggs of Dytiscidae (Coleoptera). III. The adult life and sex ratio. *Trans R. Entomol. Soc. Lond.* **118**, 23–49.

Jamieson, B.G.M. (1987) *The Ultrastructure and Phylogeny of Insect Spermatozoa,* Cambridge University Press, Cambridge. 320pp.

Janssen, A. (1989) Optimal host selection by *Drosophila* parasitoids in the field. *Funct. Ecol.* **3,** 469–479.

Janssen, A., Alphen, J.J.M. van, Sabelis, M.W. and Baliter, K. (1995) Specificity of odour-mediated avoidance of competition in *Drosophila* parasitoids. *Behav. Ecol. Sociobiol.* **36,** 229–235.

Janssen, A., Driessen, G., de Haan, M. and Roodbol, N. (1988) The impact of parasitoids on natural populations of temperate woodland *Drosophila. Neth. J. Zool.* **38,** 61–73.

Janzen, D.H. (1981) The peak of North American ichneumonid species richness lies between 38° and 42°N. *Ecology* **62,** 532–537.

Janzen, D.H. and Pond, C.M. (1975) A comparison, by sweep sampling, of the arthropod fauna of secondary vegetation in Michigan, England and Costa Rica. *Trans. R. Entomol. Soc. Lond.* **127,** 33–50.

Javahery, M. (1968) The egg parasite complex of British Pentatomoidea (Hemiptera): Taxonomy of Telenominae (Hymenoptera: Scelionidae). *Trans. R. Entomol. Soc. Lond.* **120,** 417–436.

Jell, P.A. and Duncan, P.M. (1986) Invertebrates, mainly insects, from the freshwater, Lower Cretaceous, Koonwarra fossil bed (Korumburra group), South Gippsland, Victoria. In *Plants and Invertebrates from the Lower Cretaceous Koonwarra Fossil Bed, South Gippsland, Victoria* (eds P.A. Jell and J. Roberts), Assoc. Austral. Palaeo. Mem., No. 3, Sydney, pp. 111–202.

Jervis, M.A. (1979) Courtship, mating and 'swarming' in *Aphelopus melaleucus* (Dalman) (Hymenoptera: Dryinidae). *Ent. Gaz.* **30,** 191–193.

Jervis, M.A. and Copland, M.J.W. (1996) The life cycle. In *Insect Natural Enemies* (eds M. Jervis and N. Kidd), Chapman & Hall, London. pp. 63–161.

Jervis, M.A. and Kidd, N.A.C. (1986) Host-feeding strategies in hymenopteran parasitoids. *Biol. Rev.* **61,** 395–434.

Jervis, M.A. and Kidd, N.A.C. (1991) The dynamic significance of host-feeding by insect parasitoids – what modellers ought to consider. *Oikos,* **62,** 97–99.

Jervis, M.A., Kidd, N.A.C. and Walton, M. (1992) A review of methods for determining dietary range in adult parasitoids. *Entomophaga,* **37,** 565–574.

Jervis, M.A., Kidd, N.A.C., Fitton, M.G. *et al.* (1993) Flower-visiting by hymenopteran parasitoids. *J. Nat. Hist.* **27,** 67–105.

Jervis, M.A., Kidd, N.A.C. and Almey, H.E. (1994) Post-reproductive life in the parasitoid *Bracon hebetor* (Say) (Hym., Braconidae). *J. Appl. Ent.* **117,** 72–82.

Jervis, M.A., Hawkins, B.A. and Kidd, N.A.C. (1996) The usefulness of destructive host feeding parasitoids in classical biological control: theory and observation conflict. *Ecol. Ent.* **21,** 41–46.

Johnson, J.B., Miller, T.D., Heraty, J.M. and Merickel, F.W. (1986) Observations on the biology of two species of *Orasema* (Hymenoptera: Eucharitidae). *Proc. Entomol. Soc. Wash.* **88,** 542–549.

Johnson, N.F. (1988) Midcoxal articulations and the phylogeny of the order Hymenoptera. *Ann. Entomol. Soc. Am.* **81,** 870–881.

Johnson, N.F., Rawlins, J.E. and Pavuk, D.M. (1987) Host-related antennal variation in the polyphagous egg parasite *Telenomus alsophilae* (Hymenoptera, Scelionidae). *Syst. Ent.* **12,** 437–447.

Joiner, R.L., Vinson, S.B. and Benskin, J.B. (1973) Teratocytes as source of juvenile hormone activity in a parasitoid–host relationship. *Nature New Biol.* **246,** 120–121.

Jones, D. (1985a) Parasite regulation of host insect metamorphosis: A new form of regulation in pseudoparasitized larvae of *Trichoplusia ni. J. Comp. Physiol.* B **155,** 583–590.

Jones, D. (1985b) Endocrine interraction between host (Lepidoptera) and parasite (*Chelonus*: Hymenoptera): Is the host or the parasite in control? *Ann. Entomol. Soc. Am.* **78**, 141–148.

Jones, D. (1987) Material from adult female *Chelonus* sp. directs expression of altered developmental programme of host Lepidoptera. *J. Insect Physiol.* **33**, 129–134.

Jones, D., Jones, G., Rudnicka, M. *et al.* (1986a) Pseudoparasitism of host *Trichoplusia ni* by *Chelonus* spp. as a new model system for parasite regulation of host physiology. *J. Insect Physiol.* **32**, 315–328.

Jones, D., Sreekrishna, S., Iwaya, M. and Yang, J.-N. (1986b) Comparison of viral ultrastructure and DNA banding patterns from the reproductive tracts of eastern and western hemisphere *Chelonus* spp. (Braconidae: Hymenoptera). *J. Invert. Pathol.* **47**, 105–115.

Jones, D., Gelman, D. and Loeb, M. (1992) Hemolymph concentrations of host ecdysteroids are strongly suppressed in precocious prepupae of *Trichoplusia ni* parasitized and pseudoparasitized by *Chelonus* near *curvimaculatus*. *Arch. Insect Biochem. Physiol.* **21**, 155–165.

Jones, D., Krishnan, A., Sarkari, N. and Wozniak, M. (1994) Isomeric and quaternary properties of homogenous 33 kDa protein from the venom of *Chelonus* near *curvimaculatus*. *Arch. Insect Biochem. Physiol.* **26**, 83–95.

Jones, R.L. (1996) Semiochemicals in host and mate finding behavior of *Macrocentrus grandii* Goidanich (Hymenoptera, Braconidae). *Florida Entomol.* **79**, 104–108.

Jordan, K. (1926) On a pyralid parasitic as larva on spiny saturnid caterpillars at Para. *Novit. Zool.* **33**, 367–370.

Kahn, D.M. and Cornell, H.V. (1989) Leafminers, early abscission, and parasitoids: A tritrophic interaction. *Ecology* **70**, 1219–1226.

Kainoh, Y. and Brown, J.J. (1994) Amino acids as oviposition stimulants for the egg larval parasitoid, *Chelonus* sp. near *curvimaculatus* (Hymenoptera, Braconidae). *Biol. Control.* **4**, 22–25.

Kainoh, Y. and Oishi, Y. (1993) Source of sex pheromone of the egg–larval parasitoid, *Ascogaster reticulatus* Watanabe (Hymenoptera, Braconidae). *J. Chem. Ecol.* **19**, 963–969.

Kainoh, Y., Tatsuki, S., Sugie, H. and Tamaki, Y. (1989) Host egg kairomones essential for egg–larval parasitoid, *Ascogaster reticulatus* Watanabe (Hymenoptera: Braconidae). II. Identification of internal kairomone. *J. Chem. Ecol.* **15**, 1219–1229.

Kainoh, Y., Nemoto, T. Shimizu, K. *et al.* (1991) Mating behavior of *Ascogaster reticulatus* Watanabe (Hymenoptera, Braconidae), an egg–larval parasitoid of the smaller tea tortrix, *Adoxophyes* sp. (Lepidoptera: Tortricidae). III. Identification of a sex pheromone. *Appl. Entomol. Zool.* **26**, 543–549.

Kajita, H. (1986) Role of postcopulatory courtship in insemination of two aphelinid wasps (Hymenoptera: Aphelinidae). *Appl. Ent. Zool.* **21**, 484–486.

Kaneshiro, L.N. and Johnson, M.W. (1996) Tritrophic effects of leaf nitrogen on *Liriomyza trifolii* (Burgess) and an associated parasitoid *Chrysocharis oscinidis* (Ashmead) on bean. *Biol. Control.* **6**, 186–192.

Karnavar, G.K. (1984) Studies on the influence of the parasitoid, *Apanteles glomeratus* on the metabolite levels of the host *Pieris brassicae*. *Insect Sci. Applic.* **5**, 99–100.

Kasparyan, D.R. (1980) The functional aspect of evolution of the sting in the Hymenoptera. *Entomol. Rev.* **59**, 49–54.

Kasparyan, D.R. (1981) *Fauna of the USSR. Hymenoptera 3(1). Ichneumonidae (Subfamily Tryphoninae) Tribe Tryphonini*, Oxonian Press, New Delhi. 414pp.

Kathirithamby, J. and Grimaldi, D. (1993) Remarkable stasis in some lower Tertiary parasitoids – Descriptions, new records, and review of Strepsiptera in the Oligomiocene amber of the Dominican Republic. *Entomol. Scand.* **24**, 31–41.

Kazmer, D.J., Hopper, K.R., Coutinot, D.M. and Heckel, D.G. (1995) Suitability of random amplified polymorphic DNA for genetic markers in the aphid parasitoid, *Aphelinus asychus* Walker. *Biol. Control* **5**, 503–512.

Keeler, K.H. (1978) Insects feeding at extrafloral nectaries of *Ipomoea carnea* (Convolvulaceae). *Entomol. News* **89**, 163–168.

Keilin, D. and Tate, P. (1943) The larval stages of the celery fly (*Acidia heraclei* L.) and of the braconid *Adelura apii* (Curtis), with notes upon an associated parasitic yeast-like fungus. *Parasitology*, **35**, 27–36.

Keilin, D. and Thompson, W.R. (1915) Sur le cycle evolutif des Dryinidae, hymenopteres parasites des hemipteres homopteres. *Soc. Biol. Compt. Rend.* **78**, 83–87.

Keller, M.A., Lewis, W.J. and Stinner, R.E. (1985) Biological and practical significance of movement by *Trichogramma* species: a review. *Southwestern Entomol. Suppl.* **8**, 138–155.

Kenchington, W. (1972) Variations in silk gland morphology among sawfly larvae (Hymenoptera: Symphyta). *J. Ent.* (A) **46**, 111–116.

Kenis, M. (1994) Variations in diapause among populations of *Eubazus remirugosus* (Nees) (Hym.: Braconidae), a parasitoid of *Pissodes* spp. (Col.: Curculionidae). *Norw. J. Agric. Sci. Suppl.* **16**, 77–82.

Kenis, M., Hulme, M.A. and Mills, N.J. (1996) Comparative developmental biology of populations of three European and one North American *Eubazus* spp. (Hymenoptera: Braconidae), parasitoids of *Pissodes* spp. weevils (Coleoptera: Curculionidae). *Bull. Entomol. Res.* **86**, 143–153.

Kerrich, G.J. (1969) Description of an ichneumonid (Hym.) that preys on egg-masses of weevils harmful to tea culture in Kenya. *Bull. Entomol. Res.* **59**, 469–472.

Kfir, R. and Rosen, D. (1981) Biology of the hyperparasite *Pachyneuron concolor* (Forster) (Hymenoptera: Pteromalidae) reared on *Microterys flavus* (Howard) in brown soft scale. *J. Entomol. Soc. South Afr.* **44**, 151–163.

Khasimuddin, S. and DeBach, P. (1976) Hybridization tests: A method for establishing biosystematic statuses of cryptic species of some parasitic Hymenoptera. *Ann. Entomol. Soc. Am.* **60**, 15–20.

Kidd, M.A.C. and Jervis, M.A. (1989) The effects of host-feeding behaviour on the dynamics of parasitoid–host interactions, and the implications for biological control. *Res. Popul. Ecol.* **31**, 235–274.

Kimsey, L.S. (1992) Functional morphology of the abdomen and phylogeny of chrysidid wasps (Hymenoptera: Chrysididae). *J. Hym. Res.* **1**, 165–174.

King, B.H. (1987) Offspring sex ratios in parasitoid wasps. *Q. Rev. Biol.* **62**, 367–386.

King, B.H. (1994) How do female parasitoid wasps assess host size during sex ratio manipulation? *Anim. Behav.* **48**, 511–518.

King, P.E. (1960) The passage of sperms to the spermatheca during mating in *Nasonia vitripennis* (Walker) (Hym., Pteromalidae). *Entomologist's Mon. Mag.* **96**, 136.

King, P.E. (1961) A possible method of sex ratio determination in the parasitic hymenopteran *Nasonia vitripennis*. *Nature (Lond.)* **189**, 330–331.

King, P.E. (1962a) Structure of the micropyle in eggs of *Nasonia vitripennis*. *Nature (Lond.)* **195**, 829–830.

King, P.E. (1962b) The structure and action of the spermatheca in *Nasonia vitripennis* (Walker) (Hymenoptera: Pteromalidae). *Proc. R. Entomol. Soc. Lond.* (A) **37**, 73–75.

King, P.E. (1963) The rate of egg resorption in *Nasonia vitripennis* (Walker) (Hymenoptera: Pteromalidae) deprived of hosts. *Proc. R. Entomol. Soc. Lond.* (A) **38**, 98–100.

King, P.E. and Copland, M.J.W. (1969) The structure of the female reproductive system in the Mymaridae (Chalcidoidea: Hymenoptera). *J. Nat. Hist.* **3**, 349–365.

King, P.E. and Fordy, M.R. (1970) The formation of 'accessory nuclei' in the developing oöcytes of the parasitoid hymenopterans *Ophion luteus* (L.) and *Apanteles glomeratus* (L.). *Z. Zellforsch.* **109**, 158–170.

King, P.E. and Rafai, J. (1973) A possible mechanism for initiating the parthenogenetic development of eggs in a parasitoid hymenopteran, *Nasonia vitripennis* (Walker) (Pteromalidae). *Entomologist* **106**, 118–120.

King, P.E. and Ratcliffe, N.A. (1969) The structure and possible mode of functioning of the female reproductive system in *Nasonia vitripennis* (Hymenoptera: Pteromalidae). *J. Zool. Lond.* **157**, 319–344.

King, P.E. and Richards, J.G. (1968) Oösorption in *Nasonia vitripennis* (Hymenoptera: Pteromalidae). *J. Zool.* **154**, 495–516.

King, P.E. and Richards, J.G. (1969) Oögenesis in *Nasonia vitripennis* (Walker) (Hymenoptera: Pteromalidae). *Proc. R. Entomol. Soc. Lond.* (A) **44**, 143–157.

King, P.E., Ratcliffe, N.A. and Copland, M.J.W. (1969a) The structure of the egg membranes in *Apanteles glomeratus* (L.) (Hymenoptera: Braconidae). *Proc. R. Entomol. Soc. Lond.* (A) **44**, 137–142.

King, P.E., Richards, J.G. and Copland, M.J.W. (1969b) The structure of the chorion and its possible significance during oviposition in *Nasonia vitripennis* (Walker) (Hymenoptera: Pteromalidae) and other chalcids. *Proc. R. Entomol. Soc. Lond.* (A) **43**, 13–20.

King, P.E., Ratcliffe, N.A. and Fordy, M.R. (1971) Oögenesis in a braconid, *Apanteles glomeratus* (L.) possessing an hydropic type of egg. *Zeitschr. Zellforsch. Mikrosk. Anat.* **119**, 43–57.

King, R.C. and Cassidy, J.D. (1973) Ovarian development in *Habrobracon juglandis* (Ashmead) (Hymenoptera: Braconidae) – II. observations on growth and differentiation of component cells of egg chamber and their bearing upon interpretation of radiosensitivity data from *Habrobracon* and *Drosophila*. *Int. J. Insect Morphol. Embryol.* **2**, 117–136.

Kitamura, K. (1988) Comparative studies on the biology of dryinid wasps in Japan. *Kontyû, Tokyo*, **56**, 659–666.

Kitano, H. (1975) Studies on the courtship behavior of *Apanteles glomeratus* L. 2. Role of the male wing during courtship and the releaser of mounting and copulatory behavior in males. *Kontyû* **43**, 513–521.

Kitano, H. (1982) Effect of the venom of the gregarious parasitoid *Apanteles glomeratus* on its hemocytic encapsulation by the host, *Pieris*. *J. Invert. Pathol.* **40**, 61–67.

Kitano, H. (1986) The role of *Apanteles glomeratus* venom in the defensive response of its host, *Pieris rapae crucivora*. *J. Insect Physiol.* **32**, 369–375.

Klag, J. and Bilinski, S. (1993) Oosome formation in two ichneumonid wasps. *Tissue and Cell* **25**, 121–128.

Klag, J. and Bilinski, S. (1994) Germ cell cluster formation and oogenesis in the hymenopteran *Coleocentrotus soldanskii*. *Tissue and Cell* **26**, 699–706.

Klein, J.A. and Beckage, N.E. (1990) Comparative suitability of *Trogoderma variabile* and *T. glabrum* (Coleoptera: Dermestidae) as hosts for the ectoparasite *Laelius pedatus* (Hymenoptera: Bethylidae). *Ann. Entomol. Soc. Am.* **83**, 809–816.

Klein, J.A., Ballard, D.K., Lieber, K.S. *et al.* (1991) Host developmental stage and size as factors affecting parasitization of *Trogoderma variabile* (Coleoptera: Dermestidae) by *Laelius pedatus* (Hymenoptera: Bethylidae). *Ann. Entomol. Soc. Am.* **84**, 72–78.

Klomp, H. (1981) Parasitic wasps as sleuthhounds. Response of an ichneumonid to the trail of its host. *Neth. J. Zool.* **31**, 762–772.

Klomp, H. and Teerink, B.J. (1978) The elimination of supernumary larvae of the gregarious egg-parasitoid *Trichogramma embryophagum* (Hym.: Trichogrammatidae) in eggs of the host *Ephestia kuehniella* (Lep.: Pyralidae). *Entomophaga* **23**, 153–159.

Klomp, H., Teerink, B.J. and Wei Chun Ma (1980) Discriminization between parasitized and unparasitized hosts in the egg parasite *Trichogramma embryophagum* (Hym.: Trichogrammatidae): A matter of learning and forgetting. *Neth. J. Zool.* **30**, 254–277.

Königsmann, E. (1977) Das phylogenetisce System der Hymenoptera: Teil 2: Symphyta). *Deut. Entomol.* **24**, 1–40.

Königsmann, E. (1978) Das phylogenetisce System der Hymenoptera: Teil 3: 'Terebrantes' (Unterordnung Apocrita). *Deut. Entomol.* **25**, 1–55.

Koptur, S. (1989) Mimicry of flowers by parasitoid wasp pupae. *Biotropica* **21**, 93–95.

Kovalev, O.V. (1994) Palaeontological history, phylogeny and the system of Brachycleistogastromorphs and Cynipomorphs (Hymenoptera, Brachycleistogastromorpha Infraorder n., Cynipomorpha Infraorder n.) with description of new fossil and recent families, subfamilies and genera. *Entomol. Obozren.* **73**, 385–426.

Kozlov, M.A. (1994) Renyxidae fam. n., a new remarkable family of parasitic Hymenoptera (Proctotrupoidea) from the Russian Far East. *Far Eastern Entomol.* **1**, 1–7.

Kraaijeveld, A.R. (1994) Local adaptations in a parasitoid–host system. A coevolutionary arms race? Doctoral Thesis, University of Leiden.

Kraaijeveld, A.R. and van Alphen, J.J.M. (1993) Successful invasion of North America by two Palaearctic *Drosophila* species (Diptera: Drosophilidae): A matter of immunity to local parasitoids? *Neth. J. Zool.* **43**, 235–241.

Kraaijeveld, A.R. and van Alphen, J.J.M. (1994) Geographic variation in resistance of the parasitoid *Asobara tabida* against encapsulation by *Drosophila melanogaster* larvae: The mechanism explored. *Physiol. Ent.* **19**, 9–14.

Kraaijeveld, A.R. and van Alphen, J.J.M. (1995) Variation in diapause and sex ratio in the parasitoid *Asobara tabida*. *Entomol. Exp. Appl.* **74**, 259–265.

Kraaijeveld, A.R. and van der Wel, N.N. (1994) Geographic variation in reproductive success of the parasitoid *Asobara tabida* in larvae of several *Drosophila* species. *Ecol. Ent.* **19**, 221–229.

Kraaijeveld, A.R., Nowee, B. and Najem, R.W. (1995) Adaptive variation in host-selection behaviour of *Asobara tabida*, a parasitoid of *Drosophila* larvae. *Funct. Ecol.* **9**, 113–118.

Krainska, M. (1961) A morphological and histochemical study of oogenesis in the gall-fly *Cynips folii*. *Quart. J. Microscop. Sci.* **102**, 119–129.

Krell, P.J. (1987) Replication of long virus-like particles in the reproductive tract of the ichneumonid wasp *Diadegma terebrans*. *J. Gen. Virol.* **68**, 1477–1483.

Krell, P.J. and Stoltz, D.B. (1979) Unusual baculovirus of the parasitoid wasp *Apanteles melanoscelus*: Isolation and preliminary characterisation. *J. Virol.* **29**, 1118–1130.

Krell, P.J., Summers, M.D. and Vinson, S.B. (1982) Virus with a multipartite superhelical DNA genome from the ichneumonid parasitoid *Campoletis sonorensis*. *J. Virol.* **43**, 859–870.

Krenn, H.W. and Pass, G. (1995) Morphological diversity and phylogenetic analysis of wing circulatory organs in insects, part II. Holometabola. *Zoology* **98**, 147–164.

Krespi, L., Dedryver, C.A., Rabasse, J.M. and Nénon, J.P. (1994) A morphological comparison of aphid mummies containing diapausing vs. non-diapausing larvae of *Aphidius rhopalosiphi* (Hymenoptera: Braconidae, Aphidiinae). *Bull. Entomol. Res.* **84**, 45–50.

Krishnan, A., Nair, P.N. and Jones, D. (1994) Isolation, cloning, and characterization of new chitinase stored in active form in chitin-lined venom apparatus. *J. Biol. Chem.* **269**, 20971–20976.

Kukalova-Peck, J. (1991) Fossil history and the evolution of hexapod structures. In *The Insects of Australia*, 2nd edn (CSIRO), Melbourne University Press, Melbourne. pp. 141–179.

Kumar, P. and Ballal, C.R. (1992) The effect of parasitism by *Hyposoter didymator* (Hym.: Ichneumonidae) on food consumption and utilization by *Spodoptera litura* (Lep.: Noctuidae). *Entomophaga* **37**, 197–203.

Künckel d'Herculais, J. and Langlois, L. (1891) Moeurs et métamorphoses de *Perilitus brevicollis* Haliday hyménoptère braconide parasite de l'altise de la vigne en Algerie. *Soc. Ent. Fr. Ann.* **60**, 457–466.

Kurstak, E.S. (1966) Le rôle de *Nemeritis canescens* Gravenhorst dans l'infection à *Bacillus thuringiensis* Berliner chez *Ephestia kühniella* Zeller. *Ann. Epiphyt.* **17**, 335, 451.

Kutsch, W. and Breidbach, O. (1994) Homologous structures in the nervous systems of Arthropoda. *Adv. Insect Physiol.* **24**, 1–109.

Kyeipoku, G.K. and Kunimi, Y. (1996) The effect of parasitization by *Cotesia kariyai* (Hymenoptera, Braconidae) on susceptibility of *Pseudaletia separata* (Lepidoptera, Noctuidae) larvae to an entomopoxvirus. *Appl. Ent. Zool.* **31**, 243–246.

Labeyrie, V. (1960. Contribution à l'étude de la dynamique des populations d'insectes. I. Influence stimulatrice de l'hôte *Acrolepia assectella* Z. sur la multiplication d'un Hyménoptère Ichneumonidae (*Diadromus* sp.). *Entomophaga Mém.* **1**, 1–193.

Lackie, A.M. (1988) Haemocyte behaviour. *Adv. Insect Physiol.* **21**, 85–178.

Laigo, F.M. and Tamashiro, M. (1967) Interaction between the microsporidian pathogen of the lawn-armyworm and the hymenopterous parasite *Apanteles marginiventris*. *J. Invert. Pathol.* **9**, 546–554.

Laing, D.R. and Caltagirone, L.E. (1969) Biology of *Habrobracon lineatellae* (Hymenoptera, Braconidae). *Canad. Entomol.* **101**, 135–142.

Landry, B.S., Dextraze, L. and Boivin, G. (1993) Random amplified polymorphic DNA markers for DNA fingerprinting. *Genome* **36**, 580–587.

Lanzrein, B. and Hammock, B. (1995) Degradation of juvenile hormone III *in vitro* by non-parasitized and parasitized *Spodoptera exigua* (Noctuidae) and by the endoparasitoid *Chelonus inanitus* (Braconidae). *J. Insect Physiol.* **41**, 993–1000.

LaSalle, J. (1987) New World Tanaostigmatidae (Hymenoptera, Chalcidoidea). *Contrib. Amer. Entomol. Inst.* **23**, 1–181.

LaSalle, J. (1990a) A new genus and species of Tetrastichinae (Hymenoptera: Eulophidae) parasitic on the coffee berry borer, *Hypothenemus hamperi* (Ferrari) (Coleoptera: Scolytidae). *Bull. Entomol. Res.* **80**, 7–10.

LaSalle, J. (1990b) Tetrastichinae (Hymenoptera: Eulophidae) associated with spider egg sacks. *J. Nat. Hist.* **24**, 1377–1389.

LaSalle, J. (1993) Parasitic Hymenoptera, biological control and biodiversity. In *Hymenoptera and Biodiversity* (eds J. LaSalle and I.D. Gauld), C.A.B. International, Wallingford. pp. 197–215.

LaSalle, J. (1994) North American genera of Tetrastichinae (Hymenoptera: Eulophidae). *J. Nat. Hist.* **28**, 109–236.

LaSalle, J. and Gauld, I.D. (1991) Parasitic Hymenoptera and the biodiversity crisis. *Redia*, **74**, 315–334.

LaSalle, J. and Gauld, I.D. (eds) (1993) *Hymenoptera and Biodiversity*, C.A.B. International, Wallingford.

References

LaSalle, J. and LeBeck, L.M. (1983) The occurrence of encyrtiform eggs in the Tanaostigmatidae (Hymenoptera: Chalcidoidea). *Proc. Entomol. Soc. Wash.* **85**, 397–398.

LaSalle, J. and Stage, G.I. (1985) The chalcidoid genus *Leptofoenus* (Hymenoptera: Pteromalidae). *Syst. Ent.* **10**, 285–298.

Lathrop, F.H. and Newton, R.C. (1933) The biology of *Opius melleus* Gahan. A parasite of the blueberry maggot. *J. Agric. Res.* **46**, 142–160.

Laudonia, S. and Viggiani, G. (1986) Observations on the developmental stages of *Edovum puttleri* Grissell (Hymenoptera: Eulophidae), an egg-parasitoid of Colorado potato beetles. *Boll. Lab. Entomol. Agrar. Filippo Silvestri, Portici* **43**, 97–104.

Lawrence, P.O. (1986) Host–parasite hormonal interactions: An overview. *J. Insect. Physiol.* **32**, 295–298.

Lawrence, P.O. (1988a) Superparasitism of the Caribbean fruit fly, *Anastrepha suspensa* (Diptera: Tephritidae) by *Biosteres longicaudatus* (Hymenoptera: Braconidae): Implications for host regulation. *Ann. Entomol. Soc. Am.* **81**, 233–239.

Lawrence, P.O. (1988b) *In vivo* and *in vitro* development of first instars of the parasitic wasp, *Biosteres longicaudatus* (Hymenoptera: Braconidae). In *Advances in Parasitic Hymenoptera Research* (ed. V.K. Gupta), E.J. Brill, Leiden. pp 351–365.

Lawrence, P.O. (1990a) Serosal cells of *Biosteres longicaudatus* (Hymenoptera: Braconidae): Ultrastructure and release of polypeptides. *Arch. Insect. Biochem. Physiol.* **13**, 199–216.

Lawrence, P.O. (1990b) The biochemical and physiological effects of insect hosts on the development and ecology of their insect parasites: An overview. *Arch. Insect. Biochem. Physiol.* **13**, 217–228.

Lawrence, P.O. and Akin, D. (1990) Virus-like particles from the poison glands of the parasitic wasp *Biosteres longicaudatus* (Hymenoptera: Braconidae). *Canad. J. Zool.* **68**, 539–546.

Lawrence, P.O., Baker, F.C., Tsai, L.W. *et al.* (1990) JH III levels in larvae and pharate pupae of *Anastrepha suspensa* (Diptera: Tephritidae) and in larvae of the parasitic wasp *Biosteres longicaudatus* (Hymenoptera: Braconidae). *Arch. Insect Biochem. Physiol.* **13**, 53–62.

Leal, W.S., Higuchi, H., Mizutani, N. *et al.* (1995) Multifunctional communication in *Riptortus clavatus* (Heteroptera, Alydidae) – conspecific nymphs and egg parasitoid *Ooencyrtus nezarae* use same adult attractant pheromone as chemical cue. *J. Chem. Ecol.* **21**, 973–985.

Leatemia, J.A., Laing, J.E. and Corrigan, J.E. (1995) Production of exclusively male progeny by mated, honey-fed *Trichogramma minutum* Riley (Hym., Trichogrammatidae). *J. Appl. Ent.* **119**, 561–566.

Lebeck, L.M. (1989. Extracellular symbiosis of a yeast-like microorganism within *Comperia merceti* (Hymenoptera: Encyrtidae). *Symbiosis*, **7**, 51–66.

Leclercq, J. (1951) Mise en évidence de la nature ptérinique des pigments jaunes des Hyménoptères adultes. *Bull. Ann. Soc. Entom. Belgique* **87**, 64–74.

Lecomte, C. and Pouzat, J. (1985) Réponses électroantennographiques de deux parasitoides ichneumonidés, *Diadromus pulchellus* et *D. collaris*, aux odeurs de végétaux, du phytophage hote *Acrolepiopsis assectella* et du partenaire sexuel. *Entomol. Exp. Appl.* **39**, 295–306.

Lee, P.E. and Wilkes, A. (1965) Polymorphic spermatozoa in the hymenopterous wasp *Dahlbominus*. *Science*, **147**, 1445–1446.

Leerdam, M.B. van, Smith, J.W. Jr and Fuchs, T.W. (1985) Frass-mediated, host-finding behavior of *Cotesia flavipes*, a braconid parasite of *Diatraea saccharalis* (Lepidoptera: Pyralidae). *Ann. Entomol. Soc. Am.* **78**, 647–650.

Legner, E.F. (1985) Effects of scheduled high temperature on male production in thelytokous *Muscidifurax uniraptor*. *Canad. Entomol.* **117**, 383–389.

Legner, E.F. (1987) Transfer of thelytoky to arrhenotokous *Muscidifurax raptor* Girault and Saunders (Hymenoptera: Pteromalidae). *Canad. Entomol.* **119**, 265–271.

Legner, E.F. (1991) Recombinant males in the parasitic wasp *Muscidifurax raptorellus* (Hymenoptera: Pteromalidae). *Entomophaga* **36**, 173–181.

Leiby, R.W. (1922) The polyembryonic development of *Copidosoma gelechiae* with notes on its biology. *J. Morphol.* **37**, 195–285.

Leiby, R.W. and Hill, C.C. (1924) The polyembryonic development of *Platygaster vernalis*. *J. Agric. Res.* **28**, 829–840.

Leius, K. (1961) Influence of food on fecundity and longevity of adults of *Itoplectis conquisitor* (Say) (Hymenoptera: Ichneumonidae). *Canad. Entomol.* **93**, 771–780.

Leius, K. (1963) Effects of pollens on fecundity and longevity of adult *Scambus buolianae* (Htg.) (Hymenoptera: Ichneumonidae). *Canad. Entomol.* **95**, 202–207.

Lelej, A.S. (1994) Female description of *Renyxa incredibilis* Kozlov (Hymenoptera, Proctotrupoidea, Renyxidae). *Far East. Ent.* **4**, 1–7.

Leluk, J. and Jones, D. (1989) *Chelonus* sp. near *curvimaculatus* venom proteins: Analysis of their potential role and processing during development of host *Trichoplusia ni*. *Arch. Insect Biochem. Physiol.* **10**, 1.

Leluk, J., Schmidt, J. and Jones, D. (1989) Comparative studies on the protein composition of hymenopteran venom reservoirs. *Toxicon*, **27**, 105.

Lenteren, J.C. van, Bakker, K. and Alphen, J.J.M. van (1978) How to analyse host discrimination. *Ecol. Ent.* **3**, 71–75.

Le Ralec, A. (1991) Les hymenopteres parasitoides: Adaptations de l'appareil reproducteur femelle. Morphologie et ultrastrucure de l'ovaire, de l'oeuf et de l'ovipositeur. Unpublished doctoral thesis, University of Rennes I.

Le Ralec, A. (1995) Egg contents in relation to host-feeding in some parasitic Hymenoptera. *Entomophaga* **40**, 87–93.

Le Ralec, A., Barbier, R., Nénon, J.P. and Rabasse, J.M. (1986) Anatomie et fonctionnement de l'ovaire et de l'ovipositeur d'*Aphidius uzbekistanicus* (Hymenoptera: Aphidiidae) parasitoide des pucerons des cereales. *Bull. Soc. Zool. Fr.* **111**, 211–220.

Le Ralec, A. and Rabasse, J.M. (1988) Structure, sensory receptors and operation of the ovipositor of three Aphidiidae. In *Ecology and Effectiveness of Aphidophaga* (eds E. Niemczyk and A.F.G. Dixon), SPB Academic Publishing, The Hague. pp. 83–88.

Le Ralec, A. and Wajnberg, E. (1990) Sensory receptors of the ovipositor of *Trichogramma maidis* (Hym.: Trichogrammatidae). *Entomophaga*, **35**, 293–299.

Le Ralec, A., Rabasse, J.M. and Wajnberg, E. (1996) Comparative morphology of the ovipositor of some parasitic Hymenoptera in relation to characteristics of their hosts. *Canad. Entomol.* **128**, 413–433.

Lester, L.J. and Selander, R.K. (1979) Population genetics of haplodiploid insects. *Genetics*, **92**, 1329–1345.

Levin, D.B., Laing, J.E., Jaques, R.P. and Corrigan, J.E. (1983) Transmission of the granulosis virus of *Pieris rapae* (Lepidoptera, Pieridae) by the parasitoid *Apanteles glomeratus* (Hymenoptera, Braconidae). *Environ. Ent.* **12**, 166–170.

Levine, L. and Sullivan, D.J. (1983) Intraspecific tertiary parasitoidism in *Anaphes lucens* (Hymenoptera: Pteromalidae), an aphid hyperparasitoid. *Canad. Entomol.* **115**, 1653–1658.

Lewis, W.J. (1970) Life history and anatomy of *Microplitis croceipes* (Hymenoptera: Braconidae), a parasite of *Heliothis* spp. (Lepidoptera: Noctuidae). *Ann. Entomol. Soc. Am.* **63**, 67–70.

Lewis, W.J. and Snow, J.W. (1971) Fecundity, sex ratios, and egg distribution by *Microplitis croceipes*, a parasite of *Heliothis*. *J. Econ. Entomol.* **64**, 6–8.

Lewis, W.J. and Takasu, K. (1990) Use of learned odours by a parasitic wasp in accordance with host and food needs. *Nature*, **348**, 635–636.

Lewis, W.J., Snow, J.W. and Jones, R.L. (1971) A pheromone trap for studying populations of *Cardiochiles nigriceps*, a parasite of *Heliothis virescens*. *J. Econ. Entomol.* **64**, 1417–1421.

Li, X.S. and Webb, B.A. (1994) Apparent functional role for a cysteine-rich polydnavirus protein in suppression of the insect cellular immune response. *J. Virol.* **68**, 7482–7489.

Li, Y. and Steiner, W.W.M. (1995) A wing colour mutant (*cw*) in the parasitic wasp, *Microplitis croceipes* (Hymenoptera, Braconidae). *Heredity* **86**, 158–163.

Liepert, C. and Dettner, K. (1993) Recognition of aphid parasitoids by honeydew collecting ants: The role of cuticular lipids in a chemical mimicry system. *J. Chem. Ecol.* **19**, 2143–2153.

Lim, K.P., Yule, W.N. and Stewart, R.K. (1980) A note on *Pelecinus polyturator* (Hymenoptera: Pelecinidae), a parasite of *Phyllophaga anxia* (Coleoptera: Scarabaeidae). *Canad. Entomol.* **112**, 219–220.

Liu, S.-S. and Carver, M. (1982) The effect of temperature on the adult integumental coloration of *Aphidius smithi*. *Entomol. Exp. Appl.* **32**, 54–60.

Loan, C.C. (1967) Studies on the taxonomy and biology of the Euphorinae (Hymenoptera: Braconidae). I. Four new Canadian species of *Microctonus*. *Ann. Entomol. Soc. Am.* **60**, 230–235.

Lucas, F. and Rudall, K.M. (1968) Extracellular fibrous proteins: The silks. In *Comparative Biochemistry*, vol. 26B (eds M. Florkin and E.H. Stotz), Elsevier, Amsterdam, pp. 475–558.

Luck, R.F., Stouthamer, R. and Nunney, L.P. (1992) Sex determination and sex ratio patterns in parasitic Hymenoptera. In *Evolution and Diversity of Sex Ratio in Haplodiploid Insects and Mites* (eds D.L. Wrensch and M. Ebbert), Chapman & Hall, New York, pp. 442–476.

MacBride, D.H. (1946) Failure of sperm of *Habrobracon* diploid males to penetrate the eggs. *Genetics* **31**, 224.

Macedo, M.V. de and Monteiro, R.F. (1989) Seed predation by a braconid wasp, *Allorhogas* sp. (Hymenoptera). *J. N.Y. Ent. Soc.* **97**, 358–362.

Mackauer, M. (1969) Sexual behaviour of and hybridization between three species of *Aphidius* Nees parasitic on the pea aphid (Hymenoptera: Aphidiidae). *Proc. Entomol. Soc. Wash.* **71**, 339–351.

Mackauer, M. (1986) Growth and developmental interactions in some aphids and their hymenopteran parasites. *J. Insect Physiol.* **32**, 275–280.

Madden, J.L. (1968) Behavioural responses of parasites to the symbiotic fungus associated with *Sirex noctilio*. *Nature*, **218**, 189–190.

Madden, J.L. (1982) Avian predation of the woodwasp, *Sirex noctilio* F., and its parasitoid complex in Tasmania. *Austr. Wildlife Res.* **9**, 135–144.

Maetô, K. and Thornton, I.W.B. (1993) A preliminary appraisal of the braconid (Hymenoptera) fauna of the Krakatau Islands, Indonesia, in 1984–1986, with comments on the colonising abilities of parasitoid modes. *Jpn. J. Ent.* **61**, 787–801.

Maingay, H.M., Bugg, R.L., Carlson, R.W. and Davidson, N.A. (1991) Predatory and parasitic wasps (Hymenoptera) feeding at flowers of sweet fennel (*Foeniculum vulgare* Miller var *dulce* Battandier and Trabut, Apiaceae) and spearmint (*Mentha spicata* L., Lamiaceae) in Massachusetts. *Biol. Agric. Hortic.* **7**, 363–383.

Maki, T. (1938) Thoracic musculature of insects. *Mem. Fac. Sci. Agric. Taihoku Imp. Univ.* **24**, 1–343.

Makino, S. and Sayama, K. (1994) Bionomics of *Elasmus japonicus* (Hymenoptera, Elasmidae), a parasitoid of a paper wasp, *Polistes snelleni* (Hymenoptera, Vespidae). *Jpn. J. Ent.* **62**, 377–383.

Malyshev, S.I. (1964) A comparative study of the life and development of primitive gasteruptiids (Hymenopera, Gasteruptiidae). *Entomol. Rev.* **43**, 267–271.

Malyshev, S.I. (1968) *Genesis of the Hymenoptera and the Phases of their Evolution*, Methuen and Co., London. 319 pp. (Originally published in Russian in 1966.)

Mao, H. and Kunimi, Y. (1994) Longevity and fecundity of *Brachymeria lasus* (Walker) (Hymenoptera: Chalcidoidea), a pupal parasitoid of the oriental tea tortrix, *Homona magnanima* Diakonoff (Lepidoptera: Tortricidae) under laboratory conditions. *Appl. Entomol. Zool.* **29**, 237–243.

Maple, J.D. (1937) The biology of *Ooencyrtus johnsoni* (Howard), and the role of the egg shell in the respiration of certain encyrtid larvae (Hymenoptera). *Ann. Entomol. Soc. Am.* **30**, 123–154.

Maple, J.D. (1947) The eggs and first instar larvae of Encyrtidae and their morphological adaptations for respiration. *Univ. Calif. Publs. Ent.* **8**, 25–122.

Marchal, P. (1904) Recherches sur la biologie et le développement des Hyménoptères parasites. I. La polyembryonie specifique ou germinogonie. *Arch. Zool. Exp. Gen.* **2**, 257–335.

Marle, J. van (1977) Structure and histochemistry of the venom glands of the wasps *Microbracon hebetor* Say and *Philanthus triangulum* F. *Toxicon* **15**, 529–539.

Marle, J. van and Piek, T. (1986) Morphology of the venom apparatus. In *Venoms of the Hymenoptera* (ed. T. Piek), Academic Press, London, pp. 17–44.

Marris, G.C. and Casperd, J. (1996) The relationship between conspecific superparasitism and the outcome of *in vitro* contests staged between different larval instars of the solitary endoparasitoid *Venturia canescens*. *Behav. Ecol. Sociobiol.* **39**, 61–69.

Marris, G.C., Hubbard, S.F. and Scrimgeour, C. (1996) The perception of genetic similarity by the solitary parthenogenetic parasitoid *Venturia canescens*, and its effects on the occurrence of superparasitism. *Entomol. Expt. Appl.* **78**, 167–174.

Marsh, P.M. (1991) Description of a phytophagous doryctine braconid from Brazil (Hymenoptera: Braconidae). *Proc. Entomol. Soc. Wash.* **93**, 92–95.

Martínez-Martínez, L., Leyva-Vázquez, J.L. and Mojica, H.B. (1993) Utilizacion del esperma en hembras de *Diachasmimorpha longicaudata*. *Southwest. Entomol.* **18**, 293–299.

Mashhood Alam, S. (1952) Studies on 'skeleto-muscular mechanism' of the male genitalia in *Stenobracon deesae* Cameron (Hymenoptera: Braconidae). *Beitr. Ent.* **2**, 620–634.

Mashhood Alam, S. (1954) A preliminary survey of the alimentary canal in *Stenobracon deesae* Cam. (Braconidae, Hymenoptera). *Indian J. Ent.* **26**, 55–66.

Mashhood Alam, S. (1960) The skeleto-muscular mechanism of *Stenobracon deesae* Cameron (Braconidae, Hymenoptera) – An ectoparasite of sugarcane and juar borers in India. Part II. Abdomen and internal anatomy. In *On Indian Insect Types* (ed. M.B. Mirza), Aligarh Muslim Univ. Publ. (Zool. Ser.). 75 pp. + 7 plates.

Mason, W.R.M. (1964) Regional colour patterns in the parasitic Hymenoptera. *Canad. Entomol.* **96**, 132–134.

Mason, W.R.M. (1967) Specialization in the egg structure of *Exenterus* (Hymenoptera: Ichneumonidae) in relation to distribution and abundance. *Canad. Entomol.* **99**, 375–384.

Mason, W.R.M. (1984) Structure and movement of the abdomen of female *Pelecinus polyturator* (Hymenoptera: Pelecinidae). *Canad. Entomol.* **116**, 419–426.

Masutti, L., Battisti, A., Milani, N. *et al.* (1993) *In vitro* rearing of *Ooencyrtus pityocampae* (Hym., Encyrtidae), an egg parasitoid of *Thaumetopoea pityocampa* (Lep., Thaumetopoeidae). *Entomophaga* **38**, 327–333.

Mathur, K.C. (1967) Notes on *Apistephialtes* sp., an ichneumonid larval parasite of *Hypsipyla robusta* Moore in India. *Tech. Bull. Commonwealth Inst. Biol. Contr.* **9**, 133–135.

Matsuda, R. (1965) Morphology and evolution of the insect head. *Mem. Am. Ent. Inst.* **4**, 1–334.

Matsuda, R. (1970) Morphology and evolution of the insect thorax. *Mem. Entomol. Soc. Canad.* **76**, 1–431.

Matsuda, R. (1976) The Hymenoptera. In *Morphology and Evolution of the Insect Abdomen* (ed. R. Matsuda), Pergamon Press, New York. 534 pp.

Matthews, R.W. (1974) Biology of Braconidae. *Annu. Rev. Ent.* **19**, 15–32.

Matthews, R.W. (1982) Courtship in parasitic wasps. In *Evolutionary Strategies of Parasitic Insects and Mites* (ed. P.W. Price), Plenum Press, New York. pp. 66–86.

Matthews, R.W., Matthews, J.R. and Crankshaw, O. (1979) Aggregation in male parasitic wasps of the genus *Megarhyssa*: I. Sexual discrimination, tergal stroking behaviour, and description of associated anal structures. *Florida Entomol.* **62**, 3–8.

Mattiacci, L., Dicke, M. and Posthumus, M.A. (1994) Induction of parasitoid attracting syndrome in Brussels sprouts plants by feeding of *Pieris brassicae* larvae – Role of mechanical damage and herbivore elicitor. *J. Chem. Ecol.* **20**, 2229–2247.

Mazanec, Z. (1990) Immature stages and life history of *Enytus* sp. (Hymenoptera: Ichneumonidae), a parasitoid of *Perthida glyphopa* Common (Lepidoptera: Incurvariidae). *J. Austr. Entomol. Soc.* **29**, 57–66.

McAllister, B.F. (1995) Isolation and characterisation of a retroelement from B–chromosome (PSR) in the parasitic wasp *Nasonia vitripennis*. *Insect Molec. Biol.* **4**, 253–262.

McAllister, M.K. and Roitberg, B.D. (1987) Adaptive suicidal behaviour in pea aphids. *Nature* **328**, 797–799.

McBrien, H. and Mackauer, M. (1990) Heterospecific larval competition and host discrimination in two species of aphid parasitoids: *Aphidius ervi* and *Aphidius smithi*. *Entomol. Exp. Appl.* **56**, 145–153.

McBrien, H. and Mackauer, M. (1991) Decision to superparasitise based on larval survival: Competition between larval parasitoids *Aphidius ervi* and *Aphidius smithi*. *Entomol. Exp. Appl.* **59**, 145–150.

McConnell, H.S. (1938) Additional notes on *Oocenteter tomostethae*. *Proc. Entomol. Soc. Wash.* **40**, 23–24.

McInnis, D.O., Wang, T.T.Y. and Nishimoto, J. (1986) The inheritance of a black body mutant in *Biosteres longicaudatus* (Hymenoptera; Braconidae) from Hawaii. *Proc. Hawaii Entomol. Soc.* **27**, 37–40.

McLaughlin, R.E. and Adams, C.H. (1966) Infection of *Bracon mellitor* (Hymenoptera: Braconidae) by *Mattesia grandis* (Protozoa: Neogregarinida). *Ann. Entomol. Soc. Am.* **59**, 800–802.

Meer, R.K. van der, Jouvenaz, D.P. and Wojcik, D.P. (1989) Chemical mimicry in a parasitoid (Hymenoptera: Eucharitidae) of fire ants (Hymenoptera: Formicidae). *J. Chem. Ecol.* **15**, 2247–2261.

Mellini, E. (1994) Elementi per un confronto tra il parassitoidismo degli Imenotteri e quello dei Ditteri. *Boll. Ist. Ent. G. Grandi Univ. Bologna* **49**, 41–100.

Memmott, J. and Godfray, H.C.J. (1993) Parasitoid webs. In *Hymenoptera and Biodiversity* (eds J. LaSalle and I.D. Gauld), C.A.B. International, Wallingford. pp. 217–234.

Memmott, J. and Godfray, H.C.J. (1994) The use and construction of parasitoid webs. In *Parasitoid Community Ecology* (eds B.A. Hawkins and W. Sheehan), Oxford University Press, Oxford. pp. 300–318.

Meng, C. (1970) Autoradiographische Untersuchungen am Öosom in der Öocyte von *Pimpla turionellae* L. (Hymenoptera). *Wilhelm Roux Arch* **165**, 35–52.

Menzel, J.G. and Tautz, J. (1994) Functional morphology of the subgenual organ of the carpenter ant. *Tissue and Cell* **26**, 735–746.

Mertens, J.W. (1980) Life-history and behaviour of *Laelius pedatus* (Hymenoptera: Bethylidae), a gregarious bethylid ectoparasitoid of *Anthrenus verbasci* (Coleoptera: Der.). *Ann. Entomol. Soc. Am.* **73**, 686–693.

Messner, B. and Taschenberger, D. (1981) Zur Funktionsmorphologie des Atembandes von *Agriotypus armatus* Walk. *Dtsch. Ent. Z., N.F.* **28**, 7–9.

Mettus, R.V. and Petters, R.M. (1981) Genetic investigation of a behaviour in response to mechanical shock in the parasitic wasp, *Bracon hebetor*. *Genetics* **97** (Suppl.), s72–s74.

Meyerdirk, D.E. and Moratorio, M.S. (1987) Biology of *Anagrus giraulti* (Hymenoptera: Myrmaridae), an egg parasitoid of the beet leafhopper, *Circulifer tenellus* (Homoptera: Cicadellidae). *Ann. Entomol. Soc. Am.* **80**, 272–277.

Meyers, D.M. and Deonier, D.L. (1993) A behavioral–ecological study of *Kleidotoma parydrae* Beardsley (Hymenoptera: Eucoilidae), with notes of *Anaphes* sp. (Hymenoptera: Mymaridae) parasites of *Parydra* spp. (Diptera: Ephydridae). *Contrib. Amer. Entomol. Inst.* **27**, 7–29.

Meyhöfer, R., Casas, J. and Dorn, S. (1994) Host location by a parasitoid using leafminer vibrations – characterizing the vibrational signals produced by the leafmining host. *Physiol. Ent.* **19**, 349–359.

Meyhöfer, R., Casas, J. and Dorn, S. (in press) Mechano- and chemoreceptors and their possible role in the host location behaviour of *Sympiesis sericeicornis* (Hymenoptera: Eulophidae). *Ann. Entomol. Soc. Am.*

Micha, S.G., Stammel, J. and Höller, C. (1993) 6-methyl-5-heptene-2-one, a putative sex and spacing pheromone of the aphid hyperparasitoid, *Alloxysta victrix* (Hymenoptera, Alloxystidae). *Eur. J. Entomol.* **90**, 439–442.

Michaud, J.P. (1994) Differences in foraging behaviour between virgin and mated aphid parasitoids (Hymenoptera, Aphidiidae). *Canad. J. Zool.* **72**, 1597–1602.

Middeldorf, J. and Ruthman, A. (1984) Yeast-like endosymbionts in an ichneumonid wasp. *Z. Naturforsch.* **39**, 322–326.

Miller, G.L. and Lambdin, P.L. (1985) Observations on *Anacharis melanoneura* (Hymenoptera: Figitidae), a parasite of *Hemerobius stigma* (Neuroptera: Hemerobiidae). *Entomol. News* **96**, 93–97.

Miller, J.C. (1982) Life history of insect parasitoids involved in successful multiparasitism. *Oecologia* **54**, 8–9.

Mills, N.J. (1993) Species richness and structure in the parasitoid complexes of tortricid hosts. *J. Anim. Ecol.* **62**, 45–58.

Mills, N.J. (1994a) Parasitoid guilds: Defining the structure of the parasitoid communities of endopterygote insect hosts. *Environ. Ent.* **23**, 1066–1083.

Mills, N.J. (1994b) Parasitoid guilds: A comparative analysis of the parasitoid communities of tortricids and weevils. In *Parasitoid Community Ecology* (eds B.A. Hawkins and W. Sheehan), Oxford University Press, Oxford. pp. 30–46.

Mills, N.J., Kruger, K. and Schlup, J. (1991) Short-range host location mechanisms of bark beetle parasitoids. *J. Appl. Ent.* **111**, 33–43.

Mimouni, F. (1992) Genetic variations in host infestation efficiency in two *Trichogramma* species from Morocco. *Redia* **74**, 393–400.

Minkenberg, P.M.J. (1990) The leafminers *Liriomyza trifolii* and *L. bryoniae*, their parasitoids and host plants: A review. PhD thesis, Agricultural University, Wageningen. 230 pp.

Mitchell, W.C. and Mau, R.F.L. (1971) Response of the female southern green stink bug and its parasite, *Trichopoda pannipes*, to male stink bug sex pheromones. *J. Econ. Ent.* **64**, 856–859.

Miura, K. (1992) Aggressive behavior in *Paracentrobia andoi* (Hymenoptera, Trichogrammatidae), an egg parasitoid of the green rice leafhopper. *Jpn. J. Ent.* **60**, 103–107.

Mollema, C. (1991) Heritability estimates of host selection behaviour by the *Drosophila* parasitoid *Asobara tabida*. *Neth. J. Zool.* **41**, 174–183.

Monconduit, H. and Prevost, G. (1994) Avoidance of encapsulation by *Asobara tabida*, a larval parasitoid of *Drosophila* species. *Norw. J. Agric. Sci. Suppl.* **16**, 301–309.

Mook, J.H. (1961) Observations on the oviposition behaviour of *Polemon liparae* Gir. (Hym., Braconidae). *Archs Neerl. Zool.* **14**, 423–430.

Morales-Ramos, J.A., Rojas, M.G. and King, E.G. (1995) Venom of *Catolaccus grandis* (Hymenoptera: Pteromalidae) and its role in parasitoid development and host regulation. *Ann. Entomol. Soc. Am.* **88**, 800–808.

Moran, V.C., Brothers, D.J. and Case, J.J. (1969) Observations on the biology of *Tetrastichus flavigaster* Brothers and Moran (Hymenoptera: Eulophidae) parasitic on psyllid nymphs (Homoptera). *Trans. R. Entomol. Soc. Lond.* **121**, 41–58.

Morgan, D.J.W. and Cook, J.M. (1994) Extremely precise sex ratios in small clutches of a bethylid wasp. *Oikos* **71**, 423–430.

Morris, K.R.S. (1937) The prepupal stage in Ichneumonidae, illustrated by the life-history of *Exenterus abruptorius*, Thb. *Bull. Entomol. Res.* **28**, 525–534.

Morrison, G., Auerbach, M. and McCoy, E.D. (1979) Anomalous diversity of tropical parasitoids: A general phenomenon? *Amer. Nat.* **114**, 303–307.

Morse, D.H. (1988) Interactions between the crab spider *Misumena vatia* (Clerck) (Araneae) and its ichneumonid egg predator *Trychosis cyperia* Townes (Hymenoptera). *J. Arachnol.* **16**, 132–135.

Mouzaki, D.G. and Margaritis, L.H. (1994) The eggshell of the almond wasp, *Eurytoma amygdali* (Hymenoptera, Eurytomidae). 1. Morphogenesis and fine structure of the eggshell layers. *Tissue and Cell* **26**, 559–568.

Mück, O., Hassan, S., Huger, A.M. and Krieg, A. (1981) Zur Wirkung von *Bacillus thuringiensis* Berliner auf die parasitischen Hymenopteren *Apanteles glomeratus* L. (Braconidae) und *Pimpla turionella* (L.) (Ichneumonidae). *Z. Angew. Entomol.* **92**, 303–314.

Mudd, A.R., Fisher, C. and Smith, M.C. (1982) Volatile hydrocarbons in the Dufour's gland of the parasite *Nemeritis canescens* (Grav.) (Hymenoptera: Ichneumonidae). *J. Chem. Ecol.* **8**, 1035–1042.

Münster-Swendsen, M. (1994) Pseudoparasitism: Detection and ecological significance in *Epinota tedella* (Cl.) (Tortricidae). *Norw. J. Agric. Sci. Suppl.* **16**, 329–335.

Murray, D.A.H., Monsour, C.J., Teakle, R.E. *et al.* (1995) Interactions between nuclear polyhedrosis virus and three larval parasitoids of *Helicoverpa armigera* (Hubner) (Lepidoptera, Noctuidae). *J. Austr. Entomol. Soc.* **34**, 319–322.

Myers, J.H. and Smith, J.N.M. (1978) Head flicking by tent caterpillars: A defensive response to parasitoid sounds. *Canad. J. Zool.* **56**, 1628–1631.

Nadel, H. (1987) Male swarms discovered in Chalcidoidea (Hymenoptera: Encyrtidae, Pteromalidae). *Pan-Pacific Entomol.* **63**, 242–246.

Nadel, H. and Luck, R.F. (1985) Span of female emergence and male sperm depletion in the female-biased, quasi-gregarious parasitoid, *Pachycrepoideus vindemiae* (Hymenoptera: Pteromalidae). *Ann. Entomol. Soc. Am.* **78**, 410–414.

Nagarkatti, S. (1970) The thelytokous hybrid in an interspecific cross between two species of *Trichogramma* (Hym.: Trichogrammatidae). *Curr. Sci.* **4**, 76–78.

Nahif, A.A. and Madel, G. (1990) Zur Biologie des Hyperparasitoiden *Alloxysta fuscicornis* (= *A. ancylocera*) (Hym.: Charipidae). *Entomophaga* **35**, 641–651.

Naito, T. (1982) Chromosome number differentiation in sawflies and its systematic implication (Hymenoptera, Tenthredinidae). *Kontyû, Tokyo* **50**, 569–587.

Naumann, I.D. (1991) Hymenoptera (wasps, bees, ants, sawflies). In *The Insects of Australia* (CSIRO), 2nd edn, Melbourne University Press, Melbourne. pp. 916–1000.

Naumann, I.D. and Reid, C.A.M. (1990) *Ausasaphes shiralee* sp. n. (Hymenoptera, Pteromalidae, Asaphinae), a brachypterous wasp phoretic on a flightless chrysomelid beetle (Coleoptera, Chrysomelidae). *J. Austr. Entomol. Soc.* **29**, 319–325.

Navasero, R.C. and Elzen, G.W. (1991) Sensilla on the antennae, foretarsi and palpi of *Microplitis croceipes* (Cresson) (Hymenoptera: Braconidae). *Proc. Entomol. Soc. Wash.* **93**, 737–747.

Nealis, V.G., Oliver, D. and Tchir, D. (1996) The diapause response to photoperiod in Ontario populations of *Cotesia melanoscela* (Ratzeburg) (Hymenoptera: Braconidae). *Canad. Entomol.* **128**, 41–46.

Nelson, W.A. and Farstad, C.W. (1953) Biology of *Bracon cephi* (Gahan) (Hymenoptera: Braconidae), an important native parasite of the wheat stem sawfly, *Cephus cinctus* Nort. (Hymenoptera: Cephidae), in Western Canada. *Canad. Ent.* **85**, 103–107.

Nemec, V. and Starý, P. (1983) Elpho-morph differentiation in *Aphidius ervi* Hal. biotype on *Microlophium carnosum* (Bckt.) related to parasitization on *Acyrthosiphon pisum* (Harr.) (Hym., Aphidiidae). *Zeitschr. Angew. Ent.* **95**, 524–530.

Nemec, V. and Starý, P. (1984) Population diversity of *Diaeretiella rapae* (McInt.) (Hym., Aphidiidae), an aphid parasitoid in agroecosystems. *Zeitschr. Angew. Ent.* **97**, 223–233.

Nemec, V. and Starý, P. (1985) Genetic diversity in relation to host specificity of aphid parasitoids (Hymenoptera: Aphidiidae). *Entomol. Gener.* **10**, 241–251.

Nemoto, T., Shibuya, M., Kuwahara, Y. and Suzuki, T. (1987) New 2-acylcyclo-hexane-1,3-diones – kairomone against a parasitic wasp, *Venturia canescens*, from faeces of the almond moth, *Cadra cautella*, and the Indian meal moth, *Plodia interpunctella*. *Agric. and Biol. Chem.* **51**, 1805–1810.

Nénon, J.P. and Biossangama, A. (1985) Mise en évidence du siphon respiratoire de l'Hyménoptère *Leptomastix dactylopii* (Encyrtidae) lors de son développement dans la cochenille *Planococcus citri* (Pseudococcidae). *Comptes Rendu Acad. Sci. Paris* **300**, 519–524.

Nénon, J.P., Boivin, G., LeLannic, J. and Baaren, J. van (1995a) Functional morphology of the mymariform and sacciform larvae of the egg parasitoid, *Anaphes victus* Huber (Hymenoptera, Mymaridae). *Canad. J. Zool.* **73**, 996–1000.

Nénon, J.P., Le Lannic, J., Kacem, N. *et al.* (1995b) Micromorphologie de l'oviposi-teur des hyménoptères et évolution des symphytes phytophages aux apocrites parasitoïdes. *Comptes Rendu Acad. Sci. Paris* **318**, 1045–1051.

Neser, S. (1973) Biology and behaviour of *Euplectus* species near *laphygmae* Ferrière (Hymenoptera: Eulophidae). *Rep. South Afr. Dep. Agr. Tech. Serv. Entomol. Mem.* **32**, 1–31.

Nettles, W.C. Jr (1990) *In vitro* rearing of parasitoids: Role of host factors in nutrition. *Arch. Insect Biochem. Physiol.* **13**, 167–175.

Nettles, W.C. Jr, Morrison, R.K., Xie, Z.N. *et al.* (1982) Synergistic action of potassium chloride and magnesium sulfate on parasitoid wasp oviposition. *Science* **218**, 164–166.

Newman, E. (1867) Description of the larva of *Xanthia gilvago*. *Entomologist* **3**, 342.

Nguyen, R. and Sailer, R.I. (1987) Facultative hyperparasitism and sex determination of *Encarsia smithi* (Silvestri) (Hymenoptera: Aphelinidae). *Ann. Entomol. Soc. Am.* **80**, 713–719.

Nielsen, E. (1923) Contributions to the life history of the pimpline spider parasites (*Polysphincta, Zaglyptus, Tromatobia*). *Ent. Meddel.* **14**, 137–205.

Nishida, T. (1958) Extrafloral glandular secretions, a food source for certain insects. *Proc. Hawaii Entomol. Soc.* **16**, 379–386.

Nobel, N.S. (1940) *Trichilogaster acaciae-longifoliae* (Froggatt) (Hymenopt., Chalcidoidea), a wasp causing galling of the flower-buds of *Acacia longifoliae* Willd., *A. floribunda* Sieber and *A. sophorae* R. Br. *Trans. R. Entomol. Soc. Lond.* **90**, 13–38.

Noguchi, H., Hayakawa, Y. and Downer, R.G.H. (1995) Elevation of dopamine levels in parasitized insect larvae. *Insect Biochem. Molec. Biol.* **25**, (197–201.

Noirot, C. and Quennedey, A. (1974) Fine structure of insect epidermal glands. *Ann. Rev. Entomol.* **19**, 61–80.

Noirot, C. and Quennedey, A. (1991) Glands, gland cells, glandular units: Some comments on terminology and classification. *Annls Soc. Entomol. Fr. (N.S.)* **27**, 123–128.

Noldus, L.P.J.J. (1989) Chemical espionage by parasitic wasps. How Trichogramma species exploit moth sex pheromones. Doctoral thesis, Landbouwuniversteit, Wageningen.

Noldus, L.P.J.J., Potting, R.P.J. and Barendregt, H.E. (1991) Moth sex pheromone adsorption to leaf surface – Bridge in time for chemical spies. *Physiol. Ent.* **16**, 329–344.

Nordlund, D.A. (1981) A glossary of terms used to describe chemicals that mediate intra- and interspecific interactions. In *Management of Insect Pests with Semiochemicals* (ed. E.R. Mitchell), Plenum Press, New York.

Norton, W.N. and Vinson, S.B. (1974a) Antennal sensilla of three parasitic Hymenoptera. *Int. J. Insect Morphol. Embryol.* **3**, 305–316.

Norton, W.N. and Vinson, S.B. (1974b) A comparative ultrastructural and behavioral study of the antennal sensory sensilla of the parasitoid *Cardiochiles nigriceps* (Hymenoptera: Braconidae). *J. Morphol.* **142**, 329–349.

Norton, W.N. and Vinson, S.B. (1977) Encapsulation of a parasitoid egg within its natural host: An ultrastructural investigation. *J. Invertebr. Pathol.* **30**, 55–67.

Norton, W.N. and Vinson, S.B. (1982) Synthesis of the vitelline and chorionic membranes of an ichneumonid parasitoid. *J. Morphol.* **174**, 185–195.

Norton, W.N. and Vinson, S.B. (1983) Correlating the initiation of virus replication with a specific pupal developmental phase of an ichneumonid parasitoid. *Cell. Tiss. Res.* **231**, 387–398.

Noyes, J.S. (1982) Collecting and preserving chalcid wasps (Hymenoptera: Chalcidoidea). *J. Nat. Hist.* **16**, 315–334.

Noyes, J.S. (1989) The diversity of Hymenoptera in the tropics with a special reference to parasitica in Sulawesi. *Ecol. Ent.* **14**, 197–207.

Noyes, J.S. (1990) A word on chalcidoid classification. *Chalcid Forum* **13**, 6,7.

Noyes, J.S. (1994) The reliability of published host–parasitoid records: A taxonomist's view. *Norw. J. Agric. Sci. Suppl.* **16**, 59–69.

Numata, H. (1993) Induction of adult diapause and of low and high reproductive states in a parasitoid wasp, *Ooencyrtus nezarae*, by photoperiod and temperature. *Entomol. Exp. Appl.* **66**, 127–134.

Nuttall, M.J. (1980) Insect parasites of *Sirex* (Hymenoptera: Ichneumonidae, Ibaliidae, and Orussidae). *Forest Timb. Ind. N.Z. No. 47*, 12pp.

Obara, M. and Kitano, H. (1974) Studies on the courtship behaviour of *Apanteles glomeratus* L. I. Experimental studies of wing-vibrating behaviour in the male. *Kontyû, Tokyo* **42**, 208–214. (In Japanese.)

Odebiyi, J.A. and Oatman, E.R. (1972) Biology of *Agathis gibbosa* (Hymenoptera: Braconidae), a primary parasite of the potato tuberworm. *Ann. Entomol. Soc. Am.* **65**, 1104–1114.

Odierna, G., Baldanza, F., Aprea, G. and Olmo, E. (1993) Occurrence of G-banding in metaphase chromosomes of *Encarsia berlesei* (Hymenoptera, Aphelinidae). *Genome* **36**, 662–667.

O'Donnell, D.J. (1987) Larval development and the determination of the number of instars in aphid parasitoids (Hym., Aphidiidae). *Int. J. Insect Morphol. Embryol.* **16**, 3–15.

O'Donnell, D.J. (1989) A morphological and taxonomic study of first instar larvae of Aphidiinae (Hymenoptera: Braconidae). *Syst. Ent.* **14**, 197–219.

Oeser, R. (1961) Vergleickend-Morphologische Untersuchungen über den Ovipositor der Hymenopteren. *Mitt. Zool. Mus. Berlin*, **37**, 1–119.

Oeser, R. (1962) Der reduzierte Ovipositor von *Pseudogonalos hahni* (Spin.) nebst Bemerkungen über die systematische stellung der Trigonalidae. *Wandersamm. Deutsch. Entomol. Ber.* **9**, 153–157.

Ohbayashi, T., Iwabuchi, K. and Mitsuhashi, J. (1994) *In vitro* rearing of a larval endoparasitoid, *Venturia canescens* (Gravenhorst) (Hymenoptera, Ichneumonidae). 1. Embryonic development. *Appl. Ent. Zool.* **29**, 123–126.

Ohno, K. (1983) Reproductive behaviour in scelionid wasps, egg parasitoids of pentatomids. *Pl. Prot., Tokyo* **37**, 45–49. (In Japanese.)

Ohsaki, N. and Sato, Y. (1990) Avoidance mechanisms of three pierid butterfly species against the parasitoid wasp *Apanteles glomeratus*. *Ecol. Ent.* **15**, 169–176.

Okuda, M.S. and Yeargan, K.V. (1988) Intra- and interspecific host discrimination in *Telenomus podisi* and *Trissolcus euschisti* (Hymenoptera: Scelionidae). *Ann. Entomol. Soc. Am.* **81**, 1017–1020.

Olmi, M. (1994) The Dryinidae and Embolemidae (Hymenoptera: Chrysidoidea) of Fennoscandia and Denmark. In *Fauna Entomologia Scandinavica*, Vol. 30, E.J. Brill, Leiden. 100pp.

Olson, D.L. and Nechols, J.R. (1995) Effects of squash leaf trichome exudates and honey on adult feeding, survival, and fecundity of the squash bug (Heteroptera, Coreidae) egg parasitoid *Gryon pennsylvanicum* (Hymenoptera, Scelionidae). *Env. Ent.* **24**, 454–458.

Olson, D.M. and Andow, D.A. (1993) Antennal sensilla of female *Trichogramma nubiale* (Ertle and Davis) (Hymenoptera, Trichogrammatidae) and comparisons with other parasitic Hymenoptera. *Int. J. Insect Morphol. Embryol.* **22**, 507–520.

O'Neill, W.L. (1973) Biology of *Trichopria popei* and *T. atrichomelinae* (Hymenoptera: Diapriidae), parasitoids of the Sciomyzidae (Diptera). *Ann. Entomol. Soc. Am.* **66**, 1043–1051.

Orr, C.J., Obrycki, J.J. and Flanders, R.V. (1992) Host-acceptance behaviour of *Dinocampus coccinellae* (Hymenoptera: Braconidae). *Ann. Entomol. Soc. Am.* **85**, 722–730.

Ortel, J. (1995) Accumulation of Cd and Pb in successive stages of *Galleria mellonella* and metal transfer to the pupal parasitoid *Pimpla turionellae*. *Entomol. Exp. Appl.* **77**, 89–97.

Orzack, S.H. and Gladstone, J. (1994) Quantitative genetics of sex-ratio traits in the parasitic wasp, *Nasonia vitripennis*. *Genetics* **137**, 211–220.

Osborne, P. (1960) Observations on the natural enemies of *Meligethes aeneus* (F.) and *M. viridescens* (F.) (Coleoptera: Nitidulidae). *Parasitology* **50**, 91–110.

Osman, S.E. and Führer, E. (1979) Histochemical analysis of accessory gland secretions in female *Pimpla turionellae* L. (Hym.). *Int. J. Invert. Reprod.* **1**, 323–332.

Osten, T. (1982) Vergleichend-funktionsmorphologische Untersuchungen der Kopfkapsel und der Mundwerkzeuge ausgewählter 'Scolioidea' (Hymenoptera, Aculeata). *Stuttgart. Beitr. Naturkunde Serie A* **364**, 1–60.

Ovruski, S.M. (1994) Immature stages of *Aganaspis pelleranoi* (Brèthes) (Hymenoptera: Cynipoidea: Eucoilidae), a parasitoid of *Ceratitis capitata* (Wied.) and *Anastrepha* spp. (Diptera: Tephritidae). *J. Hym. Res.* **3**, 233–239.

Owen, D.F. and Owen, J. (1974) Species diversity in temperate and tropical Ichneumonidae. *Nature*, **249**, 583–584.

Owen, J., Townes, H. and Townes, M. (1981) Species diversity of Ichneumonidae and Serphidae (Hymenoptera) in an English suburban garden. *Biol. J. Linn. Soc.* **16**, 315–336.

Own, O.S. and Brooks, W.M. (1986) Interactions of the parasite *Pediobius foveolatus* (Hymenoptera: Eulophidae) with two *Nosema* spp. (Microsporida: Nosematidae) of the Mexican bean beetle (Coleoptera: Coccinellidae). *Environ. Ent.* **15**, 32–39.

Packer, L. and Owen, R.E. (1992) Variable enzyme systems in the Hymenoptera. *Biochem. Syst. Ecol.* **20**, 1–7.

Page, R. Jr (1986) Sperm utilisation in social insects. *Annu. Rev. Entomol.* **31**, 297–320.

Paillot, A. (1933) *L'Infection chez les Insectes*, G. Patissier, Trüvoux.

Palm, N.-B. (1949) The rectal papillae in insects. *Lunds Univ. Arsskr. N.F.* **60**, 1–29.

Pamilo, P., Varvio-Aho, S. and Pekkarinen, A. (1978) Low enzyme gene variability in Hymenoptera as a consequence of haplodiploidy. *Hereditas* **88**, 93–99.

Pampel, W. (1913) Die weiblichen Geschlechtsorgane der Ichneumoniden. *Z. Wiss. Zool.* **108**, 291–357.

Papp, R.P. (1981) Notes on the reproductive biology of the parasitoid *Doryctes palliatus* (Cameron) (Hymenoptera: Braconidae) in Hawaii. *Proc. Hawaii. Entomol. Soc.* **23**, 435–436.

Parker, E.D. and Orzack, S.H. (1985) Genetic variation for sex ratio in *Nasonia vitripennis*. *Genetics* **110**, 93–105.

Parker, H.L. (1924) Recherches sur les formes post-embryonnaires des Chalcidiens. *Ann. Soc. Ent. France* **93**, 261–379 + 39 plates.

Parker, H.L. (1931) *Macrocentrus gifuensis* Ashmead, a polyembryonic braconid parasite of the European corn borer. *US Dep. Agric. Tech. Bull.* **230**, 1–62.

Parnell, J.R. (1963) Three gall midges (Diptera: Cecidomyidae) and their parasites found in the pods of broom (*Sarothamnus scoparius* (L.) Wimmer). *Trans. R. Entomol. Soc. Lond.* **115**, 261–275.

Parrott, P.J. and Glasgow, H. (1916) The leaf-weevil (*Polydrussus impressifrons* Gyll.). *Tech. Bull., N.Y. State Agric. Exp. Stat.* **56**, 1–22.

Payne, N.M. (1937) The differential effect of environmental factors upon *Microbracon hebetor* Say and its host, *Ephestia kuhniella* Zeller. *Biol. Bull. (Woods Hole, Mass.)* **73**, 147–154.

Peakall, R. (1990) Responses of male *Zaspilothynnus trilobatus* Turner wasps to females and the sexually deceptive orchid it pollinates. *Funct. Ecol.* **4**, 159–167.

Pech, L.L., Trudeau, D. and Strand, M.R. (1995) Effects of basement membranes on the behavior of hemocytes from *Pseudoplusia includens* (Lepidoptera; Noctuidae): Development of an in vitro encapsulation assay. *J. Insect Physiol.* **41**, 801–807.

Peck, O. (1937) The male genitalia in the Hymenoptera (Insecta), especially the family Ichneumonidae. *Canad. J. Res. D.* **15**, 221–274.

Pedata, P., Isidoro, N. and Viggiani, G. [(1993)/1995] Evidence of male sex glands on the antennae of *Encarsia asterobemisiae* Viggiani and Mazzone (Hymenoptera: Aphelinidae). *Boll. Lab. Entomol. Agric. 'Filippo Silvestri' Portici* **50**, 271–280

Pemberton, C.E. and Willard, H.F. (1918) A contribution to the biology of fruit-fly parasites in Hawaii. *J. Agric. Res.* **15**, 419–466.

Pennacchio, F., Vinson, S.B. and Tremblay, E. (1992) Preliminary results on *in vitro* rearing of the endoparasitoid *Cardiochiles nigriceps* from egg to second instar. *Entomol. Exp. Appl.* **64**, 209–216.

Pennacchio, F., Digilio, M.C., Tremblay, E. and Tranfaglia, A. (1994a) Host recognition and acceptance behaviour in two aphid parasitoid species: *Aphidius ervi* and *Aphidius microlophii* (Hymenoptera: Braconidae). *Bull. Entomol. Res.* **84**, 57–64.

Pennacchio, F., Vinson, S.B. and Tremblay, E. (1994b) Morphology and ultra-structure of the serosal cells (teratocytes) in *Cardiochiles nigriceps* Viereck (Hymenoptera: Braconidae) embryos. *Int. J. Insect Morphol. Embryol.* **23**, 93–104.

Pennacchio, F., Vinson, S.B. and Tremblay, E. (1994c) Regulation of *Heliothis virescens* (F.) development and hormonal metabolism by the endophagous parasitoid *Cardiochiles nigriceps* Viereck: The role of teratocytes. *Norw. J. Agric. Sci. Suppl.* **16**, 293–300.

Periquet, G., Hedderwick, M.P., El Agoze, M. and Poirie, M. (1993) Sex determination in the Hymenoptera *Diadromus pulchellus* (Ichneumonidae): Validation of the one-locus multiallele model. *Heredity* **70**, 420–427.

Perrotminnot, M.J., Guo, L.R. and Werren, J.H. (1996) Single and double infections with *Wolbachia* in the parasitic wasp *Nasonia vitripennis* – Effects on compatibility. *Genetics* **143**, 961–972.

Peter, C. and David, B.V. (1991) Observations on the oviposition behaviour of *Goniozus sensorius* (Hymenoptera: Bethylidae) a parasite of *Diaphania indica* (Lepidoptera: Pyralidae). *Entomophaga* **36**, 403–407.

Petersen, G. and Hardy, I.C.W. (1996) The importance of being larger – Parasitioid intruder–owner contests and their implications for clutch size. *Anim. Behav.* **51**, 1363–1373.

Petters, R.M. (1977) A morphogenetic fate map constructed from *Habrobracon juglandis* gynandromorphs. *Genetics* **85**, 279–287.

Petters, R.M. and Grosch, D.S. (1976) Increased production of genetic mosaics in *Habrobracon juglandis* by cold shock of newly oviposited eggs. *J. Embryol. Exp. Morph.* **36**, 127–131.

Petters, R.M. and Mettus, R.V. (1980) Decreased diploid male viability in the parasitic wasp, *Bracon hebetor. J. Hered.* **71**, 353–356.

Petters, R.M., Mettus, R.V. and Casey, J.N. (1983) Toxic and reproductive effects of the soluble organic fraction from diesel particulate emissions on the parasitoid wasp, *Bracon hebetor. Envir. Res.* **32**, 37–46.

Petters, R.M., Kendall, M.E., Taylor, R.A.J. and Mettus, R.V. (1985) Time required for mating and the degree of genetic relatedness in the parasitic wasp, *Bracon hebetor* Say (Hymenoptera: Braconidae). *Melsheimer Entomol. Ser.* **35**, 21–27.

Pfister-Wilhelm, R. and Lanzrein, B. (1996) Precocious induction of metamorphosis in *Spodoptera littoralis* (Noctuidae) by the parasitic wasp *Chelonus inanitus* (Braconidae): Identification of the parasitoid larva as the key regulatory element and the host corpora allata as the main target. *Arch. Ins. Biochem. Physiol.* **32**, 511–525.

Phillips, W.J. and Emery, W.T. (1919) A revision of the chalcid-flies of the genus *Harmolita* of America north of Mexico. *Proc. U.S. Nat. Mus.* **55**, 433–471.

Piao, C.S. and Moriya, S. (1992) Longevity and oviposition of *Torymus sinensis* Kamijo and two strains of *T. beneficus* Yasumatsu et Kamijo (Hymenoptera, Torymidae). *Jap. J. Appl. Ent. Zool.* **36**, 113–118. (In Japanese.)

Pickering, J. (1980) Larval competition and brood sex ratios in the gregarious parasitoid *Pachysomoides stupidus. Nature (London)* **283**, 291–292.

Piek, T. (1982) Solitary wasp venoms and toxins as tools for the study of neuromuscular transmission in insects. *Ciba Found. Symp.* **88**, 275–290.

Piek, T. (ed.) (1986) *Venoms of the Hymenoptera*, Academic Press, London. 570pp.

Piek, T. and Spanjer, W. (1986) Chemistry and pharmacology of solitary wasp venoms. In *Venoms of the Hymenoptera* (ed. T. Piek), Academic Press, London. pp. 161–307.

Piek, T., Spanjer, W., Njio, K.D. *et al.* (1974) Paralysis caused by the venom of the wasp, *Microbracon gelechiae. J. Insect Physiol.* **20**, 2307–2319.

Pijls, J.W.A.M., Steenbergen, H.J. van and Alphen, J.J.M. van (1996) Asexuality cured – The relations and differences between sexual and asexual *Apoanagyrus diversicornis. Heredity* **76**, 506–513.

Pimm, S.L. (1979) Sympatric speciation: A simulation model. *Biol. J. Linn. Soc.* **11**, 131–139.

Pinto, J.D. (1994) A taxonomic study of *Brachista* (Hymenoptera: Trichogrammatidae) with a description of the two new species phoretic on robberflies of the genus *Efferia* (Diptera: Asilidae). *Proc. Entomol. Soc. Wash.* **96**, 120–132.

Pinto, J.D., Velten, R.K., Platner, G.R. and Oatman, E.R. (1989) Phenotypic plasticity and taxonomic characters in *Trichogramma* (Hymenoptera: Trichogrammatidae). *Ann. Entomol. Soc. Am.* **82**, 414–425.

Pinto, J.D., Stouthamer, R., Platner, G.R. and Oatman, E.R. (1991) Variation in reproductive compability in *Trichogramma* and its taxonomic significance. *Ann. Entomol. Soc. Am.* **84**, 37–46.

Pinto, J.D., Platner, G.R. and Sassaman, C.A. (1993) Electrophoretic study of two closely related species of North American *Trichogramma*: *T. pretiosum* and *T. deion* (Hymenoptera: Trichogrammatidae). *Ann. Entomol. Soc. Am.* **86**, 702–709.

Pintureau, B. and Babault, M. (1981) Enzymatic characterization of *Trichogramma evanescens* and *Trichogramma maidis* (Hym.: Trichogrammatidae) – Study of hybrids. *Entomophaga* **26**, 11–22.

Pintureau, B. and Daumal, J. (1995) Effects of diapause and host species on some morphometric characters in *Trichogramma* (Hym., Trichogrammatidae). *Experientia* **51**, 67–72.

Poinar, G.O. (1986) Fossil evidence of spider parasitism by Ichneumonidae. *J. Arachnol.* **14**, 399–400.

Poinar, G.O. Jr (1992) *Life in Amber*, Stanford University Press, Stanford. 350pp.

Polaszek, A. (1986) The effects of two species of hymenopterous parasitoid on the reproductive system of the pea aphid, *Acyrthosiphon pisum. Entomol. Exp. Appl.* **40**, 285–292.

Ponomarenko, N.G. (1971) Some peculiarities of development of Dryinidae. *Proc. XIII Int. Congr. Ent.*, Moscow, 2–9 August, 1968. pp. 281–282.

Ponomarenko, N.G. (1975) Peculiarities of the larval development in Dryinidae. *Entomologich. Obozr.* **54**, 534–540. (In Russian.)

Poole, H.K. (1970. The wall structure of the honey bee spermatheca with comments about its function. *Ann. Entomol. Soc. Am.* **63**, 1625–1628.

Porcelli, F. (1988) Morfologia degli stadi larvali di *Habronyx heros* Wesmael (Ichneumonidae – Anomaloninae) con note di biologia. *Entomologica* **23**, 171–189.

Potting, R.P.J., Vet, L.E.M. and Dicke, M. (1995) Host microhabitat location by the stem-borer parasitoid *Cotesia flavipes*: The role of herbivore volatiles and locally and systemically induced plant volatiles. *J. Chem. Ecol.* **21**, 525–540.

Powell, J.E. (1989) Food consumption by tobacco budworm (Lepidoptera: Noctuidae) larvae reduced after parasitization by *Microplitis demolitor* or *M. croceipes* (Hymenoptera: Braconidae). *J. Econ. Entomol.* **82**, 408–411.

Powell, W. (1994) Nemec and Starý's 'Population Diversity Centre' hypothesis for aphid parasitoids re-visited. *Norw. J. Agric. Sci. Suppl.* **16**, 163–169.

Powell, W. and Walton, M.P. (1989) The use of electrophoresis in the study of hymenopteran parasitoids of agricultural pests. In *Electrophoretic Studies of Agricultural Pests* (eds H.D. Loxdale and J. den Hollander), Clarendon Press, Oxford. pp. 443–465.

Powell, W., Hardy, I.C.W., Hick, A.J. *et al.* (1993) Responses of the parasitoid *Praon volucre* (Hymenoptera, Braconidae) to aphid sex pheromone lures in cereal fields in autumn – implications for parasitoid manipulation. *Eur. J. Entomol.* **90**, 435–438.

Presnail, J.K. and Hoy, M.A. (1996) Maternal microinjection of the endoparasitoid *Cardiochiles diaphaniae* (Hymenoptera: Braconidae). *Ann. Entomol. Soc. Am.* **89**, 576–680.

Preston-Mafham, R. and Preston-Mafham, K. (1993) *The Encyclopaedia of Land Invertebrate Behaviour*, Blandford Press, London. 320pp.

Prevost, G. and Lewis, W.J. (1990) Heritable differences in the response of the braconid wasp *Microplitis croceipes* to volatile allelochemicals. *J. Insect Behav.* **3**, 277–287.

Price, P.W. (1970) Trail odours: Recognition by insect parasitic in cocoons. *Science* **170**, 546–547.

Pringle, J.W.S. (1968) Comparative physiology of the flight motor. *Adv. Insect Physiol.* **5**, 163–227.

Prokopy, R.J. and Webster, R.P. (1978) Oviposition-deterring pheromone of *Rhagoletis pomonella*. A kairomone for its parasitoid *Opius lectus*. *J. Chem. Ecol.* **4**, 481–494.

Pungerl, N.B. (1986) Morphometric and electrophoretic study of *Aphidius* species (Hymenoptera: Aphidiidae) reared from a variety of aphid hosts. *Syst. Ent.* **11**, 327–354.

Pupedis, R.J. (1978) Tube feeding by *Sisyridivora cavigena* (Hymenoptera: Pteromalidae) on *Climacia areolaris* (Neuroptera: Sisyridae). *Ann. Entomol. Soc. Am.* **71**, 773–775.

Purvis, A. and Quicke, D.L.J. (1997) Building phylogenies: are the big easy? *Trends Ecol. Evol.*, **12**, 49–50.

Quednau, F.W. (1970) Notes on life-history, fecundity, longevity, and attack pattern of *Agathis pumila* (Hymenoptera: Braconidae), a parasite of the larch casebearer. *Canad. Entomol.* **102**, 736–745.

Quezada, J.R., DeBach, P. and Rosen, D. (1973) Biological and taxonomic studies of *Signiphora borinquensis*, new species, (Hymenoptera: Signiphoridae), a primary parasitoid of diaspine scales. *Hilgardia* **41**, 543–603.

Quicke, D.L.J. (1981) Hamuli number in the Braconinae (Hymenoptera: Braconidae): An inter- and intraspecific, size-dependent, taxonomic character. *Or. Insects* **15**, 235–240.

Quicke, D.L.J. (1984) Evidence for the function of white-tipped ovipositor sheaths in Braconinae (Hymenoptera: Braconidae). *Proc. Trans. Br. Ent. Nat. Hist. Soc.* **17**, 71–79.

Quicke, D.L.J. (1986) Preliminary notes on homeochromatic associations within and between the Afrotropical Braconinae (Hym., Braconidae) and Lamiinae (Col., Cerambycidae). *Entomol. Mon. Mag.* **122**, 97–109.

Quicke, D.L.J. (1987) The Old World genera of braconine wasps (Hymenoptera: Braconidae). *J. Nat. Hist.* **21**, 43–157.

Quicke, D.L.J. (1988) Spider venoms and the prospect of safer insecticides. *New Scientist* **1640**, 38–41.

Quicke, D.L.J. (1989) Parasitic braconine wasps of the genus *Archibracon* (Hymenoptera: Braconidae). *J. Nat. Hist.* **23**, 29–70.

Quicke, D.L.J. (1990) Tergal and inter-tergal metasomal glands of male braconine wasps (Insecta, Hymenoptera, Braconidae). *Zool. Scripta* **19**, 413–423.

Quicke, D.L.J. (1991) Ovipositor mechanics of the braconine wasp genus *Zaglyptogastra* and the ichneumonid genus *Pristomerus*. *J. Nat. Hist.* **25**, 971–977.

Quicke, D.L.J. (1992) Nocturnal Australasian Braconinae (Hym., Braconidae). *Ent. Mon. Mag.* **128**, 33–37.

Quicke, D.L.J. (1993) *Principles and Techniques of Contemporary Taxonomy*, Blackie Academic and Professional, Glasgow. 311 pp.

Quicke, D.L.J. (1994) Phylogenetics and biological transitions in the Braconidae (Hymenoptera: Ichneumonoidea). *Norw. J. Agric. Sci. Suppl.* **16**, 155–162.

Quicke, D.L.J. and Achterberg, C. van (1990) Phylogeny of the subfamilies of Braconidae (Hymenoptera). *Zool. Verh., Leiden.* **258**, 1–95.

Quicke, D.L.J. and Fitton, M.G. (1995) Ovipositor steering mechanisms in parasitic wasps of the families Gasteruptiidae and Aulacidae (Hymenoptera). *Proc. R. Soc. Lond. B.* **261**, 99–103.

Quicke, D.L.J. and Gokhman, V.E. (1996) First chromosome records for the super-family Ceraphronoidea and new data for some genera and species of Evanioidea and Chrysididae (Hymenoptera: Chrysidoidea). *J. Hym. Res.* **5**, 203–205

Quicke, D.L.J. and Huddleston, T. (1991) The extraordinary genus *Gnathobracon* Costa (Hym., Braconidae, Braconinae), with the description of a new species. *Entomol. Mon. Mag.* **127**, 191–195.

Quicke, D.L.J. and Kruft, R.A. (1995) Latitudinal gradients in north American braconid wasp species richness and biology. *J. Hym. Res.* **4**, 194–203.

Quicke, D.L.J. and Marsh, P.M. (1992) Two new species of neotropical parasitic wasps with highly modified ovipositors (Hymenoptera: Braconidae: Braconinae and Doryctinae). *Proc. Entomol. Soc. Wash.* **94**, 559–567.

Quicke, D.L.J. and Usherwood, P.N.R. (1990) Spider toxins as lead structures for novel pesticides. In *Safer Insecticides: Development and Use* (eds E. Hodgson and R.J. Kuhr), Marcel Dekker, Inc., New York. pp. 385–452.

Quicke, D.L.J., Fitton, M.G. and Ingram, S. (1992a) Phylogenetic implications of the structure and distribution of ovipositor valvilli in the Hymenoptera (Insecta). *J. Nat. Hist.* **26**, 587–608.

Quicke, D.L.J., Ingram, S.N., Baillie, H.S. and Gaitens, P.V. (1992b) Sperm structure and ultrastructure in the Hymenoptera (Insecta). *Zool. Scripta* **21**, 381–402.

Quicke, D.L.J., Ingram, S.N., Proctor, J. and Huddleston, T. (1992c) Batesian and Müllerian mimicry between species with connected life histories, with a new example involving braconid wasp parasitoids of *Phoracantha* beetles. *J. Nat. Hist.* **26**, 1013–1034.

Quicke, D.L.J., Fitton, M.G., Tunstead, J.R. *et al.* (1994) Ovipositor structure and relationships within the Hymenoptera, with special reference to the Ichneumonoidea. *J. Nat. Hist.* **28**, 635–682.

Quicke, D.L.J., Fitton, M.G. and Harris, J. (1995) Ovipositor steering mechanisms in braconid wasps. *J. Hym. Res.* **4**, 110–120.

Quicke, D.L.J., Wharton, R.A. and Sittertz-Bhatkar, H. (1996) Recto-tergal fusion in the Braconinae (Hymenoptera, Braconidae): Distribution and structure. *J. Hym. Res.* **5**, 73–79.

Quicke, D.L.J., Achterberg, C. van and Godfray, H.C.J. (in press) Venom gland and reservoir structure in the Opiinae and Alysiinae (Insecta: Hymenoptera: Braconidae). *Zool. Scripta.*

Rabb, R.L. and Bradley, J.R. (1970) Marking host eggs by *Telenomus sphingus*. *Ann. Entomol. Soc. Am.* **63**, 1053–1056.

Rabouille, A., Bigot, Y., Drezen, J.M. *et al.* (1994) A member of the Reoviridae (DPRV) has a ploidy-specific genomic segment in the wasp *Diadromus pulchellus* (Hymenoptera). *Virology*, **205**, 228–237.

Raimo, B., Reardon, R.C. and Podgwaite, J.D. (1977) Vectoring gypsy moth nuclear polyhedrosis virus by *Apanteles melanoscelus* (Hym.: Braconidae). *Entomophaga* **22**, 207–215.

Rakshpal, R. (1945) The structure and development of the female genital organs of *Tetrastichus pyrillae* Crawf. (Eulophidae – Chalcidoidea) with a comparison of the genital organs in the two sexes. *Indian J. Ent.* **7**, 65–74.

Ramachandran, R. and Norris, D.M. (1991) Volatiles mediating plant–herbivore–natural enemy interactions: Electroantennogram responses of soybean looper, *Pseudoplusia includens*, and a parasitoid, *Microplitis demolitor*, to green leaf volatiles. *J. Chem. Ecol.* **17**, 1665–1690.

Ramachandran, R., Norris, D.M., Phillips, J.K. and Phillips, T.W. (1991) Volatiles mediating plant–herbivore–natural enemy interactions – Soybean looper frass volatiles, 3-octanone and guaiacol, as kairomones for the parasitoid *Microplitis demolitor*. *J. Agric. and Food Chem.* **39**, 2310–2317.

Ramadan, M.M. and Beardsley, J.W. (1992) Supernumary moults in the 1st instar of *Diachasmimorpha tryoni* (Cameron) (Hymenoptera: Braconidae: Opiinae). *Proc. Hawaii Entomol. Soc.* **31**, 235–258.

Ramadan, M.M., Wong, T.T.Y. and Wong, M.A. (1991) Influence of parasitoid size and age on male mating success of opiines (Hymenoptera: Braconidae), larval parasitoids of fruit flies (Diptera: Tephritidae). *Biol. Control* **1**, 248–255.

Ramirez, B.W. (1969) Fig wasps: Mechanisms of pollen transfer. *Science* **163**, 580–581.

Ramírez, W. (1974) Coevolution of *Ficus* and Agaonidae. *Annls Miss. Bot. Gard.* **61**, 770–780.

Ramírez, W. (1978) Evolution of mechanisms to carry pollen in Agaonidae (Hymenoptera Chalcidoidea). *Tijdschr. Ent.* **121**, 279–293.

Ramírez, W. (1986) Oviposition behavior of *Critogaster* and *Idarnes* (Sycophaginae, Torymidae: Chalcidoidea) long-tailed parasites of the New World *Pharmacosycea* and *Urostigma* figs. *Brenesia.* **25–26**, 323–325.

Ramírez, W. (1987) The influence of the microenvironment – the interior of the syconium – in the coevolution between fig wasps (Agaonidae) and the fig (*Ficus*). In *Insects – Plants* (eds V. Labeyrie, G. Fabres and D. Lachaise), Dr W. Junk Publishers, Dordrecht. pp. 329–334.

Ramírez, W. (1991) Evolution of the mandibular appendage in fig wasps (Hymenoptera: Agaonidae). *Rev. Biol. Trop.* **39**, 87–95.

Ramirez, W. and Marsh, P.M. (1996) A review of the genus *Psenobolus* (Hymenoptera: Braconidae) from Costa Rica, an inquiline fig wasp with brachypterous males, with descriptions of two new species. *J. Hym. Res.* **5**, 64–72.

Rasch, E.M., Cassidy, J.D. and King, R.C. (1975) Estimates of genome size in haplo-diploid species of parasitoid wasps. *J. Histochem. Cytochem.* **23**, 317.

Rasch, E.M., Cassidy, J.D. and King, R.C. (1977) Evidence for dosage compensation in parthenogenetic Hymenoptera. *Chromosoma* **59**, 323–340.

Rasnitsyn, A.P. (1964) On hibernation of ichneumon-flies (Hymenoptera: Ichneumonidae). *Entomologich. Obozr.* **43**, 46–51. (In Russian.)

Rasnitsyn, A.P. (1980) Origin and evolution of Hymenoptera. *Trudy Paleont. Inst.* **174**, 1–190.

Rasnitsyn, A.P. (1988) An outline of evolution of the hymenopterous insects (Order Vespida). *Oriental Insects* **22**, 115–145.

Ratcliffe, N.A. and King, P.E. (1969) Morphological, ultrastructural, histochemical and electrophoretic studies on the venom system of *Nasonia vitripennis* Walker (Hymenoptera: Pteromalidae). *J. Morphol.* **127**, 177–204.

Rathcke, B.J. and Price, P.W. (1976) Anomalous diversity of tropical ichneumonid parasitoids: A predation hypothesis. *Am. Nat.* **110**, 889–893.

Ratner, S. and Vinson, S.B. (1983) Encapsulation reactions *in vitro* by haemocytes of *Heliothis virescens*. *J. Insect Physiol.* **29**, 855–864.

Rechav, Y. and Orion, T. (1975) The development of the immature stages of *Chelonus inanitus*. *Ann. Entomol. Soc. Am.* **68**, 457–462.

Reed, H.C., Tan, S., Reed, D.K. *et al.* (1994) Evidence for a sex attractant in *Aphidius colemani* Viereck with potential use in field studies. *Southwest. Entomol.* **19**, 273–278.

Reed-Larsen, D.A. and Brown, J.J. (1990) Embryonic castration of the codling moth, *Cydia pomonella* by an endoparasitoid, *Ascogaster quadridentata*. *J. Insect Physiol.* **36**, 111–118.

Reid, J.A. (1941) The thorax of the wingless and short-winged Hymenoptera. *Trans. R. Entomol. Soc. Lond.* **91**, 367–446.

Reymonet, C., Fabres, G. and Labeyrie, V. (1987) Étude préliminaire d'un marqueur physiologique externe de la diapause chez *Dinarmus acutus* (Hym. Pteromalidae) parasite de *Bruchus affinis* (Col. Bruchidae). *Annls Soc. Entomol. Fr. (N.S.)* **23**, 241–246.

Rice, R.E. and Jones, R.A. (1982) Collections of *Prospaltella perniciosi* Tower (Hymenoptera: Aphelinidae) on San Jose Scale (Homoptera: Diaspididae) pheromone traps. *Environ. Ent.* **11**, 876–880.

Richards, O.W. (1949) The significance of the number of wing-hooks in bees and wasps. *Proc. R. Entomol. Soc. Lond. (A)* **24**, 75–78.

Richardson, P.M., Holmes, W.P. and Saul, G.B. II (1987) The effect of tetracycline on nonreciprocal cross incompatibility in *Mormoniella* [= *Nasonia*] *vitripennis. J. Invert. Pathol.* **50**, 176–183.

Richerson, J.V. and Borden, J.H. (1971) Sound and vibration are not obligatory host finding stimuli for the bark beetle parasite *Coeloides brunneri* Viereck. *Canad. Entomol.* **94**, 748–763.

Richerson, J.V. and Borden, J.H. (1972a) Host finding behaviour of *Coeloides brunneri* (Hymenoptera: Braconidae). *Canad. Entomol.* **104**, 1235–1250.

Richerson, J.V. and Borden, J.H. (1972b) Host finding by heat perception in *Coeloides brunneri* (Hymenoptera: Braconidae). *Canad. Entomol.* **104**, 1877–1878.

Richerson, J.V., Borden, J.H and Hollingdale, J. (1972) Morphology of a unique sensillum placodeum on the antennae of *Coeloides brunneri* (Hymenoptera: Braconidae). *Canad. J. Zool.* **50**, 909–913.

Ridley, M. (1993) Clutch size and mating frequency in parasitic Hymenoptera. *Amer. Nat.* **142**, 893–910.

Rimsky-Korsakov, M. (1917) Observations biologiques sur les Hyménoptères aquatiques. *Rev. Russe d'Entomol.* **16**, 209–225.

Rimsky-Korsakov, M. (1933) Methoden zur Untersuchung von Wasserhymenopteren. *Hbch. Biologisch. Arbeitsmeth.* **9**, 227–258.

Ritter, K.S. and Johnson, J.A. (1991) Effects of host sterols on the development and sterol composition of *Microplitis demolitor* (Hymenoptera: Braconidae) in *Heliothis zea* (Lepidoptera: Noctuidae). *Ann. Entomol. Soc. Am.* **84**, 79–86.

Rivero-Lynch, A. P. (1994) Reproductive behaviour of parasitoids in patchy environments. Unpublished PhD Thesis, Imperial College, University of London.

Rivers, D.B., Hink, W.F. and Denlinger, D.L. (1993) Toxicity of venom from *Nasonia vitripennis* (Hymenoptera, Pteromalidae) towards fly hosts, nontarget insects, different developmental stages, and cultured insect cells. *Toxicon*, **31**, 755–766.

Rivers, D.B. and Denlinger, D.L. (1995) Venom-induced alterations in fly lipid metabolism and its impact on larval development of the ectoparasitoid *Nasonia vitripennis* (Walker) (Hymenoptera: Pteromalidae). *J. Invert. Pathol.* **66**, 104–110.

Rizki, R.M. and Rizki, T.M. (1984) Selective destruction of a host blood cell type by a parasitoid wasp. *Proc. Natn. Acad. Sci. USA* **81**, 6154–6158.

Rizki, R.M. and Rizki, T.M. (1990a) Parasitoid virus-like particles destroy *Drosophila* cellular immunity. *Proc. Natl. Acad. Sci. USA*, **87**, 8388–8392.

Rizki, R.M. and Rizki, T.M. (1990b) Microtubule inhibitors block morphological changes induced in *Drosophila* blood cells by a parasitoid wasp factor. *Experientia*, **46**, 311–315.

Rizki, R.M. and Rizki, T.M. (1990c) Encapsulation of parasitoid eggs in phenoloxidase-deficient mutants of *Drosophila melanogaster. J. Insect Physiol.* **36**, 523–529.

Rizki, R.M. and Rizki, T.M. (1991) Effects of lamellolysin from a parasitoid wasp on *Drosophila* blood cells *in vitro. J. Exp. Zool.* **257**, 236–244.

Rizki, T.M., Rizki, R.M. and Carton, Y. (1990) *Leptopilina heterotoma* and *L. boulardi*: Strategies to avoid cellular defence responses of *Drosophila melanogaster. Exp. Parasitol.* **70**, 466–475.

Robacker, D.C. and Hendry, L.B. (1977) Neral and geranial: Components of the sex pheromone of the parasitic wasp, *Itoplectis conquisitor*. *J. Chem. Ecol.* **3**, 563–577.

Robacker, D.C., Weaver, K.M. and Hendry, L.B. (1976) Sexual communication and associative learning in the parasitic wasp, *Itoplectis conquisitor*. *J. Chem. Ecol.* **2**, 39–48.

Robertson, B., Hillerton, J.E. and Vincent, J.F.V. (1984) The presence of zinc or manganese as the predominant metal in the mandibles of adult stored product beetles. *J. Stored Prod. Res.* **20**, 133–137.

Robertson, P.L. (1968) A morphological and functional study of the venom apparatus in representatives of some major groups of Hymenoptera. *Aust. J. Zool.* **16**, 133–166.

Rogers, D. (1972) The ichneumon wasp *Venturia canescens*: Oviposition and avoidance of superparasitism. *Entomol. Exp. Appl.* **15**, 190–194.

Rohwer, S.A. and Cushman, R.A. (1917) Idiogastra, a new suborder of Hymenoptera with notes on the immature stages of *Oryssus*. *Proc. Entomol. Soc. Wash.* **19**, 89–98.

Roitberg, B.D., Mangel, M., Lalonde, R.G. *et al.* (1992) Seasonal dynamic shifts in patch exploitation by parasitic wasps. *Behav. Ecol.* **3**, 156–165.

Roitberg, B.D., Sircom, J., Roitberg, C.A. *et al.* (1993) Life expectancy and reproduction. *Nature,* **364**, 108.

Rojas, M.G., Morales-Ramos, J.A. and King, E.G. (1996) *In vitro* rearing of the boll weevil (Coleoptera: Curculionidae) ectoparasitoid *Catolaccus grandis* (Hymenoptera: Pteromalidae) on meridic diets *J. Econ. Entomol.* **89**, 1095–1104.

Rojas-Rousse, D. (1972) Description et fonctionnement de l'appareil genital interne de *Diadromus pulchellus* Wesmael (Hymenoptera: Ichneumonidae) – I. Testicules, canaux deferents, glandes annexes, canaux collecteurs, canal ejaculateur du male. *Int. J. Insect Morphol. Embryol.* **1**, 225–232.

Rojas-Rousse, D. and Palevody, C. (1981) Structure et fonctionnement de la spermatheque chez l'endoparasite solitaire *Diadromus pulchellus* Wesmael (Hymenoptera: Ichneumonidae). *Int. J. Insect Morph. Embryol.* **10**, 309–320.

Rojas-Rousse, D. and Palevody, C. (1983) Organogenesis and ultrastructure of placoid sensilla of the antenna of *Diadromus pulchellus* Wesmael (Hymenoptera, Ichneumonidae). *Int. J. Insect Morph. Embryol.* **12**, 171–185. (In French.)

Rojas-Rousse, D., Bigot, Y. and Periquet, G. (1993) DNA insertions as a component of the evolution of unique satellite DNA families in 2 genera of parasitoid wasps – *Diadromus* and *Eupelmus* (Hymenoptera). *Mol. Biol. Evol.* **10**, 383–396.

Ronquist, F. (1994a) Morphology, phylogeny and evolution of cynipoid wasps. *Acta. Univ. Ups.* **38**, 1–29.

Ronquist, F. (1994b) Evolution of parasitism among closely related species: Phylogenetic relationships and the origin of inquilinism in gall wasps (Hymenoptera, Cynipidae). *Evolution* **48**, 241–266.

Ronquist, F. (1995) Phylogeny and early evolution of the Cynipoidea (Hymenoptera). *Syst. Ent.* **20**, 309–335.

Roques, A. (1976) Observations sur la biologie et le comportement cleptoparasite d'*Eurytoma waachtli* (Chalc.: Eurytomidae), parasite de *Pissodes validirostris* (Col.: Curculionidae) dans les cones de pin sylvestre a Fontainebleau. *Entomophaga* **21**, 289–295.

Rosen, D. (1967) On the relationship between ants and parasites of coccids and aphids on citrus. *Beitr. Ent.* **17**, 281–286.

Rosen, D. (1986) The role of taxonomy in effective biological control programs. *Agric. Ecosyst. Env.* **15**, 121–129.

Rosen, D. (1988) Parasitic Hymenoptera in biological control: the genus *Aphytis*. In *Advances in Parasitic Hymenoptera Research* (ed. V.K. Gupta), E.J. Brill, Leiden. pp. 411–416.

Rosen, D. and DeBach, P. (1990) Ectoparasites. In *The Armoured Scale Insects, Their Biology, Natural Enemies and Control*, Vol. B (ed. D. Rosen), Elsevier Science Publishers, Amsterdam. pp. 99–120.

Rosen, D. and Kfir, R. (1983) A hyperparasite of coccids develops as a primary parasite of fly puparia. *Entomophaga*, **28**, 83–87.

Rosenheim, J.A. and Rosen, D. (1992) Influence of egg load and host size on host-feeding behaviour in the parasitoid *Aphytis lingnanensis*. *Ecol. Ent.* **17**, 263–272.

Rosi, M.C., Isidoro, N., Colazza, S. and Bin, F. (1995) Functional anatomy of the female reproductive system of *Trissolcus basalis* (Woll.). In *Trichogramma and Other Egg Parasites. 3rd International Symposium, Cairo (Egypt), Oct. 4–7, 1994* (ed. INRA), Les Colloques de l'INRA No. 73, Paris, pp. 101–103.

Roskam, J.C. (1986) Biosystematics of insects living in female birch catkins. IV. Egg–larval parasitoids of the genus *Platygaster* Latreille and *Metaclistis* Förster (Hymenoptera, Platygastridae). *Tijdschr. Ent.* **129**, 125–140.

Ross, H.H. (1945) Sawfly genitalia: Terminology and study techniques. *Entomol. News* **56**, 261–268.

Ross, K.G. and Fletcher, D.J.C. (1985) Genetic origin of male diploidy in the fire ant *Solenopsis invicta* (Hymenoptera: Formicidae), and its evolutionary significance. *Evolution* **39**, 888–903.

Ross, K.G., Vargo, E.L., Keller, L. and Trager, J.C. (1993) Effect of founder event on variation in the genetic sex-determining system of the fire ant *Solenopsis invicta*. *Genetics* **135**, 843–854.

Rössler, Y. and DeBach, P. (1972) The biosystematic relations between a thelytokous and an arrhenotokous form of *Aphytis mytilaspidis* (Le Baron) (Hymenoptera: Aphelinidae). I. The reproductive relations. *Entomophaga* **17**, 391–423.

Rössler, Y. and DeBach, P. (1973) Genetic variability in a thelytokous form of *Aphytis mytilaspidis* (Le Baron) (Hymenoptera: Aphelinidae). *Hilgardia* **42**, 149–175.

Rotheram, S. (1967) Immune surface of eggs of a parasitic wasp. *Nature* **214**, 700.

Rotheram, S. (1973) The surface of the egg of a parasitic insect. 1. The surface of the egg and first-instar larva of *Nemeritis*. *Proc. R. Soc. Lond. B.* **183**, 179–194.

Rotheray, G.E. (1979) The biology and host searching behaviour of a cynipoid parasite of aphidophagous syrphid larvae. *Ecol. Ent.* **4**, 75–82.

Rotheray, G.E. (1981a) Courtship, male swarms and a sex pheromone of *Diplazon pectoratorius* (Thunberg) (Hymenoptera: Ichneumonidae). *Entomol. Gaz.* **30**, 191–193.

Rotheray, G.E. (1981b) Emergence from the host puparium by *Diplazon pectoratorius* (Gravenhorst) (Hymenoptera: Ichneumonidae), a parasitoid of aphidophagous syrphid larvae. *Entomol. Gaz.* **32**, 39–41.

Rothschild, M. (1984) Aide Memoire mimicry. *Ecol. Ent.* **9**, 311–319.

Rotundo, G. and Tremblay, E. (1993) Electroantennographic responses of chestnut moths (Lepidoptera: Tortricidae) and their parasitoid *Ascogaster quadridentatus* Wesmael (Hymenoptera: Braconidae) to volatiles from chestnut (*Castanea sativa* Miller) leaves. *Redia* **76**, 361–373.

Rotundo, G., Cavallaro, R. and Tremblay, E. (1988) *In vitro* rearing of *Lysiphlebus fabarum* (Hym.: Braconidae). *Entomophaga* **33**, 261–267.

Rousset, F., Bouchon, D., Pintureau, B. *et al.* (1992) *Wolbachia* endosymbionts responsible for various alterations of sexuality in arthropods. *Proc. R. Soc. Lond. B.* **250**, 91–98.

Ruberson, J.R., Tauber, M.J. and Tauber, C.A. (1988) Reproductive biology of two biotypes of *Edovum puttleri*, a parasitoid of Colorado potato beetle eggs. *Entomol. Exp. Appl.* **46**, 211–219.

Ruberson, J.R., Tauber, M.J. and Tauber, C.A. (1989a) Intraspecific variability in hymenopteran parasitoids: Comparative studies of two biotypes of the egg parasitoid *Edovum puttleri* (Hymenoptera: Eulophidae). *J. Kansas Entomol. Soc.* **62**, 189–202.

Ruberson, J.R., Tauber, M.J., Tauber, C.A. and Tingey, W.M. (1989b) Interactions at three trophic levels: *Edovum puttleri* Grissell (Hymenoptera: Eulophidae), the Colorado potato beetle, and insect-resistant potatoes. *Canad. Entomol.* **121**, 841–851.

Rudall, K.M. and Kenchington, W. (1971) Arthropod silks: The problem of fibrous proteins in animal tissues. *Ann. Rev. Ent.* **16**, 73–96.

Ryan, R.B. (1963) Contribution to the embryology of *Coeloides brunneri* (Hymenoptera: Braconidae). *Ann. Entomol. Soc. Am.* **56**, 639–648.

Ryan, R.B. and Saul, G.B. (1968) Post-fertilization effects of incompatibility factors in *Mormoniella*. *Mol. Gen. Genet.* **103**, 29–36.

Ryan, R.B. and Yoshimoto, C.M. (1975) Laboratory crossings with different sources of the larch casebearer parasite *Chrysocharis laricinellae* (Hymenoptera: Eulophidae). *Canad. Entomol.* **107**, 1301–1304.

Salkeld, E.H. (1959) Notes on anatomy, life-history, and behaviour of *Aphaereta pallipes* (Say) (Hymenoptera: Braconidae), a parasite of the onion maggot, *Hylemya antiqua* (Meig.). *Canad. Entomol.* **91**, 91–97.

Salkeld, E.H. (1967) Histochemistry of the excretory system of the endoparasitoid *Aphaereta pallipes* (Say) (Hymenoptera: Braconidae). *Canad. J. Zool.* **45**, 967–973.

Salt, G. (1931) Parasites of the wheat-stem sawfly *Cephus pygmaeus* Linnaeus in England. *Bull. Entomol. Res.* **22**, 479–545.

Salt, G. (1937) The egg-parasite of *Sialis lutaria*: A study of the influence of the host upon a dimorphic parasite. *Parasitology* **29**, 539–553.

Salt, G. (1941) The effects of hosts upon their insect parasites. *Biol.Rev.* **16**, 239–264.

Salt, G. (1958) Experimental studies in insect parasitism. VI. Host suitability. *Bull. Entomol. Res.* **29**, 223–246.

Salt, G. (1959) Role of glycerol in the cold-hardening of *Bracon cephi* (Gahan). *Canad. J. Zool.* **37**, 59–69.

Salt, G. (1961) Competition among insect parasitoids. *Symp. Soc. Exp. Biol.* **15**, 96–119.

Salt, G. (1965) Experimental studies in insect parasitism. XIII. The haemocytic reaction of a caterpillar to eggs of its habitual parasite. *Proc. R. Soc. Lond. B* **162**, 303–318.

Salt, G. (1966) Experimental studies in insect parasitism. XIV. The haemocytic reaction of a caterpillar to larvae of its habitual parasite. *Proc. R. Soc. Lond. B* **165**, 155–178.

Salt, G. (1968) The resistance of insect parasitoids to the defence reactions of their hosts. *Biol. Rev.* **43**, 200–232.

Salt, G. (1970) Experimental studies in insect parasitism. XV. The means of resistance of a parasitoid larva. *Proc. R. Soc. Lond. B* **176**, 105–114.

Salt, G. (1977) Problems of orientation associated with cocoon-spinning by *Nemeritis*. *Ecol. Ent.* **2**, 171–177.

Sander, K. and Feddersen, I. (1985) Developmental failure after experimental activation of insect eggs. *Int. J. Insect Repro. Devel.* **8**, 219–226.

Sanderson, A.R. (1988) Cytological investigations of parthenogenesis in gall wasps (Cynipidae, Hymenoptera). *Genetica (The Hague)* **77**, 198–216.

Sandlan, K. (1980) Host location by *Coccygomimus turionellae* (Hymenoptera: Ichneumonidae). *Entomol. Exp. and Appl.* **27**, 233–245.

Sanger, C. and King, P.E. (1971) Structure and function of the male genitalia of *Nasonia vitripennis* (Walker) (Hym.: Pteromalidae). *Entomologist* **104**, 137–149.

Sappel, N.P., Jeng, R.S., Hubbes, M. and Liu, F.H. (1995) Restriction–fragment-length polymorphisms in polymerase chain-reaction amplified ribosomal DNAs of three *Trichogramma* (Hymenoptera, Trichogrammatidae) species. *Genome* **38**, 419–425.

Saul, G.B. II (1990) Gene map of the parasitic wasp *Nasonia vitripennis* (= *Mormoniella vitripennis*) (2N=10). In *Genetic Maps. Locus Maps of Complex Genomes* (ed. S.J. O'Brien), Cold Spring Harbour Laboratory Press, New York. pp. 3.198–3.201.

Saul, G.B. II, Whiting, P.W., Saul, S.W. and Heidner, C.A. (1965) Wild-type and mutant sticks of *Mormoniella*. *Genetics* **52**, 1318–1327.

Saunders, D.S. (1965) Larval diapause of maternal origin: Induction of diapause in *Nasonia vitripennis* (Walk.) (Hymenoptera: Pteromalidae). *J. Exp. Biol.* **42**, 495–508.

Saunders, D.S. (1975) Spectral sensitivity and intensity thresholds in *Nasonia* photoperiodic clock. *Nature* **253**, 732–734.

Schaefer, P.W. (1993) Overwintering aggregations of female *Brachymeria intermedia* (Hymenoptera, Chalcididae). *Entomol. News* **104**, 133–135.

Schiff, N.M. and Feldlaufer, M.F. (1996) Neutral sterols of sawflies (Symphyta) – their relationship to other Hymenoptera. *Lipids* **31**, 441–443.

Schlinger, E.I. and Hall, J.C. (1960) The biology, behaviour, and morphology of *Praon palitans* Muesebeck, an internal parasite of the spotted alfalfa aphid, *Therioaphis maculata* (Buckton) (Hymenoptera: Braconidae, Aphidiinae). *Ann. Entomol. Soc. Am.* **53**, 44–60.

Schlinger, E.I. and Hall, J.C. (1961) The biology, behaviour, and morphology of *Trioxys utilis*, an internal parasite of the spotted alfalfa aphid, *Therioaphis maculata* (Hymenoptera: Braconidae, Aphidiinae). *Ann. Entomol. Soc. Am.* **54**, 34–45.

Schmidt, J.M. and Smith, J.J.B. (1985a) The ultrastructure of the wings and the external sensory morphology of the thorax in female *Trichogramma minutum* Riley (Hymenoptera: Trichogrammatidae). *Proc. R. Soc. Lond. B* **224**, 287–313.

Schmidt, J.M. and Smith, J.J.B. (1985b) Host volume measurement by the parasitoid wasp *Trichogramma minutum*: The roles of curvature and surface area. *Entomol. Exp. Appl.* **39**, 213–221.

Schmidt, J.M. and Smith, J.J.B. (1987) The external sensory morphology of the legs and hairplate system of female *Trichogramma minutum* Riley (Hymenoptera: Trichogrammatidae). *Proc. R. Soc. Lond. B* **232**, 323–366.

Schmidt, J.O. (1978) *Dasymutilla occidentalis*: A long-lived aposematic wasp (Hymenoptera: Mutillidae). *Entomol. News* **89**, 135–136.

Schmidt, J.O. and Blum, M.S. (1977) Adaptations and responses of *Dasymutilla occidentalis* (Hymenoptera: Mutillidae) to predators. *Entomol. Exp. Appl.* **21**, 99–111.

Schmidt, K. and Kuhbandner, B. (1983) Ontogeny of the sensilla placodea on the antennae of *Aulacus striatus* Jurine (Hymenoptera: Aulacidae). *Int. J. Insect Morphol. and Embryol.* **12**, 43–57.

Schmidt, O., Andersson, K., Will, A. and Schuchmann-Feddersen, I. (1990) Viruslike particle proteins from a hymenopteran endoparasitoid are related to a protein component of the immune system in the lepidopteran host. *Arch. Insect Biochem. Physiol.* **13**, 107–115.

Schmitz, J. and Moritz, R.F.A. (1994) Sequence analysis of the D1 and D2 regions of 28S rDNA in the hornet (*Vespa crabro*) (Hymenoptera, Vespinae). *Insect Molec. Biol.* **3**, 273–277.

Schneider, F. (1951) Einige physiologische Beziehungen zwischen Syrphidenlarven und ihren Parasiten. *Z. Angew. Entomol.* **33**, 150–162.

Schneiderman, H.A. and Horwitz, J. (1958) The induction and termination of facultative diapause in the chalcid wasps *Mormoniella vitripennis* (Walker) and *Tritneptis klugii* (Ratzeburg). *J. Exp. Biol.* **35**, 520–550.

Scholz, D. and Höller, C. (1992) Competition for hosts between two hyperparasitoids of aphids, *Dendrocerus laticeps* and *D. carpenteri* (Hymenoptera: Megaspilidae): The benefit of interspecific host discrimination. *J. Insect Behav.* **5**, 289–300.

Schönitzer, K. and Lawitzky, G. (1987) A phylogenetic study of the antenna cleaner in Formicidae, Mutillidae, and Tiphiidae (Insecta, Hymenoptera). *Zoomorphology* **107**, 273–285.

Schroder, R.F.W., Sidor, A.M. and Athana, M.M. (1996) Parental care in *Erixestus winnemana* (Hymenoptera, Pteromalidae), an egg parasite of *Calligrapha* (Coleoptera, Chrysomelidae). *Entomol. News* **107**, 161–165.

Schrott, A. (1986) Vergleichende Morphologie und Ultrastruktur des Cenchrus-Dornenfeldapparates bei Pflanzenwespen. *Ber. nat.-med. Verein, Innsbruck* **73**, 159–168.

Scudder, G.G.E. (1961a) The comparative morphology of the insect ovipositor. *Trans. R. Entomol. Soc. Lond.* **113**, 25–40.

Scudder, G.G.E. (1961b) The functional morphology and interpretation of the insect ovipositor. *Canad. Entomol.* **93**, 267–272.

Scudder, G.G.E. (1971) The comparative morphology of insect genitalia. *Annu. Rev. Ent.* **16**, 379–406.

Seitner, M. and Nötzl, P. (1925) *Pityophthorus henscheli* Seitner und sein Parasit *Cosmophorus henscheli* Ruschka. *Zeitschr. Angew. Entomol.* **11**, 187–196.

Shapiro, A.M. (1976) Beau geste? *Am. Nat.* **110**, 900–902.

Sharkey, M.J. and Mason, W.R.M. (1986) The generic validity of *Aenigmostomus* and *Asiacardiochiles* (Hymenoptera: Braconidae). *Proc. Entomol. Soc. Wash.* **88**, 300–302.

Sharkey, M.J. and Wahl, D. (1992) Cladistics of the Ichneumonoidea (Hymenoptera). *J. Hym. Res.* **1**, 15–24.

Sharkey, M.J. and Wharton, R.A. (1994) A revision of the genera of the world Ichneutinae (Hymenoptera: Braconidae). *J. Nat. Hist.* **28**, 873–912.

Shaw, M.R. (1983) On[e] evolution of endoparasitism: The biology of some genera of Rogadinae (Braconidae). *Contr. Am. Entomol. Inst.* **20**, 307–328.

Shaw, M.R. (1987) Host associations of species of *Eulophus* in Britain (Hymenoptera: Eulophidae). *Entomol. Gaz.* **38**, 59–63.

Shaw, M.R. (1994) Parasitoid host ranges. In *Parasitoid Community Ecology* (eds B.A. Hawkins and W. Sheehan), Oxford University Press, Oxford. pp. 111–144.

Shaw, M.R. (1995) Observations on the adult behaviour and biology of *Histeromerus mystacinus* Wesmael (Hymenoptera: Braconidae). *The Entomologist* **114**, 1–13.

Shaw, M.R. and Aeschlimann, J.-P. (1994) Host ranges of parasitoids (Hymenoptera: Braconidae and Ichneumonidae) reared from *Epermenia chaerophyllella* (Goeze) (Lepidoptera: Epermeniidae) in Britain, with description of a new species of *Triclistus* (Ichneumonidae). *J. Nat. Hist.* **28**, 619–629.

Shaw, M.R. and Huddleston, T. (1991) Classification and biology of braconid wasps (Hymenoptera: Braconidae). *Handbks Ident. Br. Insect* **7(11)**, 1–126.

Shaw, S.R. (1985) A phylogenetic study of the subfamilies Meteorinae and Euphorinae (Hymenoptera: Braconidae). *Entomography* **3**, 277–270.

Shaw, S.R. (1988) Euphorine phylogeny: The evolution of diversity in host-utilization by parasitic wasps (Hymenoptera: Braconidae). *Ecol. Ent.* **13**, 323–335.

Shaw, S.R. (1990) Phylogeny and biogeography of the parasitoid wasp family Megalyridae (Hymenoptera). *J. Biogeogr.* **17**, 569–581.

Shaw, S.R. (1991) An unusual manner of aggregation in the braconid *Chelonus* (*Microchelonus*) *hadrogaster* McComb (Hymenoptera). *J. Insect Behav.* **4**, 537–542.

Shaw, S.R. (1993) Observations on the ovipositional behaviour of *Neoneurus mantis*, an ant-associated parasitoid from Wyoming (Hymenoptera: Braconidae). *J. Insect Behav.* **6**, 649–658.

Sheehan, W. (1991) Host range patterns of Hymenopteran parasitoids of exophytic lepidopteran folivores. In *Insect–Plant Interactions*, Volume 3 (ed. E. Bernays), CRC Press, Boca Ralon. pp. 209–247.

Sheehan, W. and Hawkins, B.A. (1991) Attack strategy as an indicator of host range in metopiine and pimpline Ichneumonidae (Hymenoptera). *Ecol. Ent.* **16**, 129–131.

Sheppard, W.S. and Heydon, S.L. (1986) High levels of genetic variability in three male-haploid species (Hymenoptera: Argidae, Tenthredinidae). *Evolution* **40**, 1350–1353.

Shevyrev, I.Ya. (1912) *Parasites and Hyperparasites in the Insect World*, St. Petersburg, 216pp. (In Russian.)

Shimizu, T., Tatsuki, Y. and Takeda, N. (1993) Aromatic amino acids in the venom of the braconid parasitoid *Apanteles kariyai*. *Z. Naturforsch. C.* **48**, 108–109.

Short, J.R.T. (1952) The morphology of the head of larval Hymenoptera with special reference to the head of the Ichneumonoidea, including a classification of the final instar larvae of the Braconidae. *Trans. R. Entomol. Soc. Lond.* **103**, 27–84.

Short, J.R.T. (1959) On the skeleto-muscular mechanisms of the anterior abdominal segments of certain adult Hymenoptera. *Trans. R. Entomol. Soc. Lond.* **111**, 175–203.

Short, J.R.T. (1978) The final larval instars of the Ichneumonidae. *Mem. Amer. Entomol. Inst.* **25**, 1–508.

Shorthouse J.D. (1980) Modification of galls of *D. polita* by the inquiline *Periclistus pirata*. *Bull. Scot. Bot. Fr., Actual Botaniques* **127**, 79–84.

Shu, S.Q. and Jones, R.L. (1993) Evidence for a multicomponent sex-pheromone in *Eriborus terebrans* (Gravenhorst) (Hym., Ichneumonidae), a larval parasitoid of the European corn-borer. *J. Chem. Ecol.* **19**, 2563–2576.

Shuja-Uddin (1977) Observations on normal and diapausing cocoons of the genus *Lipolexis* Foerster (Hymenoptera: Aphidiidae) from India. *Boll. Lab. Entomol. Agrar. Filippo Silvestri, Portici* **34**, 51–54.

Silvers, M.J. and Nappi, A.J. (1986) *In vitro* study of physiological suppression of supernumerary parasites by the endoparasitic wasp *Leptopilina heterotoma*. *J. Parasitol.* **72**, 405–409.

Silvestri, F. (1906) Contribuzioni alla conoscenza biologica degli Imenotteri parassiti. I. Biologia del *Litomastix truncatellus* (Dalm.). *Annali R. Scuola Superiore Agricoltura Portici* **6**, 3–51.

Simmonds, F.J. (1953) Observations on the biology and mass-breeding of *Spalangia drosophilae* Ashm. (Hym., Spalangidae), a parasite of the frit fly, *Oscinella frit* (L.). *Bull. Entomol. Res.* **44**, 773–778.

Simser, D.H. and Coppel, H.C. (1980a) Aggregation behaviour of *Brachymeria lasus* (Walker) (Hymenoptera, Chalcididae) in the laboratory. *Env. Ent.* **9**, 486–488.

Simser, D.H. and Coppel, H.C. (1980b) Female-produced sex pheromone in *Brachymeria lasus* and *B. intermedia* (Hym.: Chalcididae). *Entomophaga* **25**, 373–380.

Sivinski, J. and Webb, J.C. (1989) Acoustic signals produced during courtship in *Diachasmimorpha* (= *Biosteres*) *longicaudata* (Hymenoptera: Braconidae) and other Braconidae. *Ann. Entomol. Soc. Am.* **82**, 117–120.

Skinner, S.W. (1982) Maternally inherited sex ratio in the parasitoid wasp, *Nasonia vitripennis*. *Science*, **215**, 1133–1134.

Skinner, S.W. (1985) Son-killer – a third extrachromosomal factor affecting the sex ratio in the parasitoid wasp, *Nasonia vitripennis*. *Genetics* **109**, 745–759.

Skinner, W.S., Dennis, P.A. and Quistad, G.B. (1990) Partial characterization of toxins from *Goniozus legneri* (Hymenoptera: Bethylidae). *J. Econ. Ent.* **83**, 733–736.

Slansky, F. Jr (1986) Nutritional ecology of endoparasitic insects and their hosts: An overview. *J. Insect Physiol.* **32**, 255–261.

Slifer, E.H. (1969) Sense organs on the antennae of a parasitic wasp, *Nasonia vitripennis*. *Biol. Bull.* **136**, 253–263.

Slobodchikoff, C.N. (1967) Bionomics of *Grotea californica* Cresson, with a description of the larva and pupa (Hymenoptera: Ichneumonidae). *Pan-Pac. Entomol.* **43**, 161–168.

Slovák, M. (1984) New data about developmental stages of *Exetastes cinctipes* (Hym., Ichneumonidae). *Biologiá (Bratislava)* **41**, 543–548.

Slovák, M. (1986) Mating behaviour in laboratory reared *Exetastes cinctipes* (Hym., Ichneumonidae). *Biologiá (Bratislava)* **41**, 543–548.

Sluss, R.R. and Leutenegger, R. (1968) The fine structure of the trophic cells of *Perilitus coccinellae*. *J. Ultrastr. Res.* **24**, 441–451.

Smilowitz, Z. (1974) Relationships between the parasitoid *Hyposoter exiguae* (Viereck) and cabbage looper, *Trichoplusia ni* (Hübner): Evidence for endocrine involvement in successful parasitism. *Ann. Entomol. Soc. Am.* **67**, 317–319.

Smirnoff, W.A. (1971) Susceptibility of *Dahlbominus fuscipennis* (Chalcidoidea: Eulophidae) to the microsporidian *Thelohania pristiphorae*. *Canad. Entomol.* **103**, 1165–1167.

Smith, D.R. (1970) A new Nearctic *Xyela* causing galls on *Pinus* spp. (Hymenoptera: Xyelidae). *J. Georgia Entomol. Soc.* **5**, 69–72.

Smith, E.L. (1968) Biosystematics and morphology of the Symphyta – I. Stem-galling *Euura* of the California region, and a new female genitalic nomenclature (Hymenoptera: Tenthredinidae). *Ann. Entomol. Soc. Am.* **61**, 1389–1407.

Smith, E.L. (1969) Evolutionary morphology of external insect genitalia. 1. Origin and relationships to other appendages. *Ann. Entomol. Soc. Am.* **62**, 1051–1079.

Smith, E.L. (1970) Evolutionary morphology of external insect genitalia. 2. Hymenoptera. *Ann. Entomol. Soc. Am.* **63**, 1–27.

Smith, E.L. (1972) Biosystematics and morphology of the Symphyta – III. External genitalia of *Euura* (Hymenoptera: Tenthredinidae): Sclerites, sensilla, musculature, development and oviposition behaviour. *Int. J. Insect Morphol. Embryol.* **1**, 321–365.

Smith, L. (1988) Dispersal behaviour of two pteromalid parasitoids of house fly pupae in a dairy environment (Hymenoptera: Chalcidoidea). In *Advances in Parasitic Hymenoptera Research* (ed. V.K. Gupta), E.J. Brill, Leiden. pp 333–344.

Smith, O.J. (1952) Biology and behavior of *Microctonus vittatae* Muesebeck (Braconidae). *Univ. Cal. Publ. Entomol.* **9**, 315–344,

Smith, R.H. and Shaw, M.R. (1980) Haplodiploid sex ratios and the mutation rate. *Nature*, **287**, 728–729.

Smith, S.G. and Wallace, D.R. (1971) Allelic sex determination in lower Hymenoptera, *Neodiprion nigroscutum*. *Canad. J. Genet. Cytol.* **13**, 617–621.

Smith Trail, D.R. (1980) Behavioural interactions between parasites and hosts: Host suicide and the evolution of complex life cycles. *Am. Nat.* **116**, 77–91.

Snodgrass, R.E. (1933) Morphology of the insect abdomen. Part II. *Smithsonian Misc. Coll.* **89**, 1–148.

Snodgrass, R.E. (1941) Male genitalia of Hymenoptera. *Smithson. Misc. Coll.* **99**, 1–86.

Soenjaro, E. (1979) Effect of labelling with the radioisotope zinc-65 on the performance of the eulophid, *Colpoclypeus florus*, a parasite of Tortricidae. *Entomol. Exp. Appl.* **25**, 304–310.

Soldán, T. and Starý, P. (1981) Parasitogenic effects of *Aphidius smithi* (Hymenoptera, Aphidiidae) on the reproductive organs of the pea aphid, *Acyrthosiphon pisum* (Homoptera, Aphidiidae). *Acta. Ent. Bohemoslov.* **78**, 243–253.

Sommerman, K.M. (1956) Parasitization of nymphal and adult psocids (Psocoptera). *Proc. Entomol. Soc. Wash.* **58**, 149–152.

Southwood, T.R.E. (1957) Observations on swarming in Braconidae (Hymenoptera) and Coniopterygidae (Neuroptera). *Proc. R. Entomol. Soc. Lond. (A)* **32**, 80–82.

Spanjer, W., Grosu, L. and Piek, T. (1977) Two different paralyzing preparations obtained from a homogenate of the wasp *Microbracon hebetor* (Say). *Toxicon* **15**, 413–421.

Speicher, B.R. (1936) Oögenesis, fertilization and early cleavage in *Habrobracon. J. Morphol.* **59**, 401–421.

Speicher, B.R. (1937. Oögenesis in a thelytokous wasp, *Nemeritis canescens* (Grav.). *J. Morphol.* **61**, 453–467.

Speicher, B.R. and Speicher, K.G. (1940) The occurrence of diploid males in *Habrobracon brevicornis. Am. Nat.* **74**, 379–382.

Speicher, K.G. (1934) Impaternate females in *Habrobracon. Biol. Bull.* **67**, 277.

Speicher, K.G. and Speicher, B.R. (1938) Diploids from unfertilized eggs in *Habrobracon. Biol. Bull.* **74**, 247–252.

Speirs, D.C., Sherratt, T.N. and Hubbard, S.F. (1991) Parasitoid diets: Does superparasitism pay? *Trends Ecol. Evol.* **6**, 22–25.

Spradbery, J.P. (1969) The biology of *Pseudorhyssa sternata*, a cleptoparasite of siricid wasps. *Bull. Entomol. Res.* **59**, 291–297.

Spradbery, J.P. (1970a) The biology of *Ibalia drewseni* Borries (Hymenoptera: Ibaliidae), a parasite of siricid woodwasps. *Proc. R. Entomol. Soc. Lond.* **45**, 104–113.

Spradbery, J.P. (1970b) Host finding by *Rhyssa persuasoria* (L.), an ichneumonid parasite of siricid woodwasps. *Anim. Behav.* **18**, 103–114.

Spradbery, J.P. (1973) A comparative study of the phytotoxic effects of siricid woodwasps on conifers. *Ann. Appl. Biol.* **75**, 309–320.

Spradbery, J.P. (1974) The responses of *Ibalia* species (Hymenoptera: Ibaliidae) to the fungal symbionts of siricid woodwasp hosts. *J. Entomol. (A)* **48**, 217–222.

Spradbery, J.P. and Kirk, A. (1978) Aspects of ecology of siricid woodwasps (Hymenoptera; Siricidae) in Europe, North Africa and Turkey with special reference to the biological control of *Sirex noctilio* F. in Australia. *Bull. Entomol. Res.* **68**, 341–359.

Starý, P. (1970) *The Biology of Aphid Parasites*, Dr W. Junk, The Hague. 643pp.

Starý, P. and Völkl, W. (1988) Aggregations of aphid parasitoid adults (Hymenoptera, Aphidiidae). *J. Appl. Ent.* **105**, 270–279.

Stavraki-Paulopoulou, H.G. (1966) Contribution a l'étude de la capacité reproductrice et de la fécondité réelle d'*Opius concolor* Szepl. (Hymenoptera–Braconidae). *Ann. Epiphyties* **17**, 391–435.

Stechmann, D.H., Völkl, W. and Starý, P. (1996) Ant attendance as a critical factor in the biological control of the banana aphid, *Pentalonia nigronervosa* Coq. (Hom.: Aphididae) in Oceania. *J. Appl. Ent.* **120**, 119–123.

Steffan, J.R. (1961. Comportment de *Lasiochalcidia igiliensis* Ms., chalcidide parasite de fourmilions. *Comptes Rendus*, **253**, 2401–2403.

Steffensen, D. and LaChance, L.E. (1960) Radioisotopes and the genetic mechanism: Cytology and genetics of divalent metals in nuclei and chromosomes. *Symp. Radioisot. Biosphere*, pp. 132–145.

Steinberg, S., Podoler, H. and Rosen, D. (1987) Competition between two parasites of the Florida red scale in Israel. *Ecol. Ent.* **12**, 299–310.

Steinberg, S., Prag, H. and Rosen, D. (1993) Host plant affects fitness and host acceptance in the aphid parasitoid *Lysiphlebus testaceipes* (Cresson). *IOBC WPRS Bull.* **16**, 161–164.

Steiner, A.L. (1986) Stinging behaviour of solitary wasps. In *Venoms of the Hymenoptera* (ed. T. Piek), Academic Press, London. pp. 63–160.

Steinkraus, D.C. and Cross, E.A. (1993) Description and life history of *Acarophenax mahunkai*, n. sp. (Acari, Tarsonemina: Acarophenacidae), an egg parasite of the lesser mealworm (Coleoptera: Tenebrionidae). *Ann. Entomol. Soc. Am.* **86**, 239–249.

Stepper, J., Becker, C. and Schmidt, K. (1983) Feinbau und Ontogenese der Porenplatten auf den Antenna von of *Pimpla turionellae* (Hymenoptera, Ichneumonidae). *Zoomorphologie* **102**, 11–32.

Stern, V.M., Schlinger, E.I. and Bowen, W.R. (1965) Dispersal studies of *Trichogramma semifumatum* (Hymenoptera: Trichogrammatidae) tagged with radioactive phosphorus. *Ann. Entomol. Soc. Am.* **58**, 234–240.

Sternlicht, M. (1973) Parasitic wasps attracted by the sex pheromone of the coccid host. *Entomophaga* **18**, 339–342.

Stille, B. and Dävring, L. (1980) Meiosis and reproductive strategy in the partheno-genetic gall wasp *Diplolepis rosae* (L.) (Hymenoptera, Cynipidae). *Hereditas* **92**, 353–362.

Stillwell, M.A. (1965) Hypopleural organs of the woodwasp larva *Tremex columba* (L.) containing the fungus *Daedalea unicolor* Bull. ex Fries. *Canad. Entomol.* **97**, 783–784.

Stoltz, D.B. (1981) A putative baculovirus in the ichneumonid parasitoid *Mesoleius tenthredinis*. *Canad. J. Microbiol.* **27**, 116–122.

Stoltz, D.B. (1990) Evidence for chromosomal transmission of polydnavirus DNA. *J. Gen. Virol.* **71**, 1051–1056.

Stoltz, D.B. (1993) The polydnavirus life cycle. In *Parasites and Pathogens of Insects. Volume 1: Parasites* (eds N.E. Beckage, S.N. Thompson and B.A. Federici), Academic Press, San Diego. pp 167–187.

Stoltz, D.B. and Cook, D.I. (1983) Inhibition of host phenoloxidase activity by parasitic Hymenoptera. *Experientia* **39**, 1022–1024.

Stoltz, D.B. and Faulkner, G. (1978) A putative baculovirus in the ichneumonid parasitoid, *Mesoleius tenthredinis*. *Canad. J. Microbiol.* **27**, 116–122.

Stoltz, D.B. and Guzo, D. (1986) Apparent haemocytic transformations associated with parasitoid-induced inhibition of immunity in *Malacosoma disstria* larvae. *J. Insect Physiol.* **32**, 377–388.

Stoltz, D.B. and Vinson, S.B. (1977) Baculovirus-like particles in the reproductive tracts of female parasitoid wasps. II. The genus *Apanteles*. *Canad. J. Microbiol.* **23**, 28–37.

Stoltz, D.B. and Vinson, S.B. (1979) Viruses and parasitism in insects. *Adv. Virol. Res.* **24**, 125–171.

Stoltz, D. and Whitfield, J.B. (1992) Viruses and virus-like entities in the parasitic Hymenoptera. *J. Hym. Res.* **1**, 125–139.

Stoltz, D.B., Krell, P. and Vinson, S.B. (1981) Polydisperse viral DNAs in ichneumonid ovaries: A survey. *Canad. J. Microbiol.* **27**, 123–130.

Stoltz, D.B., Krell, P., Summers, M.D. and Vinson, S.B. (1984) Polydnaviridae – a proposed family of insect viruses with segmented, double-stranded, circular DNA genomes. *Intervirology*, **21**, 1–4.

Stoltz, D.B., Guzo, D., Belland, E.R. *et al.* (1988a) Venom promotes uncoating *in vitro* and persistence *in vivo* of DNA from a braconid polydnavirus. *J. Gen. Virol.* **69**, 903–907.

Stoltz, D.B., Krell, P., Cook, D. *et al.* (1988b) An unusual virus from the parasitic wasp *Cotesia melanoscela*. *Virology* **162**, 311–320.

Stouthamer, R. (1991. Effectiveness of several antibiotics in reverting thelytoky to arrhenotoky in *Trichogramma* spp. In *Trichogramma and Other Egg Parasitoids* (eds E. Wajnberg and S.B. Vinson), Les Colloques de l'INRA No. 56, pp.119–122.

Stouthamer, R. (1993) The use of sexual versus asexual wasps in biological control. *Entomophaga* **38**, 3–6.

Stouthamer, R. and Kazmer, D.J. (1994) Cytogenetics of microbe-associated parthenogenesis and its consequences for gene flow in *Trichogramma* wasps. *Heredity* **73**, 317–327.

Stouthamer, R. and Luck, R.F. (1991) Transition from bisexual to unisexual cultures in *Encarsia perniciosi* (Hymenoptera, Aphelinidae) – New data and a reinterpretation. *Ann. Entomol. Soc. Am.* **84**, 150–157.

Stouthamer, R. and Werren, J.H. (1993) Microbes associated with parthenogenesis in wasps of the genus *Trichogramma*. *J. Invert. Pathol.* **61**, 6–9.

Stouthamer, R., Luck, R.F. and Hamilton, W.D. (1990) Antibiotics cause parthenogenetic *Trichogramma* (Hymenoptera/Trichogrammatidae) to revert to sex. *Proc. Nat. Acad. Sci. USA* **87**, 2424–2427.

Stouthamer, R., Luck, R.F. and Werren, J.H. (1992) Genetics of sex determination and the improvement of biological control using parasitoids. *Environ. Ent.* **21**, 427–435.

Stouthamer, R., Breeuwer, J.A.J., Luck, R.F. and Werren, J.H. (1993) Molecular identification of microorganisms associated with parthenogenesis. *Nature*, **361**, 66–68.

Stouthhamer, R., Lükö, S. and Mak, F. (1994) Influence of parthenogenesis *Wolbachia* on host fitness. *Norw. J. Agric. Sci. Suppl.* **16**, 117–122.

Strand, M.R. (1986) The physiological interactions of parasitoids with their hosts and their influence on reproductive strategies. In *Insect Parasitoids* (eds J. Waage and D. Greathead), Academic Press, London. pp. 97–136.

Strand, M.R. (1994) *Microplitis demolitor* polydnavirus infects and expresses in specific morphotype of *Pseudoplusia includens* hemocytes. *J. Gen. Virol.* **75**, 3007–3020.

Strand, M.R. and Godfray, H.C.J. (1989) Superparasitism and ovicide in parasitic Hymenoptera: A case study of the ectoparasitoid *Bracon hebetor*. *Behav. Ecol. Sociobiol.* **24**, 421–432.

Strand, M.R. and Noda, T. (1991) Alterations in the haemocytes of *Pseudoplusia includens* after parasitism by *Microplitis demolitor*. *J. Insect Physiol.* **37**, 839–850.

Strand, M.R. and Ode, P.J. (1990) Chromosome number of the polyembryonic parasitoid *Copidosoma floridanum* (Hymenoptera: Encyrtidae). *Ann. Entomol. Soc. Am.* **84**, 834–837.

Strand, M.R. and Pech, L.L. (1995) *Microplitis demolitor* polydnavirus induces apoptosis of a specific haemocyte morphotype in *Pseudoplusia includens*. *J. Gen. Virol.* **76**, 283–291.

Strand, M.R. and Vinson, S.B. (1982) Source and characterization of an egg recognition kairomone of *Telenomus heliothidis*, a parasitoid of *Heliothis virescens*. *Physiol. Ent.* **7**, 83–90.

Strand, M.R. and Vinson, S.B. (1983) Host acceptance behavior of *Telenomus heliothidis* (Hymenoptera: Scelionidae) towards *Heliothis virescens* (Lepidoptera: Noctuidae). *Ann. Entomol. Soc. Am.* **76**, 781–785.

Strand, M.R., Ratner, S. and Vinson, S.B. (1980) Maternally induced host regulation by egg parasitoid *Telenomus heliothidis*. *Physiol. Ent.* **8**, 469–475.

Strand, M.R., Johnson, J.A. and Culin, J.D. (1988) Developmental interactions between the parasitoid *Microplitis demolitor* (Hymenoptera: Braconidae) and its host *Heliothis virescens* (Lepidoptera: Noctuidae). *Ann. Entomol. Soc. Am.* **81**, 822–830.

Strand, M.R., Johnson, J.A. and Culin, J.D. (1990a) Intrinsic interspecific competition between the polyembryonic parasitoid *Copidosoma floridanum* and the solitary endoparasitoid *Microplitis demolitor* in *Pseudoplusia includens*. *Entomol. Exp. Appl.* **55**, 275–284.

Strand, M.R., Roitberg, B.D. and Papaj, D.R. (1990b) Acridine orange: A potentially useful internal marker of Hymenoptera and Diptera. *J. Kansas Entomol. Soc.* **63**, 634–637.

Strand, M.R., Baehrecke, E.H. and Wong, E.P. (1991) The role of host endocrine factors in the development of polyembryonic parasitoids. *Biol. Control* **1**, 144–152.

Strand, M.R., McKenzie, D.I., Grassl, V. *et al.* (1992) Persistence and expression of *Microplitis demolitor* polydnavirus in *Pseudoplusia includens*. *J. Gen. Virol.* **73**, 1627–1635.

Streams, F.A. (1971) Encapsulation of insect parasites in superparasitized hosts. *Entomol. Exp. Appl.* **14**, 484–490.

Streams, F.A. and Greenberg, L. (1969) Inhibition of the defence reaction of *Drosophila melanogaster* parasitized simultaneously by the wasps *Pseudeucoila bochei* and *Pseudeucoila mellipes*. *J. Invert. Pathol.* **13**, 371–377.

Strickland, E.H. (1923) Biological notes on parasites of prairie cutworms. *Bull. Dep. Agric. Domin. Can., Ent. Brch.* **22**, 1–40.

Strien-van Liempt, W.T.F.H. van and Alphen, J.J.M. van (1981) The absence of interspecific host discrimination in *Asobara tabida* Nees and *Leptopilina heterotoma* (Thomson) coexisting larval parasitoids of *Drosophila* species. *Neth. J. Zool.* **31**, 701–712.

Stuart, M.K. and Burkholder, W.E. (1991) Monoclonal antibodies specific to *Laelius pedatus* (Bethylidae) and *Bracon hebetor* (Braconidae), two hymenopterous parasitoids of stored-product pests. *Biol. Control* **1**, 302–308.

Styer, E.L., Hamm, J.J. and Nordlund, D.A. (1987) A new virus associated with the parasitoid *Cotesia marginiventris* (Hymenoptera: Braconidae): Replication in noctuid host larvae. *J. Invert. Pathol.* **50**, 302–309.

Sugimoto, T., Imoarai, T. and Tsuji, H. (1983) Oosorption in eulophid wasp, *Chrysocharis pentheus* Walker (Hymenoptera: Eulophidae). *Appl. Ent. Zool.* **18**, 287–289.

Sugonjaev, E.S. (1963) On seasonal cyclic adaptations of the chalcid-wasp *Blastothrix confusa* Erd. to its host, *Parthenolecanium corni* Bouche. *Zool. Zhurn.* **42**, 1732–1735. (In Russian.)

Sugumaran, M. and Kanost, M.R. (1993) Regulation of insect hemolymph phenoloxidases. In *Parasites and Pathogens of Insects*, Vol. 1 (eds N.E. Beckage, S.N. Thompson and B.A. Federici), Academic Press, San Diego. pp. 317–342.

Sullivan, D.J. (1987) Insect hyperparasitism. *Annu. Rev. Ent.* **32**, 49–70.

Suomalainen, E. (1950) Parthenogenesis in animals. *Adv. Genet.* **3**, 193–253.

Suzuki, Y., Tsuji, H. and Sasakawa, M. (1984) Sex allocation and effects of superparasitism on secondary sex ratios in the gregarious parasitoid, *Trichogramma chilonis* (Hymenoptera: Trichogrammatidae). *Anim. Behav.* **32**, 478–484.

Svoboda, J.A., Schiff, N.M. and Feldlaufer, M.F. (1995) Sterol composition of three species of sawflies (Hymenoptera: Symphyta) and their dietary plant material. *Experientia* **51**, 150–152.

Swedenborg, P.D. and Jones, R.L. (1992a) Multicomponent sex pheromone in *Macrocentrus grandii* Goidanich (Hymenoptera: Braconidae). *J. Chem. Ecol.* **18**, 1901–1912.

Swedenborg, P.D. and Jones, R.L. (1992b) (Z)-4-tredecenal, a pheromonially active air oxidation product from a series of (Z,Z)-9,13-dienes in *Macrocentrus grandii* (Goidanich) (Hymenoptera: Braconidae). *J. Chem. Ecol.* **18**, 1913–1931.

Swedenborg, P.D., Jones, R.L., Liu, H.-W. and Krick, T.P. (1993) (3R*,5S*,6R*)-3,5-dimethyl-6-(methylethyl)-3,4,5,6-tetrahydropyran-2-one, a third sex pheromone component for *Macrocentrus grandii* (Goidanich) (Hymenoptera: Braconidae) and evidence for its utility at eclosion. *J. Chem. Ecol.* **19**, 485–502.

Swedenborg, P.D., Jones, R.L., Zhou, H.Q. *et al.* (1994) (3R*,5S*,6R*)-3,5-dimethyl-6-(methylethyl)-3,4,5,6-tetrahydropyran-2-one and (3S*,5R*,6S*)-3,5-dimethyl-6-(methylethyl)-3,4,5,6-tetrahydropyran-2-one, a pheromone of *Macrocentrus grandii* (Goidanich) (Hymenoptera: Braconidae). *J. Chem. Ecol.* **20**, 3373–3380.

Syvertsen, T.C., Jackson, L.L., Blomquist, G.J. and Vinson, S.B. (1995) Alkadienes mediating courtship in the parasitoid *Cardiochiles nigriceps* (Hymenoptera, Braconidae). *J. Chem. Ecol.* **21**, 1971–1989.

Szelenyi, G. (1941) Notes on the tetrastichine genus *Myiomisa* Rond. (Hymenoptera: Chalcidoidea) with the redescription of the genotype and with description of a new species parasitising in the galls of *Eriophyes phloeocoptes*. *Nal. Noev. Evk.* **1**, 89–97.

Szklarzewicz, T., Bilinski, S.M., Klag, J. and Jablonska, A. (1993) Accessory nuclei in the oocytes of the cuckoo wasp, *Chrysis ignita* (Hymenoptera, Aculeata). *Folia Histochem. Cytobiol.* **31**, 227–231.

Tagawa, J. (1977) Localization and histology of the sex pheromone-producing gland in the parasitic wasp, *Apanteles glomeratus*. *J. Insect Physiol.* **23**, 49–56.

Tagawa, J. (1983) Female sex pheromone glands in the parasitic wasps, genus *Apanteles*. *Appl. Entomol. Zool.* **18**, 416–427.

Tagawa, J. (1996) Function of the cocoon of the parasitoid wasp, *Cotesia glomerata* L. (Hymenoptera: Braconidae): Protection against desiccation. *Appl. Entomol. Zool.* **31**, 99–103.

Tagawa, J. and Kitano, H. (1981) Mating behaviour of the braconid wasp, *Apanteles glomeratus* L. (Hymenoptera: Braconidae) in the field. *Appl. Ent. Zool.* **16**, 345–450.

Takabayashi, J., Noda, T. and Takahashi, S. (1991) Plants produce attractants for *Apanteles kariyai*, a parasitoid of *Pseudaletia separata*; Cases of 'communication' and 'misunderstanding' in parasitoid–plant interactions. *Appl. Ent. Zool.* **26**, 237–243.

Takadera, K., Yamashita, M., Hatakeyama, M. and Oishi, K. (1996) Similarities in vitellin antigenicity and vitellogenin mRNA nucleotide sequence among sawflies (Hymenoptera: Symphyta: Tenthredinoidea). *J. Insect Physiol.* **42**, 417–422.

Takasu, K. and Hirose, Y. (1988) Host discrimination in the parasitoid *Ooencyrtus nezarae*: The role of the egg stalk as an external marker. *Entomol. Exp. Appl.* **47**, 45–48.

Takasu, K. and Lewis, W.J. (1993) Host- and food-foraging of the parasitoid *Microplitis croceipes*: Learning and physiological state effects. *Biol. Control.* **3**, 70–74.

Tamashiro, M. (1971) A biological study of the venoms of two species of *Bracon. Hawaii Agric. Exp. Stn. Tech. Bull.* **70**, 52pp.

Tanada, Y. (1955) Field observations on a microsporidian parasite of *Pieris rapae* L. and *Apanteles glomeratus* L. *Proc. Hawaii Entomol. Soc.* **15**, 609–616.

Tanaka, M. (1985a) Early embryogenic development of the parasitic wasp, *Trichogramma chilonis* (Hymenoptera, Trichogrammatidae). In *Recent Advances in Insect Embryology in Japan* (eds H. Ando and K. Miya), ISEBU Co., Tsukuba. pp. 171–179.

Tanaka, M. (1985b) Embryogenic and early post-embryonic development of the parasitic wasp, *Trichogramma chilonis* (Hymenoptera, Trichogrammatidae). In *Recent Advances in Insect Embryology in Japan* (eds H. Ando and K. Miya), ISEBU Co., Tsukuba. pp. 181–189.

Tanaka, T. (1987) Morphology and functions of calyx fluid filaments in the reproductive tracts of endoparasitoid, *Microplitis mediator* (Hym.: Braconidae). *Entomophaga* **32**, 9–17.

Tanaka, T. and Vinson, S.B. (1991) Depression of thoracic gland activity of *Heliothis virescens* by venom and calyx fluids from the parasitoid, *Cardiochiles nigriceps*. *J. Insect Physiol.* **37**, 139–144.

Tanaka, T. and Wago, H. (1990) Ultrastructural and functional maturation of tera-tocytes of *Apanteles kariyai*. *Arch. Insect Biochem. Physiol.* **13**, 187–197.

Tanaka, T., Tagashira, E. and Sakurai, S. (1994) Reduction of testis growth of *Pseudaletia separata* larvae after parasitization by *Cotesia kariyai*. *Arch. Insect Biochem. Physiol.* **26**, 111–122.

Tanaka, T., Yagi, S. and Nakamatu, Y. (1992) Regulation of parasitoid sex allocation and host growth by *Cotesia* (*Apanteles*) *kariyai* (Hymenoptera, Braconidae). *Ann. Entomol. Soc. Am.* **85**, 310–316.

Tarasco, E. (1995) Morfologia larvale e biologia di *Coelichneumon rudis* (Boyer de Fonscolombe) (Hymenoptera: Ichneumonidae), endoparassitoide delle crialidi della *Thaumetopoea pityocampa* (Denis et Schiffermüller) (Lepidoptera: Thaumetopoeidae). *Entomologica* **29**, 5–51.

Tardieux, I. and Rabasse, J.M. (1988) Induction of athelytokous reproduction in the *Aphidius colemani* (Hymenoptera: Aphidiidae) complex. *J. Appl. Ent.* **106**, 58–61.

Tauber, M.J., Tauber, C.A., Nechols, J.R. and Obrycki, J.J. (1983) Seasonal activity of parasitoids: Control by external, internal and genetic factors. In *Diapause and Life Cycle Strategies in Insects* (eds V.K. Brown and I. Hodek), Dr W. Junk Publishers, The Hague. pp. 87–108.

Temerak, S.A. (1983) Host preferences of the parasitoid *Bracon brevicornis* Wesmael (Hym., Braconidae) and host sensitivity to its venom. *Z. Ang. Ent.* **96**, 37–41.

Teraoka, T. and Numata, H. (1995) Induction of adult diapause in a parasitoid wasp, *Ooencyrtus nezarae* under natural conditions. *Entomol. Exp. Appl.* **76**, 329–332.

Tersac, J. and Guerdoux, J. (1981) Élevage de *Pimpla instigator* (Hym.: Ichneumonidae) dans des conditions artificielles: I. Ponte dans un leurre. *Entomophaga* **26**, 221–232.

Thelen, E. and Farish, D. J. (1977) An analysis of the grooming behavior of wild and mutant strains of *Bracon hebetor*. *Behaviour* **62**, 70–102.

Theopold, U., Krause, E. and Schmidt, O. (1994) Cloning of a VLP-protein coding gene from a parasitoid wasp *Venturia canescens*. *Arch. Insect Biochem. Physiol.* **26**, 137–145.

Thomas, J.A. and Elmes, G.W. (1993) Specialized searching and the hostile use of allomones by a parasitoid whose host, the butterfly *Maculinea rebeli*, inhabits ants nests. *Animal Behav.* **45**, 593–602.

Thompson, S.N. (1975) Defined meridic and holidic diets and aseptic feeding pro-cedures for artificial rearing of the ectoparasitoid *Exeristes roborator* (Fabricius). *Ann. Entomol. Soc. Am.* **68**, 220–226.

Thompson, S.N. (1979) The effects of dietary carbohydrate on larval development and lipogenesis in the parasite, *Exeristes roborator*. *J. Parasitol.* **65**, 849–854.

Thompson, S.N. (1981a) Essential amino acid requirements of four species of par-asitic Hymenoptera. *Comp. Biochem. Physiol.* **69A**, 173–174.

Thompson, S.N. (1981b) *Brachymeria lasus* and *Pachycrepoideus vindemiae*: sterol requirements during larval growth of two hymenopterous insect parasites reared *in vitro* on chemically defined media. *Exp. Parasitol.* **51**, 220–235.

Thompson, S.N. (1982) Effects of the insect parasite, *Hyposoter exiguae* on the total body glycogen and lipid level of its host, *Trichoplusia ni*. *Comp. Biochem. Physiol.* **72B**, 233–237.

Thompson, S.N. (1983) Biochemical and physiological effects of metazoan endoparasites on their hosts species. *Comp. Biochem. Physiol.* **74B**, 183–211.

Thompson, S.N. (1986a) The metabolism of insect parasites (parasitoids): An overview. *J. Insect Physiol.* **32**, 421–423.

Thompson, S.N. (1986b) Nutrition and *in vitro* culture of insect parasitoids. *Annu. Rev. Entomol.* **31**, 197–219.

Thompson, S.N. and Barlow, J.S. (1970) The change in fatty acid pattern of *Itoplectis conquisitor* (Say) reared on different hosts. *J. Parasit.* **56**, 845–846.

Thompson, S.N. and Barlow, J.S. (1974) The fatty acid composition of parasitic Hymenoptera and its possible biological significance. *Ann. Entomol. Soc. Am.* **67**, 627–632.

Thompson, W.R. and Parker, H.L. (1930) The morphology and biology of *Eulimneria crassifemur*, an important parasite of the European corn borer. *J. Agric. Res.* **40**, 321–345.

Thorpe, W.H. (1932a) Experiments upon respiration in the larvae of certain parasitic Hymenoptera. *Proc. R. Soc. Lond. (B)* **109**, 450–471.

Thorpe, W.H. (1932b) The primary larvae of three ophionine ichneumonids, parasitic on *Rhyacionia buoliana*. *Parasitology* **24**, 107–110 plus 1 plate.

Thorpe, W.H. (1936) On a new type of respiratory interrelation between an insect (chalcid) parasitite and its host (Coccidae). *Parasitology* **28**, 517–540.

Tilden, R.L. and Ferkovich, S.M. (1987) Regulation of protein synthesis during egg development of the parasitic wasp, *Microplitis croceipes*. *Insect Biochem.* **17**, 783–792.

Tillman, P.G. (1994) Age-dependent parasitization and production of female progeny for *Microplitis croceipes* (Hymenoptera, Braconidae). *Southwest. Ent.* **19**, 335–338.

Timberlake, P.H. (1916) Note on an interesting case of two generations of a parasite reared from the same individual host. *Canad. Entomol.* **48**, 89–91.

Tobias, V.I. (1993) Dependence of wing venation in the Hymenoptera on ecological environment of their habitats. *Entomol. Obozr.* **72**, 497–506. (In Russian.)

Tobias, V.I. and Belokobyl'skij, S.A. (1983) An aberrant venation in Braconidae (Hymenoptera) and its importance in the investigation of the phylogeny of the family. *Entomol. Obozr.* **62**, 341–347.

Togashi, I. (1965) Preliminary report on the rectal papillae of the Symphyta (Hymenoptera). *Kontyû* **33**, 230–234.

Togashi, I. (1970) The comparative morphology of the internal reproductive organs of the Symphyta (Hymenoptera). *Mushi*, Supplement **43**, 1–114.

Tonapi, G.T. (1958a) A comparative study of spiracular structure and mechanisms in some Hymenoptera. *Trans. R. Entomol. Soc. Lond.* **110**, 489–520.

Tonapi, G.T. (1958b) A comparative study of the respiratory system of some Hymenoptera Part I: Symphyta. *Indian J. Ent.* **20**, 108–120.

Tonapi, G.T. (1958c) A comparative study of the respiratory system of some Hymenoptera Part II: Apocrita Parasitica. *Indian J. Ent.* **20**, 203–220.

Tonapi, G.T. (1958d) A comparative study of the respiratory system of some Hymenoptera Part III: Apocrita-Aculeata. *Indian J. Ent.* **20**, 245–269.

Torchio, P.F. (1972) *Sapyga pumila* Cresson, a parasite of *Megachile rotundata* (F.) (Hymenoptera: Sapygidae; Megachilidae). I: Biology and description of immature stages. *Melanderia* **10**, 1–22.

Tothill, J.D. (1922) The natural control of the fall webworm (*Hyphantria cunea* Drury) with an account of its several parasites. *Bull. Dep. Agric. Can., Ent. Brch.* **19**, 1–107.

Townes, H. (1958) Some biological characteristics of the Ichneumonidae (Hymenoptera) in relation to biological control. *J. Econ. Ent.* **51**, 650–652.

Townes, H. (1971) Ichneumonids as biological control agents. *Proceedings of the Tall Timbers Conference on Ecological Animal Control by Habitat Management.* pp. 235–248.

Townes, H. (1975) The parasitic Hymenoptera with the longest ovipositors, with descriptions of two new Ichneumonidae. *Entomol. News.* **86**, 123–127.

Townes, H.K. Jr (1939) Protective odours among the Ichneumonidae (Hymenoptera). *Bull. Brooklyn Entomol. Soc.* **34**, 29–30.

Tremblay, E. (1966) Ricerche sugli imenotteri parassiti. II. Osservazioni sull'origine e sul destino sell'involucro embrionale degli Afidiini (Hymenoptera: Braconidae: Aphidiinae) e considerazioni sul significato generale delle membrane embrionali. *Boll. Lab. Entomol. Agrar. Filippo Silvestri, Portici* **24**, 119–166.

Tremblay, E. (1991) Embryonic strategies in koinobiont endoparasitoid Hymenoptera. *Redia* **74**, 439–443.

Tremblay, E. and Caltagirone, L.E. (1973) Fate of polar bodies in insects. *Annu. Rev. Ent.* **18**, 421–444.

Tremblay, E. and Calvert, D. (1971) Embryosystematics in the aphidiines (Hymenoptera: Braconidae). *Boll. Lab. Entomol. Agrar. Filippo Silvestri, Portici* **29**, 223–247.

Tremblay, E. and Calvert, D. (1972) New cases of polar nuclei utilization in insects. *Annls Soc. Entomol. Fr. (N.S.)* **8**, 495–498.

Tremblay, E. and Laccarino, F.M. (1971) Notizie sull'ultrastruttura dei trofociti di *Aphidius matricariae* Hal. (Hymenoptera: Braconidae). *Boll. Lab. Entomol. Agrar. Filippo Silvestri, Portici* **29**, 305–313 + 3plates.

Triltsch, H. (1996) On the parasitization of the ladybird *Coccinella septempunctata* L. (Col., Coccinellidae). *J. Appl. Ent.* **120**, 375–378.

Tripp, H.A. (1961) The biology of a hyperparasite *Euceros frigidus* Cress. (Ichneumonidae) and description of the planidial stage. *Canad. Entomol.* **93**, 40–58.

Trjapitzin, V.A. (1978) Superfamily Chalcidoidea. In *Keys to the Identification of Insects of the European Part of the USSR* (ed. G.S. Medvedev), Nauka Publishers, Leningrad. pp. 39–58.

Turlings, T.C.J. (1994) The active role of plants in the foraging successes of entomophagous insects. *Norw. J. Agric. Sci. Suppl.* **16**, 211–219.

Turlings, T.C.J. and Tumlinson, J.H. (1991) Do parasitoids use herbivore-induced plant chemical defences to locate hosts? *Florida. Entomol.* **74**, 42–50.

Turlings, T.C.J. and Tumlinson, J.H. (1992) Systematic release of chemical signals by herbivore-injured corn. *Proc. Natn. Acad. Sci.* **89**, 8399–8402.

Turlings, T.C.J., Battenburg, F.D.H. van and Strein-van Liempt, W.T.F.H. van (1985) Why is there no interspecific host discrimination in the two coexisting larval parasitoids of *Drosophila* species, *Leptopilina heterotoma* (Thomson) and *Asobara tabida* (Nees). *Oecologia* 67, 352–359.

Turlings, T.C.J., Tumlinson, J.H. and Lewis, W.J. (1990) Exploitation of herbivore-induced plant odours by host-seeking parasitic wasps. *Science* **250**, 1251–1253.

Turlings, T.C.J., McCall, P.J., Alborn, H.T. and Tumlinson, J.H. (1993a) An elicitor in caterpillar oral secretions that induces corn seedlings to emit chemical signals attractive to parasitic wasps. *J. Chem. Ecol.* **19**, 411–425.

Turlings, T.C.J., Wäckers, F.L., Vet, L.E.M. *et al.* (1993b) Learning of host-finding cues by hymenopterous parasitoids. In *Insect Learning* (eds D.R. Papaj and A.C. Lewis), Chapman & Hall, New York. pp. 51–78.

Uematsu, H. and Sakanoshita, A. (1987) Effects of venom from an external parasitoid, *Euplectrus kuwanae* (Hymenoptera: Eulophidae) on larval ecdysis of *Argyrogramma albostriata* (Lepidoptera: Noctuidae). *Appl. Ent. Zool.* **22**, 139–144.

Ueno, T. (1994) Self-recognition by the parasitic wasp *Itoplectis naranyae* (Hymenoptera: Ichneumonidae). *Oikos* **70**, 333–339.

Ueno, T. (1995) Abdominal tip movements during oviposition by two parasitoids (Hymenoptera: Ichneumonidae) as an index of predicting the sex of depositing eggs. *Appl. Entomol. Zool.* **30**, 588–590.

Unruh, T.R., White, W., González, D. *et al.* (1983) Heterozygosity and effective size in laboratory populations of *Aphidius ervi* (Hym.: Aphidiidae). *Entomophaga* **28**, 245–258.

Unruh, T.R., White, W., González, D. and Woolley, J.B. (1989) Genetic relationships among seventeen *Aphidius* (Hymenoptera: Aphidiidae) populations, including six species. *Ann. Entomol. Soc. Am.* **82**, 754–768.

Urano, T. and Hijii, N. (1995) Resource utilization and sex allocation in response to host size in two ectoparasitoid wasps on subcortical beetles. *Entomol. Exp. Appl.* **74**, 23–35.

Vagina, N.P. (1982) Morphology of the neuroendocrine system of *Alysia manducator* Panz. (Hymenoptera, Braconidae). *Entomol. Rev.* **66**, 82–90.

Vagina, N.P. (1987) Role of neurosecretory cells of the brain in regulation of larval development of *Alysia manducator* Panz. (Hymenoptera, Braconidae) larvae. *Entomolog. Obozr.* **61**, 245–251. (In Russian.)

Vance, A.M. and Smith, H.D. (1933) The larval head of parasitic Hymenoptera and nomenclature of its parts. *Ann. Entomol. Soc. Am.* **26**, 86–94.

Vander Meer, R.K., Jouvenaz, D.P. and Wojcik, D.P. (1989) Chemical mimicry in a parasitiod (Hymenoptera: Eucharitidae) of fire ants (Hymenoptera: Formicidae). *J. Chem. Ecol.* **15**, 2247–2261.

Vanlerberghe-Masutti, F. (1994a) Molecular identification and phylogeny of parasitic wasp species (Hymenoptera: Trichogrammatidae) by mitochondrial DNA RFLP and RAPD markers. *Insect Molec. Biol.* **3**, 229–237.

Vanlerberghe-Masutti, F. (1994b) Detection of genetic variability in *Trichogramma* populations using molecular markers. *Norw. J. Agric. Sci. Suppl.* **16**, 171–176.

Varley, G.C. (1937) Description of the eggs and larvae of four species of chalcidoid Hymenoptera parasitic on the knapweed gall-fly. *Proc. R. Entomol. Soc. Lond. B* **6**, 122–130.

Varley, G.C. (1964) A note on the life history of the ichneumon fly *Euceros unifasciatus* Voll. with a description of its planidium larva. *Entomologist's Mon. Mag.* **100**, 113–116.

Varley, G.C. and Butler, C.G. (1933) The acceleration of development of insects by parasitism. *Parasitology*, **25**, 263–268.

Varricchio, P., Gargiulo, G., Graziani, F. *et al.* (1995) Characterization of *Aphidius ervi* (Hymenoptera, Braconidae) ribosomal genes and identification of site-specific insertion elements belonging to the non-LTR retrotransposon family. *Insect Biochem. Molec. Biol.* **25**, 603–612.

Vaughn, T.T., Antolin, M.F. and Bjostad, L.B. (1996) Behavioral and physiological responses of *Diaretiella rapae* to semiochemicals. *Entomol. Exp. Appl.* **78**, 187–196.

Veen, J.C. van (1981) The biology of *Poecilostictus cothurnatus* (Hymenoptera, Ichneumonidae) an endoparasite of *Bupalus pinarius* (Lepidoptera, Geometridae). *Ann. Entomol. Fenn.* **47**, 77–93.

Veen, J.C. van and Wijk, M.E.L. van (1985) The unique structure and functions of the ovipositor of the non-paralysing ectoparasitoid *Colpoclypeus florus* Walk. (Hym., Eulophidae) with special reference to antennal sensilla and immature stages. *Z. Angew. Entomol.* **99**, 511–531.

Veen, J.C. van and Wijk, M.E.L. van (1987) Parasitization strategy in the non-paralysing ectoparasitoid *Colpoclypeus florus* (Hym., Eulophidae) towards the common summer host *Adoxophyes orana* (Lep., Tortricidae). *Z. Angew. Entomol.* **104**, 402–417.

Veerkamp, F.A. (1980) Behavioural differences between two strains and the hybrids of the wasp *Pseudocoila bochei* Weld. (Hym.: Cynipidae), a parasite of *Drosophila melanogaster. Neth. J. Zool.* **30**, 431–449.

Venkatraman, T.V. and Subba Rao, B.R. (1954) The mechanism of oviposition in *Stenobracon deesae* (Cam.) (Hymenoptera: Braconidae). *Proc. R. Entomol. Soc. Lond (A)* **29**, 1–8.

Vereschagina, V.V. (1961) *Tetrastichus* (*Myiomisa*) *sajoi* Szelenyi, a predator of the plum shoot mite, *Eriophyes phloeocoptes. Nal. Trud. Mold. Nauch.-issl. Inst. Sadov. Vinog.* **7**, 31–33.

Vet, L.E.M. and Dicke, M. (1992) Ecology of infochemical use by natural enemies in a tritrophic context. *Annu. Rev. Ent.* **37**, 141–172.

Vet, L.E.M., Wackers, F.L. and Dicke, M. (1991) How to hunt for hiding hosts – The reliability-detectability problem in foraging parasitoids. *Neth. J. Zool.* **41**, 202–213.

Viggiani, G. (1964) La specializzazione entomoparasitica in alcuni Eulofidi (Hym., Chalcioidea). *Entomophaga* **9**, 111–119.

Viggiani, G. (1984) Bionomics of the Aphelinidae. *Annu. Rev. Entomol.* **29**, 257–276.

Viggiani, G. (1988) Advances in taxonomy and biology of the Aphelinidae (Hymenoptera: Chalcidoidea). In *Advances in Parasitic Hymenoptera Research* (ed. V.K. Gupta), E.J. Brill, Leiden. pp 79–84.

Viggiani, G. and Garonna, A.P. [(1986)/1987] Preliminary observations on the biology of *Archenomus orientalis* Silvestri (Hymenoptera: Aphelinidae), parasite of white peach scale (*Pseudaulacaspis pentagona* Targ.-Tozz.). *Boll. Lab. Entomol. Agrar. Filippo Silvestri, Portici* **43**, 223–227.

Viggiani, G. and Mazzone, P. (1982) Antennal sensilla of some *Encarsia* Foerster (Hymenoptera: Aphelinidae), with particular reference to sensorial complexes of the male. *Res. Hym. Chalc.* **78**, (19–26.

Vincent, J.F.V. and King, M.J. (1996) The mechanism of drilling by wood wasp ovipositors. *Biomimetics* **3**, 187–201.

Vinson, S.B. (1969) General morphology of the digestive and internal reproductive system of adult *Cardiochiles nigriceps* (Hymenoptera: Braconidae). *Ann. Entomol. Soc. Am.* **62**, 1414–1419.

Vinson, S.B. (1972) Courtship behavior and evidence for a sex pheromone in the parasitoid *Campoletis sonorensis* (Hymenoptera: Ichneumonidae). *Env. Ent.* **1**, 409–415.

Vinson, S.B. (1975) Source of material in the tobacco budworm which initiates host searching by the egg–larval parasitoid, *Chelonus texanus*. *Ann. Entomol. Soc. Am.* **68**, 381–384.

Vinson, S.B. (1976) Host selection by insect parasitoids. *Annu. Rev. Entomol.* **21**, 109–133.

Vinson, S.B. (1978) Sexual behavior and source of a sexual pheromone from *Cardiochiles nigriceps*. *Ann. Entomol. Soc. Am.* **71**, 832–837.

Vinson, S.B. (1985) The behaviour of parasitoids. In *Comprehensive Insect Physiology, Biochemistry and Physiology* (eds G.A. Kerkut and L.I. Gilbert), Pergamon Press, New York. pp. 417–469.

Vinson, S.B. (1988) Comparison of host characteristics that elicit host recognition behavior of parasitic Hymenoptera. In *Advances in Parasitic Hymenoptera Research* (ed. V.K. Gupta), E.J. Brill, Leiden. pp. 285–291.

Vinson, S.B. (1990) How parasitoids deal with the immune system of their host: An overview. *Arch. Insect Bioch. Physiol.* **13**, 3–27.

Vinson, S.B. (1991) Chemical signals used by parasitoids. *Redia* **74**, 15–42.

Vinson, S.B. and Guillot, F.S. (1972) Host marking source of a substance that results in host discrimination in insect parasitoids. *Entomophaga* **17**, 241–245.

Vinson, S.B. and Jang, H.-S. (1989) Activation of *Campoletis sonorensis* (Hymenoptera: Ichneumonidae) eggs by artificial means. *Ann. Entomol. Soc. Am.* **80**, 486–489.

Vinson, S.B. and Lewis, W.J. (1973) Teratocytes: Growth and numbers in the hemocoel of *Heliothis virescens* attacked by *Microplitis croceipes*. *J. Invert. Pathol.* **22**, 351–355.

Vinson, S.B. and Scarborough, T.A. (1991) Interactions between *Solenopsis invicta* (Hymenoptera: Formicidae), *Rhopalosiphum maidis* (Homoptera: Aphididae), and the parasitoid *Lysiphlebus testaceipes* Cresson (Hymenoptera: Aphidiidae). *Ann. Entomol. Soc. Am.* **84**, 158–164.

Vinson, S.B., Edson, K.M. and Stoltz, D.B. (1979) Effect of a virus associated with the reproductive system of the parasitoid wasp, *Campoletis sonorensis*, on host weight gain. *J. Invert. Pathol.* **34**, 133–137.

Vinson, S.B., Bin, F. and Strand, M.R. (1986) The role of the antennae and host factors in host selection behavior of *Trissolcus basalis* (Woll.) (Hym.: Scelionidae). In *Trichogramma and Other Egg Parasites* (INRA), Les Colloques de l'INRA no. 43. Paris.

Vinson, S.B., Mourad, A.K. and Sebesta, D.K. (1994a) Sources of possible host regulatory factors in *Cardiochiles nigriceps* (Hymenoptera: Braconidae). *Arch. Insect Biochem. Physiol.* **26**, (197–209).

Vinson, S.B., Williams, H.J. and Lu, J. (1994b) Identification of different compounds from different plants responsible for the orientation of *Campoletis sonoriensis* to potential host sites. *Norw. J. Agric. Sci. Suppl.* **16**, 207–210.

Visser, B.J., Labruyère, W.T., Spanjer, W. and Piek, T. (1983) Characterization of two paralysing toxins (A-MTX and B-MTX), isolated from a homogenate of the wasp *Microbracon hebetor* (Say). *Comp. Biochem. Physiol. B* **75B**, 523–538.

Visser, M.E., Luyckx, B., Nell, H.W. and Boskamp, Ge.J.F. (1992) Adaptive superparasitism in solitary parasitoids: Marking of parasitized hosts in relation to the pay-off from superparasitism. *Ecol. Ent.* **17**, 76–82.

Voegele, J., Brun, P. and Daumal, J. (1974) Les trichogrammes. 1. Modalités de la prise de possession et de l'elimination de l'hôte chez le parasite embryonnaire *Trichogramma brasiliensis. Ann. Soc. Entomol. Fr.* **10**, 757.

Vogt, E.A. and Nechols, J.R. (1991) Diel activity patterns of the squash bug egg parasitoid *Gryon pennsylvanicum* (Hymenoptera: Scelionidae). *Ann. Entomol. Soc. Am.* **84**, 303–308.

Völkl, W. and Kranz, P. (1995) Nocturnal activity and resource utilization in the aphid hyperparasitoid, *Dendrocerus carpenteri. Ecol. Ent.* **20**, 293–297.

Völkl, W. and Stadler, B. (1996) Colony orientation and successful defence behaviour in the conifer aphid, *Schizolachnus pineti. Entomol. Expt. Appl.* **78**, 197–200.

Völkl, W., Hübner, G. and Dettner, K. (1994) Interactions between *Alloxysta brevis* (Hymenoptera, Cynipoidea, Alloxystidae) and honeydew-collecting ants: How an aphid hyperparasitoid overcomes ant aggression by chemical defence. *J. Chem. Ecol.* **20**, 2901–2915.

Völkl, W., Liepert, C., Birnbach, R. *et al.* (1996) Chemical and tactile communications between the root aphid parasitoid *Paralipsis enervis* and trophobiotic ants – consequences for parasitoid survival. *Experientia* **52**, 731–738.

Volkoff, A.-N. and Colazza, S. (1992) Growth patterns of teratocytes in the immature stages of *Trissolcus basalis* (Woll.) (Hymenoptera: Scelionidae), an egg parasitoid of *Nezara viridula* (L.) (Hemiptera: Pentatomidae). *Int. J. Insect Morphol. Embryol.* **21**, 323–336.

Volkoff, A.-N. and Daumal, J. (1994) Ovarian cycle in immature and adult stages of *Trichogramma cacoeciae* and *T. brassicae* (Hym.: Trichogrammatidae). *Entomophaga* **39**, 303–312.

Volkoff, A.-N., Daumal, J., Barry, P. *et al.* (1995a) Development of *Trichogramma cacoeciae* Marchal (Hymenoptera: Trichogrammatidae): Time table and evidence for a single larval instar. *Int. J. Insect Morphol. Embryol.* **24**, 459–466.

Volkoff, A.-N., Ravallec, M., Bossy, J.-P. *et al.* (1995b) The replication of *Hyposoter didymator* polydnavirus: Cytopathology of the calyx cells in the parasitoid. *Biol. Cell.* **83**, 1–13.

Volkoff, N., Vinson, S.B., Wu, Z.X. and Nettles, W.C. Jr (1992) *In vitro* rearing of *Trissolcus basalis* [Hym., Scelionidae], an egg parasitoid of *Nezara viridula* (Hem., Pentatomidae). *Entomophaga* **37**, 141–148.

Waage, J.K. (1982) Sib-mating and sex ratio strategies in scelionid wasps. *Ecol. Entomol.* **7**, 103–112.

Waage, J. and Greathead, D. (eds) (1986) *Insect Parasitoids*, Academic Press, London. 389 pp.

Wäckers, F.L. (1994) Visual cues in food and host-foraging by hymenopterous parasitoids. *Norw. J. Agric. Sci. Suppl.* **16**, 347–352.

Wäckers, F.L. and Lewis, W.J. (1994) Olfactory and visual learning and their combined influence on host site location by the parasitoid *Microplitis croceipes* (Cresson). *Biol. Control* **4**, 105–112.

Wahrman, M.Z. and Zhu, S. (1993) Haploid and diploid-cell cultures from a haplodiploid insect. *Invert. Repro. Devel.* **24**, 79–86.

Wajnberg, E. (1993) Genetic variation in sex allocation in a parasitic wasp: Variation in sex pattern within sequences of oviposition. *Entomol. Exp. Appl.* **69**, 221–229.

Waldvogel, M.G. and Brown, M.W. (1978) An overwintering site of the gypsy moth parasite, *Brachymeria intermedia*. *Env. Ent.* **7**, 782.

Walker, I. (1979) Some British species of *Anagrus* (Hymenoptera: Mymaridae). *Zool. J. Linn. Soc.* **67**, 181–202.

Waller, J.B. (1965) The effect of venom of *Bracon hebetor* on the respiration of the wax moth *Galleria mellonella*. *J. Insect Physiol.* **11**, 1595–1599.

Waloff, N. (1974) Biology and behaviour of some species of Dryinidae (Hymenoptera). *J. Ent. (A)* **49**, 97–109.

Walter, G.H. (1983a) Divergent male ontogenies in Aphelinidae (Hymenoptera: Chalcidoidea): A simplified classification and a suggested evolutionary sequence. *Biol. J. Linn. Soc.* **16**, 63–82.

Walter, G.H. (1983b) Differences in host relationships between male and female heteronomous parasitoids (Aphelinidae: Chalcidoidea): A review of host location, oviposition and pre–imaginal physiology and morphology. *J. Entomol. Soc. South Afr.* **46**, 261–282.

Walter, G.H. (1986) Suitability of a diphagous parasitoid, *Coccophagus bartletti* Annecke and Insley (Hymenoptera, Aphelinidae), for sex ratio studies: Ovipositional and host-feeding behaviour. *J. Entomol. Soc. South Afr.* **49**, 141–152.

Walter, G.H. (1988a) Heteronomous host relationships in aphelinids – Evolutionary pathways and adaptive significance (Hymenoptera: Chalcidoidea). In *Advances in Parasitic Hymenoptera Research* (ed. V.K. Gupta), E.J. Brill, Leiden. pp. 313–326.

Walter, G.H. (1988b) Activity patterns and egg production in *Coccophagus bartletti*, an aphelinid parasitoid of scale insects. *Ecol. Ent.* **13**, 95–105.

Walther, C. and Rathmayer, W. (1974) The effect of *Habrobracon* venom on excitatory neuromuscular transmission in insects. *J. Comp. Physiol.* **89**, 23–38.

Walther, C. and Rathmayer, W. (1983) Block of synaptic vesicle exocytosis without block of Ca^{2+}-influx. An ultrastructural analysis of the paralysing action of *Habrobracon* venom on locust motor nerve terminals. *Neuroscience* **9**, 213–224.

Walton, G.A. (1948) A minute bethylid wasp of medical interest. *Proc. R. Entomol. Soc. (A)* **23**, 98.

Wang, T. and Laing, J.E. (1989) Polyembryony in *Holcothorax testaceipes* (Hymenoptera: Encyrtidae). *Ann. Entomol. Soc. Am.* **82**, 725–729.

Ware, A.B. and Compton, S.G. (1992) Repeated evolution of elongate multiporous plate sensilla in female fig wasps (Hymenoptera: Agaonidae: Agaoninae). *Proc. Kon. Ned. Akad. v. Wetensch.* **95**, 275–292.

Wardle, A.R. and Borden, J.H. (1990) Learning of host microhabitat form by *Exeristes roborator* (F.) (Hymenoptera: Ichneumonidae). *J. Insect Behav.* **3**, 251–263.

Webb, B.A. and Summers, M.D. (1990) Venoms and viral expression products of the endoparasitic wasp *Campoletis sonorensis* share epitopes and related sequences. *Proc. Natn. Acad. Sci. USA* **87**, 4961–4965.

Webb, B.A. and Summers, M.D. (1992) Stimulation of polydnavirus replication by 20-hydroxyecdysone. *Experientia* **48**, 1018–1022.

Weber, C.A., Smilanik, I.M., Ehler, L.E. and Zalom, F.G. (1996) Ovipositional behavior and host discrimination in three scelionid egg parasitoids of stink bugs. *Biol. Control* **6**, 245–252.

Weinstein, P. and Austin, A.D. (1991) The host relationships of trigonalyid wasps (Hymenoptera: Trigonalyidae), with a review of their biology and a catalogue to world species. *J. Nat. Hist.* **25**, 399–433.

Weinstein, P. and Austin, A.D. (1995) Primary parasitism, development and adult biology in the wasp *Taeniogonalys venatoria* Riek (Hymenoptera: Trigonalyidae). *Aust. J. Zool.* **43**, 541–555.

Weinstein, P. and Austin, A.D. (in press) Thelytoky in *Taeniogonalys venatoria* Riek (Hymenoptera: Trigonalyidae), with notes on its distribution and first description of the male sex. *J. Austr. Entomol. Soc.*

Weis-Fogh, T. (1973) Quick estimates of flight fitness in hovering animals, including novel mechanisms for lift production. *J. Exp. Biol.* **59**, 169–230.

Weisser, W.W. and Houston, A.I. (1993) Host discrimination in parasitic wasps: When is it advantageous? *Funct. Ecol.* **7**, 27–39.

Went, D.F. (1982) Egg activation and parthenogenetic reproduction in insects. *Biol. Rev.* **57**, 319–344.

Went, D.F. and Krause, G. (1973) Normal development of mechanically activated, unlaid eggs of an endoparasitic hymenopteran. *Nature,* **244**, 454–455.

Went, D.F. and Krause, G. (1974a) Egg activation in *Pimpla turionellae* (Hym.). *Naturwissenschaften,* **61**, 407–408.

Went, D.F. and Krause, G. (1974b) Alteration of egg architecture and egg activation in an endoparasitic hymenopteran as a result of natural or imitated oviposition. *Wilhelm Roux' Arch.* **175**, 173–184.

Werren, J.H. (1983) Sex ratio evolution under local mate competition in a parasitic wasp. *Evolution* **37**, 116–124.

Werren, J.H. and Assem, J. van den (1986) Experimental analysis of a paternally inherited extrachromosomal factor. *Genetics* **114**, 217–233.

Werren, J.H., Skinner, S.W and Charnov, E.L. (1981) Paternal inheritance of a daughterless sex ratio factor. *Nature,* **293**, 467–468.

Werren, J.H., Skinner, S.W. and Huger, A.M. (1986) Male-killing bacteria in a parasitic wasp. *Science,* **231**, 990–992.

Werren, J.H., Zhang, W. and Guo, L.R. (1995) Evolution and phylogeny of *Wolbachia*: reproductive parasites of arthropods. *Proc. R. Soc. Lond. B* **261**, 55–71.

Weseloh, R.M. (1972) Sense organs of the hyperparasite *Cheiloneurus noxius* (Hymenoptera: Encyrtidae) important in host selection processes. *Ann. Entomol. Soc. Am.* **65**, 41–46.

Weseloh, R.M. (1976a) Dufours gland: Source of sex pheromone in a hymenopterous parasitoid. *Science,* **193**, 695–697.

Weseloh, R.M. (1976b) Reduced effectiveness of the gypsy moth parasite, *Apanteles melanoscelus*, in Connecticut due to poor seasonal synchronization with its host. *Environ. Ent.* **5**, 743–746.

Weseloh, R.M. (1977) Behavioural responses of the parasite, *Apanteles melanoscelus,* to gypsy moth silk. *Environ. Entomol.* **5**, 1128–1132.

Weseloh, R.M. (1980) Sex pheromone gland of the gypsy moth parasitoid, *Apanteles melanoscelus*: Revaluation and ultrastructural survey. *Ann. Entomol. Soc. Am.* **73**, 576–580.

Weseloh, R.M. (1981) Host location by parasitoids. In *Semiochemicals: Their Role in Pest Control* (eds D.A. Nordlund, R.L. Jones and W.J. Lewis), John Wiley, New York. pp. 79–96.

Weseloh, R.M. (1986) Artificial selection for host suitability and development length of the gypsy moth (Lepidoptera: Lymantriidae) parasite, *Cotesia melanoscela* (Hymenoptera: Braconidae). *J. Econ. Ent.* **79**, 1212–1216.

Whalley, P.E.S. (1969) The mymarid (Hym.) egg parasites of *Tettigella viridis* L. (Hem., Cicadellidae) and embryo–parasitism. *Entomol. Mon. Mag.* **105**, 239–243.

Wharton, R.A. (1993) Bionomics of the Braconidae. *Annu. Rev. Entomol.* **38**, 121–143.

Wheeler, D.E. and Krutzsch, P.H. (1992) Internal reproductive system in adult males of the genus *Camponotus* (Hymenoptera: Formicidae: Formicinae). *J. Morphol.* **211**, 307–317.

White, M.J.D. (1954) *Animal Cytology and Evolution*, 2nd edn, Cambridge University Press, Cambridge. 454 pp.

Whitfield, J.B. (1987) Male swarming by a microgastrine braconid, *Apanteles coniferae* (Haliday) (Hymenoptera). *Proc. Trans. Br. Ent. Nat. Hist. Soc.* **20**, 133–135.

Whitfield, J.B. (1988) Two new species of *Paradelius* (Hymenoptera: Braconidae) for North America with biological notes. *Pan-Pacific Entomologist* **64**, 313–319.

Whitfield, J.B. (1992) Phylogeny of the non-aculeate Apocrita and the evolution of parasitism in the Hymenoptera. *J. Hym. Res.* **1**, 3–14.

Whitfield, J.B., Johnson, N.F. and Hamerski, M.R. (1989) Identity and phylogenetic significance of the metapostnotum in nonaculeate Hymenoptera. *Ann. Entomol. Soc. Am.* **82**, 663–673.

Whiting, P.W. (1943) Multiple alleles in complementary sex determination in *Habrobracon*. *Genetics* **28**, 365–382.

Whiting, P.W. (1954) Comparable mutant eye colors in *Mormoniella* and *Pachycrepoideus* (Hymenoptera: Pteromalidae). *Evolution* **8**, 135–147.

Whiting, P.W. (1960) Polyploidy in *Mormoniella*. *Genetics* **45**, 949–970.

Whiting, P.W., Greb, R.J. and Speicher, B.R. (1934) A new type of sex intergrade. *Biol. Bull.* **66**, 152–165.

Wiebes, J.T. (1982) The phylogeny of the Agaonidae (Hymenoptera, Chalcidoidea). *Neth. J. Zool.* **32**, 395–411.

Wigglesworth, V.B. (1953) *The Principles of Insect Physiology*, 5th edn, Methuen, London.

Wilkes, A. (1964) Inherited male-producing factor in an insect that produces its males from unfertilised eggs. *Science* **144**, 305–307.

Wilkes, A. (1965) Sperm transfer and utilization by the arrhenotokous wasp *Dahlbominus fuscipennis* (Zett.) (Hymenoptera: Eulophidae). *Canad. Entomol.* **97**, 647–657.

Wilkes, A. (1966) Sperm utilization following multiple insemination in the wasp *Dahlbominus fuscipennis*. *Canad. J. Genet. Cytol.* **8**, 451–461.

Wilkes, A. and Lee, P.E. (1965) The ultrastructure of dimorphic spermatozoa in the hymenopteran *Dahlbominus fuscipennis* (Zett.) (Eulophidae). *Canad. J. Gen. Cytol.* **7**, 609–619.

Wilkinson, M. (1995) Coping with missing entries in phylogenetic inference using parsimony. *Syst. Biol.* **44**, 501–514.

Wilkinson, M. and Benton, M.J. (1996) Sphenodontid phylogeny and the problems of multiple trees. *Phil. Trans. R. Soc. Lond.* B **351**, 1–16.

Willers, D. (1980) Untersuchung über die physiologische Eignung einiger Schmetterlingsarten als Wirte der polyphagen Puppenparasiten *Pimpla turionellae* L. und *Itoplectis conquisitor* Say unter besonderer Berücksichtigung der Larvenmortalität der Parasiten. Diss. Forst. Fak. Univ. Göttingen.

Willers, D. and Lehmann-Danzinger, H. (1984) Survival of endo-parasitic Hymenoptera in pupae of Lepidoptera by inhibition of phenol oxidase. *Zeitschr. Parasitenkunde* **70**, 403–414.

Williams, H., Wong, M.A., Wharton, R.A. and Vinson, S.B. (1988) Hagen's gland morphology and chemical content analysis for three species of parasitic wasps (Hymenoptera: Braconidae). *J. Chem. Ecol.* **14**, 1727–1736.

Williams, R.N., Fickle, D.S. and Galford, J.R. (1992) Biological studies of *Brachyserphus abruptus* (Hym.: Proctotrupidae), a nitidulid parasite. *Entomophaga* **37**, 91–98.

Williams, T. and Polaszek, A. (1996) A re-examination of host relations in the Aphelinidae (Hymenoptera: Chalcidoidea). *Biol. J. Linn. Soc.* **57**, 35–45.

Wilson, D.D. and Ridgway, R.L. (1974) Cocoon spinning behaviour of the parasitoid *Campoletis sonorensis*. *Env. Ent.* **3**, 714–717.

Wilson, F. and Woolcock, L.T. (1960) Temperature determination of sex in a parthenogenetic parasite, *Ooencyrtus submetallicus* (Howard) (Hymenoptera: Encyrtidae). *Australian J. Zool.* **8**, 153–169.

Wishart, G. and Steenburgh, W.E. van (1934) Contribution to the technique of propogation of *Chelonus annulipes* Wesm.; an imported parasite of the European corn borer. *Canad. Entomol.* **66**, 121–125.

Wiskerke, J.S.C., Dicke, M. and Vet, L.E.M. (1993) Larval parasitoid uses aggregation pheromone of adult hosts in foraging behaviour: A solution to the reliability-detectability problem. *Oecologia*, **93**, 145–148.

Wolf, R. and Wolf, D. (1988) Activation by calcium ionophore injected into unfertilized ovarian eggs explanted from *Pimpla turionellae* (Hymenoptera). *Zool. J. Physiol.* **92**, 501–512.

Woods, P.E. and Guttmann, S.I. (1987) Genetic variation in *Neodiprion* (Hymenoptera: Symphyta: Diprionidae) sawflies and a comment on low levels of genetic diversity within the Hymenoptera. *Ann. Entomol. Soc. Am.* **80**, 590–599.

Woolley, J.B. (1988) Phylogeny and classification of the Signiphoridae (Hymenoptera: Chalcidoidea). *Syst. Ent.* **13**, 465–501.

Woolley, J.B. and Vet, L.E.M. (1981) Postovipositional webspinning behaviour in a hyperparasite, *Signiphora coquilletti* Ashmead (Hymenoptera: Signiphoridae). *Neth. J. Zool.* **31**, 627–633.

Wylie, (1985) Posterior dispersal of eggs and larvae of *Microctonus vittatae* (Hymenoptera: Braconidae) in crucifer-infesting flea beetles (Coleoptera: Chrysomelidae). *Canad. Entomol.* **117**, 541–545.

Xu, D.M. and Stoltz, D. (1991) Evidence for a chromosomal location of polydnavirus DNA in the ichneumonid parasitoid *Hyposoter fugitivus*. *J. Virol.* **65**, 6693–6704.

Yamada, Y. (1987) Characteristics of the oviposition of a parasitoid, *Chrysis shanghaiensis* (Hymenoptera, Chrysididae). *Appl. Ent. Zool.* **22**, 456–464.

Yamada, Y. (1991) Role of the teeth on the abdominal end in *Praestochrysis shanghaiensis* (Hymenoptera, Chrysididae). *Jpn. J. Ent.* **59**, 99–103.

Yang, J.C., Chu, Y.I. and Talekar, N.S. (1993) Biological studies of *Diadegma semiclausum* (Hym., Ichneumonidae), a parasite of diamondback moth. *Entomophaga* **38**, 579–586.

Yasuda, K. and Tsurumachi, M. (1995) The influence of male adults of the leaf-footed plant bug, *Leptoglossus australis* (Fabricius) (Heteroptera, Coreidae), on host-searching of the egg parasitoid, *Gryon pennsylvanicum* (Ashmead) (Hymenoptera, Scelionidae). *Appl. Ent. Zool.* **30**, 139–144.

Yazgan, S. (1972) A chemically defined synthetic diet and larval nutritional requirements of the endoparasitoid *Itoplectis conquisitor* (Hymenoptera). *J. Insect Physiol.* **18**, 2123–2141.

Yazgan, S. and House, H.L. (1970) An hymenopterous insect, the parasitoid *Itoplectis conquisitor*, reared axenically on a chemically-defined diet. *Canad. Entomol.* **102**, 1304–1306.

Yazlovetsky, I.G. and Ageeva, L.I. (1995) Artificial medium for ectoparasitic larvae of *Elasmus albipennis* (Hymenoptera, Elasmidae). *Zool. Zhurn.* **74**, 90–96.

Yazlovetsky, I.G. and Nepomnyashchaya, A.M. (1989) Culturing larvae of *Bracon hebetor* (Hymenoptera, Braconidae) on an artificial nutrient medium. *Zool. Zhurn.* **68**, 120–125.

Yeargan, K.V. and Braman, S.K. (1986) Life history of the parasite *Diolcogaster facetosa* (Weed) (Hymenoptera: Braconidae) and its behavioural adaptation to the defensive response of a lepidopteran host. *Ann. Entomol. Soc. Am.* **79**, 1029–1033.

Yeargan, K.V. and Braman, S.K. (1989) Life history of the hyperparasitoid *Mesochorus discitergus* (Hymenoptera: Ichneumonidae) and tactics used to overcome the defensive behavior of the green cloverworm (Lepidoptera: Noctuidae). *Ann. Entomol. Soc. Am.* **82**, 393–398.

Yoshimoto, C.M. (1978) Voucher specimens for entomology in North America. *Bull. Entomol. Soc. Am.* **24**, 141–142.

Young, D.K. (1990) Distribution of *Pelecinus polyturator* in Wisconsin (Hymenoptera: Pelecinidae), with speculations regarding geographical parthenogenesis. *Great Lakes Entomol.* **23**, 1–4.

Zaki, F.N. (1985) Reactions of the egg parasitoid *Trichogramma evanescens* Westw. to certain insect sex pheromones. *Z. Ang. Ent.* **99**, 448–453.

Zarani, F.E. and Margaritis, L.H. (1994) The eggshell of the almond wasp, *Eurytoma amygdali* (Hymenoptera, Eurytomidae). 2. The micropylar appendage. *Tissue and Cell*, **26**, 569–577.

Zchori-Fein, E., Geden, C.J. and Rutz, D.A. (1992a) Microsporidioses of *Muscidifurax raptor* (Hymenoptera: Pteromalidae) and other pteromalid parasitoids of muscoid flies. *J. Invert. Pathol.* **60**, 292–298.

Zchori-Fein, E., Roush, R.T. and Hunter, M.S. (1992b) Male production induced by antibiotic treatment in *Encarsia formosa* (Hymenoptera: Aphelinidae), an asexual species. *Experientia*, **48**, 102–105.

Zchori-Fein, E., Faktor, O., Zeidan, M. *et al.* (1995) Parthenogenesis-inducing microorganisms in *Aphytis* (Hymenoptera: Aphelinidae). *Insect Mol. Biol.* **4**, 173–178.

Zessin, W. (1985) Neue oberliassische Apocrita und die Phylogenie der Hymenoptera (Insecta, Hymenoptera). *Dtsch. Entomol. Z.,* N.F. **32**, 129–142.

Zhang, D. and Dahlman, D.L. (1989) *Microplitis croceipes* teratocytes cause developmental arrest of *Heliothis virescens* larvae. *Arch. Insect Biochem. Physiol.* **12**, 51–61.

Zhang, D., Dahlman, D.L. and Gelman, D.B. (1992) Juvenile hormone esterase activity and ecdysteroid titer in *Heliothis virescens* larvae injected with *Microplitis croceipes* teratocytes. *Arch. Insect Biochem. Physiol.* **20**, 231–242.

Zhang, D., Dahlman, D.L., Järlfors, U.E. *et al.* (1994) Ultrastructure of *Microplitis croceipes* (Cresson) (Braconidae: Hymenoptera) teratocytes. *Int. J. Insect Morphol. Embr.* **23**, 173–187.

Ziegler, I and Harmsen, R. (1969) The biology of pteridines in insects. *Advances in Insect Physiology.* **6**, 139–203.

Zinna, G. (1955) Un nuovo parassita della *Dioryctria splendidella* H.S. *Crataepoides Russoi* n.sp. rappresentante di un nuovo genere. *Boll. Lab. Entomol. Agrar. Filippo Silvestri, Portici* **14**, 65–82.

Zinna, G. (1960) Ricerche sugli insetti entomofagi. I. Specializzazione ento-moparassitica negli Encyrtidae: Studio morfologico, etologico e fisiologico del *Leptomastix dactylopii* Howard. (Con note del Dr. D. C. Lloyd, Commonwealth Inst. Biol. Control, Fontana, California, USA.) *Boll. Lab. Entomol. Agrar. Filippo Silvestri, Portici* **18**, 1–148.

Zinna, G. (1961) Ricerche sugli insetti entomofagi. II. Specializzazione ento-moparassitica negli Aphelinidae: Studio morfologico, etologico e fisiologico del *Coccophagus bivittatus* Compere, nuovo parassita del *Coccus hesperidum* L. per l'Italia. *Boll. Lab. Entomol. Agrar. Filippo Silvestri, Portici* **19**, 301–358.

Zinna, G. (1962) Ricerche sugli insetti entomofagi. III. Specializzazione ento-moparassitica negli Aphelinidae: Interdependenze biocenotiche tra due specie associate. Studio morfologico, etologico e fisiologico del *Coccophagoides similis* (Masi) e *Azotus matritensis*. *Boll. Lab. Entomol. Agrar. Filippo Silvestri, Portici* **20**, 73–182.

Zinnert, K.-D. (1969) Vergleichende Untersuchungen zur Morphologie und Biologie der Larvenparasiten (Hymenoptera: Ichneumonidae und Braconidae) mitteleuropäischer Blattwespen aus der Subfamilie Nematinae (Hymenoptera: Tenthredinidae). *Zeitschr. Angew. Entomol.* **64**, 180–217, 277–306.

Index